12

GENERAL MATHEMATICS

PNG UPPER SECONDARY

Christine McRae
Esther Gawac Ehava
Phyl Haydock
John Watson
Rory Barrett

OXFORD

Oxford University Press is a department of the University of Oxford. It furthers the University's objective of excellence in research, scholarship, and education by publishing worldwide. Oxford is a registered trademark of Oxford University Press in the UK and in certain other countries.

Published in Australia by
Oxford University Press
Level 8, 737 Bourke Street, Docklands, Victoria 3008, Australia

First published 2013
Reprinted 2014 (twice), 2017 (twice), 2018, 2019, 2021, 2022, 2023

This book was originally published by ESA Publications, Auckland, New Zealand. This edition, specially adapted for the Grade 11 syllabus in Papua New Guinea, is published by arrangement with ESA Publications. Authors of the original work were Phyl Haydock, John Watson and Rory Barrett and this adaptation has been produced by Christine McRae with the assistance of Esther Gawac Ehava, Mac Dandava and John Gesa Junior.

ISBN 978 0 19 557870 6

Illustrated by diacriTech, Chennai, India
Typeset by diacriTech, Chennai, India
Printed in China by Golden Cup Printing Co. Ltd

Oxford University Press Australia & New Zealand is committed to sourcing paper responsibly.

Contents

Introduction

This book has been published to provide the information required by students in order to successfully complete the Upper Secondary course in General Mathematics at Grade 12 in Papua New Guinea.

The book is written in a manner that best develops an overall understanding of the mathematical concepts and processes set out in the PNG Grade 12 Syllabus. The order of Units in this book follows the order of Units presented in the General Mathematics syllabus for Grade 12:

Unit 12.1 Measurement

Unit 12.2 Managing Money 2

Unit 12.3 Probability and Statistics

Unit 12.4 Algebra and Graphs

Unit 12.5 Applying Geometry in Papua New Guinea Arts

Within each Unit there are a number of Topics which follow the main sub-headings and bullet points set out within each Unit in the syllabus. The aim is to provide structure and content in a concise and compact format for students to use as an effective resource to support the classroom experience. It is acknowledged that Mathematics is more interesting and meaningful when students are exposed to a variety of resources and materials and we encourage students and teachers not to rely on this book as a sole source of information.

If you have any suggestions about how this book might be improved in future editions, please make contact with Oxford University Press:

Fax: 00 61 3 99349 100

Customer Service Email: exportsales.au@oup.com

We wish you every success in your studies.

The authors

General Mathematics and Advanced Mathematics – similarities in content

General Mathematics and Advanced Mathematics have some content that is the same or similar. In Grade 12 General Mathematics some Topics within Unit 12.3 Probability and Statistics have content that is the same as or similar to Grade 11 Advanced Mathematics. However, in the *Save Buk* series, our aim is to publish separate books at Grades 11 and 12, in every subject, and for each book to cover the content of the Grade 11 or the Grade 12 syllabus exactly.

Acknowledgments

There are many people to be thanked for their help and assistance in enabling the publication of this series to happen. First, we acknowledge the cooperation and generosity of Mark Sayes at ESA Publications in New Zealand, who responded with interest and support when the proposal to adapt his Study Guide series was put to him.

The main author of this book is Christine McRae, with input from Esther Gawac Ehava. With the approval of ESA Publications, the authors drew on material from *NCEA Level 1 Mathematics* by Phyl Haydock and John Watson, and *NCEA Level 2 Mathematics* and *NCEA Level 3 Statistics and Modelling* by Rory Barrett. The objective has been to provide Grade 12 students in PNG with a book that they can use as a compact summary of content and skills related to the PNG Grade 12 General Mathematics syllabus.

We would also like to acknowledge many other individuals who have been happy to advise and assist in different ways: Vagi Hanua Bino, who read the manuscript, Betty Pulpulis, Joy Sahumlal, Greg Kapanombo, Anne Sangi, and Safak Deliismail. John Gesa Junior and Mac Dandava were also involved in assessing the titles from ESA Publications, in the early stages. For information about insurance in PNG we thank Jason Denham at Aon Risk Services and Jason R McIlvena. The publishers wish to make it clear that acknowledging these individuals does not imply any endorsement of this book by any of them; all responsibility for the content rests with the authors and the publishers.

Unit 12.1 Measurement

Topic 1: Scales and dimension – perimeter and area (revision)

This Unit focuses on mathematics that deals with scales and dimension. Mathematical skills should be developed in an applied context and the local environment should be used as the context. Topic 1 covers:

- Making measurements of everyday objects based on circles, triangles, rectangles, cuboids, cylinders and triangular prisms, or any combination of these.
- Perimeter, area (including surface area), volume and capacity.
- Conversion between units, eg m^2 to ha, cm^3 to litres.

Introduction

In order to solve problems involving **measurement** of everyday objects, the **formulae** for the perimeter, area, **surface area** and volume of standard shapes need to be known.

Perimeter

The **perimeter** of a figure is the total distance around the outside of the figure.

Example A

For each shape shown, the perimeter is calculated by adding the lengths of the sides.

1.

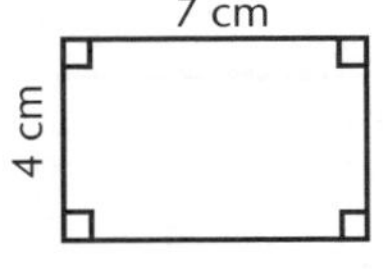

Perimeter $= 4 + 7 + 4 + 7$
$= 22$ cm

2.

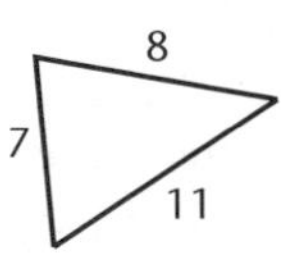

Perimeter $= 7 + 8 + 11$
$= 26$ units

3.

Perimeter $= 11 + 11 + 6$
$= 28$ cm

Note the use of marks on two sides of the triangle in part **3** to indicate sides of equal length.

When finding the perimeter of a **composite** or **compound** shape, use the properties of the figures making up the shape to **simplify** the calculation or to find the unknown lengths.

Example B

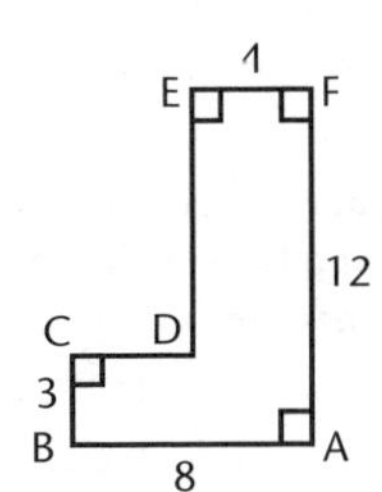

Perimeter = AB + BC + CD + DE + EF + FA

= AB + (BC + DE) + (CD + EF) + FA [rearranging]

= AB + FA + AB + FA
[BC + DE = FA and CD + EF = AB]

$= 8 + 12 + 8 + 12$

$= 40$ units

Alternatively, side CD is $8 - 4 = 4$ units and side DE is $12 - 3 = 9$ units, etc.

Circumference

The perimeter of a **circle** is called the **circumference**.

In any circle, the circumference divided by the **diameter** gives a **constant** value known as π (the Greek letter pi) which is an **irrational number** between 3 and 4. This gives rise to the formula for the circumference of a circle (where C is the circumference, d is the diameter and r is the **radius** of the circle):

$$C = \pi d \text{ or } C = 2\pi r$$

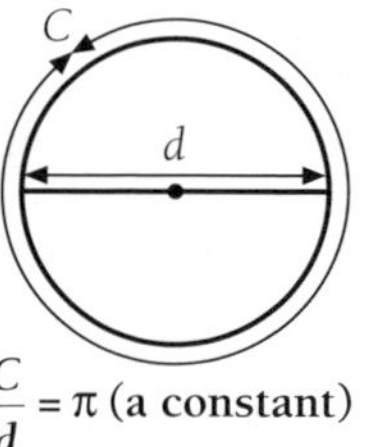

$\frac{C}{d} = \pi$ (a constant)

Since π is an infinite non-recurring **decimal**, an **approximate** value of π is used. The calculator gives an approximation of about 3.141592654.

Example C

The circumference of the circle shown is calculated as follows:

$C = 2\pi r$ [use this formula since r is given]

$= 2 \times \pi \times 2.4$ [substituting $r = 2.4$]

$= 15.1$ cm (1 dp)

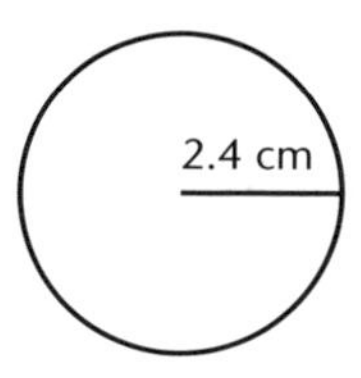

Note: If using a calculator, use the π value on the calculator and round sensibly at the *end* of the calculation.

Inverse problems are solved by substituting into the appropriate formula.

Example D

Q. Find the diameter of a circle with a circumference of 24 cm.

A. $C = \pi d$ [formula involving d]

$\therefore 24 = \pi d$ [substituting $C = 24$]

$\frac{24}{\pi} = d$ [dividing by π]

$\therefore d = 7.6$ cm (1 dp) [using the calculator, and **rounding**]

Area

Area is a measure of the amount of two-dimensional space covered by a surface.

Area is a *square* measure. The basic metric unit is the **square metre** (symbol **m²**). Commonly used metric units are **square centimetre** (**cm²**), **square kilometre** (**km²**) and **hectare** (**ha**).

$$1 \text{ ha} = 10\,000 \text{ m}^2$$

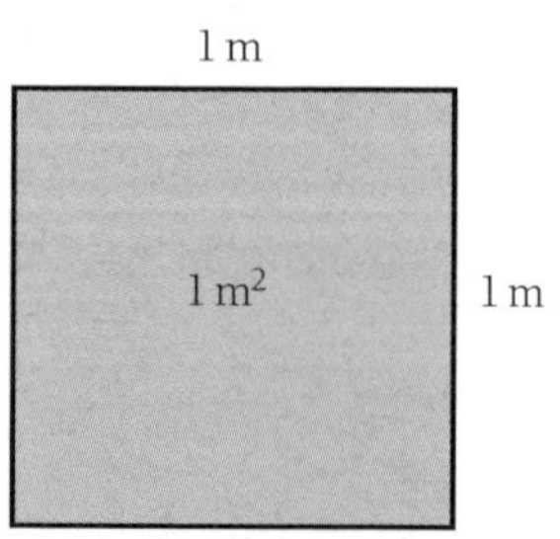

A 1 m by 1 m square covers 1 square metre

Conversion between square units

Square centimetres and square metres (square measure) do not relate to each other in the same way as centimetres and metres (**linear** measure).

The figure shows a square 1 m by 1 m (or 100 cm by 100 cm).

The area, using the units m^2 or cm^2 in turn, shows that:

$$1\ m^2 = 10\ 000\ cm^2$$

Similarly:

$$1\ cm^2 = 100\ mm^2$$

$$1\ km^2 = 1\ 000\ 000\ m^2 = 100\ ha$$

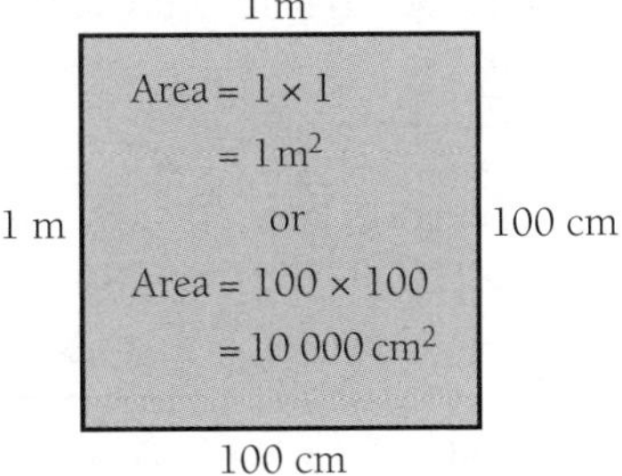

Area formulae

The **table** shows the formulae required to find the area of some common shapes.

Note: *h* always refers to the *perpendicular* height.

Square	Rectangle	Parallelogram
Area = length × length $A = a^2$ 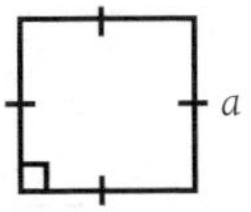	Area = length × breadth $A = l \times b$ 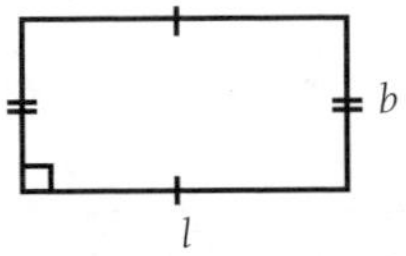	Area = base × height $A = bh$ 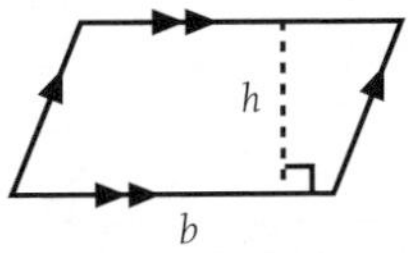

Trapezium
Area = $\frac{1}{2}$ × (sum of lengths of parallel sides) × (distance between parallel sides) $A = \frac{1}{2}(a + b)h$ 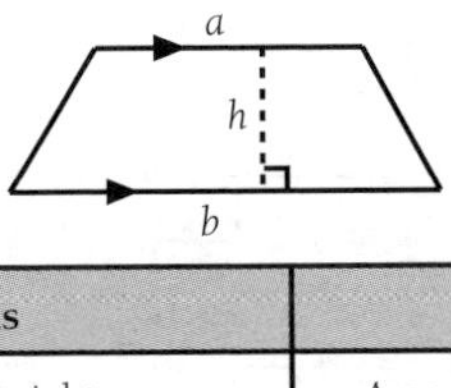 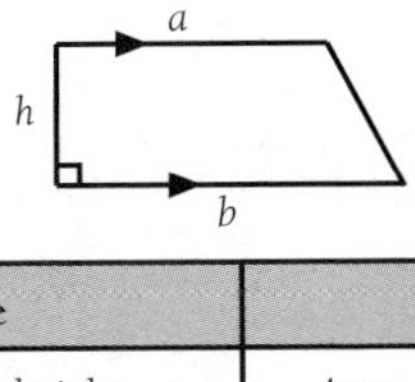

Rhombus	Triangle	Circle
Area = base × height $A = bh$ or $A = \frac{1}{2}$ × **diagonal** 1 × diagonal 2 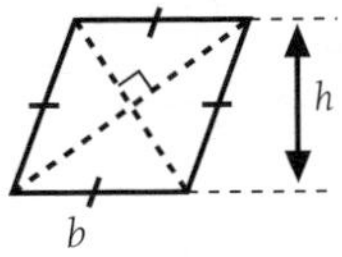	Area = $\frac{1}{2}$ × base × height $A = \frac{1}{2}bh$ 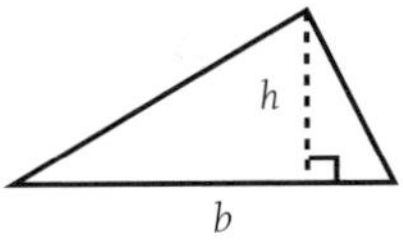	Area = π × (radius)^2 $A = \pi \times r^2$ 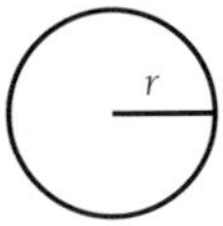

Sectors of circles

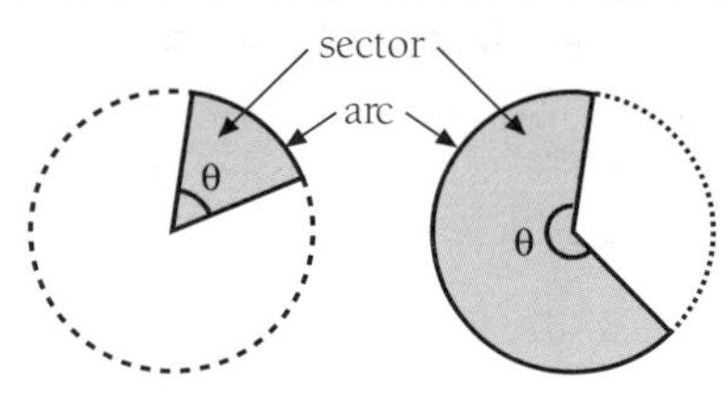

A **sector** of a circle is a **fraction** of the whole circle formed by 2 **radii** and an **arc**.

If the **angle** between the 'arms' of the sector is θ°, then the fraction of the whole circle taken up by the sector is $\frac{\theta}{360}$.

The **arc length** of a sector is a fraction of the circumference:

$$\text{arc length} = \frac{\theta}{360} \times \pi d \text{ or } \frac{\theta}{360} \times 2\pi r$$

The **area of a sector** is a fraction of the area of the circle:

$$\text{area of sector} = \frac{\theta}{360} \times \pi r^2$$

Example E

Q. For the sector shown, find:

1. The arc length, *s*.
2. The area shaded.

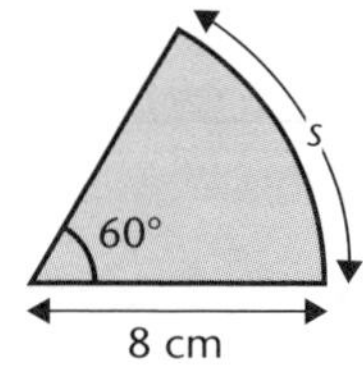

A. $\theta° = 60°$, $r = 8$ cm.

1. arc length $= \frac{\theta}{360} \times 2\pi r$

$= \frac{60}{360} \times \frac{2 \times \pi \times 8}{1}$

$= 8.4$ cm (1 dp)

2. area of sector $= \frac{\theta}{360} \times \pi r^2$

$= \frac{60}{360} \times \pi \times 8^2$

$= 33.5$ cm^2 (1 dp)

Areas of composite shapes

Many **composite** shapes are formed using a combination of common shapes. The area of the composite shape can be found by finding the area of each of the common shapes, and then adding them together.

Example F

Q. The diagram shows a factory floor plan. Find the total floor area of the factory.

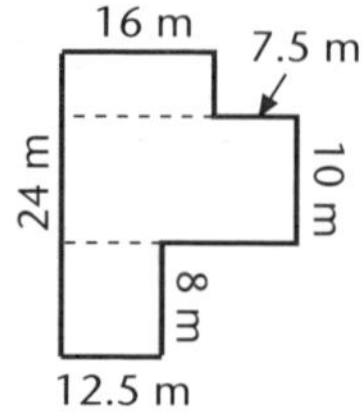

A. The floor can be divided into three sections, as shown.

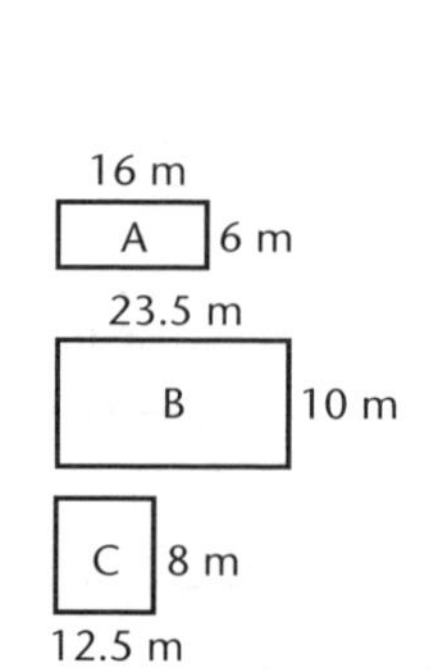

Total area = Area A + Area B + Area C

$= 16 \times 6 + 23.5 \times 10 + 12.5 \times 8$

$= 96 + 235 + 100$

$= 431$ m^2

Some compound shapes can be viewed as shapes from which one or more other shapes have been removed. For example, in the figure alongside, the shaded area is a rectangle with a rhombus removed.

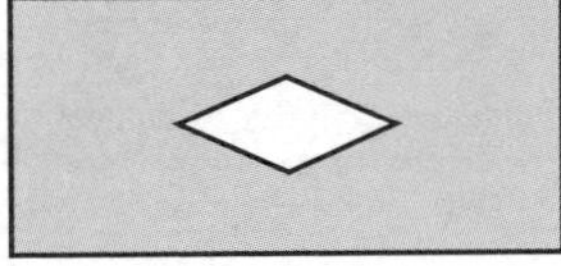

The area of the resulting shape is found by subtracting the area(s) of the removed shape(s) from the area of the original shape.

Example G

Q. A semicircular archway is formed as shown. Calculate the area of the shaded **cross-section** in square metres.

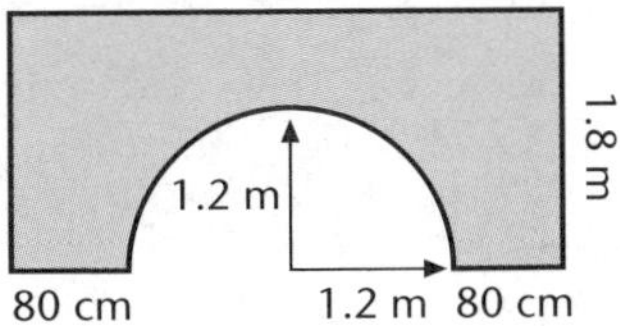

A. Shaded area = area rectangle – area of semicircle

$= b \times h - \frac{1}{2}\pi r^2$ [formulae]

$= 4 \times 1.8 - \frac{1}{2}\pi(1.2)^2$ [changing all lengths to m ($b = 0.8 + 2 \times 1.2 + 0.8$)]

$= 4.94\ \text{m}^2$ (3 sf)

Areas by measurement

In some problems side/angle sizes are not given and the student is required to make any measurements necessary in order to calculate the area. Identify the shape and measure only the lengths or angles used in the formula for the area of the shape.

For example, to find the area of the trapezium shown, measure the two parallel sides (a and c) and the distance between them (h). The lengths of b and d are not required for the area formula. Ensure all measurements use the same units.

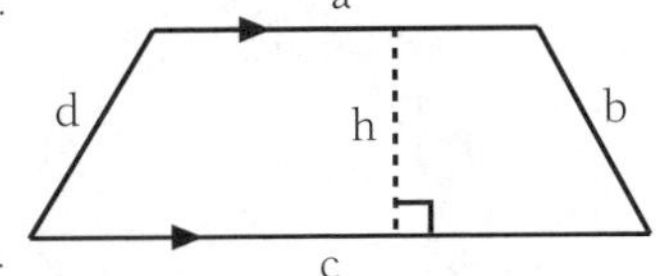

Unit 12.1 Activity 1: Perimeter, circumference and area

1. For each of the following shapes, find: **i.** The perimeter. **ii.** The area.

a. 8 cm, 5 cm

b.

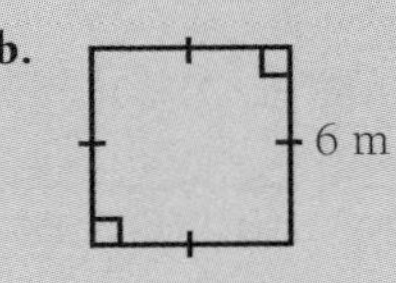

c.

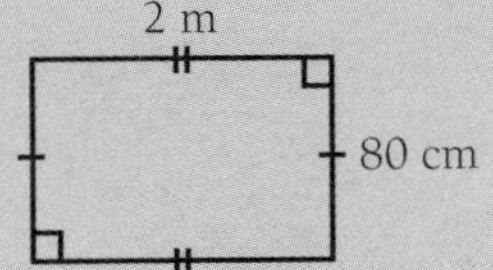

d. 2.5 m, 3.75 m, 1.25 m

e.

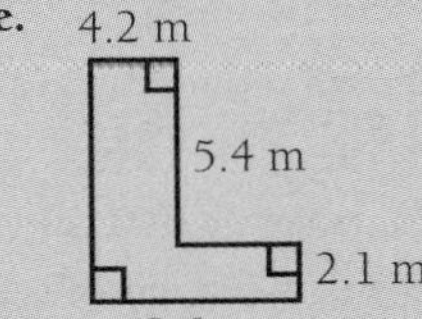

f.

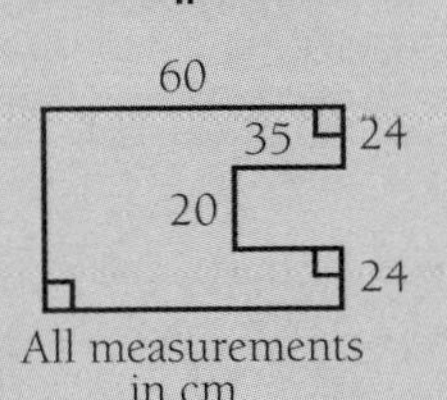

All measurements in cm.

2. Find the area of each the following:

a.

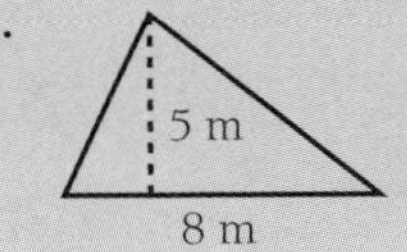

b.

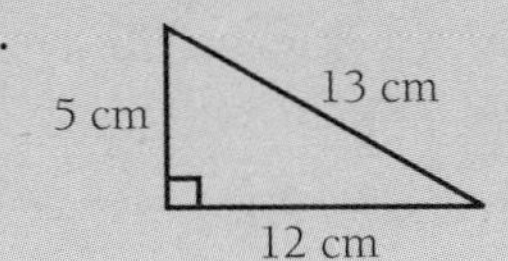

c.

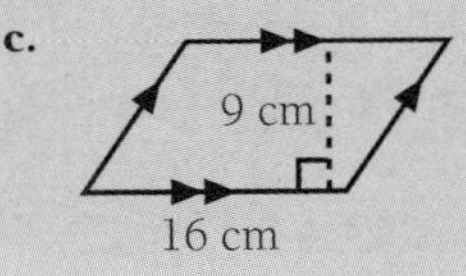

d.

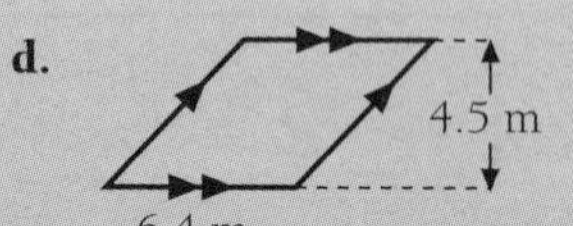

e.

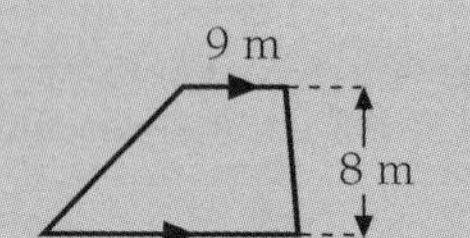

f.

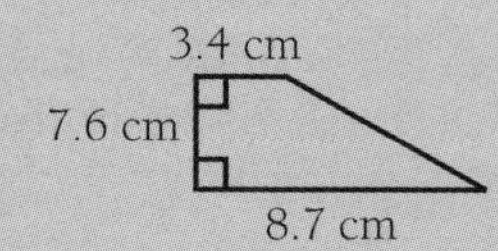

3. For each of the following, find (correct to 1 decimal place):

i. The circumference or arc length. **ii.** The area.

a.

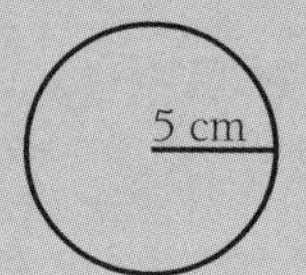

b.

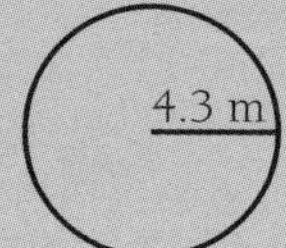

c.

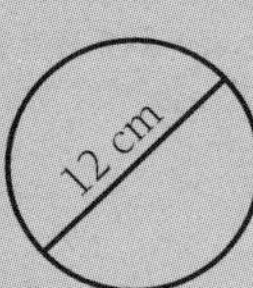

d.

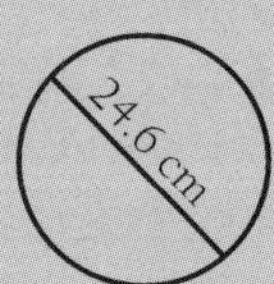

e.

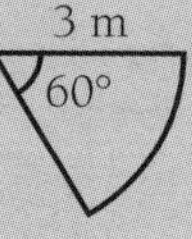

f.

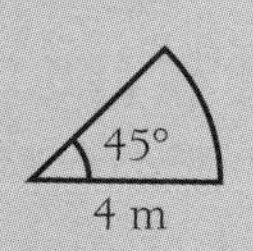

4. Use a ruler to make any necessary measurements then calculate the areas shaded. Give answers in cm^2 and mm^2 to 3 sf.

a.

b.

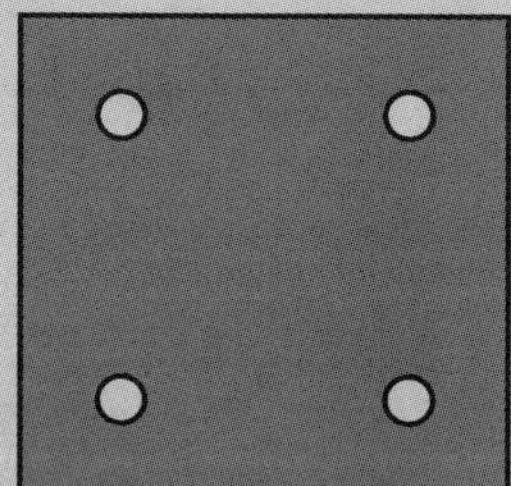

5. A property developer bought three adjoining properties measuring 2 400 m^2, 4 600 m^2, and 5 350 m^2 respectively.

a. Calculate the total area, in hectares, that the property developer bought.

b. How many sections of 650 m^2 each can the property developer subdivide the total area into?

6. A circular pool has a 25 m fence around it. What is the diameter of the pool?

7. A regular hexagonal garden of side length 2 m is to be made. Use the scale drawing of a regular hexagon alongside to take any measurements you need to find the area of the garden, to 2 sf.

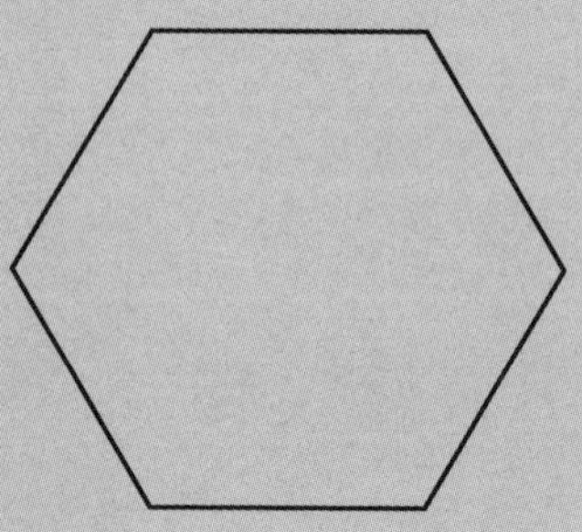

8. Matthias draws a square border (line) around his drawing. If the total length of the border is 2 m, what area is enclosed by the border? Give answer in square centimetres and in square metres.

9. A rectangular pool, 30 m by 12 m, has a small area cut off by a rope for children to play in. The rope joins the midpoint of a width (A) to the midpoint of a length (B) of the pool. What is the area of the remainder of the pool (shown shaded)?

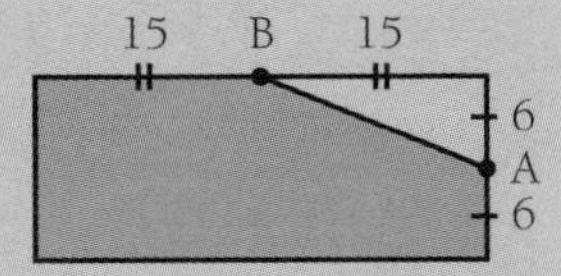

For the following questions, *show all working and state clearly what is being calculated at each step.*

10. A cylindrical wire framework for a lampshade is constructed by joining two circles of diameter 20 cm by eight straight wires each 24 cm long.

Find the weight of the framework if the wire weighs 0.25g per cm.

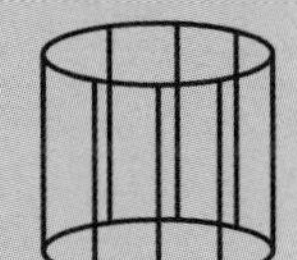

11. A villager has a garden, as shown in the diagram (not to scale). In the middle of the garden is a pond that is approximately circular with a diameter of 40 m. The villager wishes to fertilise the garden. The recommended rate of application of the fertiliser is 25 g/m^2. How much fertiliser will be needed? (Round your answer appropriately, stating the units used.)

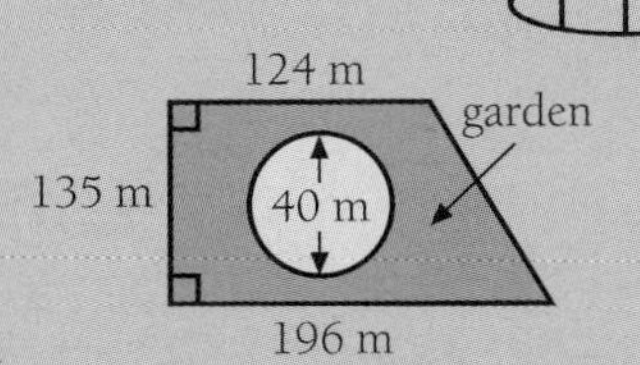

12. A security firm bought 36 metres of fencing to place around a semicircular compound. What area of compound will this length of fencing enclose?

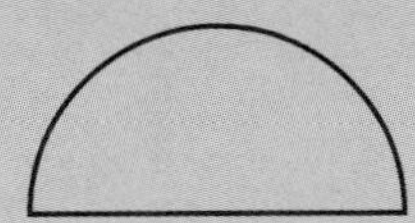

Unit 12.1 Measurement

Topic 2: Scales and dimension – triangles and formulae (revision)

In Topic 2 we continue to focus on measurement by revising the properties of triangles and the formulae that apply to triangles, in line with the Syllabus, p. 20 'Perimeter and area of triangles', in particular Pythagoras' theorem and the use of the sine ratio.

Properties of triangles and formulae that apply to triangles

Triangles are closed figures with three straight sides.

The three angles of a triangle sum to 180°

$a + b + c = 180°$

The side lengths, and angles, of a **scalene triangle** are all different.

The longest side is opposite the largest angle and similarly the shortest side is opposite the smallest angle.

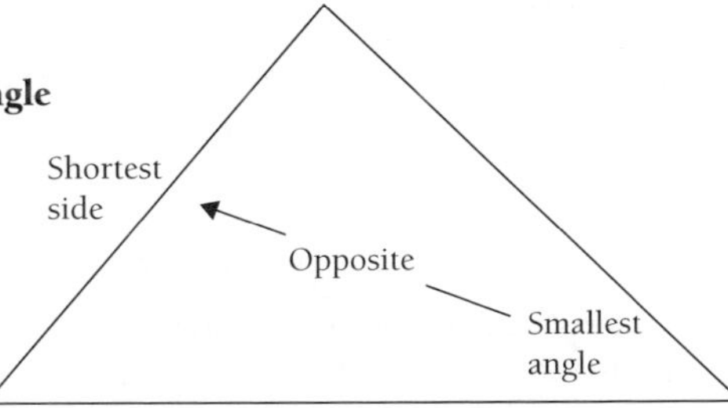

An **equilateral triangle** has all sides of equal length and all angles of equal size. All angles are 60°.

In an **isosceles triangle** two of the sides have the same length. The angles opposite these equal sides are the same size.

The line joining the **vertex** (formed by the sides of equal length) to the third side bisects the third side at right angles.

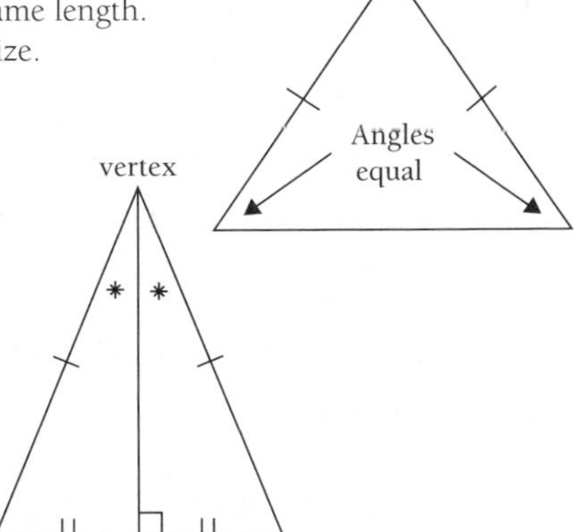

In a **right-angled triangle** the largest angle is a **right angle** (90°). The other two angles sum to 90°.

The side opposite the right angle is called the **hypotenuse**.

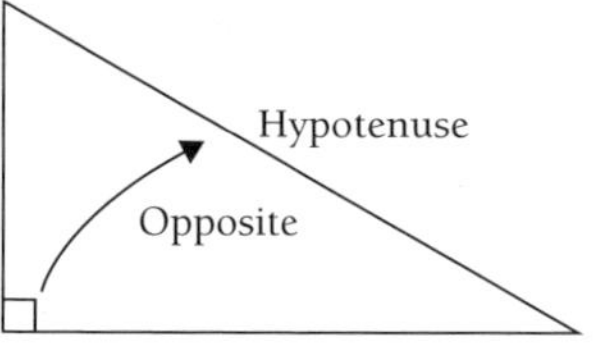

Pythagoras' rule applies only to right-angled triangles.

If c represents the length of the hypotenuse, and a and b represent the lengths of the other two sides, Pythagoras' theorem can be stated as:

$c^2 = a^2 + b^2$

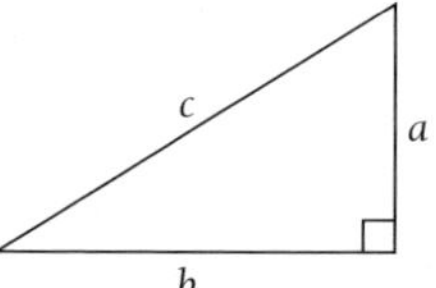

SOH CAH TOA is a helpful way to remember the **trigonometric ratios sine**, **cosine** and **tangent**, that apply only to right-angled triangles:

$$\sin\theta = \frac{\text{opposite}}{\text{hypotenuse}}$$

$$\cos\theta = \frac{\text{adjacent}}{\text{hypotenuse}}$$

$$\tan\theta = \frac{\text{opposite}}{\text{adjacent}}$$

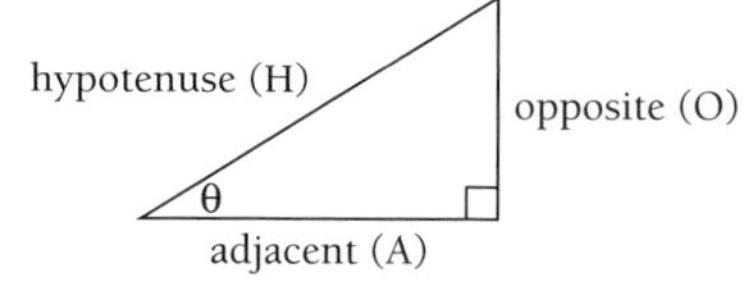

These ratios can be used to find unknown side-lengths and angles in right-angled triangles.

The **sine rule** and/or the **cosine rule** can be applied to triangles that are not right-angled, to find unknown side-lengths and angles. To use the sine and the cosine rule we use a conventionally labelled triangle.

In the triangle, the size of the angles is represented by the capital letters A, B and C and the lengths of the sides, opposite these angles, by the lower case letters a, b and c.

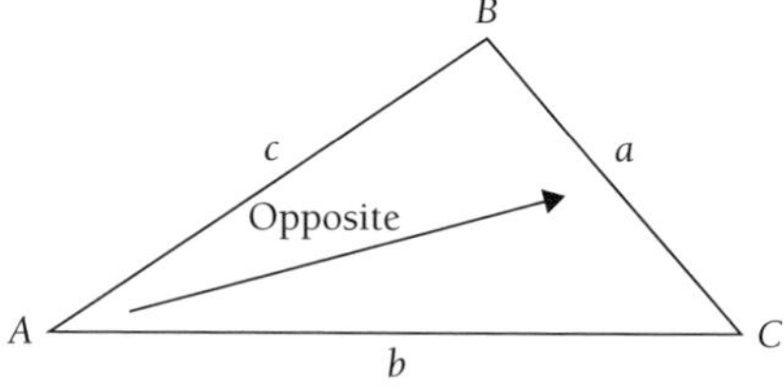

The **sine rule** is used to find unknown side lengths and angles in a triangle where:

i. Two angles (and hence three angles) and one side are known.

$$\frac{a}{\sin A} = \frac{b}{\sin B} = \frac{c}{\sin C}$$

Two parts of this formula are used to find the length of another side.

ii. Two side lengths and a non-included angle are known.

$$\frac{\sin A}{a} = \frac{\sin B}{b} = \frac{\sin C}{c}$$

Two parts of this formula are used to find the size of another angle.

The **cosine rule** is used:

i. To find the length of the third side when the length of two sides is known and the size of the included angle is known. (The included angle is 'opposite' the unknown side.)

In the case where side lengths b and c and angle A are known, the cosine rule to find the length of side a is:

$$a^2 = b^2 + c^2 - 2bc \cos A$$

ii. To find any of the angles in a triangle when the length of all three sides is known.

The cosine rule to find angle A is:

$$\cos A = \frac{b^2 + c^2 - a^2}{2bc}$$

The **perimeter of a triangle** is the sum of the lengths of the three sides of the triangle.

Example A

Q. Find the lengths of the unmarked sides and hence the perimeter of the following triangles:

a.

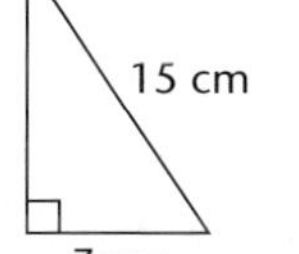

b.

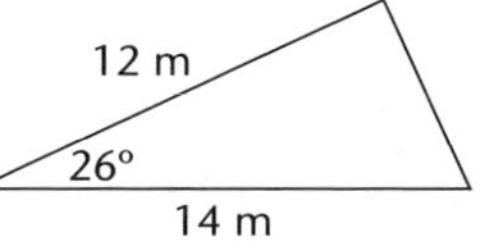

c.

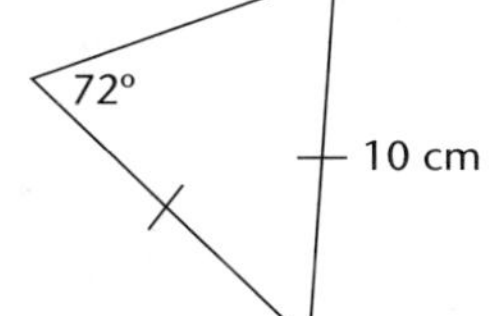

d.

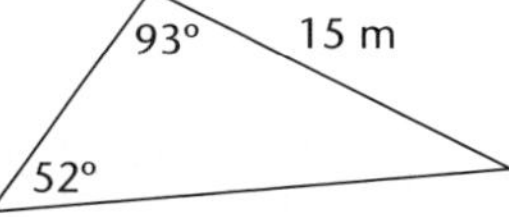

e.

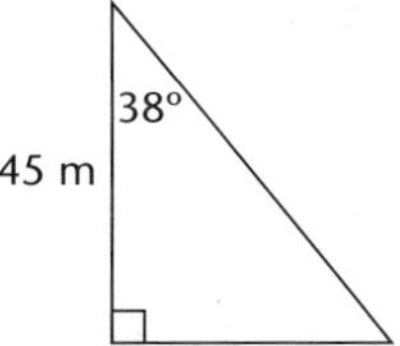

A. **a.** The length of the third side, x, of this right-angled triangle can be found using Pythagoras' rule:

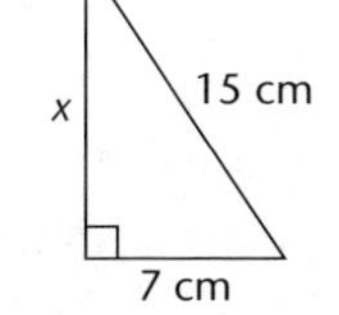

$$x^2 + 7^2 = 15^2$$

$$x^2 = 15^2 - 7^2$$

$$x = \sqrt{176} \approx 13.27 \text{ cm}$$

The perimeter of the triangle = (15 + 7 + 13.27) cm = 35.27 cm

b. The length of the third side, y, in this triangle can be found using the cosine rule:

$$y^2 = 12^2 + 14^2 - 2 \times 12 \times 14 \cos 26°$$

$$y^2 = 38.005$$

$$y \approx 6.16$$

The perimeter of the triangle = 12 + 14 + 6.16 = 32.16 m

c. This is an isosceles triangle, so there are two sides of length 10 cm and their opposite angles are both 72°. The included angle, θ, can be found:

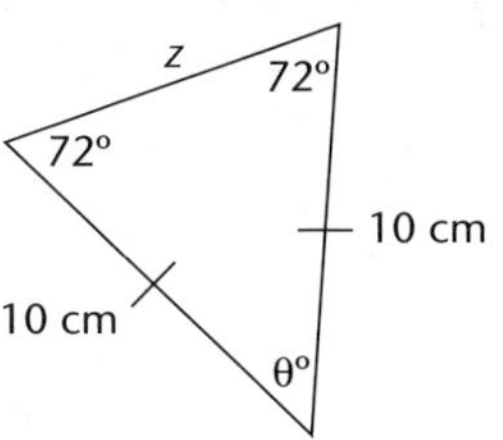

$$\theta = 180° - 2 \times 72° = 36°$$

The length of the third side, z, can be found using the cosine rule:

$$z^2 = 10^2 + 10^2 - 2 \times 10 \times 10 \cos 36°$$

$z = \sqrt{38.1966} \approx 6.18$

The perimeter of the triangle = 10 + 10 + 6.18 = 26.18 cm

d. The third angle in the triangle can be found:

$180° - 93° - 52° = 35°$

The length of the other two sides in the triangle can be found using the sine rule:

$$\frac{p}{\sin 35°} = \frac{15}{\sin 52°}$$

$$p = \frac{15 \sin 35°}{\sin 52°} \approx 10.92 \text{ m}$$

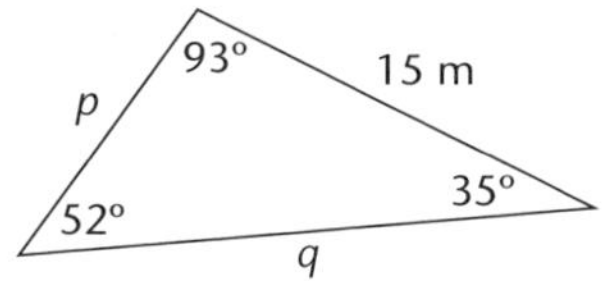

$$\frac{q}{\sin 93°} = \frac{15}{\sin 52°}$$

$$p = \frac{15 \sin 93°}{\sin 52°} \approx 19.01 \text{ m}$$

The perimeter of the triangle = 15 + 10.92 + 19.01 = 44.93 m

e. The lengths of the two sides, *s* and *t*, can be found using the SOHCAHTOA ratios:

i. $\tan 38° = \frac{t}{45}$

$t = 45 \tan 38° \approx 35.16$ m

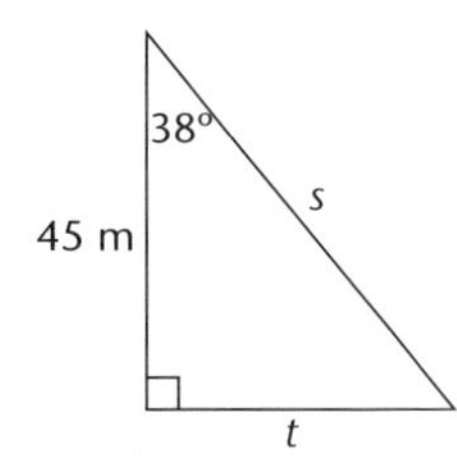

ii. $\cos 38° = \frac{45}{s}$

$$s = \frac{45}{\cos 38°} \approx 57.11 \text{ m}$$

The perimeter of the triangle = 45 + 35.16 + 57.11 = 137.27 m

Unit 12.1 Activity 2: Solving triangles

Give your answers correct to two decimal places, where appropriate.

1. Find the size of the angle θ in each of the following triangles:

a.

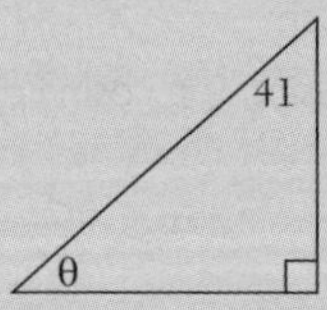

b.

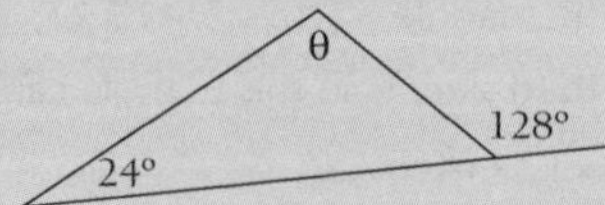

c.

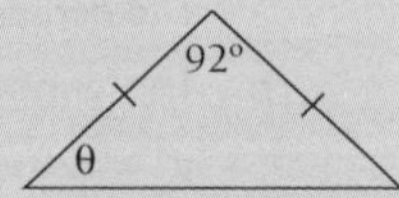

2. Find the length of the side marked *x* in each of the following triangles:

a.

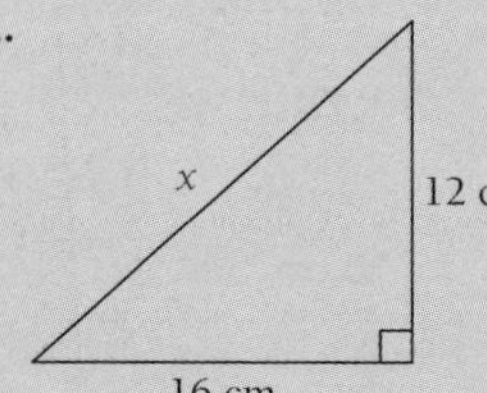

b.

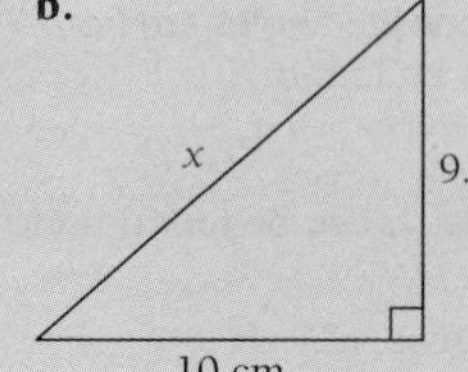

c.

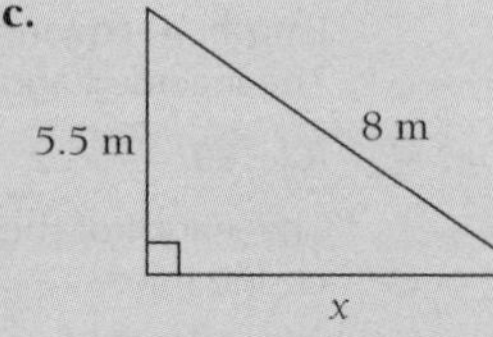

3. Find the lengths of the unmarked sides and hence the perimeter of the following triangles:

a.

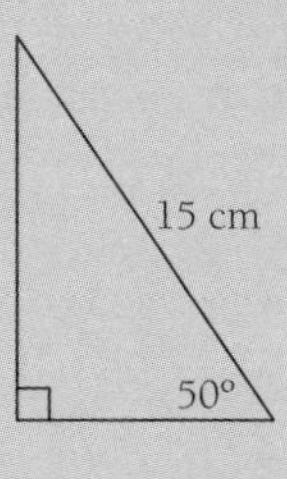

b.

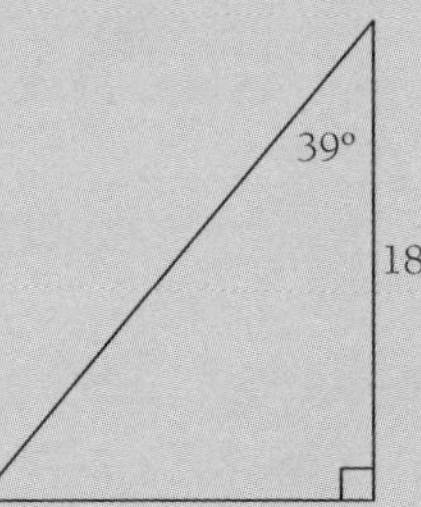

c.

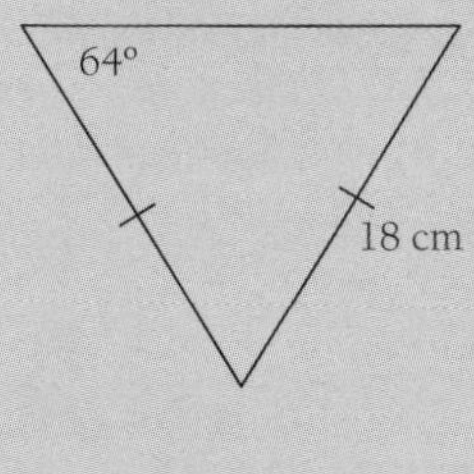

4. Find the length of the side marked x in each of the following triangles:

a.

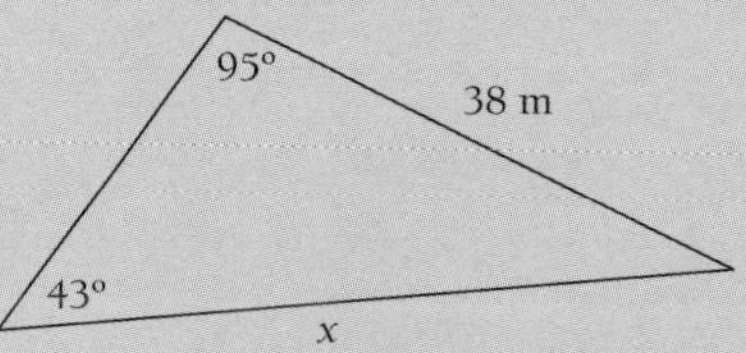

b.

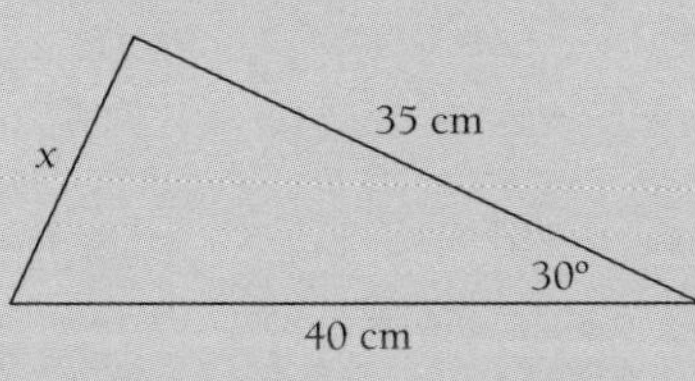

c.

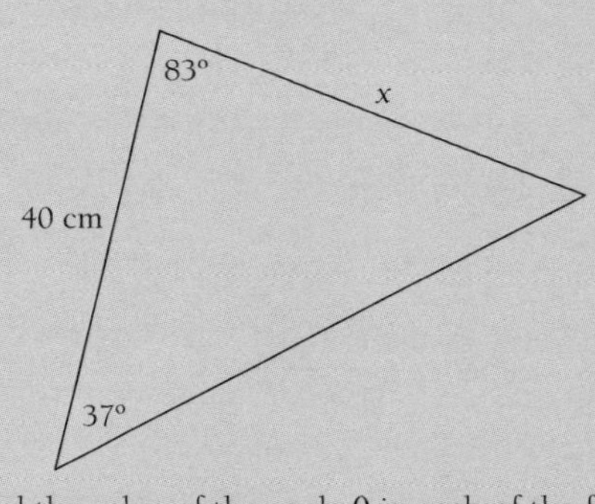

d.

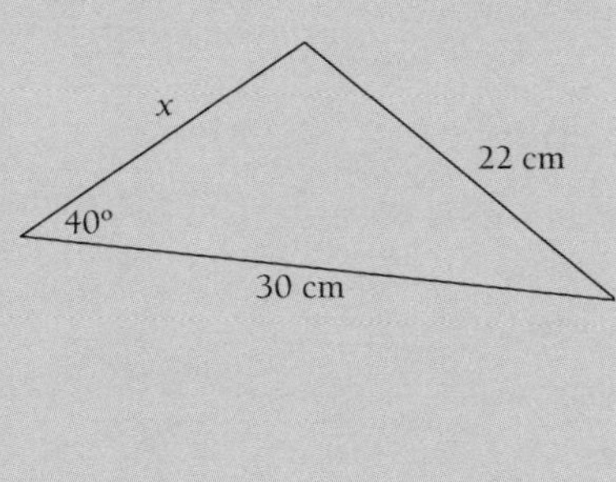

5. Find the value of the angle θ in each of the following triangles:

a.

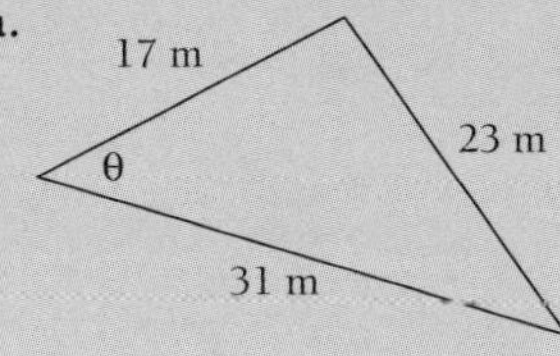

b.

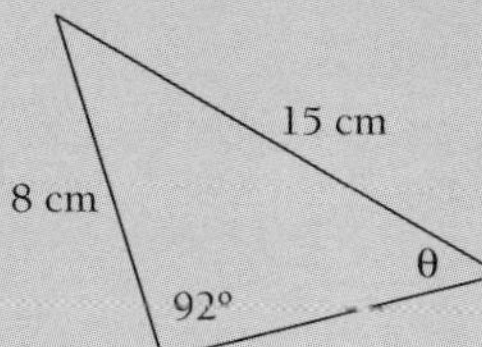

c.

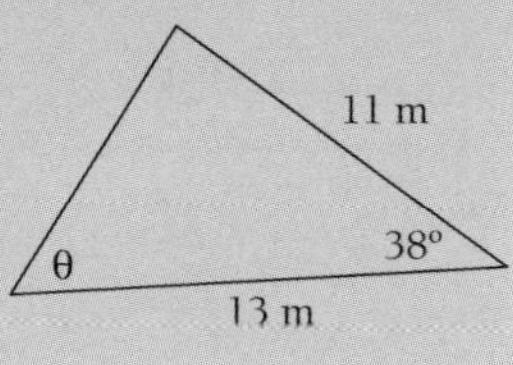

Unit 12.1 Measurement

Topic 3: Scales and dimension – area of triangle

In line with the Syllabus (p. 20), Topic 3 continues the focus on triangles, in particular calculating the area of a triangle using Heron's formula.

Formulae for finding the area of a triangle

There are three formulae for finding the area of a triangle and the one that is used depends on the given measurements of the triangle:

1. The base measurement and the height of an **altitude** are given.

If the length of the base, b, and the height, h (perpendicular to the base) are known, then the area of the triangle is given by the rule:

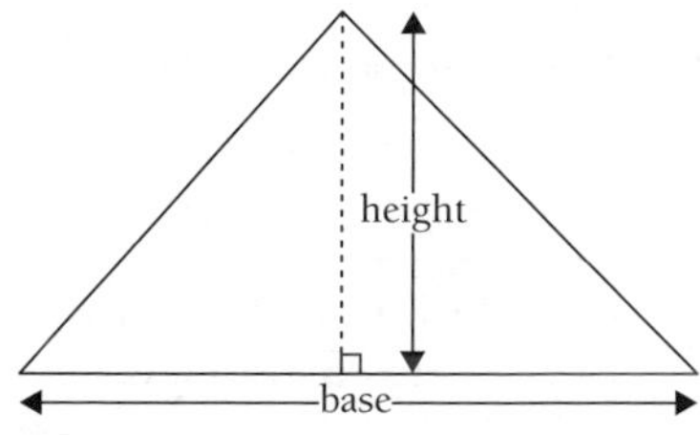

Area of triangle $= \frac{1}{2} \times$ base $\times$ height

$$\text{Area} = \frac{1}{2} b \times h$$

A **proof** of this formula is given below. If the rectangle is completed around the triangle then it can be seen that the triangle occupies half the area of the rectangle.

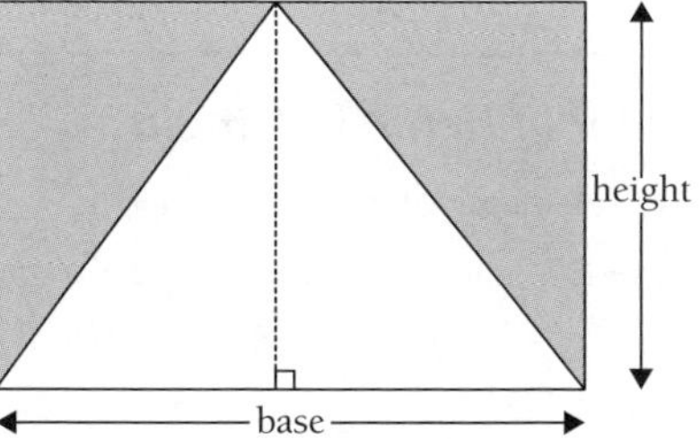

Area of rectangle = base × height

So:

$$\text{Area of triangle} = \frac{1}{2} \times \text{base} \times \text{height} = \frac{1}{2} b \times h$$

2. The length of two sides and the size of an angle are given. A formula for the area of a triangle can be found for these cases.

Consider a conventionally labelled triangle with the perpendicular height, h, shown:

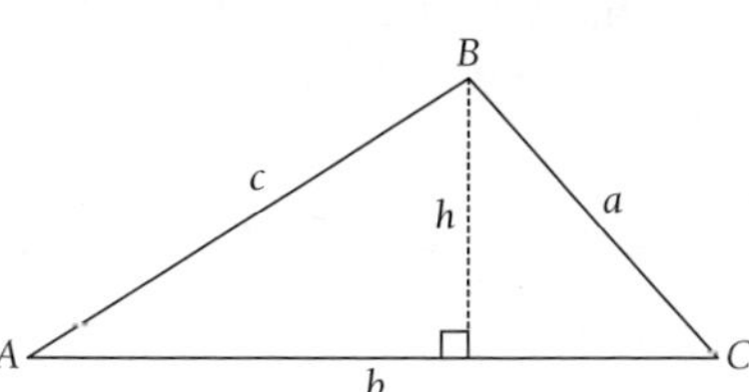

$$\text{Area of triangle} = \frac{1}{2} b \times h \qquad (1)$$

but

$$\sin A = \frac{h}{c} \Leftrightarrow h = c \sin A \qquad (2)$$

Substituting (2) into (1) gives:

$$\text{Area of triangle} = \frac{1}{2} bc \sin A$$

This formula could be used to find the area of the triangle when the length of two of the sides, b and c, and the **magnitude** of the included angle, A, is known.

Similarly:

If a, b and C are known then the formula is:

$$\text{Area of triangle} = \frac{1}{2}ab \sin C$$

and if a, c and B are known then the formula is:

$$\text{Area of triangle} = \frac{1}{2}ac \sin B$$

In all cases it is necessary to know the length of two of the sides and the magnitude of the *included angle* to use this formula to find the area.

If an angle other than the included angle is known, then the sine rule can be used to find the included angle.

3. **Heron's formula** for the area of a triangle.

This formula is named after a Greek mathematician and can be used to calculate the area of any triangle when given the lengths of the three sides.

> If the lengths of the three sides of a triangle are a, b and c and $s = \frac{1}{2}(a+b+c)$ then **Heron's formula** for the area of the triangle is:
>
> $$Area = \sqrt{s(s-a)(s-b)(s-c)}$$
>
> s is called the semi-perimeter of the triangle.

Proof of Heron's formula

Area of any $\triangle ABC = \frac{1}{2}$ base × height (1)

Heron's formula does not involve the altitude, therefore we need to express h in terms of a, b, c and s.

By Pythagoras' theorem:

In $\triangle ABD$, $h^2 = c^2 - (a-x)^2$ (2)

In $\triangle ADC$, $h^2 = b^2 - x^2$ (3)

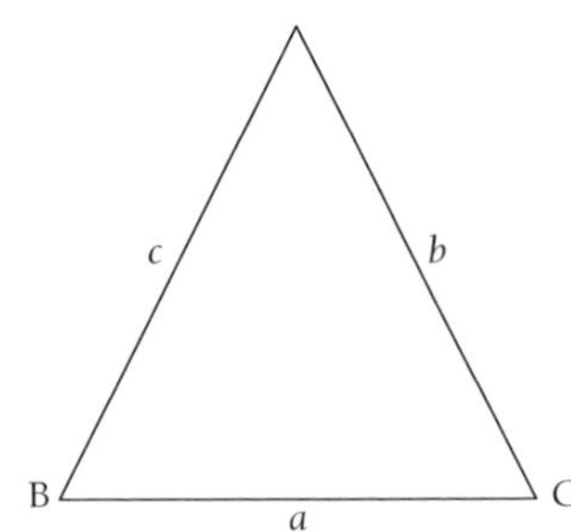

Equate the two values of h^2 to one another:

$$c^2 - (a-x)^2 = b^2 - x^2$$

$$c^2 - (a^2 - 2ax + x^2) = b^2 - x^2$$

$$c^2 - a^2 + 2ax - x^2 = b^2 - x^2$$

$$2ax = a^2 + b^2 - c^2$$

$$x = \frac{a^2 + b^2 - c^2}{2a}$$

Substitute x into equation 3:

$$h^2 = b^2 - \left(\frac{a^2 + b^2 - c^2}{2a}\right)^2$$

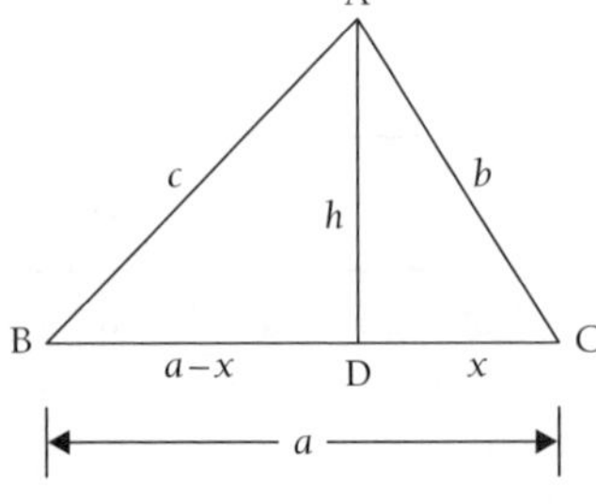

$$h^2 = \left(b + \frac{a^2 + b^2 - c^2}{2a}\right)\left(b - \frac{a^2 + b^2 - c^2}{2a}\right)$$

$$h^2 = \left(\frac{2ab + a^2 + b^2 - c^2}{2a}\right)\left(\frac{2ab - a^2 - b^2 + c^2}{2a}\right)$$

$$h^2 = \frac{\left[\left(2ab + a^2 + b^2\right) - c^2\right]\left[c^2 - \left(a^2 + b^2 - 2ab\right)\right]}{4a^2}$$

$$h^2 = \frac{[(a+b)^2 - c^2][c^2 - (a-b)^2]}{4a^2}$$

$$h^2 = \frac{\left[(a+b+c)(a+b-c)\right]\left[(c+a-b)(c-(a-b)\right]}{4a^2}$$

$$h^2 = \frac{\left[(a+b+c)(a+b-c)\right]\left[(a-b+c)(c+b-a)\right]}{4a^2}$$

Use $2s = a + b + c$ to express the **factors** in terms of s, a, b and c.

$a + b = 2s - c$, $b + c = 2s - a$, $a + c = 2s - b$

$$h^2 = \frac{(a+b+c)(a+b-c)(a+c-b)(b+c-a)}{4a^2}$$

$$h^2 = \frac{2s(2s-c-c)(2s-b-b)(2s-a-a)}{4a^2}$$

$$h^2 = \frac{2s(2s-2c)(2s-2b)(2s-2a)}{4a^2}$$

$$h^2 = \frac{16s(s-a)(s-b)(s-c)}{4a^2}$$

$$h^2 = \frac{4s(s-a)(s-b)(s-c)}{a^2}$$

$$h = \frac{\sqrt{4s(s-a)(s-b)(s-c)}}{\sqrt{a^2}}$$

$$h = \frac{2\sqrt{s(s-a)(s-b)(s-c)}}{a}$$

Substitute for h in (1), also taking into consideration that the base of the ΔABC above is a.

Area of ABC $= \frac{1}{2} \times b \times h$

$$= \frac{1}{2} \times a \times \frac{2\sqrt{s(s-a)(s-b)(s-c)}}{a}$$

$$= \sqrt{s(s-a)(s-b)(s-c)}$$

Example A

Q. Find the area of the following triangles:

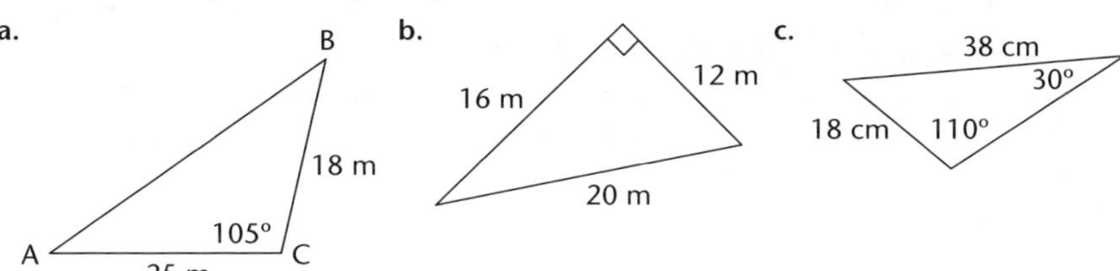

A. **a.** We know the length of two sides: $a = 18$ m and $b = 25$ m, and the magnitude of the included angle $C = 105°$, so we can use the formula:

$$\text{Area of triangle} = \frac{1}{2}\,ab \sin C$$

$$= \frac{1}{2} \times 18 \times 25 \times \sin 105°$$

$= 217.33\ \text{m}^2$, correct to two **decimal places**.

b. This triangle is a right-angled triangle so can be considered to have a base (the side of length 12 m) and a perpendicular height (the side of length 16 m). Hence:

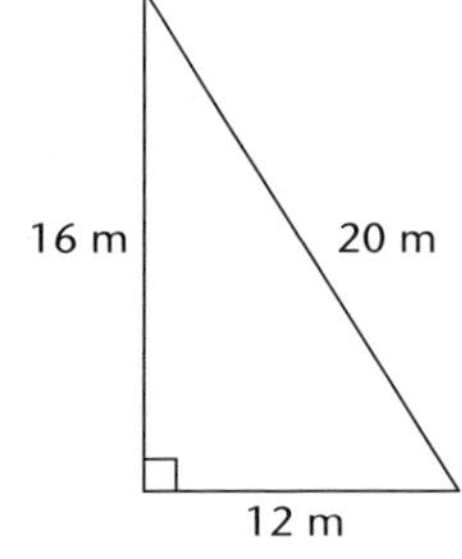

$$\text{Area of triangle} = \frac{1}{2}\,b \times h$$

$$= \frac{1}{2} \times 12 \times 16$$

$$= 96\ \text{m}^2$$

c. We are given two side lengths for this triangle but not the included angle. The included angle can be found using the rule: 'the angles of a triangle sum to 180°':

Included angle $= 180° - 30° - 110°$

$= 40°$

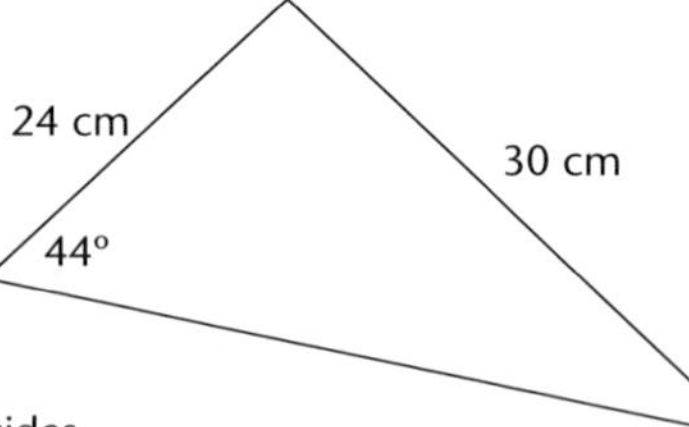

$$\text{Area of triangle} = \frac{1}{2} \times 38 \times 18 \times \sin 40°$$

$= 219.83\ \text{cm}^2$, correct to two decimal places.

Example B

Q. Use the sine rule to find the other angles in this triangle and hence the area of the triangle:

A. The sine rule can be used to find θ and then the angle that is the included angle for the two given sides.

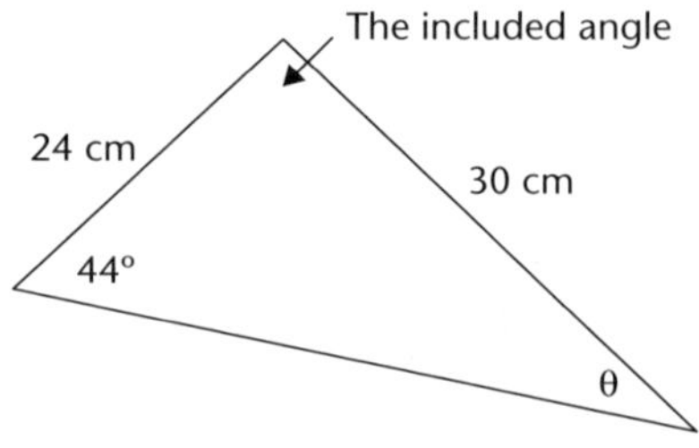

$$\frac{24}{\sin\theta} = \frac{30}{\sin 44°}$$

$$\sin\theta = \frac{24 \sin 44°}{30}$$

$$\sin\theta = 0.5557$$

$$\theta = 33.76°$$

Included angle $= 180° - 44° - 33.76° = 102.24°$

$$\text{Area of triangle} = \frac{1}{2} \times 24 \times 30 \times \sin 102.24° = 351.82\ \text{cm}^2$$

Example C

Q. Find the area of a triangular fish pond whose sides are 4, 5 and 7 metres.

A. Using Heron's formula:

$s = \frac{1}{2}$ (the perimeter of the triangle)

$s = \frac{a+b+c}{2} = \frac{4+5+7}{2} = \frac{16}{2} = 8$

Substituting this value and the length of the three sides in the formula:

$\text{Area} = \sqrt{s(s-a)(s-b)(s-c)}$

$\text{Area} = \sqrt{8(8-4)(8-5)(8-7)}$

$= \sqrt{8 \times 4 \times 3 \times 1}$

$= \sqrt{96}$

$= \sqrt{16 \times 6}$

$= 4\sqrt{6}\ \text{m}^2 \approx 9.80\ \text{m}^2$

Note: The area of this triangle can be found by first using the cosine rule to find the size of one of the angles and then using this formula: Area $= \frac{1}{2} bc \sin A$

Unit 12.1 Activity 3: Area of a triangle

Give your answers correct to two decimal places.

1. Find the area of each of the following triangles:

a.

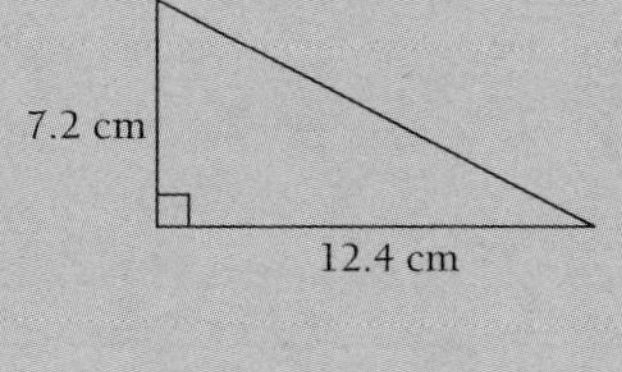

b.

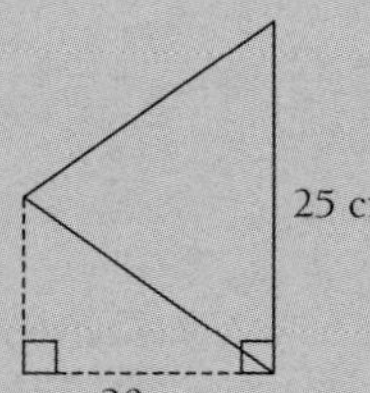

c.

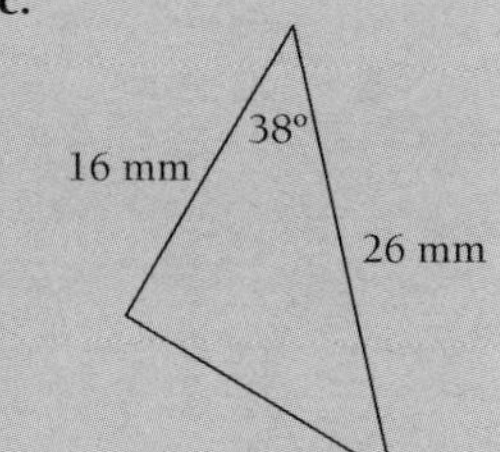

d.

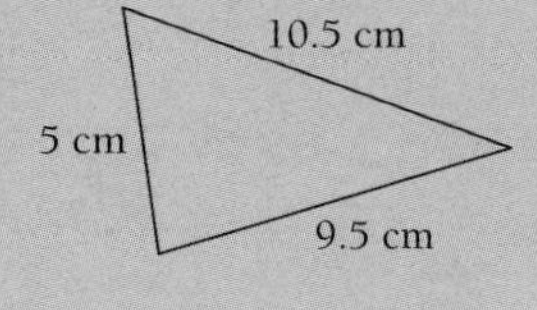

e.

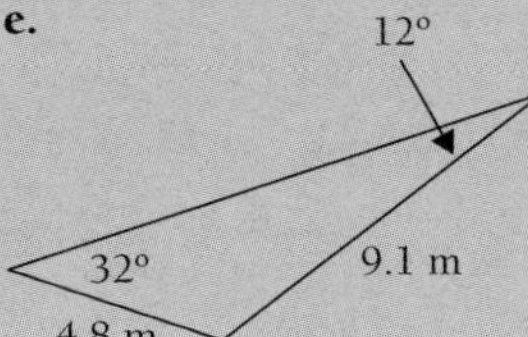

f.

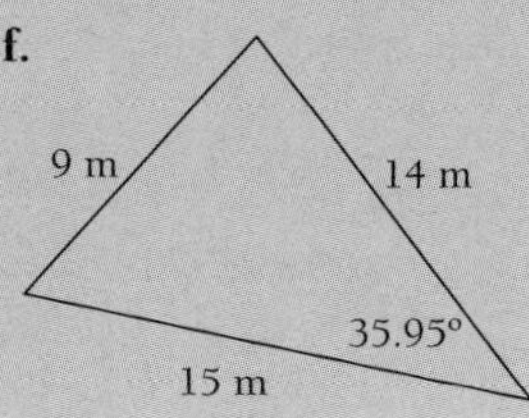

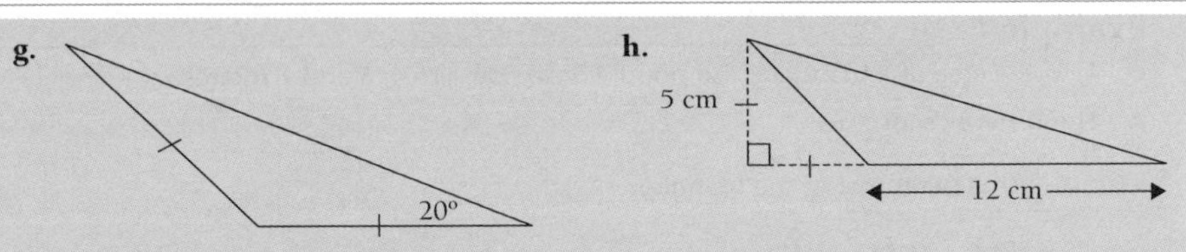

2. Find the area of each of the following triangles. You may need to find an appropriate side length or angle, using a trigonometric ratio or the sine or cosine rule, before you can calculate the area.

a.

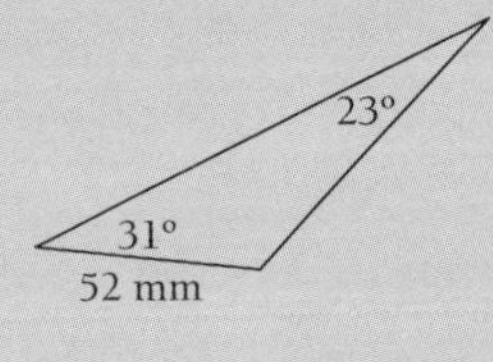

b.

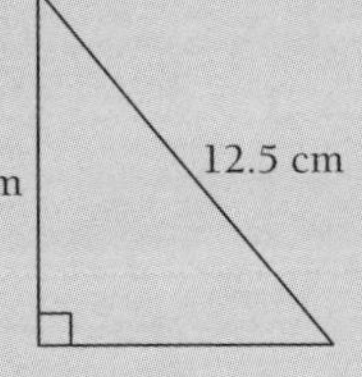

c.

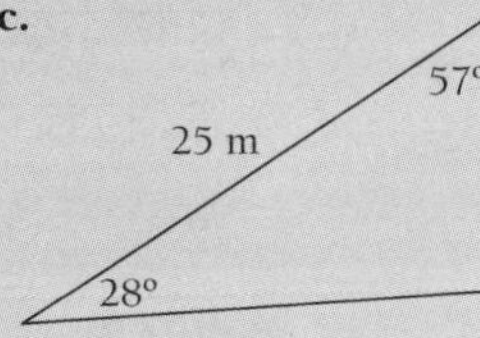

d.

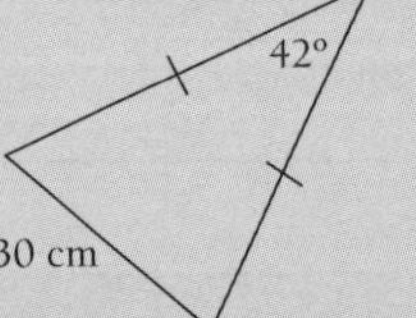

e.

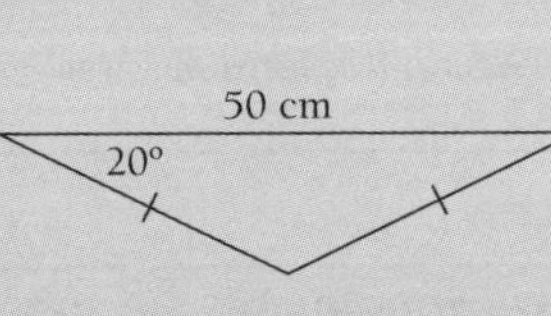

f.

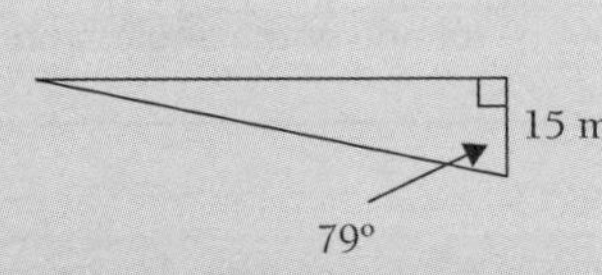

3. Which one of the following would be a calculation for the area of the triangle given, using Heron's formula?

A. $\sqrt{56 \times 32 \times 37 \times 43}$

B. $\sqrt{56 \times 4 \times 9 \times 15}$

C. $28\sqrt{4 \times 9 \times 15}$

D. $\sqrt{28 \times 4 \times 9 \times 15}$

E. $28\sqrt{32 \times 37 \times 43}$

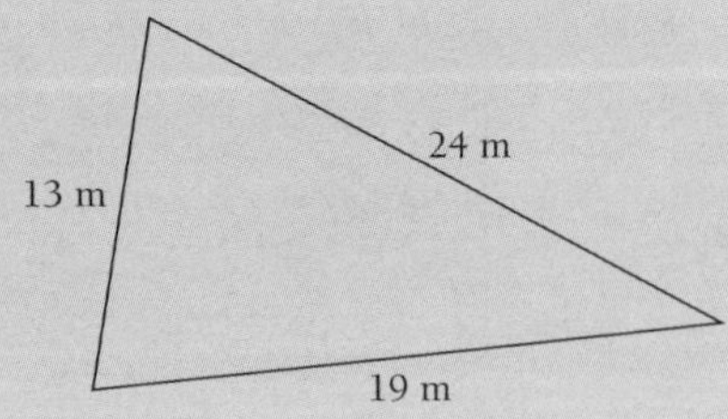

Unit 12.1 Measurement

Topic 4: Scales and dimension – polygons

In line with the Syllabus (p. 20) we now turn our attention to perimeter and area of polygons:

- Calculating the area of a polygon by identifying triangular shapes.

About polygons

A **polygon** is a many-sided, closed figure with straight sides. There are commonly used names that apply to some polygons:

3 sides: triangle
4 sides: **quadrilateral**
5 sides: **pentagon**
6 sides: hexagon
7 sides: heptagon
8 sides: **octagon**
9 sides: nonagon
10 sides: decagon
20 sides: icosagon

The **area** of a polygon can be found by dividing the polygon into triangles, finding the area of the triangles and then summing these areas.

Quadrilateral: 2 triangles

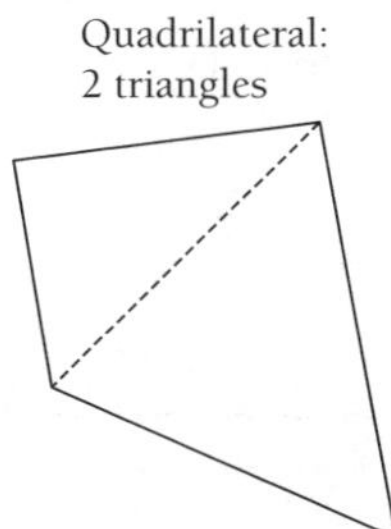

Pentagon : 3 triangles

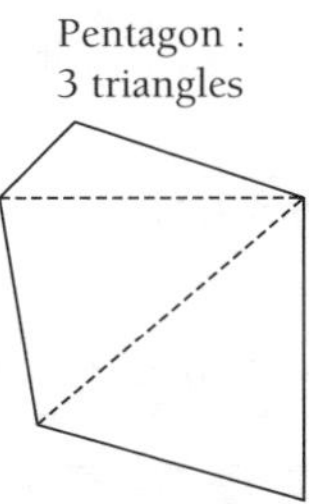

Hexagon : 4 triangles

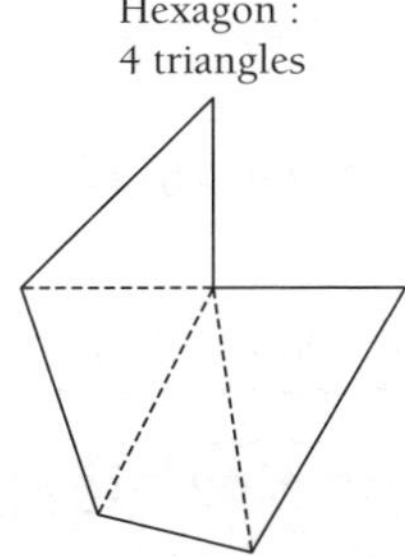

Nonagon : 7 triangles

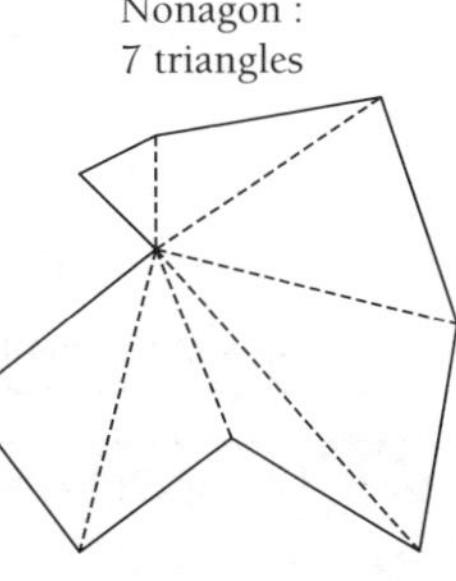

An n-sided polygon can be divided into $(n - 2)$ triangles.

Example A

Q. Find the area of this pentagon:

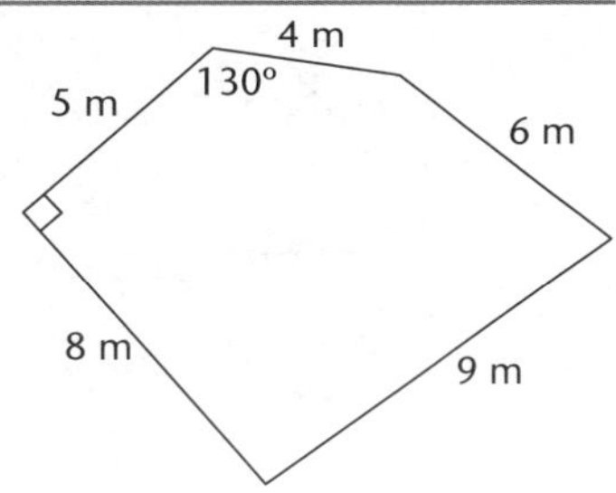

A. The pentagon can be divided into three triangles:

Triangle 1 is a right-angled triangle, so:

Area triangle 1 = $\frac{1}{2} \times 5 \times 8 = 20\text{ m}^2$

In triangle 1:

$\tan\theta = \frac{8}{5}$; $\theta = 57.99°$

Length of the hypotenuse $= \sqrt{5^2 + 8^2} = 9.434$ m

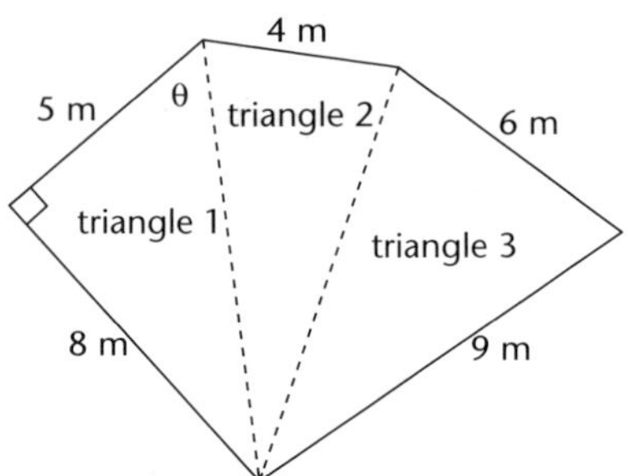

Triangle 2

Area of triangle 2 $= \frac{1}{2} \times 4 \times 9.434 \sin 72.01°$

$= 17.946$ m^2

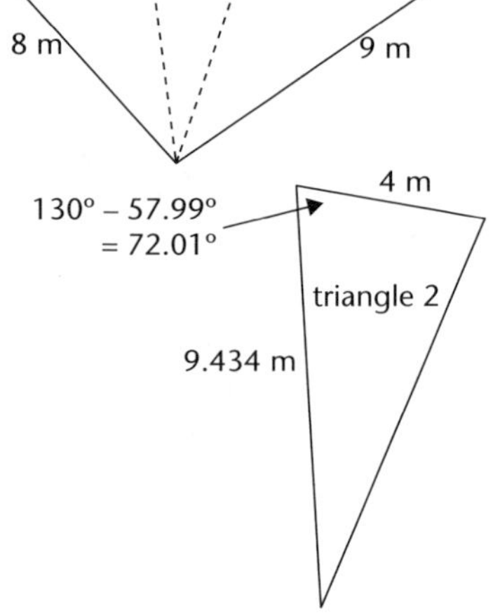

The length of the third side in triangle 2 can be found using the cosine rule:

Length $= \sqrt{4^2 + 9.434^2 - 2 \times 4 \times 9.434 \cos 72.01°}$

$= 9.038$ m

Triangle 3

The length of the three sides in triangle 3 are known, so Heron's formula can be used to find the area:

$s = \frac{1}{2}(6 + 9 + 9.038) = 12.019$

Area $= \sqrt{12.019(12.019 - 6)(12.019 - 9)(12.019 - 9.038)}$

$= 25.516$ m^2

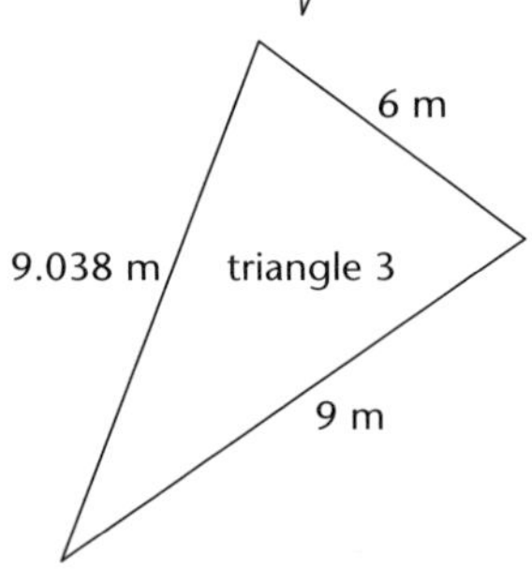

Area of the pentagon = 20 + 17.946 + 25.516 = 63.46 m^2

Unit 12.1 Activity 4: Area of a polygon

Give your answers correct to two decimal places.

1. Find the area of each of the following figures:

a. A parallelogram

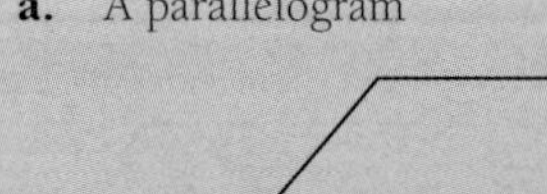

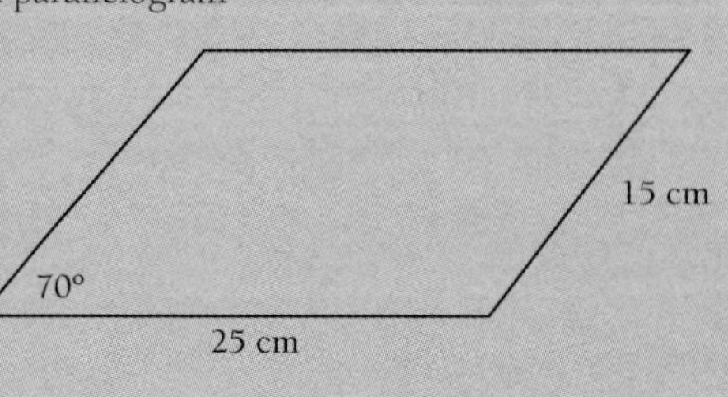

b.

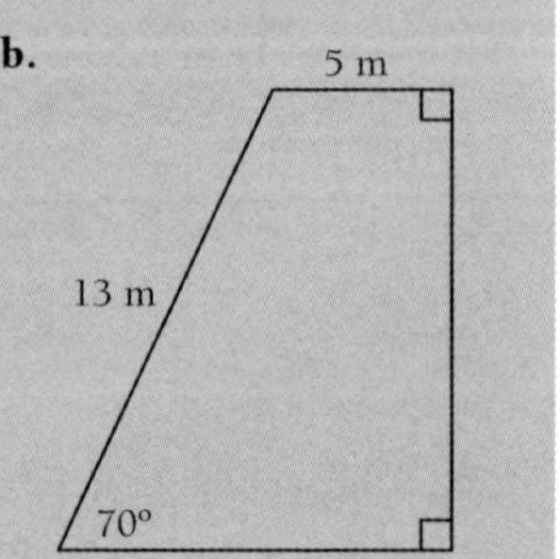

c.

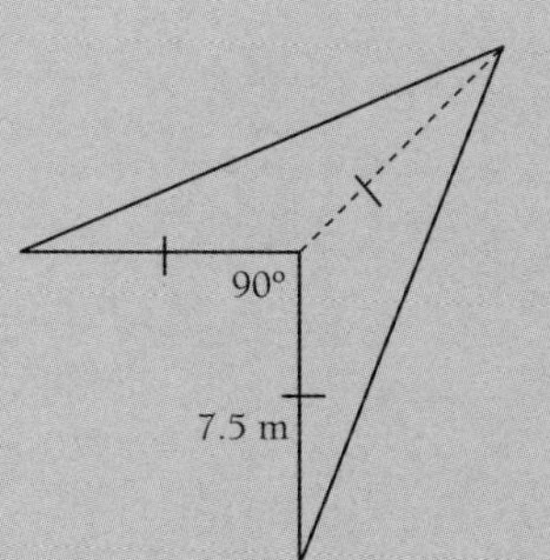

d.

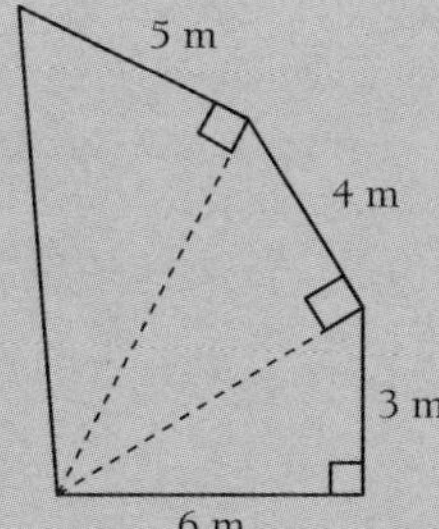

e.

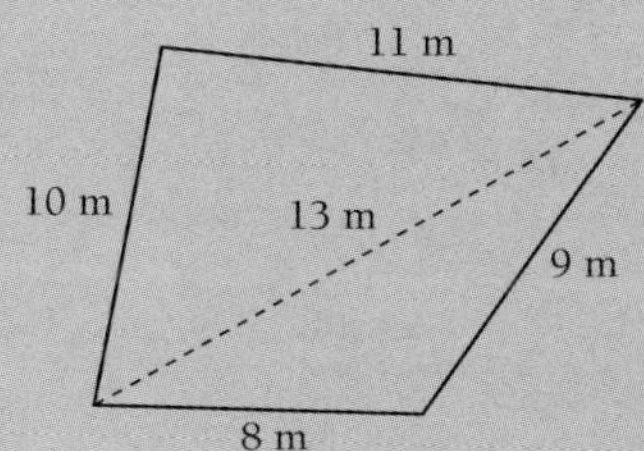

f. A rhombus

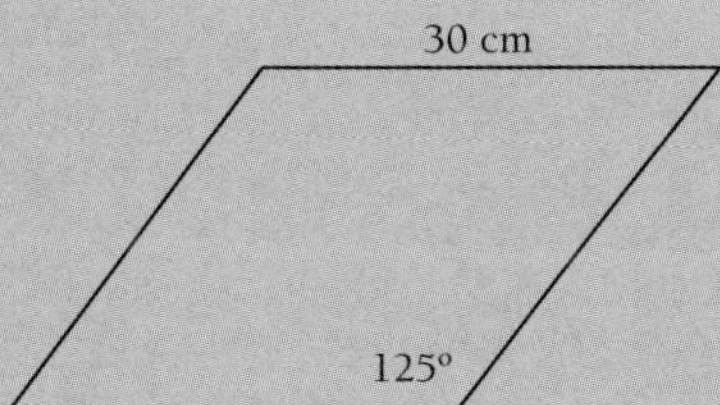

2. Find the area of each of the following polygons:

a.

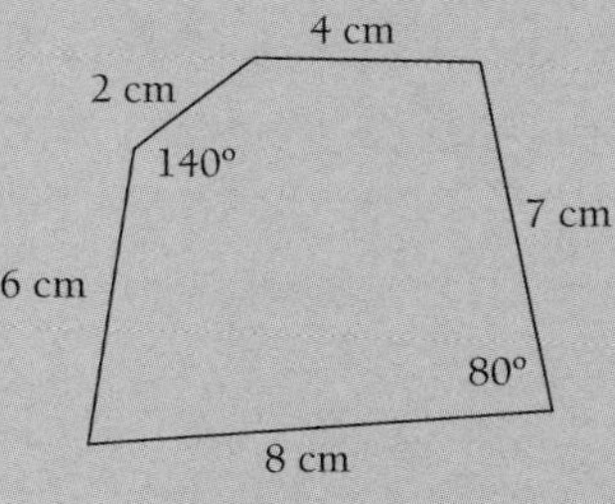

b.

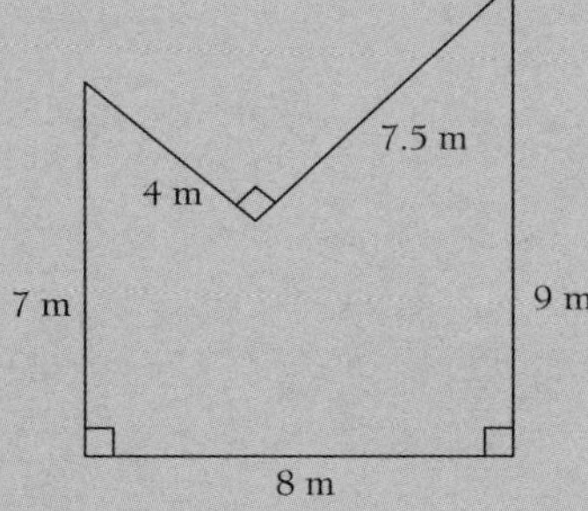

c.

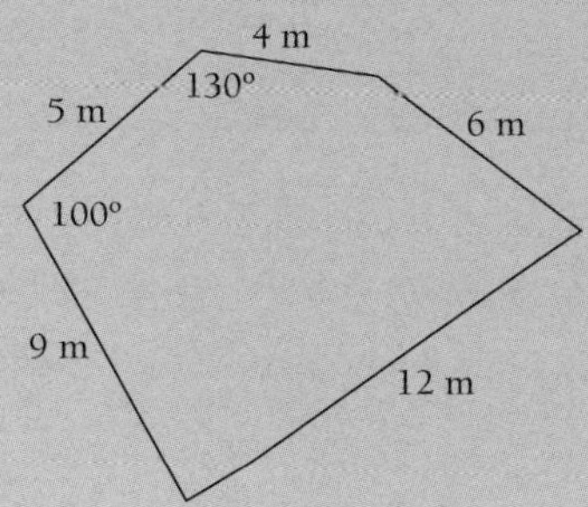

d.

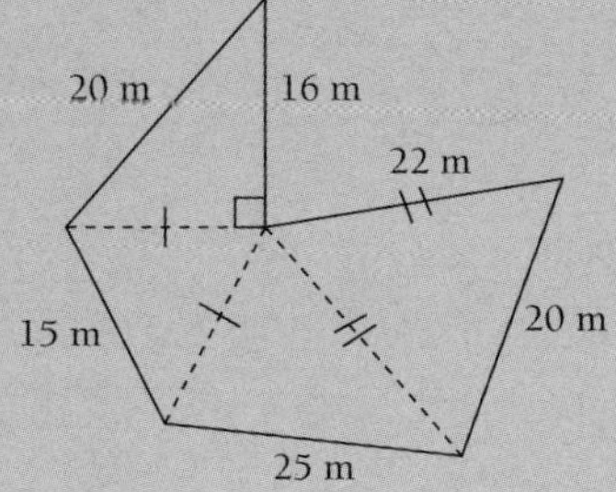

3. Find the perimeter and area of this pentagonal field:

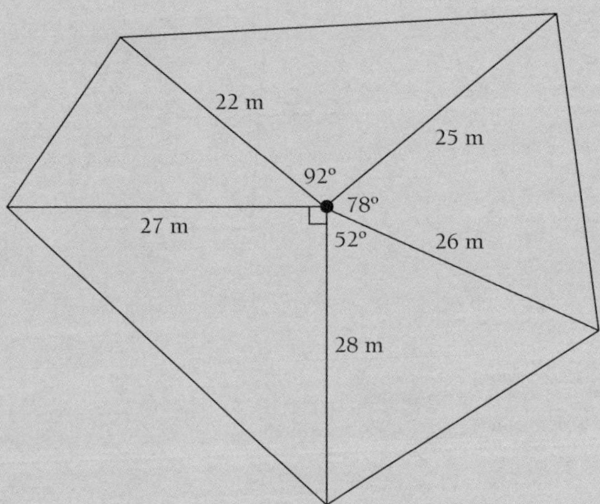

4. Find the perimeter and area of the following pentagon:

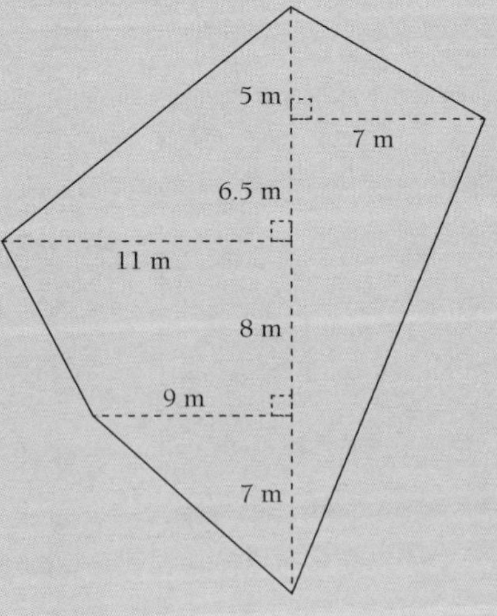

Unit 12.1 Measurement

Topic 5: Scales and dimension – regular polygons

Topic 5 builds on the information in the previous Topic and looks at regular polygons, focusing on:

- Calculating the area of a polygon by identifying triangular shapes.
- Properties of regular polygons.

A **regular** polygon has sides of equal length, and interior angles of the same size.

A regular polygon of n sides can be divided into n isosceles triangles by joining the vertices to the '**centre**'.

For example, this regular eight-sided polygon (octagon) is divided into eight isosceles triangles:

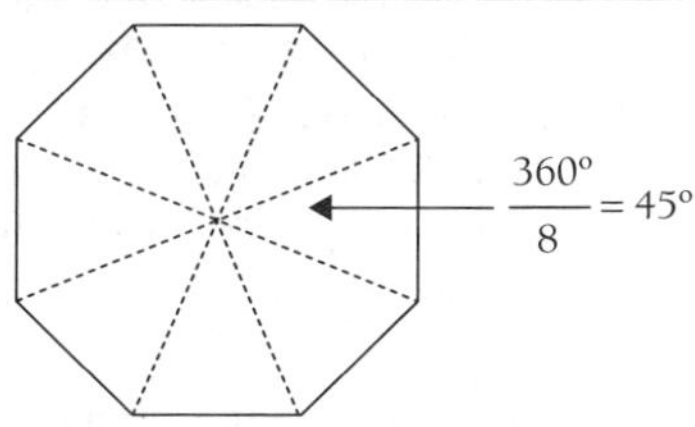

The angle at the centre, 360°, is divided equally into n parts, hence:

> The angle in the triangle at the 'centre' of an n-sided regular polygon = $\left(\frac{360}{n}\right)^{\circ}$

An **exterior angle** of a regular polygon is the angle between the extension of one of the sides and an adjacent side.

For an n-sided regular polygon the n exterior angles are all the same size:

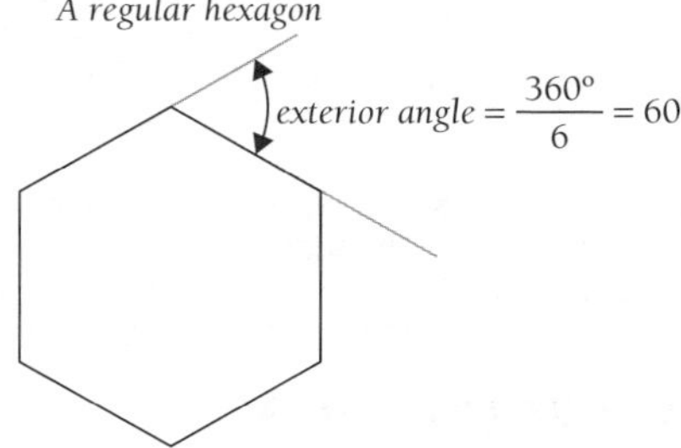

The sum of the exterior angles of a regular polygon is 360°.

> For an n-sided regular polygon each **exterior angle** = $\left(\frac{360}{n}\right)^{\circ}$

An **interior angle** of a regular polygon is the angle between two adjacent sides, on the interior of the polygon. The interior angles are all the same size in a regular polygon:

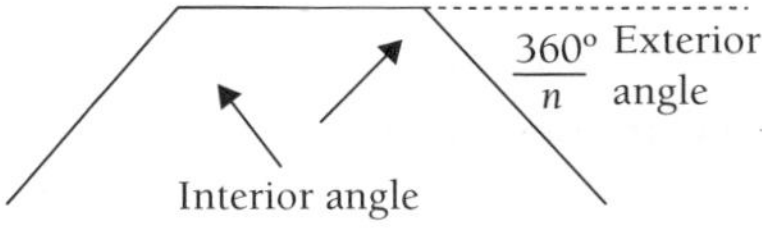

The interior angle and the exterior angle of a polygon form a **straight angle** and therefore sum to 180°.

> The **interior angle** of a regular n-sided polygon = 180° – the exterior angle
> $= 180° - \left(\frac{360}{n}\right)^{\circ}$

Example A

Q. **1.** Find the exterior angle and interior angle of a regular pentagon (5 sides).

2. Find the angle in the triangle, at the centre, when the polygon is divided into triangles.

A. **1.** The sum of the exterior angles of any regular polygon is 360°, so if there are five exterior angles on a pentagon, all of equal size, then each angle is:

$$\frac{360°}{5} = 72°$$

The sum of the interior and exterior angles is 180°, so the interior angle is 180° – 72° = 108°.

Exterior angle
Interior angle
108°
72°

2. The angle at the 'centre' is a full revolution, 360°, so the angle at the centre of each of the triangles is

$$\frac{360°}{5} = 72°$$

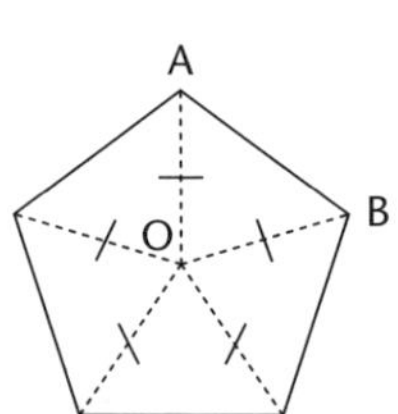

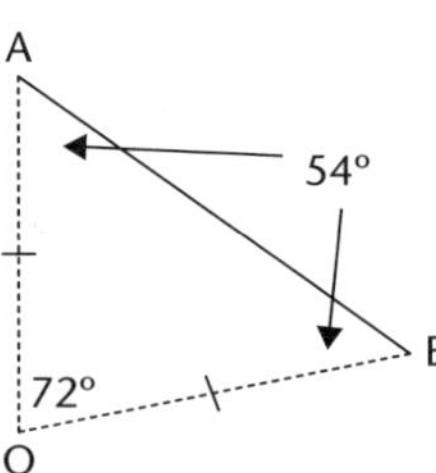

Note: The angle at the centre in the triangle has the same size as the exterior angle in a regular polygon. The other two angles in the triangle will be equal in size because this is an isosceles triangle. They are each (180° – 72°)/2 = 54° (half the size of the interior angle).

The area of a regular polygon

The size of the angle at the centre in a regular polygon is known if the number of sides is known. The only measurement that is needed to find the area is the length of the side of the polygon.

Example B

Q. Find the area enclosed by a regular polygon with nine sides if the sides are of length 5 metres.

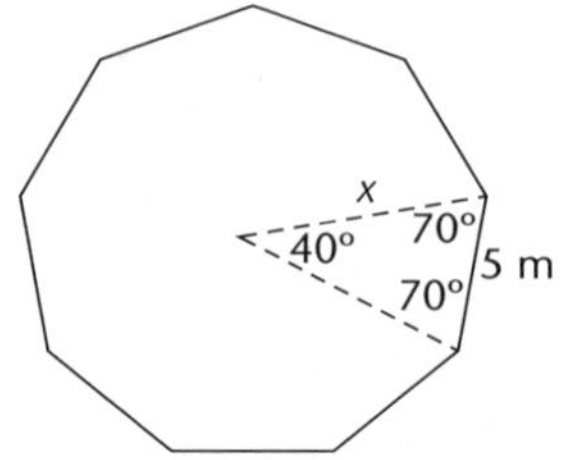

A. The angle at the centre of the nonagon is $\frac{360°}{9} = 40°$

so the other two angles in the isosceles triangle are:

$$\frac{180° - 40°}{2} = 70°$$

Using the sine rule to find the length, x metres, of the other sides of the triangle:

$$\frac{x}{\sin 70°} = \frac{5}{\sin 40°}$$

$$x = \frac{5\sin 70°}{\sin 40°} = 7.310 \text{ m}$$

We now have the length of two sides (both 7.31 m) and the included angle (40°).

The area of one of the nine triangles $= \frac{1}{2} \times 7.31 \times 7.31 \times \sin 40° = 17.172 \text{ m}^2$

The area of the whole nonagon $= 9 \times 17.172 = 154.55 \text{ m}^2$

Unit 12.1 Activity 5: Regular polygons

1. Find the size of the interior angles of:

a. A regular triangle.

b. A regular hexagon.

c. A regular polygon with 12 sides.

2. Find the angles p, q, r and s in each of the following regular polygons (O is the centre of each polygon):

a. A square.

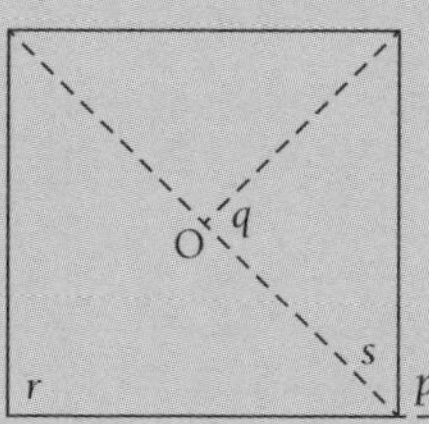

b. A pentagon.

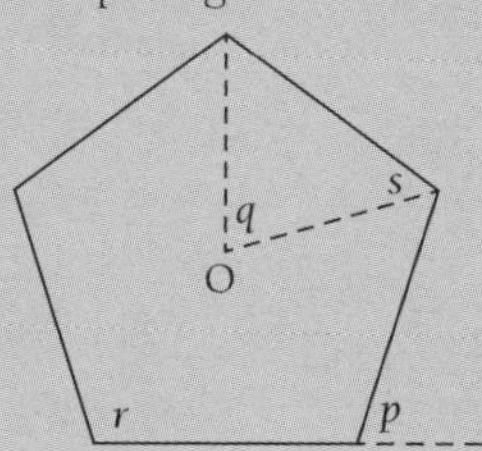

c. An octagon.

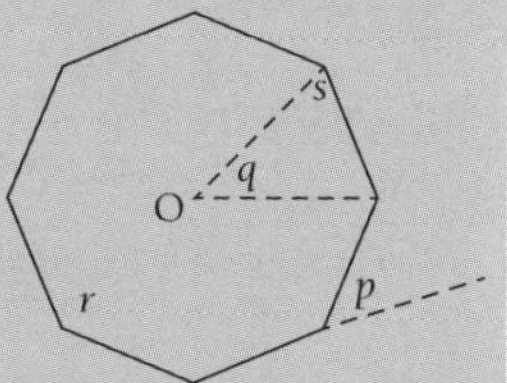

3. Find the area of:

a. A regular pentagon of side length 6 m.

b. A regular octagon of side length 4.5 cm.

c. A regular 16-side polygon of side length 12 m.

d. A regular heptagon of side length 10 cm.

4. A five-pointed star is formed by extending the sides of a regular pentagon.

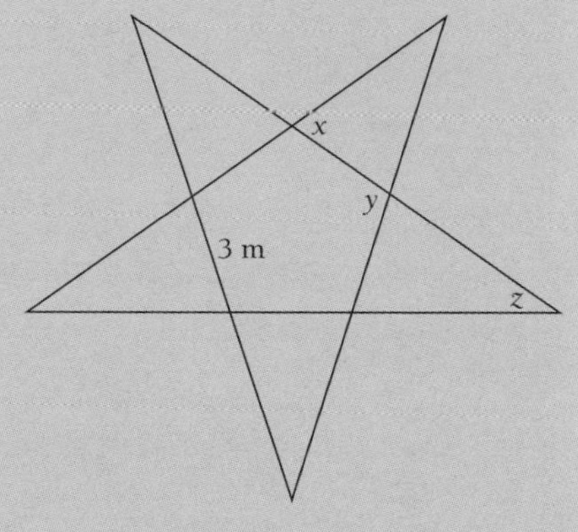

a. Find the size of the angles:

i. x

ii. y

iii. z

b. Find the area of the five-pointed star.

5. A regular polygon has an interior angle of 160°. Find the number of sides of this polygon.

Unit 12.1 Measurement

Topic 6: Scales and dimension – volume and total surface area (TSA)

Following the Syllabus (p. 20) the next few Topics examine volume and surface area:

- Calculating the surface area and volumes of pyramids, spheres and truncated solids.

The **volume** of a solid is the amount of space that the solid occupies.

Units for volume are **cubic metres** (m^3), **cubic centimetres** (cm^3) or **cubic millimetres** (mm^3).

1 cubic metre = 100 × 100 × 100 = 1 000 000 cubic centimetres.

1 cubic centimetre = 10 × 10 × 10 = 1 000 cubic **millimetres**.

The liquid **capacity** (volume) of a container is usually quoted in **millilitres** (1 cm^3 = 1 mL), litres (1 000 cm^3 = 1 L), kilolitres (1 000 L = 1 kL) or megalitres (1 000 kL = 1 ML)

The **total surface area (TSA)** of a solid refers to the entire area on its outside.

Units for TSA are the same as units for area: square metres, square centimetres, square millimetres and square kilometres.

1 square metre = 100 × 100 = 10 000 square centimetres

1 square kilometre = 1 000 × 1 000 = 1 000 000 square metres

1 square centimetre = 10 × 10 = 100 square millimetres

The total surface area of solids that have **plane** (flat) surfaces can be found by opening out the figure so that the surfaces are flat. This is called the **net** of the figure.

The volume and total surface area of some solids

Figure	Total surface area (TSA)	Volume
Cuboid All faces are rectangles Length, l Width, w Height, h h w l	**Net** w h l TSA = $2(l \times w + l \times h + w \times h)$	Volume = $l \times w \times h$

Cylinder Radius of base r, height h 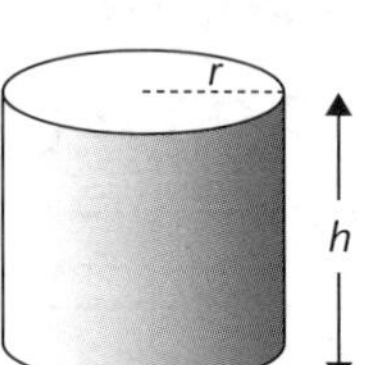	**Net** 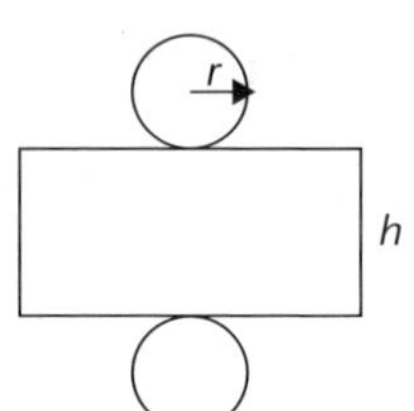 TSA = Area of rectangle + 2 × area of circle $= 2\pi rh + 2\pi r^2$	Volume $= \pi r^2 h$
Sphere Radius r 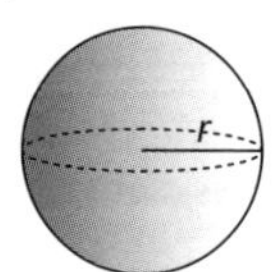	**No net** TSA $= 4\pi r^2$	Volume $= \frac{4}{3}\pi r^3$
Cone Base radius r, height h, slant edge s 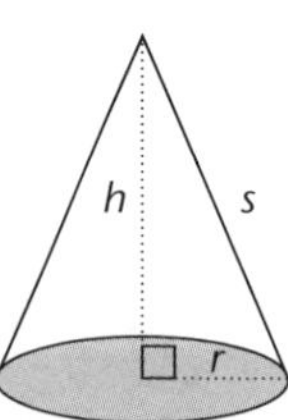	**Net** 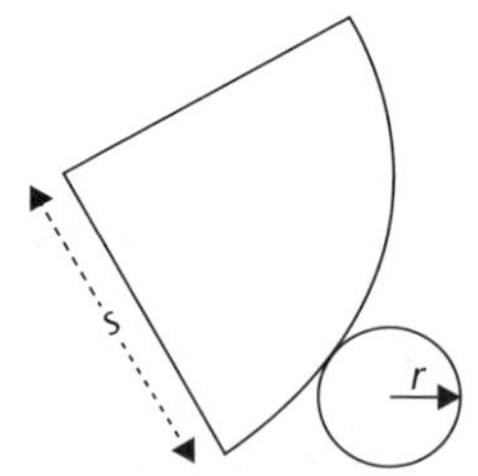 TSA $= \pi rs + \pi r^2$	Volume $= \frac{1}{3}\pi r^2 h$
Right-square pyramid The base is a square and the vertex is above the centre of the base. h is the vertical height of the pyramid and a is the altitude of the sloping side. 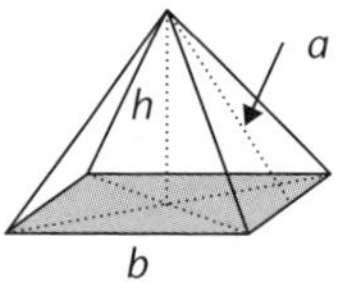	**Net** 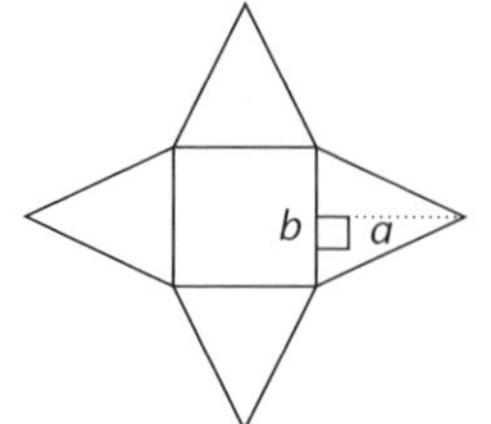 TSA = Area of base + 4 × (area of triangle) $= b^2 + 4 \times (\frac{1}{2}ba)$ $= b^2 + 2ba$	Volume $= \frac{1}{3}b^2 h$ **Note:** The volume of any pyramid $= \frac{1}{3}Ah$ where A is the area of the base and h is the height of the pyramid.

<table>
<tr>
<td>

Solids with a constant cross-section

A **prism** has a constant cross-section and rectangular sides, eg, a triangular prism:

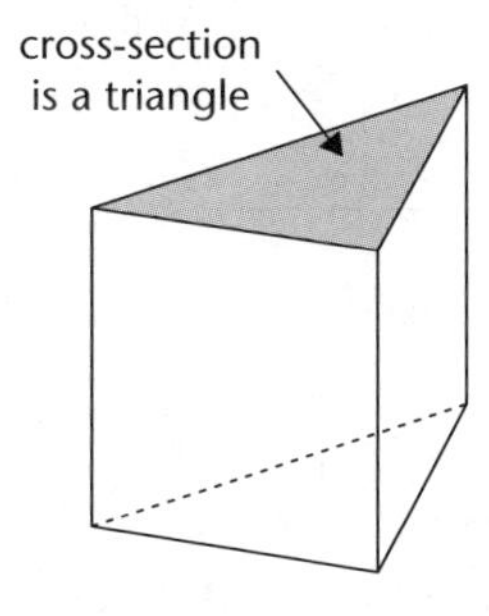

</td>
<td>

Net for a triangular prism

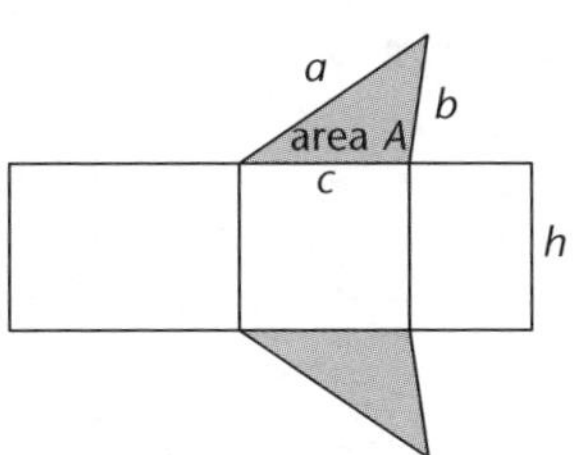

TSA = 2 × area of cross-section + area of rectangular sides.
$= 2 \times A + h(a + b + c)$

</td>
<td>

Volume = area of cross-section × height of sides
$= A \times h$

</td>
</tr>
</table>

Example A

Q. For the composite solid shown, find:

a. The volume.

b. The total surface area.

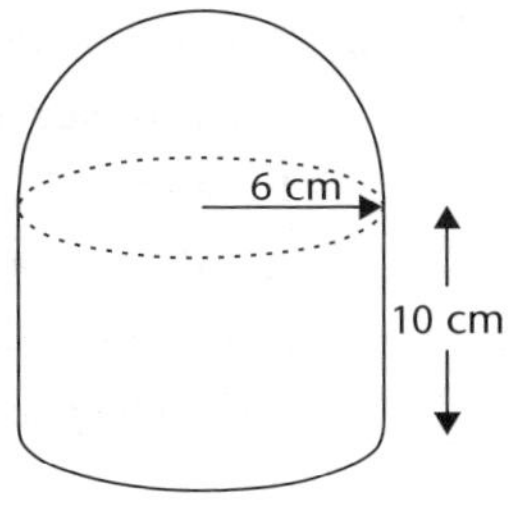

A. a. The solid is a composition of a hemisphere with radius 6 cm and a cylinder of radius 6 cm and height 10 cm.

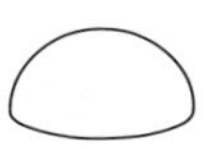

The volume of the hemisphere is $\frac{1}{2} \times \left(\frac{4}{3}\pi \times 6^3\right) = 144\pi$

The volume of the cylinder is $\pi \times 6^2 \times 10 = 360\pi$

Hence the volume of the solid is

$144\pi + 360\pi = 504\pi \approx 1\ 583.36\ \text{cm}^3$

b. The total surface area of the composite figure

= (half the TSA of a sphere) + (the TSA of a cylinder) – (the area of the top of the cylinder)

$$\text{TSA} = \frac{1}{2}\left(4\pi r^2\right) + \left(2\pi rh + 2\pi r^2\right) - \pi r^2$$

$$= 3\pi r^2 + 2\pi rh$$

$$= 3 \times \pi \times 6^2 + 2 \times \pi \times 6 \times 10$$

$$= 716.28\ \text{cm}^2$$

Example B

Q. For the triangular prism shown at right, find:

a. The total surface area.

b. The volume.

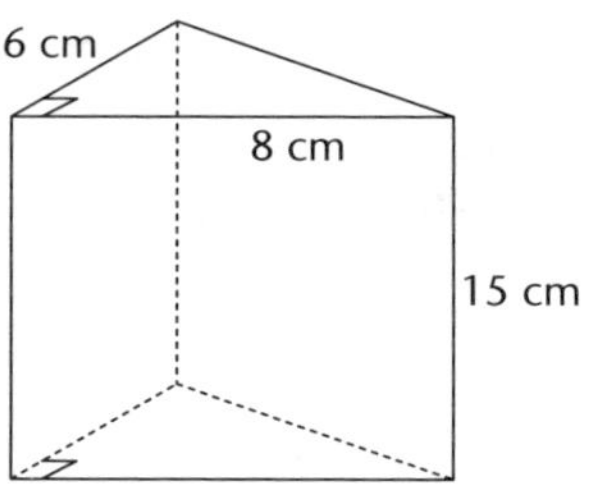

A. a. A net of the five surfaces is given below:

6 cm | 10 cm | 6 cm | 8 cm | 10 cm | 15 cm

Area 1 $6 \times 15 = 90\text{ cm}^2$ | **Area 2** $8 \times 15 = 120\text{ cm}^2$ | **Area 3** $10 \times 15 = 150\text{ cm}^2$

Total surface area

$= 2 \times (\text{area of triangle}) + \text{Area 1} + \text{Area 2} + \text{Area 3}$

$= 2 \times (\frac{1}{2} \times 8 \times 6) + 6 \times 15 + 8 \times 15 + 10 \times 15$

$= 48 + 90 + 120 + 150$

$= 408$

The total surface area of the prism is 408 cm^2

b. Volume of a prism = area of cross-section × height

The **cross-section** is triangular:

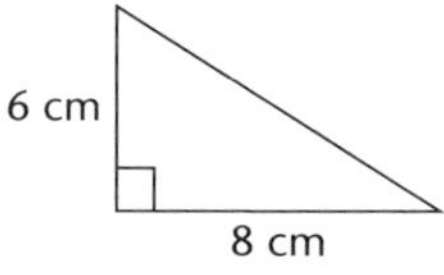

The area of the cross-section is

$\frac{1}{2} \times 8 \times 6 = 24\text{ cm}^2$

and the height is 15 cm.

Substituting in: volume of a prism = area of cross-section × height

$= 24 \times 15$

$= 360$

The volume of the prism is 360 cm^3

Unit 12.1 Activity 6: Volume and surface area

Give your answers correct to the nearest **whole number**.

1. Find the volume of the following solids:

a.

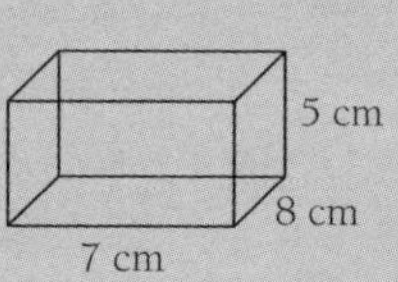

b.

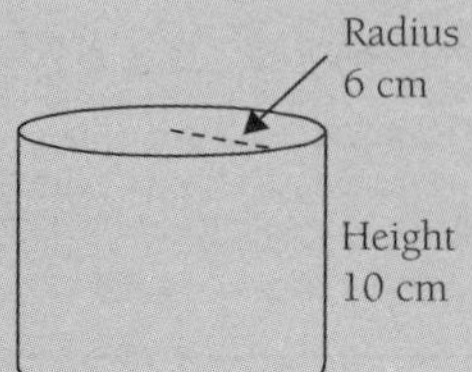

c.

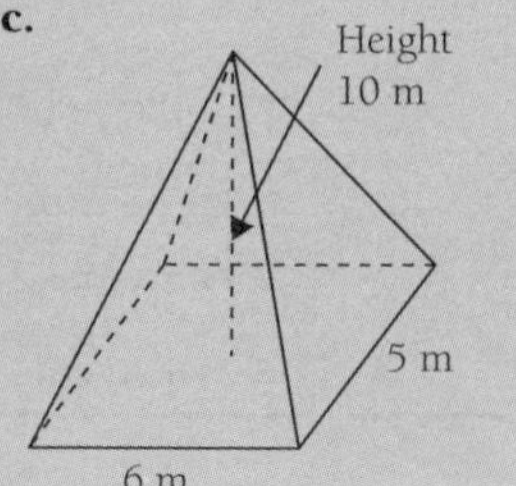

d. Cone.

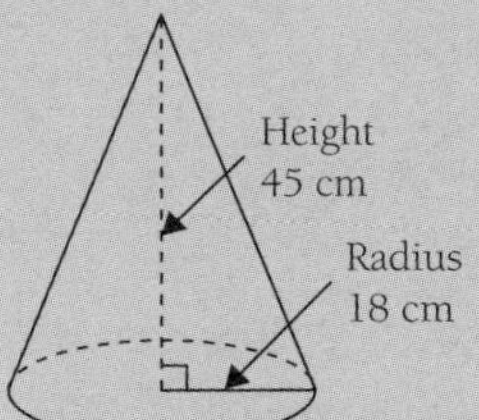

e. Hemisphere.

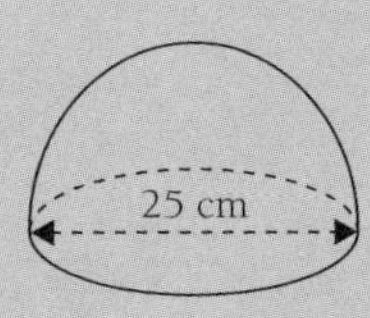

f. Triangular prism.

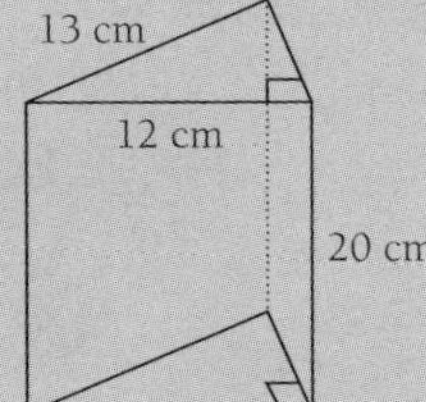

g. Face: semicircles.

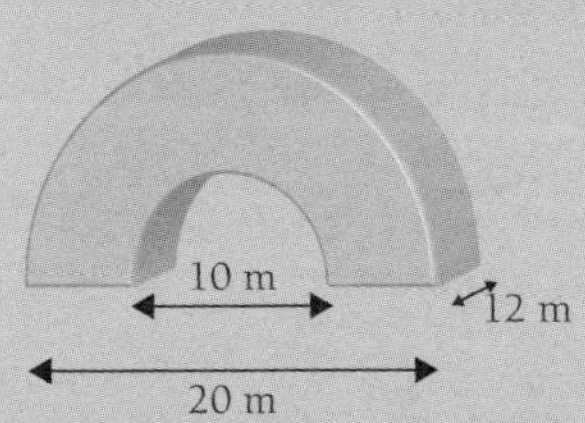

h. Ends: equilateral triangles of side length 2 metres and length 2.5 metres.

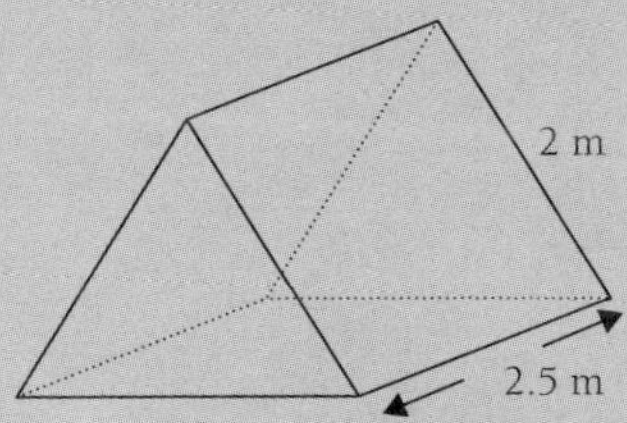

2. Find the total surface area and the volume of the following solids:

a. Cuboid: 15 × 12 × 10 cm.

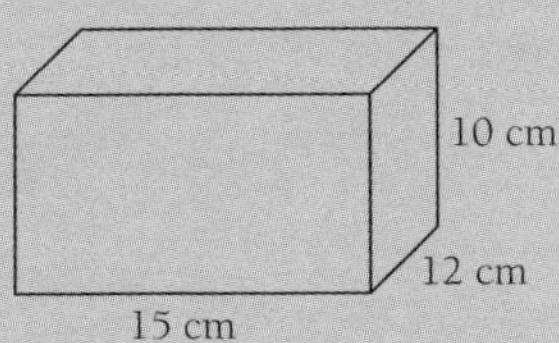

b. Cylinder: radius 7 cm, height 18 cm.

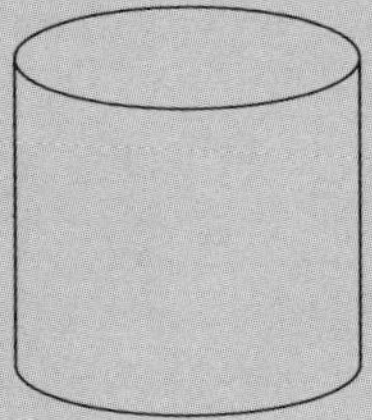

c. Hemisphere.

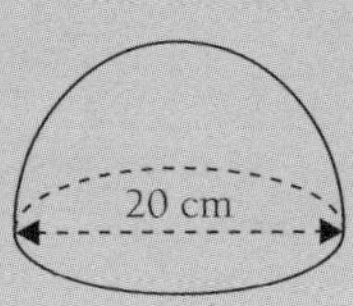

d. Triangular prism.

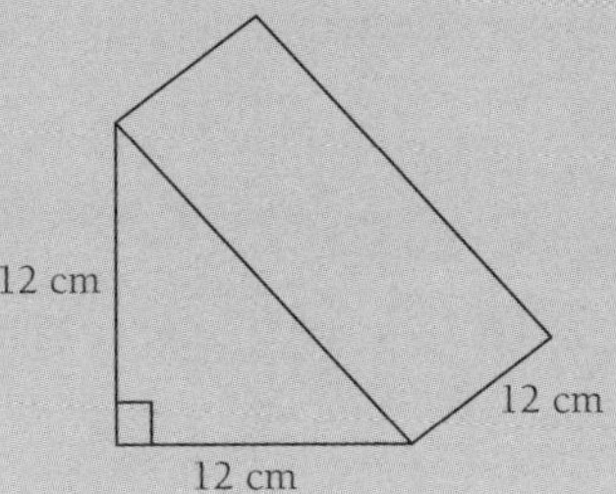

e. Prism.

Cross-section is a regular pentagon of area 688.2 cm^2

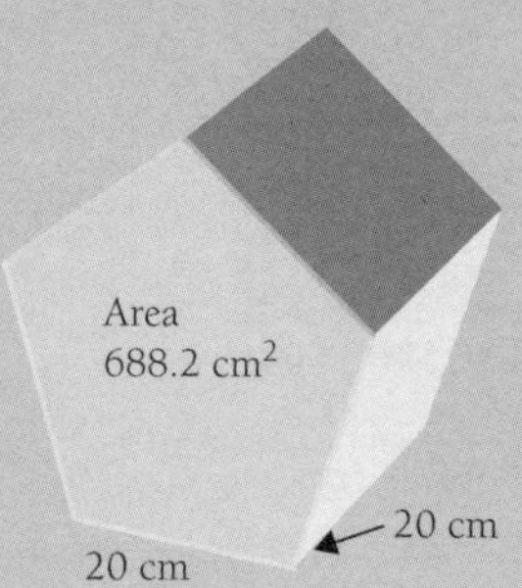

f. Cone.

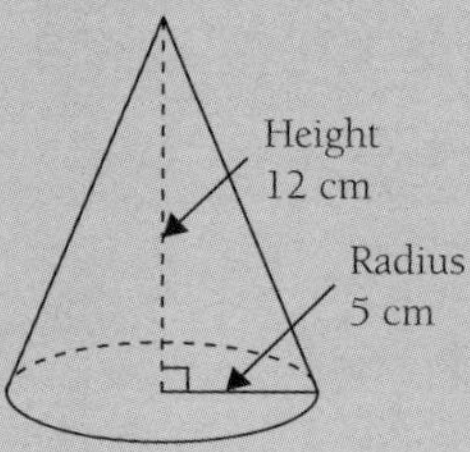

g. Pyramid.

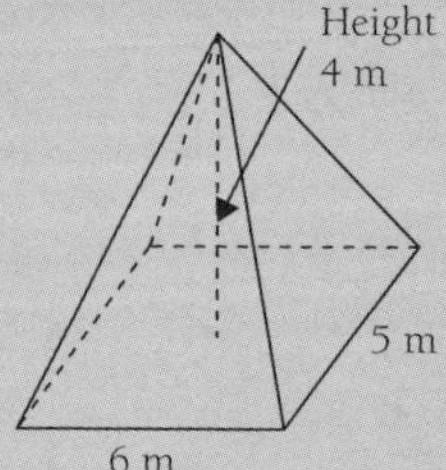

3. Grain poured at a certain speed from a chute forms a conical pile that is as high as the diameter of the conical pile. Find the volume and TSA of such a conical pile of grain that is 3.2 metres high.

4. A bamboo pole has an inside diameter of 4 cm. What is the length of pole that will hold 2 litres of liquid?

5. A container for water has been made from half of a large cylindrical container. The radius of the circular end is 305 mm and the height (length) of the cylinder is 880 mm. Find the maximum amount of water, in litres, that can be held in this half-cylinder container.

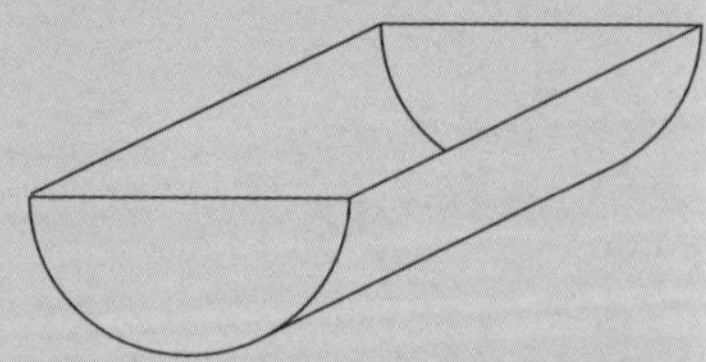

6. A cylindrical water tank holds 10 000 litres of water. If the height of the tank is 1.8 m, find:

a. The diameter of the tank.

b. The TSA of the tank.

Give your answers correct to two decimal places.

7. A food container is in the shape of the pyramid shown at right. The top of the container is open and is a square. The height of the container is 14 cm. Find:

a. The capacity of the container in cm^3.

b. The area of material needed to make the container in cm^2 (no top required).

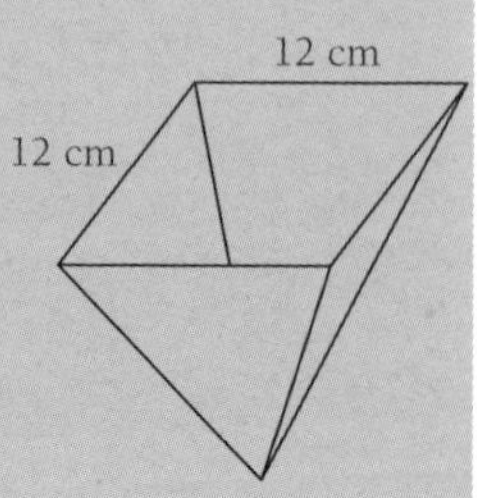

8. A drink container is in the shape of a tetrahedron where each of the four faces is an equilateral triangle of side length 11 cm. The height of this triangular pyramid is 8.98 cm. Find:

a. The capacity of the container in cubic cm.

b. The area of material needed to make the container in cm^2.

11 cm

9. A drink container is in the shape shown at right. The container is 12 cm high at its highest point and 9 cm high at the other corners. Find:

a. The capacity of the container in cubic cm.

b. The area of material needed to make the container in cm^2.

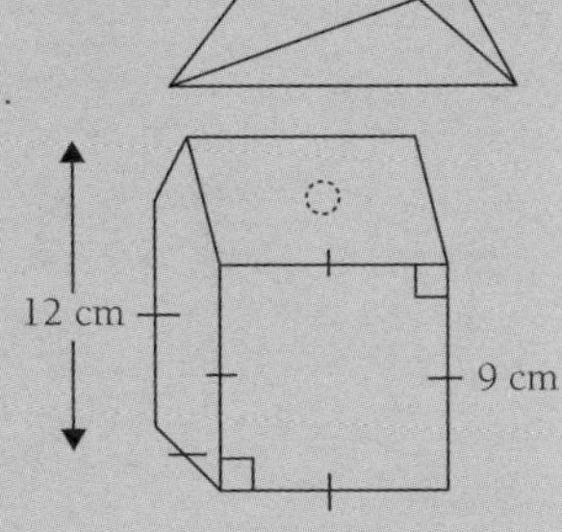

10. The diagram shows a shallow trough. The ends are in the shape of a trapezium. How long will it take to fill the trough from a hose which delivers water at a rate of 24 litres per minute?

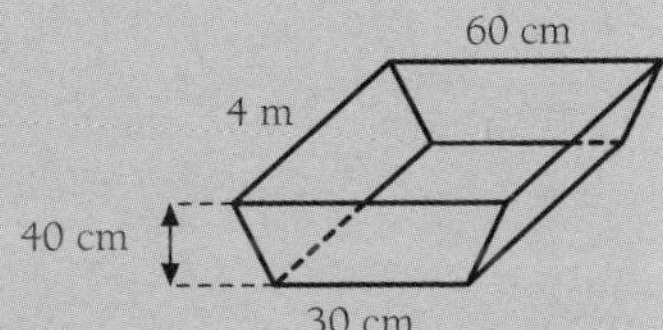

11. The diagram shows the swimming pool for a new hotel in Port Moresby. The pool is filled at the rate of 2 000 L per minute. The pool is filled to within 10 cm of the top of the pool.

a. How long does it take to fill the pool?

b. An old hotel paid water charges of K176.25 to fill its pool which has a capacity of 150 000 L. At this rate of water charges, how much will it cost to fill the swimming pool at the new hotel?

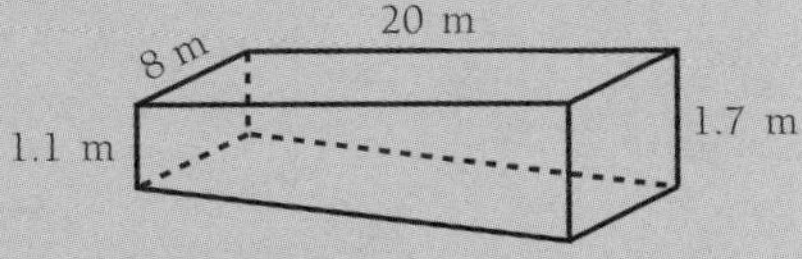

c. Comment on how exact your answers in parts **a.** and **b.** are likely to be.

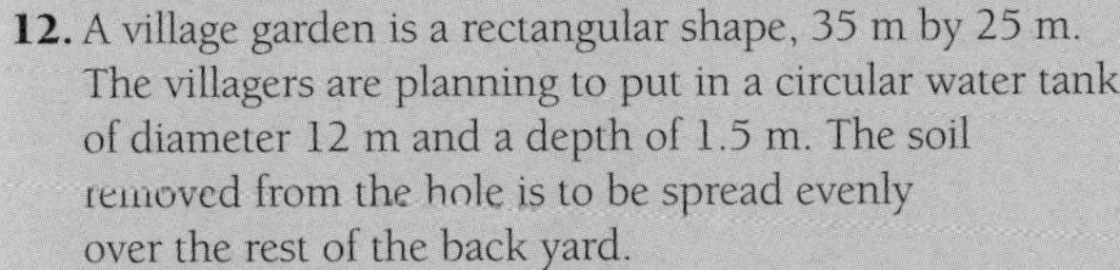

12. A village garden is a rectangular shape, 35 m by 25 m. The villagers are planning to put in a circular water tank of diameter 12 m and a depth of 1.5 m. The soil removed from the hole is to be spread evenly over the rest of the back yard.

If the hole is dug to a depth of 1.5 m, by how many centimetres would the height of the garden increase after the spreading of the soil?

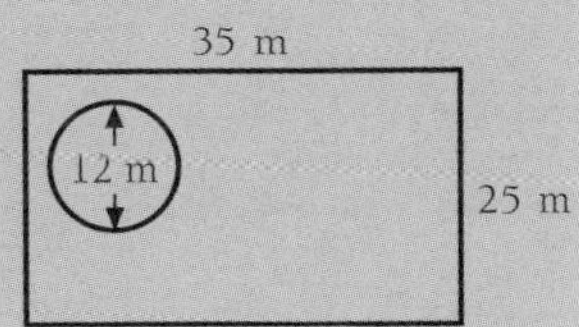

Unit 12.1 Measurement

Topic 7: Scales and dimension – similar figures and area

Topic 7 continues the focus on volume and surface area:

- Calculating the surface area and volumes of similar figures.

Similar figures and area

Consider these similar right-angled triangles:

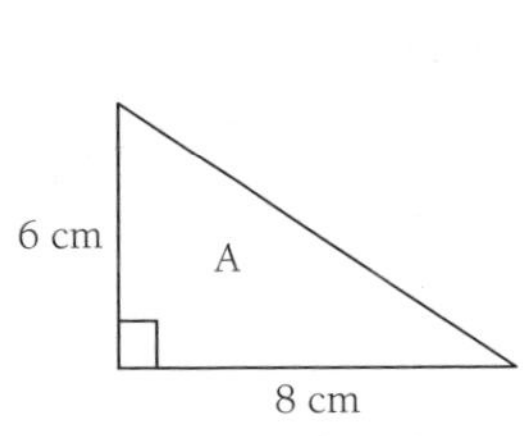

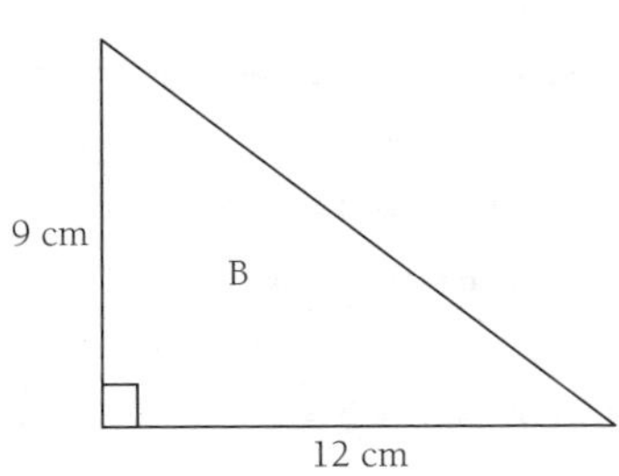

Base length triangle A : Base length triangle B = 8:12

$= 2:3$

Area A : Area B = $\frac{1}{2} \times 8 \times 6 : \frac{1}{2} \times 12 \times 9$

$= 24:54$

$= 4:9$

$= 2^2:3^2$

A similar exercise with circles:

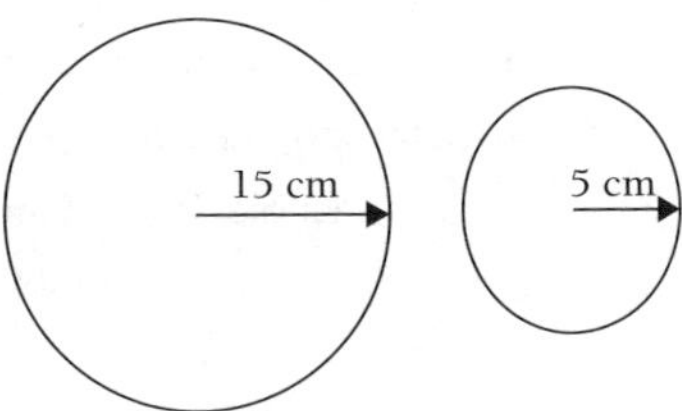

Radius length: 15:5 = 3:1

Area: $\pi \times 15^2 : \pi \times 5^2 = 225\pi : 25\pi$ (**common factor** 25π)

$= 9:1$

$= 3^2:1^2$

> *In general:*
>
> If two figures are similar and the ratio of corresponding **lengths** is a:b then the ratio of corresponding **areas** in the figures is a^2:b^2

If the **scale factor, *k***, of two similar figures is known then:

> *In general:*
>
> If two figures are similar and the scale factor of these figures is k then the scale factor for the corresponding areas of these figures is k^2

Example A

Q. Triangles A and B are similar. If the area of triangle A is 12 cm^2 find the area of triangle B.

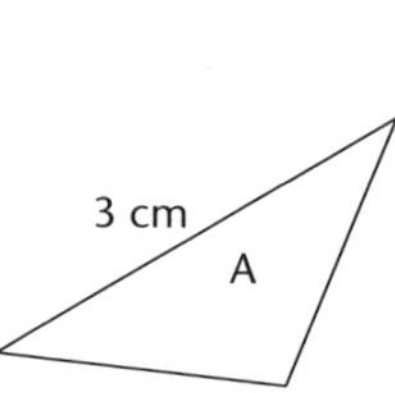

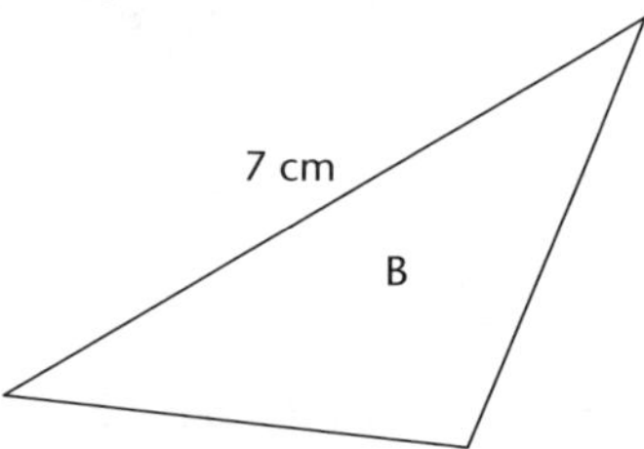

A. **Ratio** of lengths = 3:7

Therefore we know that the ratio of areas will be $3^2:7^2 = 9:49$

The area of triangle A is 12 cm^2 so we can form the ratio equation:

12 : area of triangle B = 9:49

$$\frac{\text{Area of triangle B}}{12} = \frac{49}{9}$$

$$\text{Area of triangle B} = \frac{49 \times 12}{9}$$

$$= 65\frac{1}{3}$$

The area of triangle B is $65\frac{1}{3}$ cm^2

Example B

Q. For the triangle at right find:

a. The shaded area, given that the unshaded area is 5 cm^2.

b. The ratio of shaded area to unshaded area.

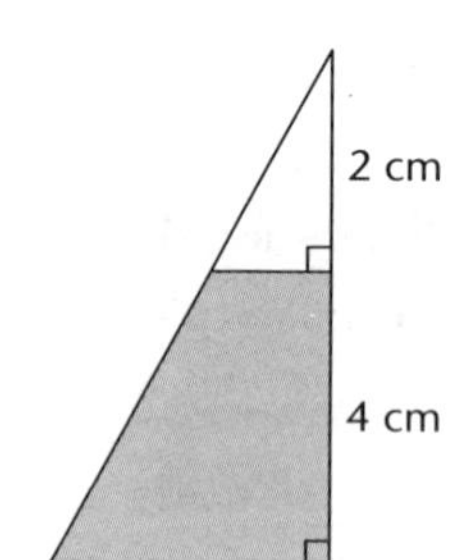

A. **a.** Drawing the similar triangles separately:

The triangles A and B are similar, hence:

Ratio of side lengths = 6:2

= 3:1

The ratio of areas will be $3^2:1^2 = 9:1$

area of triangle A : area of triangle B = 9:1

area of triangle A : 5 = 9:1

$$\frac{\text{area of triangle A}}{5} = \frac{9}{1}$$

$$\text{Hence, area of triangle A} = \frac{9 \times 5}{1} = 45$$

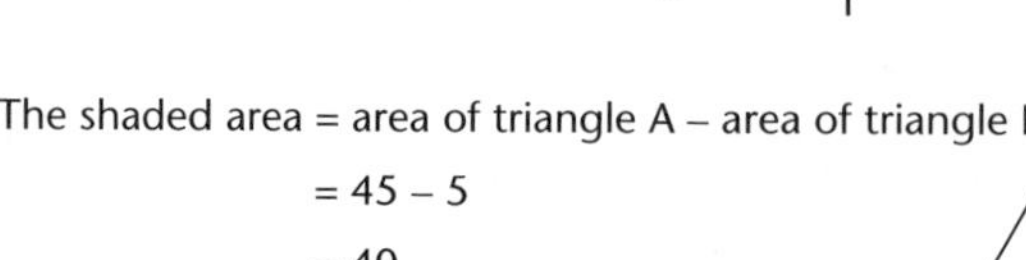

The shaded area = area of triangle A – area of triangle B

= 45 – 5

= 40

The area of triangle A is 40 cm^2

b. The ratio of shaded area to unshaded area is 40:5 = 8:1

Example C

Q. A set of house plans has a scale factor of $\frac{1}{50}$. If a living space on the plans has an area of 56.8 cm^2, what is the actual area of this living space in square metres?

A. The (length) scale factor for the house plans is $\frac{1}{50}$.

This means that any actual length in the house is 50 times the length on the plan.

The scale factor for corresponding areas is $\left(\frac{1}{50}\right)^2 = \frac{1}{2\,500}$ which means that any area on the actual house will be 2 500 times the area on the plan.

So the actual area of the living space = 2 500 × 56.8 = 142 000 square centimetres.

There are 100 × 100 = 10 000 square centimetres in a square metre, so dividing 142 000 by 10 000 gives the actual area of the living space as 14.2 square metres.

Finding the ratio of lengths given the ratio of areas

The principle linking similar figures, length and area can be used in reverse. If we know the ratio of corresponding areas then we can find the ratio of lengths.

For similar figures where corresponding areas are in the ratio A:B, the corresponding lengths are in the ratio $\sqrt{A} : \sqrt{B}$.

Example D

Q. Find the height h for the cone in the diagram.

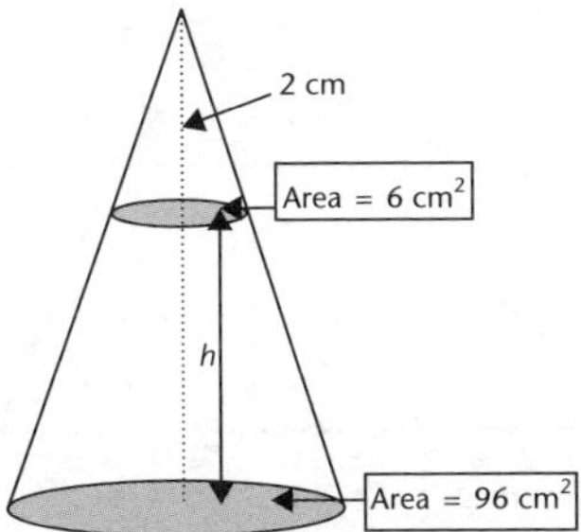

A. The cone can be drawn as two similar cones:

The ratio of areas is 96:6 = 16:1

Hence the ratio of lengths is $\sqrt{16} : \sqrt{1} = 4 : 1$

The heights are corresponding lengths in the cones and will be in the ratio 4:1, hence:

$(h + 2) : 2 = 4:1$

$$\frac{h+2}{2} = \frac{4}{1}$$

$$h + 2 = \frac{4 \times 2}{1}$$

$$h + 2 = 8$$

$$h = 6$$

The height h for the cone is 6 cm.

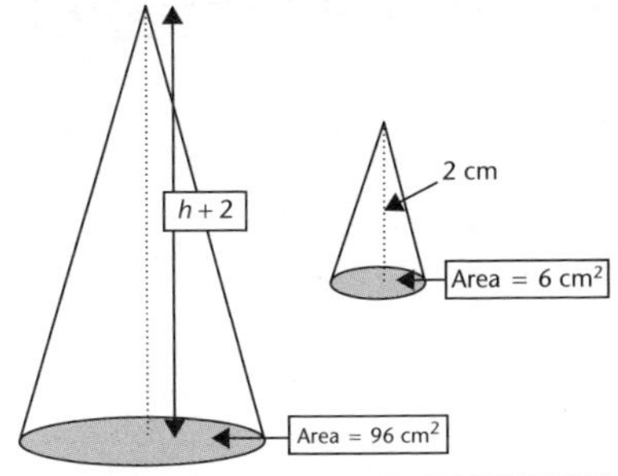

Unit 12.1 Activity 7: Similar figures and areas

1. The diameters of two circles are 6 cm and 10 cm respectively. Determine the ratio of their corresponding areas.
2. Find the ratio of the corresponding areas of these two similar triangles:

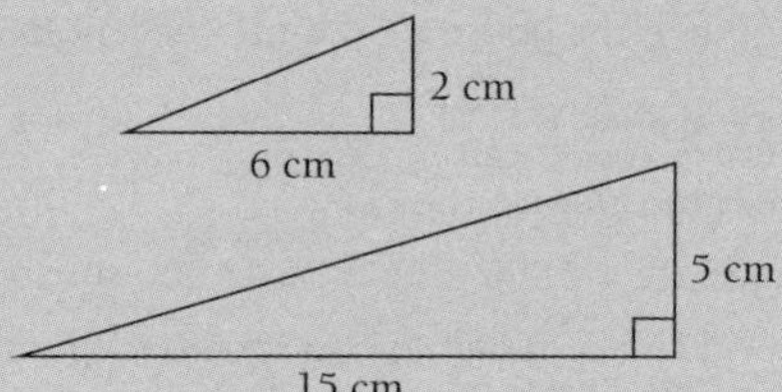

3. The scale drawing of an object has the scale factor of $\frac{4}{15}$ and an area of 32 cm^2. What is the area of the actual object in cm^2?
4. The ratio of the radius of two similar circular flower beds is 3:5. If the area of the smaller flower bed is approximately 706.9 cm^2, find the area of the larger flower bed.
5. One centimetre on a map represents 250 metres. If a field on the map has an area of 4.62 cm^2, what is the area of the actual field, in hectares? (1 hectare = 10 000 m^2)
6. The triangle below is partially shaded. Find the ratio of the shaded portion to the unshaded portion.

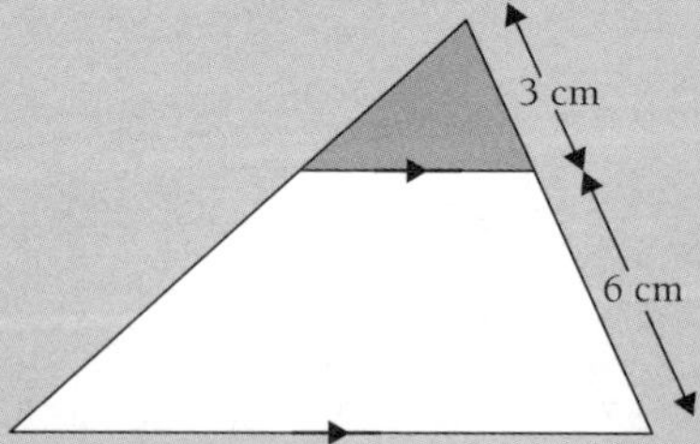

7. a. Find the area of the quadrilateral.
   b. The quadrilateral is a scale drawing of an actual field where the scale factor is 750. Find, in metres, the length of the longest side of the actual field.
   c. Use your answer from part **a**. to find the area of the actual field in square metres.

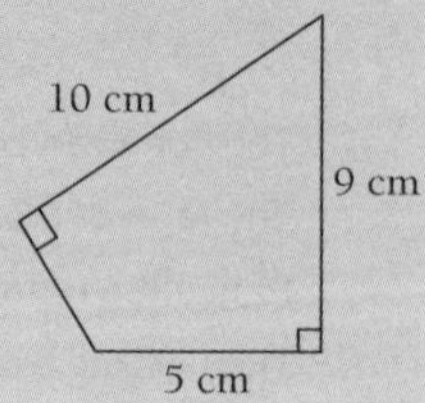

Unit 12.1 Measurement

Topic 8: Scales and dimension – similar solids, length, area and volume

Topic 8 continues the focus on volume and surface area:

- Calculating the surface area and volumes of similar solids, length, area and volume.

Similar solids, length, area and volume

i. Consider the following spheres; all spheres are similar.

The ratio of corresponding **lengths** (radius) = 3:4

The ratio of corresponding shaded **areas**

$= \pi \times 3^2 : \pi \times 4^2$

$= 9\pi : 16\pi$

$= 9:16$

$= 3^2:4^2$

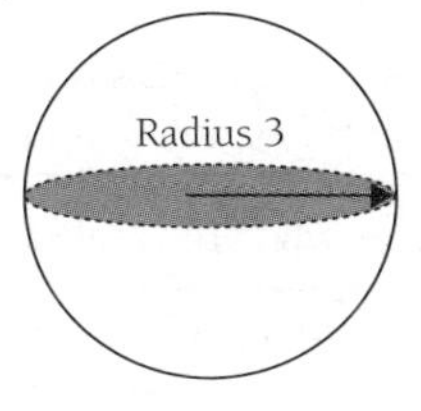

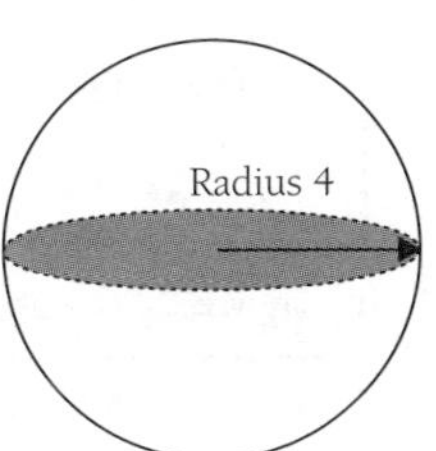

The ratio of **volumes** $= \frac{4}{3}\pi \times 3^3 : \frac{4}{3}\pi \times 4^3$ (area of a sphere is $\frac{4}{3}\pi r^3$)

$= 3^3:4^3$ (dividing by the common factor $\frac{4}{3}\pi$)

For the given spheres the following ratios apply:

Length $3:4$

Area $3^2:4^2$

Volume $3^3:4^3$

ii. Consider the following similar cones:

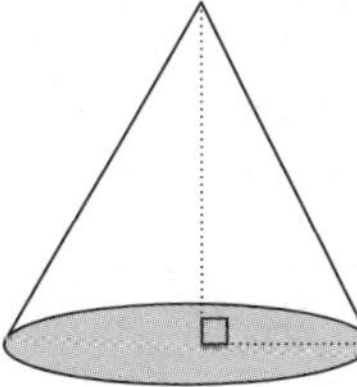

Radius 6 cm
Height 15 cm

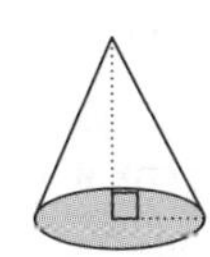

Radius 4 cm
Height 10 cm

The ratio of corresponding lengths $6:4 = 3:2$ (using the radii)

The ratio of corresponding areas $= \pi \times 3^2 : \pi \times 2^2$ (using the area of the base)

$= 9\pi:4\pi$

$= 9:4$

$= 3^2:2^2$

The ratio of volumes $= \frac{1}{3}\pi \times 6^2 \times 15 : \frac{1}{3}\pi \times 4^2 \times 10$ (volume of a cylinder is $\frac{1}{3}\pi r^2 h$)

$= 36 \times 15 : 16 \times 10$ (dividing by common factor $\frac{1}{3}\pi$)

$= 540:160$

$= 27:8$

$= 3^3:2^3$

For the given cones the following ratios apply:

Ratio of corresponding lengths $3:2$

Ratio of corresponding areas $3^2:2^2$

Ratio of corresponding volumes $3^3:2^3$

These two examples illustrate the relationship between length, area and volume for similar figures:

> If two figures are similar and a pair of corresponding lengths are in the ratio $a:b$ then any corresponding areas on the figures will be in the ratio $a^2:b^2$ and any corresponding volumes will be in the ratio $a^3:b^3$

Similarly we can establish rules when the scale factor of similar figures is known:

> If two figures are similar and a pair of corresponding lengths have the scale factor of k then corresponding areas on the figures will have a scale factor of k^2 and any corresponding volumes will have a scale factor of k^3

Example A

Q. If a cone of height 8 cm can hold 160 cm^3 of liquid, how much liquid can a similar cone, of height 16 cm, hold?

A. The scale factor of the cones is $\frac{16}{8} = 2$

so the scale factor of the volumes will be $2^3 = 8$

The volume of the smaller cone is 160 cm^3, hence:

volume of larger cone $= 8 \times 160$

$= 1\ 280$

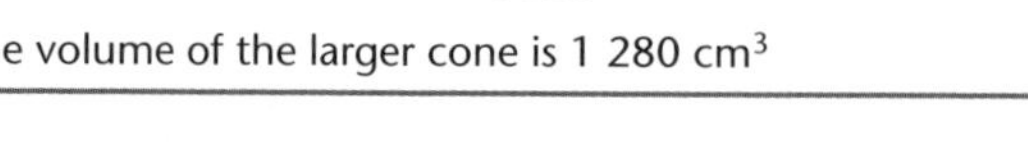

8 cm

16 cm

The volume of the larger cone is 1 280 cm^3

Example B

Q. The shaded bases of the two pyramids are 100 cm^2 and 64 cm^2 respectively.

a. Find the ratio of the heights of the pyramids in simplest form.

b. Find the volume of the smaller pyramid if the volume of the large one is $266\frac{2}{3}$ cm^3

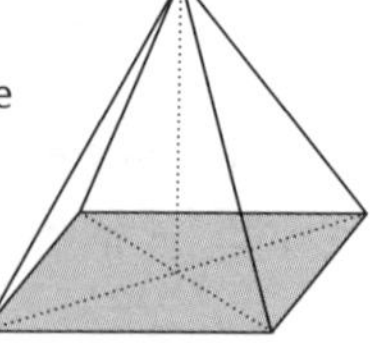

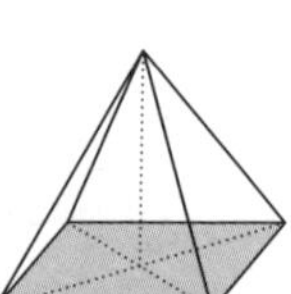

A. a. The areas of the bases are in the ratio 100:64 so the ratio of corresponding lengths is $\sqrt{100} : \sqrt{64} = 10 : 8 = 10:8$.

Hence the ratio of the heights of the pyramids is $10:8 = 5:4$

b. If corresponding lengths are in the ratio 5:4 then corresponding volumes are in the ratio $5^3:4^3 = 125:64$.

The volume of the large pyramid is $266\frac{2}{3}$ cm^3 and let the volume of the smaller pyramid be V cm^3.

Equating ratios: $266\frac{2}{3} : V = 125:64$

$$\frac{800}{3} : V = 125 : 64$$

$$\frac{V}{\frac{800}{3}} = \frac{64}{125}$$

$$V = \frac{64 \times \frac{800}{3}}{125}$$

$$V = 136\frac{8}{15}$$

The volume of the smaller pyramid is 136 8/15 cm^3 ≈ 136.53 cm^3

Example C

Q. If the shaded portion of the triangular pyramid has volume 120 cm^3, what is the volume of the unshaded portion?

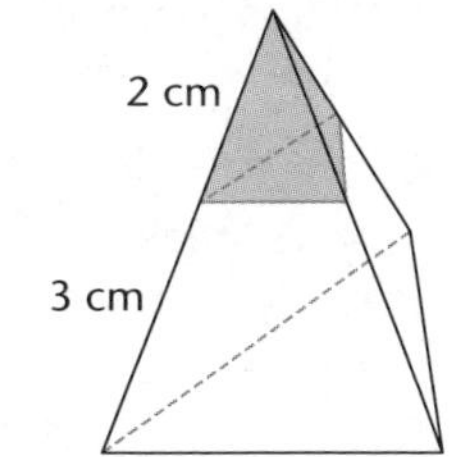

A. The similar figures are shown in the diagrams.

Length, area, and volume ratios are:

Length 5:2

Area $5^2:2^2 = 25:4$

Volume $5^3:2^3 = 125:8$

The volume of the smaller figure is 120 cm^3, hence:

(volume of larger figure):120 = 125:8

$$\frac{\text{Volume of larger figure}}{120} = \frac{125}{8}$$

$$\text{Volume of larger figure} = \frac{125 \times 120}{8}$$

$$= 1\,875\,\text{cm}^3$$

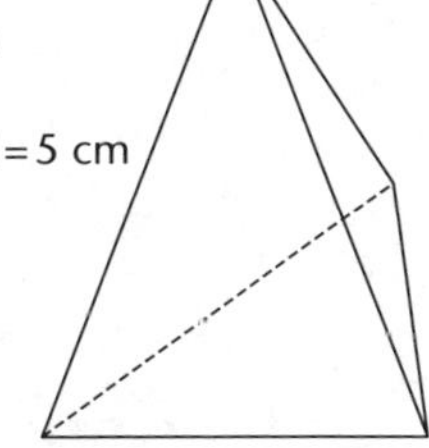

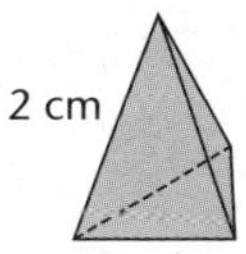

The volume of the unshaded portion of the original diagram:

= volume of the larger figure – volume of the smaller figure

= 1 875 – 120

= 1 755

The volume of the unshaded portion is 1 755 cm^3.

Unit 12.1 Activity 8: Similar solids, length, area and volume

1. Two spheres have radii 4 cm and 8 cm respectively. What is the ratio of:
   a. Their diameters?
   b. Their surface areas?
   c. Their volumes?

2. These two cones are similar. What is the scale factor of:
   a. Their heights?
   b. Their base areas?
   c. Their volumes?

Radius 3 cm

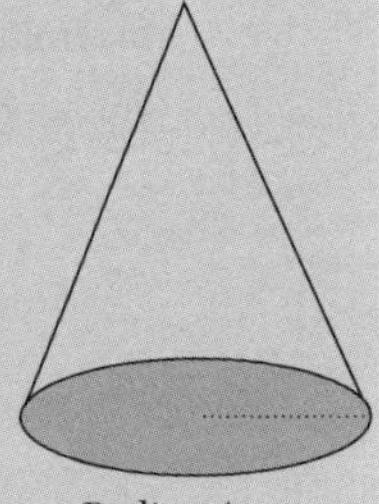
Radius 4 cm

3. The radii of the two spheres are 5 cm and 3 cm. Find the scale factor of:
   a. Their radii.
   b. The shaded areas.
   c. Their volumes.

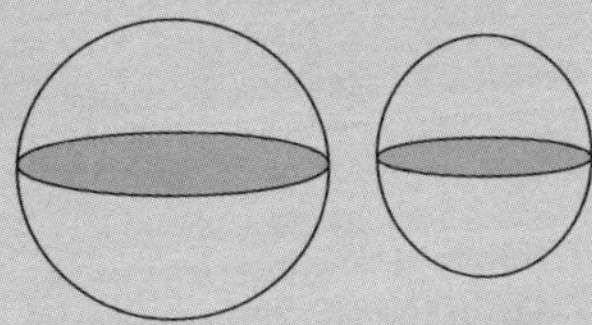

4. If these two square pyramids are similar, and the ratio of the shaded areas is 25:16, find:
   a. The ratio of their corresponding sloping edges.
   b. The ratio of their volumes.

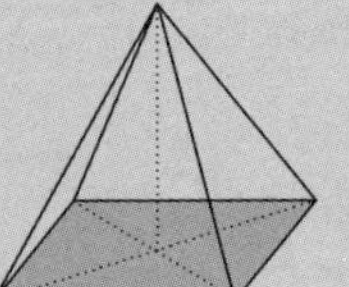
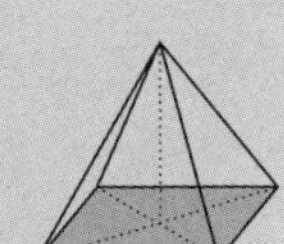

5. The ratio of the volumes of these two cubes is 64:343. Find the ratio of their corresponding diagonals.

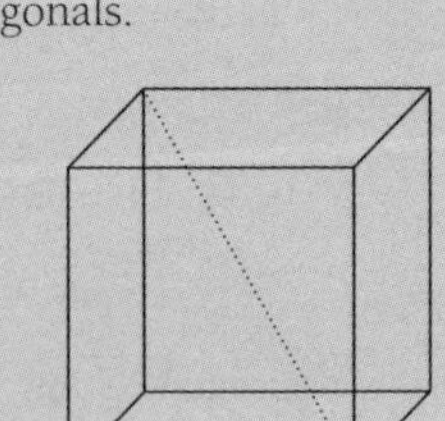

6. The height of a model of an ice cream cone has a scale factor of $\frac{3}{5}$. If the model has a volume of 18π cm^3, what is the volume of the actual ice cream cone?

7. 

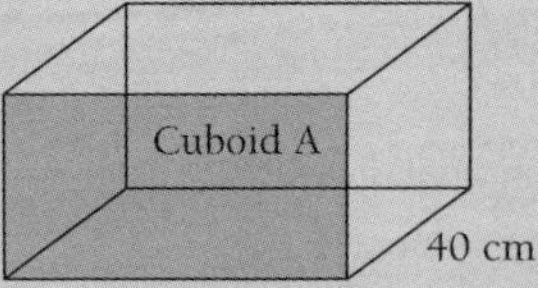

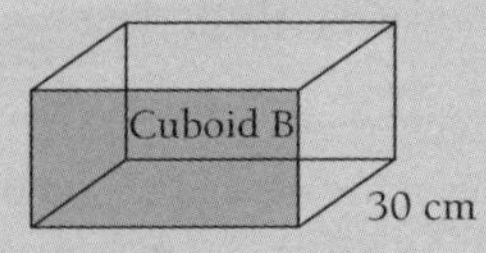

The cuboids above are similar. The shaded area in cubiod A is 2 100 cm^2. Find:
   a. The scale factor of the cuboids.
   b. The shaded area in cuboid B.
   c. The scale factor of their volumes.

8. A 1:10 scale model of a statue is to be made. If the original statue required 5 litres of modelling material, what volume of material, in cubic centimetres, is required to make the model?

9. A 1:20 model of a house is made. If the model has the volume of 15 000 cm^3, find the volume of the real house, in cubic metres.

10. The volumes of two similar solids are 320 cm^3 and 1 080 cm^3 respectively.

a. Find the scale factor of their:

i. Volumes.

ii. Heights.

b. If the height of the first-mentioned solid is 24 cm, what is the height of the other solid?

11. For the similar tetrahedrons given, find the ratio of edge lengths x:y.

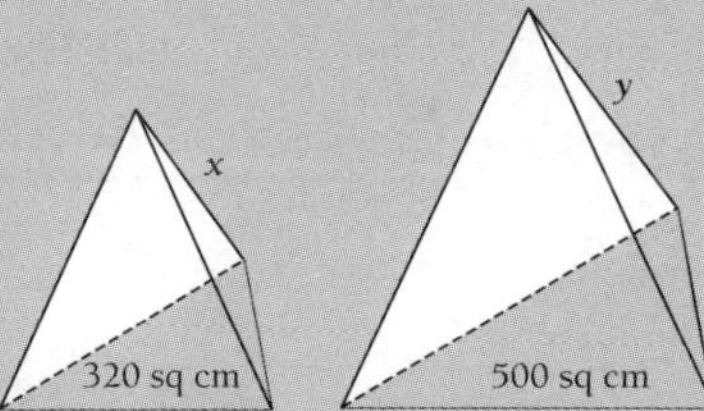

12. If two similar solids have total surface areas of 1 017 cm^2 and 2 825 cm^2 respectively, what is the ratio of their volumes?

13. For the square pyramid given, find:

a. The ratio of the smaller shaded area to the larger shaded area.

b. The size of the smaller area, given that the larger area is 400 cm^2.

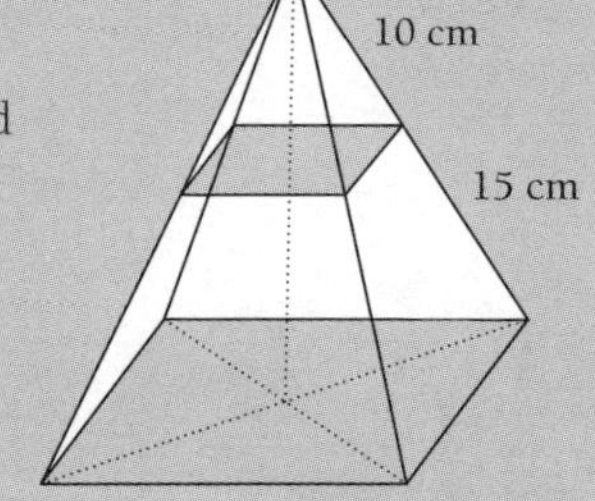

14. If the shaded part of the cone has volume 6.25 cm^3 find:

a. The volume of the whole cone.

b. The ratio of the volume of the shaded part of the cone to the volume of the unshaded part.

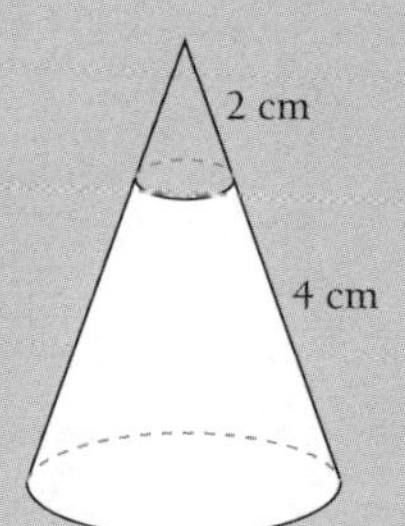

15. Two similarly shaped bottles have heights 30 cm and 20 cm respectively. If the smaller bottle holds 240 mL of liquid, what is the capacity of the larger bottle?

16. A hexagonal platform is positioned 25 cm above the hexagonal base of a glass container – as shown in the diagram.

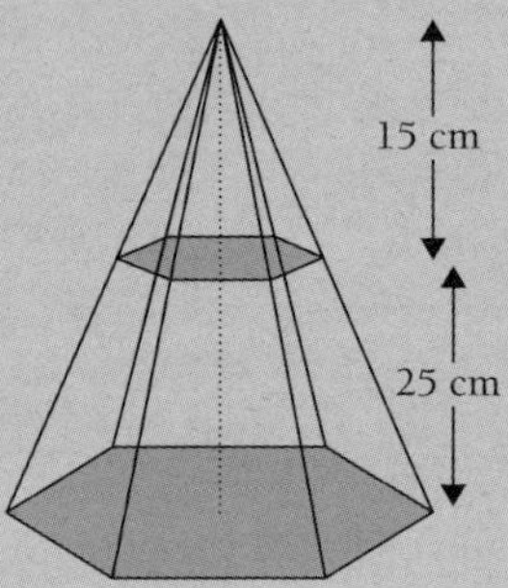

a. Find the ratio of the area of the platform to the area of the base.

b. If the base edge is 24 cm, find the length of the edge of the platform.

c. Find the ratio of the volume above the platform to the volume below the platform.

Unit 12.1 Measurement

Topic 9: Scales and dimension – total surface area and volume of truncated solids

In this Topic the focus is on truncated solids (Syllabus p. 20):

- Calculating the surface area and volumes of truncated solids.

Truncated solids

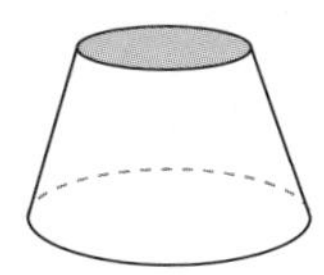

Truncated means 'cut off'.

A truncated cone is illustrated at right. The cut surface is in the shape of a circle, so the cut has been made parallel to the base.

To find the total surface area and the volume of the truncated cone we need to look at the whole cone and the part that is cut off from the whole cone.

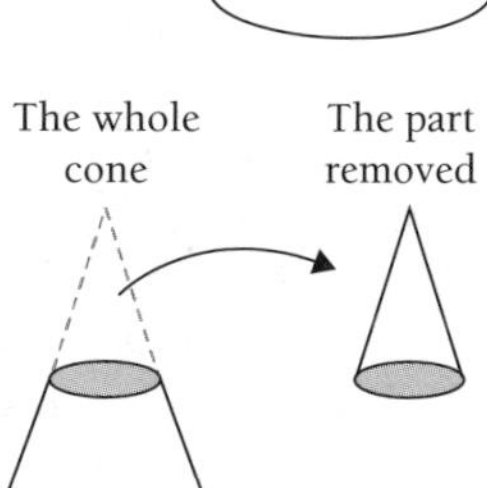

We can consider truncated figures as being cut off by a plane surface, the orientation of the **plane** surface dictating the shape of the cut surface:

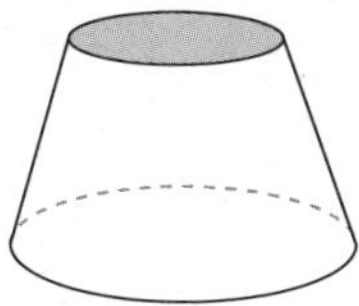

A truncated cone: The cut surface is parallel to the base so is a circle.

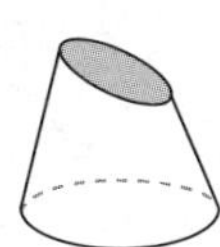

A truncated cone: The cut surface is not parallel to the base so is an ellipse.

All cut surfaces of a sphere are circles.

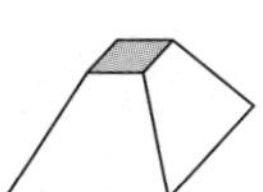

A square pyramid: The cut surface is parallel to the base so is a square.

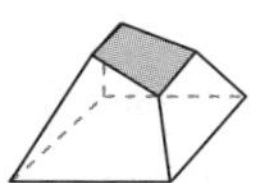

A square pyramid: The cut surface is not parallel to the base and in this case is a trapezium.

Example A

Q. A cone that has a base radius of 8 cm and a height of 25 cm is truncated parallel to the base, so that its height is now 15 cm. For this truncated cone find:

1. The volume.
2. The total surface area.

A. 1. There are *three* ways (**a**, **b**, **c** below) in which the volume of the truncated cone can be found:

a. Using the formula for the volume of a cone:

$$\text{Volume of the whole cone} = \frac{1}{3}\pi r^2 h$$
$$= \frac{1}{3} \times \pi \times 8^2 \times 25$$
$$= 1\,675.52 \text{ cm}^2$$

The part removed from the whole cone has a height of $25 - 15 = 10$ cm. We need to find the radius of this smaller cone that has been removed.

At right is a cross-section of half of the cone. The smaller triangle with base r cm and height 10 cm is similar (AAA) to the large triangle with base 8 cm and height 25 cm.

We can use ratio to find lengths in these similar triangles:

$r:8 = 10:25$

$$\frac{r}{8} = \frac{10}{25}$$

$$r = \frac{10 \times 8}{25} = 3.2 \text{ cm}$$

The volume of the small cone

$$= \frac{1}{3}\pi r^2 h$$

$$= \frac{1}{3} \times \pi \times 3.2^2 \times 10$$

$$= 107.23 \text{ cm}^2$$

Hence, the volume of the truncated cone = 1 675.52 – 107.23 = 1 568.29 cm^2

b. Using a specific formula:

Formula for the volume of a truncated cone or pyramid

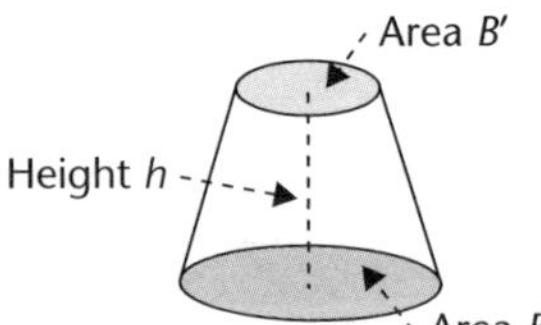

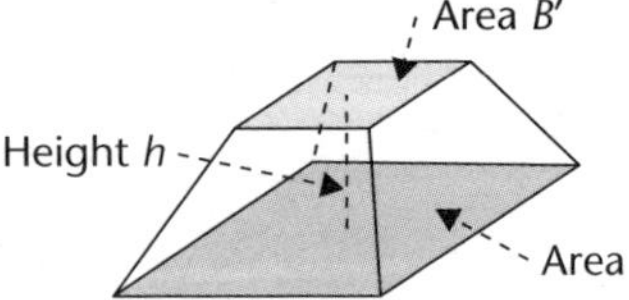

If B is the area of the lower base, B' is the area of the upper base and h is the vertical height of the truncated solid, volume $= \frac{1}{3}(B + B' + \sqrt{BB'}) \times h$

Note: The cut surface, B', and the base surface, B, are parallel.

Using this formula for this example:

$$\text{Volume} = \frac{1}{3}(B + B' + \sqrt{BB'}) \times h$$

$$= \frac{1}{3}\left(\pi \times 8^2 + \pi \times 3.2^2 + \sqrt{\pi \times 8^2 \times \pi \times 3.2^2}\right) \times 15 = 1\,568.28 \text{ cm}^2$$

c. Using similar figures, the ratio of lengths and hence the ratio of volumes:

We have already found that the volume of the whole cone

$$= \frac{1}{3}\pi r^2 h$$

$$= \frac{1}{3} \times \pi \times 8^2 \times 25$$

$$= 1\,675.52 \text{ cm}^2$$

The ratio of the height of the whole cone to the height of the smaller cone = 25:10 = 5:2

Using the rule for similar figures and the ratio of volumes, if the ratio of lengths is 5:2 then the ratio of volumes is $5^3:2^3$ = 125:8

Hence we get the ratio equation x: 1 675.52 = 8:125 where x is the volume of the smaller cone.

Solving this equation:

$$\frac{x}{1\,675.52} = \frac{8}{125}$$

$$x = \frac{8 \times 1\,675.52}{125} = 107.23$$

Hence, the volume of the truncated cone = 1 675.52 – 107.23 = 1 568.29 cm^2

2. The formula for the **total surface area** (TSA) of a cone is:

TSA = (area of the base) + (area of the curved surface)

$= \pi r^2 + \pi r s$

where s is the length of the slant edge of the cone.

For the large cone, s can be found using Pythagoras' rule: $s = \sqrt{25^2 + 8^2} = 26.249$ cm

Similarly for the smaller cone:

$s = \sqrt{10^2 + 3.2^2} = 10.500$ cm

The total surface area of the truncated cone

= (TSA of the large cone) – (area of the curved surface of the smaller (removed) cone) + (area of the base of the smaller cone)

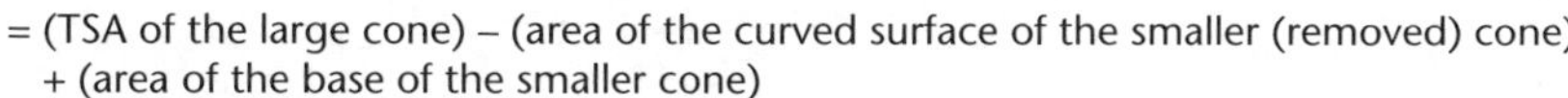

$= (\pi \times 8^2 + \pi \times 8 \times 26.249) - (\pi \times 3.2 \times 10.500) + (\pi \times 3.2^2)$

= 787.38 cm^2

Example B

Q. An eating container is in the shape of a truncated pyramid as shown in the diagram. The top is open and is a square of side length 10 cm; the bottom surface is a square of side length 4 cm and the vertical height is 12 cm. Find:

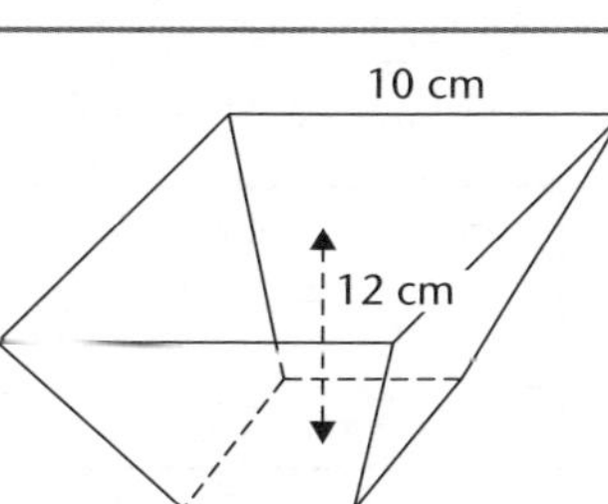

1. The total surface area.

2. The volume of the pyramid.

A. 1. The total surface area consists of a square base and four sides in the shape of trapeziums. To find the area of the trapeziums we need to know the 'height', h, on the surface of the trapezium; it is *not* 12 cm.

The height, h, can be found using Pythagoras:

$h = \sqrt{12^2 + 3^2} \approx 12.37$ cm

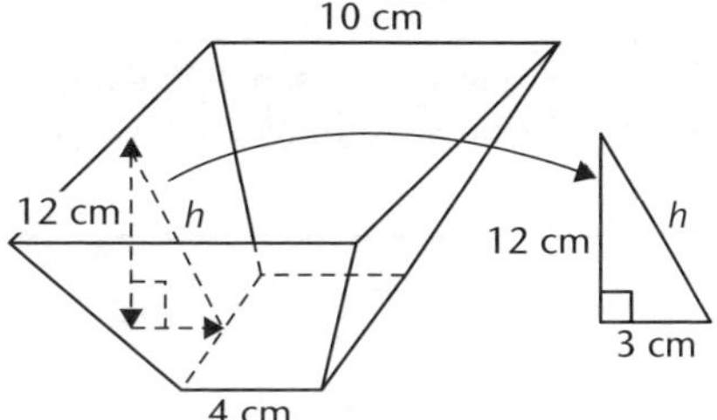

The relevant dimensions of the trapezium sides are given on the diagram below:

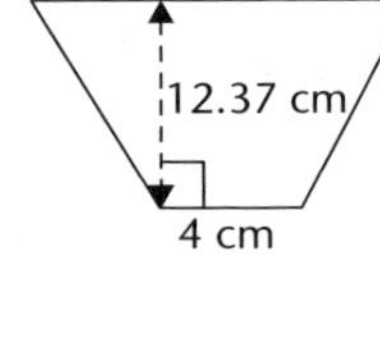

So the area of the trapezium, using the formula $Area = \frac{1}{2}$ (*Sum of parallel sides*) × *height* is:

$\frac{1}{2}(10+4) \times 12.37 = 86.59 \text{ cm}^2$

The TSA of the container is $4 \times 86.59 + 4 \times 4 = 362.36 \text{ cm}^2$

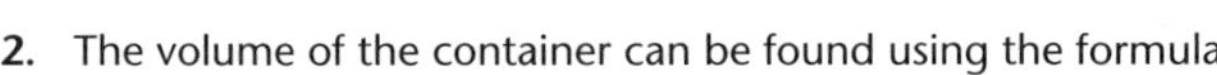

2. The volume of the container can be found using the formula

   $\text{Volume} = \frac{1}{3}(B + B' + \sqrt{BB'}) \times h$

   where $B = 10 \times 10 = 100$; $B' = 4 \times 4 = 16$ and $h = 12$

   $\text{Volume} = \frac{1}{3}(100 + 16 + \sqrt{100 \times 16}) \times 12$

   $= 624 \text{ cm}^3$

Formulae for the volume and surface area of a spherical segment

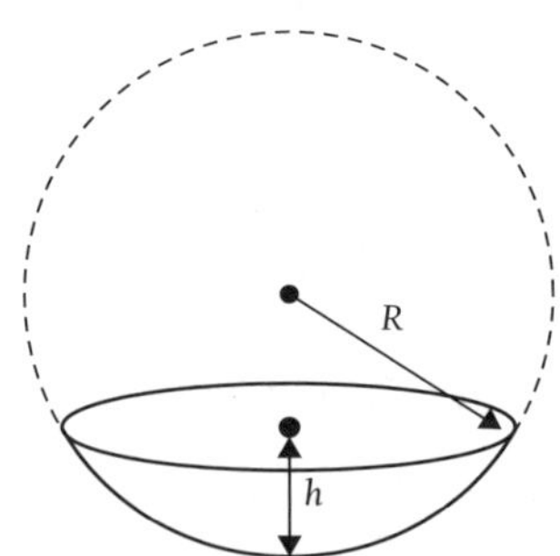

1. A formula for the volume of a spherical **segment**.

> If the radius of the sphere is R units and the height of the spherical segment is h units, then the volume, V, of the spherical segment is given by the formula:
>
> $$V = \frac{\pi h^2}{3}(3R - h)$$

2. A formula for the surface area of a spherical segment.

> If the radius of the sphere is R units and the height of the spherical segment is h units, then the surface area (curved part only) of the spherical segment is given by the formula:
>
> $$\text{Surface area} = 2\pi Rh$$

Unit 12.1 Activity 9: Truncated solids

Give your answers correct to two decimal places.

1. For the truncated cuboid illustrated at right, find:
   a. The volume.
   b. The TSA.

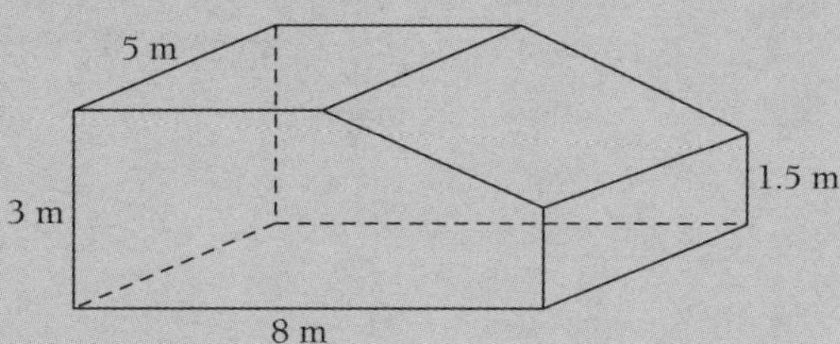

2. A solid in the shape of one-quarter of a sphere of radius 20 cm is shown in the diagram at right. Find:
   a. The volume.
   b. The TSA.

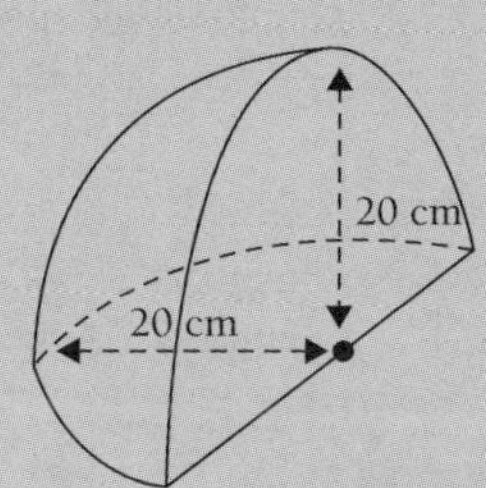

3. An ice cream container is in the shape of a truncated cone. If its base is 13 cm in diameter, its top is 14 cm in diameter and its height is 15 cm, find the capacity of the container to the nearest mL.

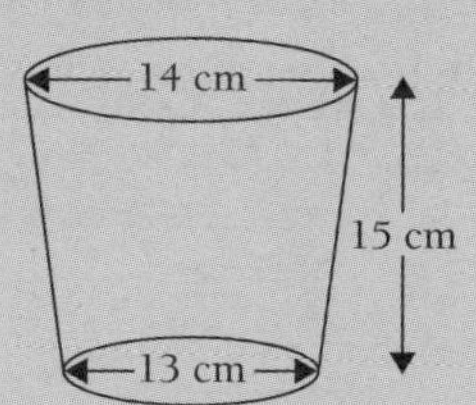

4. A truncated pyramid has a rectangular base of dimensions 34 × 18 cm and the square top, of side length 7 cm, is parallel to the base. The vertical height of the pyramid is 10 cm. Find the volume of the pyramid.

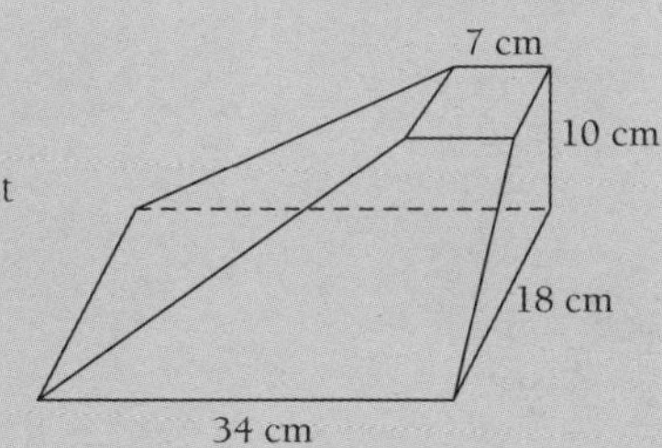

5. A sphere of radius 18 cm is truncated so that the remainder forms a bowl of height 9 cm. Find:
   a. The radius of the top rim of the bowl.
   b. The area of the top surface of the bowl.
   c. The maximum capacity of the bowl, in litres, using the formula $V = \frac{\pi h^2}{3}(3R - h)$
   d. The area of the curved surface of the bowl using the formula: Surface area = $2\pi Rh$

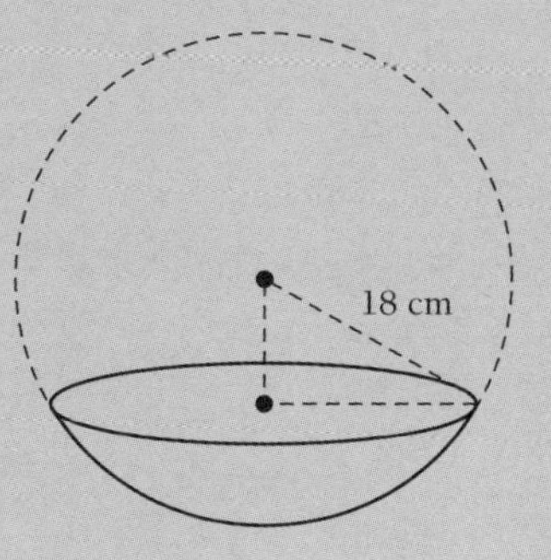

6. A bowl is in the shape of a truncated sphere, as shown in the diagram. The radius of the sphere is 16 cm and the height is 21 cm. Find:

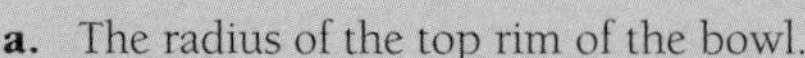

a. The radius of the top rim of the bowl.

b. The area of the top surface of the bowl.

c. The maximum capacity of the bowl, in litres, using the formula $V = \frac{\pi h^2}{3}(3R - h)$

d. The area of the curved surface of the bowl using the formula Surface area = $2\pi Rh$

21 cm

7. A right, square pyramid has been truncated so that its height is 2/3 of the original height of 4.5 m. Find:

a. The length x.

b. The volume of the truncated pyramid.

c. The TSA of the pyramid.

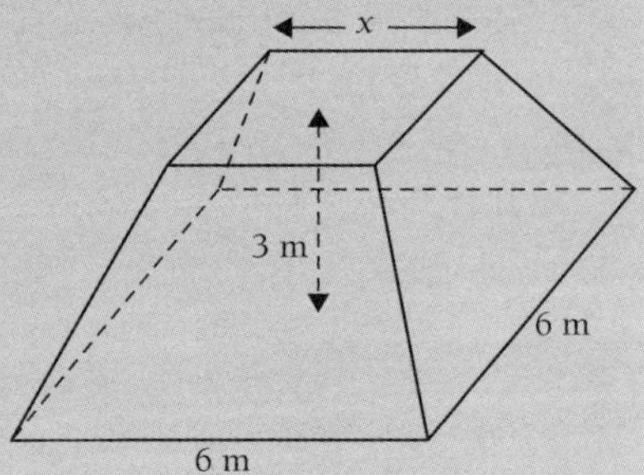

Unit 12.1 Measurement

Topic 10: Scales and dimension – modelling measurement problems

The Syllabus (p. 20) encourages the development and assessment of skills in an applied context. In Topic 10 that is what we will do, so that students learn how to apply mathematical procedures to solve practical problems. This Topic covers:

- Choosing and making appropriate measurements to solve a measuring problem.
- Solving measurement problems using estimation.
- Devising and using with justification a mathematical model as a problem-solving strategy, and commenting on the rationale for the choice of the model.
- Commenting on the limitations or effectiveness of a model to solve measurement problems.

Introduction

Everyday objects are often irregular in shape, and not made up of standard shapes such as the rectangle, cylinder or cuboid. In order to work out the perimeter, area or volume of an irregular object in a practical context, an estimate is often made by using an appropriate model (made up of standard shapes) to approximate the object's shape.

Estimating perimeters

Various techniques can be applied to find an **estimate** of the perimeter of a region.

One technique for estimating a perimeter involves drawing a polygon around the region and measuring the length of each side of the polygon. If the diagram is drawn to **scale**, the scale is then used to convert the measure to an actual distance.

Example A

A **map** of a lake is drawn to a scale of 1:100 000, ie 1 cm represents 1 km.

The approximate length of the shoreline of the lake can be found by drawing a polygon ABCDEFGH as shown, around the perimeter of the lake.

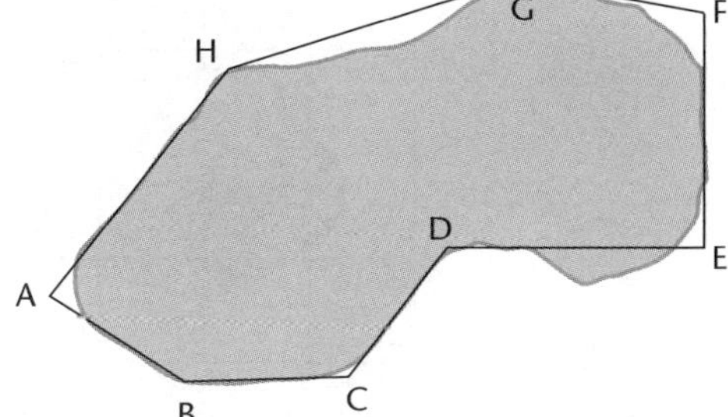

Measure each side of the polygon.

AB + BC + CD + DE + EF + FG + GH + HA

= 14 + 14 + 14 + 22 + 20 +16 + 25 + 24

= 149 mm

= 14.9 cm

Using the scale of 1 cm to 1 km, the length of the shore of the lake is approximately 15 km.

Note: By using more lines, a better approximation would be gained.

An approximation of the perimeter of a region can also be found by taking a piece of string and running the string around the perimeter of the region. The string is straightened out and measured against a ruler. The scale is then used to find the actual measure. The perimeter of the lake could be found this way.

A third technique for estimating a perimeter is to approximate shapes within the region using standard shapes such as circles, triangles or rectangles.

Example B

Q. Find an approximate perimeter for the oil stain shown.

A. The shape of the oil stain is approximately circular.

A circle of diameter 3.5 cm can be drawn as shown.

Using $C = \pi d$, the perimeter of the oil stain is approximately 11 cm.

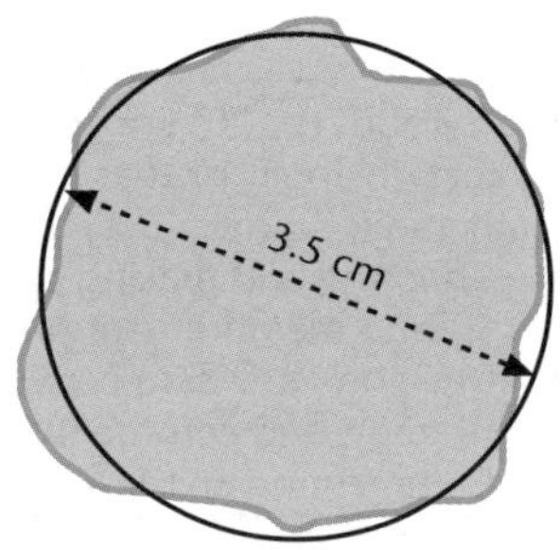

Estimating areas

Areas can be estimated by drawing a grid over the region. Squares of side length 1 cm are often (but not always) the best to use. The approximate area can be found by counting all the squares completely covering the region and adding to it one-half of all the squares covering part of the region.

Example C

To find the approximate area of a raunwara, draw a grid of squares to cover the region.

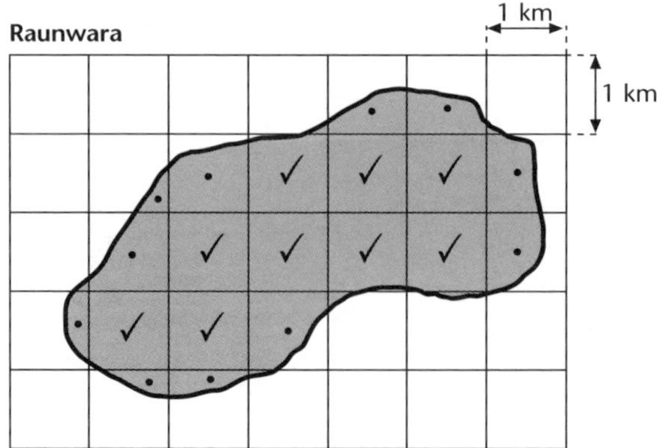

By counting: there are 9 squares (✓) completely covering the raunwara and 11 squares (•) partially covering the raunwara.

The area of the raunwara is approximately $(9 + \frac{11}{2}) \times 1\text{ km}^2 = 14.5\text{ km}^2$, since each square represents an area 1 km by 1 km.

Note: Some squares which almost completely cover the region can be classed as 'full' squares and others that hardly cover any of the region at all can be ignored.

Often an area can be 'divided' approximately into standard shapes then the area is estimated by finding the area of the standard shapes.

Example D

A lake, shown below, can be divided into a triangle, a trapezium and a rectangle:

1 cm : 10 km

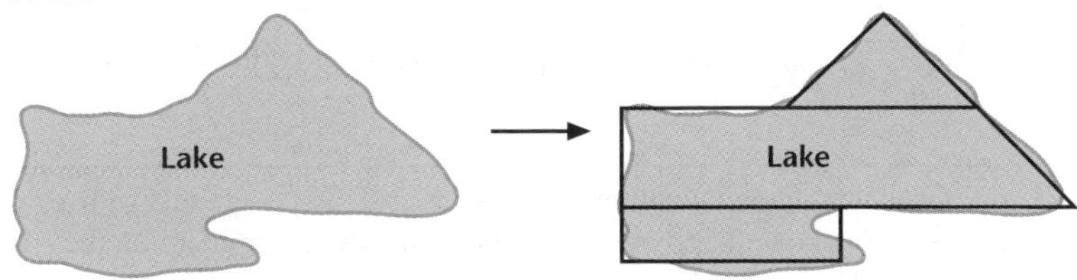

By taking appropriate measurements and using the scale, an approximation to the area of the lake can be found using standard area formulae.

Estimating volumes

In some practical situations the volume of an irregular shape can be estimated by comparing it with the volume of a familiar object. For example a 2L container of ice cream takes up 2 000 cm^3 and an 8 L bucket has a volume of 8 000 cm^3. Alternatively, the dimensions of familiar objects may be used to estimate a volume.

Example E

Q. A photo in a newspaper shows a man walking past some large reels of paper. Estimate the volume of a reel of paper.

A. The **average** height of a man is about 1.7 m.

Each reel is about three times this height, ie about $1.7 \times 3 = 5$ m approximately.

The diameter of the reels is $\frac{1.4}{7.6}$ of the height [dividing diameter by height]

$= 0.184 \times 5$ m

$= 0.9$ m approximately

The volume, V, of a reel is therefore approximately $V = \pi \times \left(\frac{0.9}{2}\right)^2 \times 5$ [using $V = \pi r^2 h$]

$= 3.2\ \text{m}^3$

Standard shapes for which there is a volume formula may be used to model an irregular shape. More than one shape may be required for the model. Be ready to comment on the strengths or limitations of your model.

Example F

Ice cream is served in containers made from corrugated plastic, as shown here.

1. Jane takes measurements to determine the amount of ice cream in a serving. She decides to model the container as a square-based pyramid with its pointed end cut off, as shown.

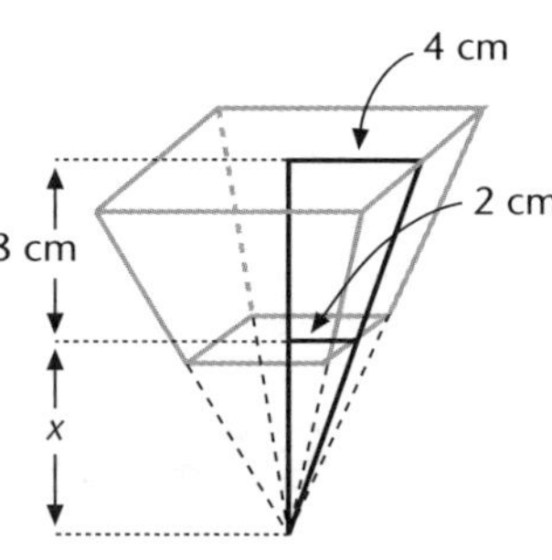

Jane measures the container:

- The base is a square of side length 4 cm.
- The top is a square of side length 8 cm.
- The height is 8 cm.

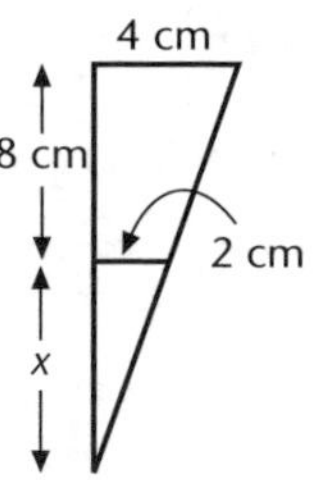

The triangle alongside shows some dimensions of the pyramids.

By similar triangles (or scale drawing) it can be seen that the length marked x is 8 cm. [the large triangle is double the size of the small triangle]

Therefore the volume of the container is:

$V_1 = \frac{1}{3} \times 8^2 \times 16 - \frac{1}{3} \times 4^2 \times 8$ [volume pyramid = $\frac{1}{3}$ × area base × height]

$= 298.7\ \text{cm}^3$

The ice cream comes 4 cm above the top of the container.

Jane models this extra icy-slush as a hemisphere of radius 4 cm. This gives a volume:

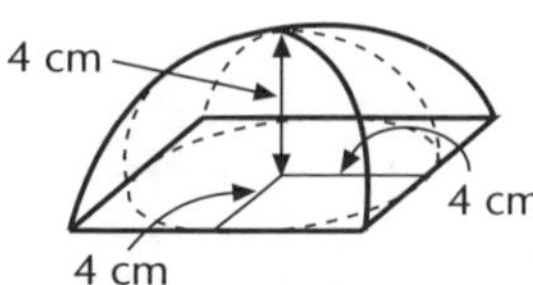

$V_2 = \frac{1}{2} \times \frac{4}{3} \times \pi \times 4^3$ [volume sphere = $\frac{4}{3}\pi r^3$]

$= 134.0\ \text{cm}^3$

Jane's estimate of the total volume of ice cream is therefore:

$V = V_1 + V_2 = 298.7 + 134.0 = 432.7\ \text{cm}^3$

So Jane estimates that each serve is approximately 430 mL (2 sf). [1 cm^3 holds 1 mL]

2. Jane now evaluates her model.

   Jane decides that her model is a close one for the container (aside from corrugations in the plastic) but that the model of a hemisphere for the top part of the ice cream is not a very accurate one. A square-based pyramid may be a better model if the ice cream forms more of a peaked shape than a rounded shape. In fact, if a pyramid model (with a square base of side length 8 cm and height 4 cm) is used, the volume of the ice cream above the container would be $\frac{1}{3} \times 8^2 \times 4 = 85.3$ or 85 mL, a significantly smaller result than 134 mL.

In some situations, an object whose volume is required may be immersed in water. By measuring the millilitres or litres of water displaced, the equivalent volume of water can be calculated, and therefore the volume of the object found.

Unit 12.1 Activity 10: Estimation

1. The diagram shows the path taken on a map by an orienteer travelling from A to B. The scale of the map is 1 cm to 40 m.

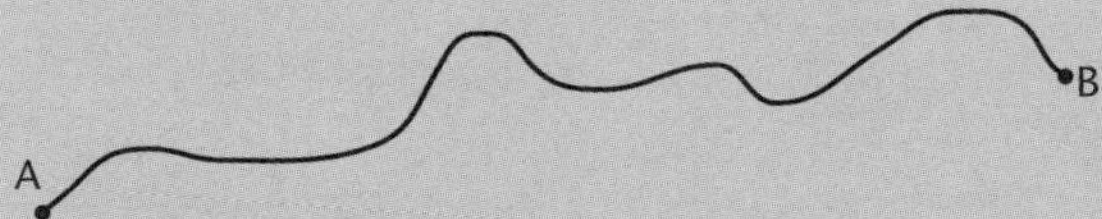

   a. Explain how you could find the approximate distance travelled by the orienteer in going from A to B.

   b. What it the approximate distance?

   c. Give a reason why the actual distance travelled by the orienteer could be significantly further than your estimated answer.

2. The map, drawn to scale, shows a small island in Milne Bay Province.

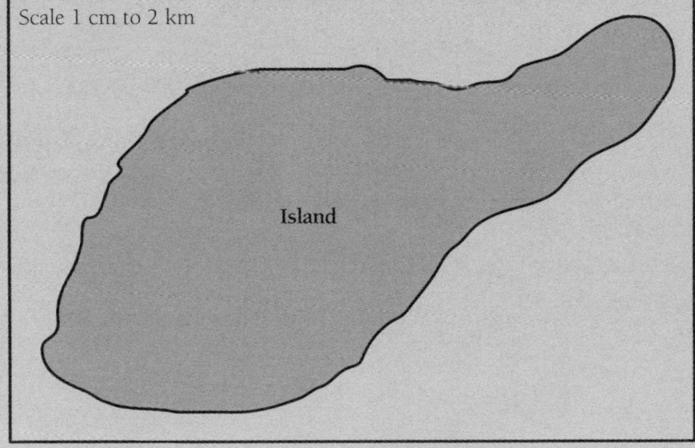

a. Find, approximately, the perimeter of the island.

b. Find, approximately, the area of the island.

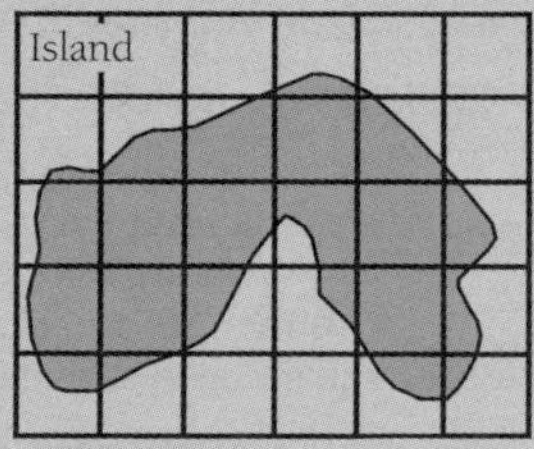

3. Use the grid markings to calculate an approximate area for another island shown. Each square on the grid represents an area 50 m by 50 m.

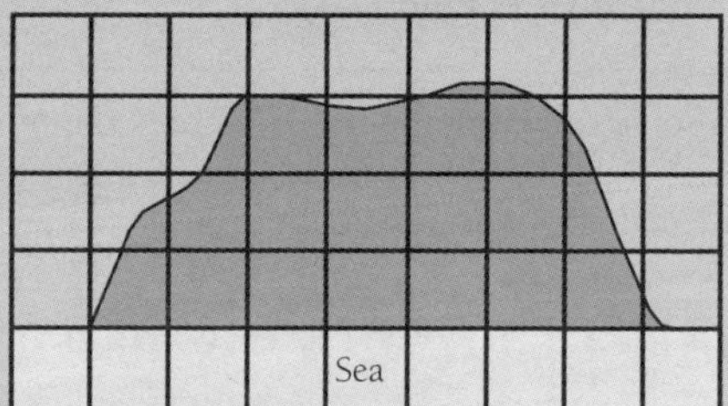

4. An island is in the approximate shape of a prism with a trapezium-shaped cross-section. (The cross-section is shown in the diagram. Each square on the grid represents an area 100 m by 100 m.) If the island is 1.2 km long, find the approximate volume of above-sea-level land.

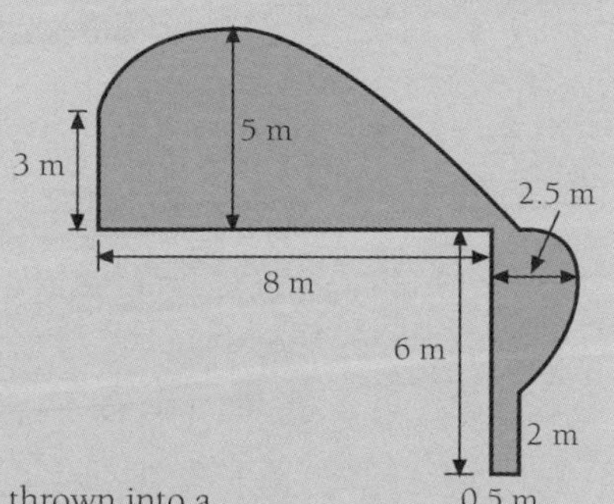

5. The Kasi family wish to spread good soil over their garden to a depth of 15 cm. They draw a **sketch** plan of their garden, as shown. Calculate the approximate amount of soil they will need.

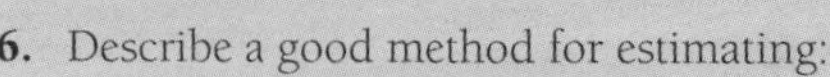

6. Describe a good method for estimating:

a. The **mass**   b. The volume

of a grain of rice.

7. To estimate the volume of a large rock, the rock is thrown into a cylindrical water tank of diameter 1.5 m which is nearly full of water. It is noted that the water level in the tank rises by 1.5 cm.

a. What is the volume of the rock? Give answer to 3 sf.

b. If the rock is roughly spherical, what is its diameter?

c. Comment on the above method's usefulness in estimating volume.

8. Bunu needs to fill a 3 m by 4 m rectangular garden with topsoil to a depth of 15 cm. He will bring the topsoil home in his trailer. Bunu wants to estimate the volume of topsoil in a 500 kg trailer load. He fills up a 12.5 L bucket with topsoil and weighs it. The mass of the topsoil in the bucket is 10.75 kg.

a. What volume of soil does Bunu have on the trailer?

b. How many trailer loads does Bunu need to fill his garden with topsoil?

9. A cubic metre of sand is ordered. The sand is poured into a sandpit but there is too much. If the sandpit is 2 m square and 18 cm deep and the extra sand makes a 'mountain' above the edges (as shown alongside):

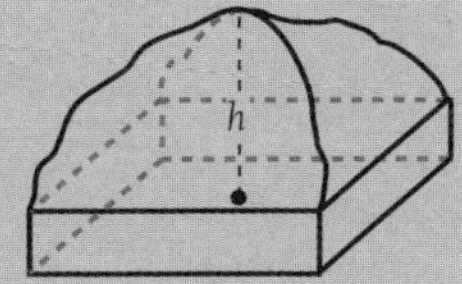

a. Estimate h, the height of the top of the sand above the ground. State any assumptions you have made.

b. Sand is removed from the sandpit until the top of the sand is uniformly 3 cm below the top of the sandpit sides. A second sandpit is to be built to accommodate the extra sand that was removed. This sandpit is to be a shape other than a cuboid, but approximately the same depth as the original sandpit. Describe and give the dimensions of a suitable design for the second sandpit.

10. Ian has a trailer with tray dimensions of 1.4 m by 2.5 m. The sides of the trailer are 30 cm high. He can load dry soil to a maximum depth of 0.5 metres.

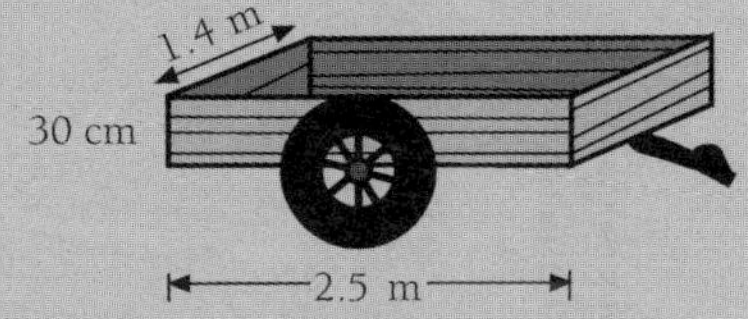

a. Find the approximate volume of soil the trailer can carry.

b. What assumption have you made in your calculations?

c. What are the limitations of your model?

Unit 12.1 Measurement

Topic 11: Surveying – using pace length to estimate distances

Surveying is the second section of Unit 12.1 (Syllabus p. 21), beginning with 'surveying on level ground without obstacles'. Topic 11 covers:

- Determining pace length by measuring 100, using a tape measure.
- Determining pace length and estimating distances and areas using pace length.

Introduction

Land surveying involves the measuring, **sketching** and recording of lengths and areas, and showing this information using a suitable scale on a map or a plan. The process of sketching and recording measurements is called **field book entry**.

The lengths, widths or distances of a surveyed field or region may be measured using a tape measure or trundle wheel, or may be estimated by pacing. Pace is the length of one large step and is one of the earliest forms of measuring length. Pacing is not an accurate method but it is convenient as it does not require modern equipment – just the human body.

Investigation: pace length

1. Measure a length of 20 m or more, using a tape measure, a metre ruler or a trundle wheel. Pace the distance by starting your count with the first step after the starting point and ending the count with the first step after the finishing point. The paces done should be natural.

An individual's pace length is calculated by:

$$\text{Pace length} = \frac{\textit{number of metres}}{\textit{number of paces}}$$

Repeat the above step several times and average the results to find the pace length.

2. Use the pace length above to estimate the distances and the area of the school's playing field

Example A

Q. A person takes an average 25 paces to cover a 20 m distance. Calculate his pace length.

A. $\text{Pace length} = \frac{20 \text{ m}}{25 \text{ paces}} = 0.8 \text{ m/pace}$

Example B

Q. Yamanduo estimates that she takes 25 paces to cover 20 metres. She finds that a rectangular room has a length that is 11 paces and a width that is 10 paces. Calculate:

1. The pace length.
2. The length of the room in metres.

3. The width of the room in metres.
4. The area of the room.

A. 1. Pace length $= \frac{20 \text{ m}}{25 \text{ paces}} = 0.8$ m/pace
2. Length = 0.8 m/pace × 11 paces = 8.8 m
3. Width = 0.8 m/pace × 10 paces = 8 m
4. Area = $l \times w$ = 8.8 m × 8 m = 70.4 m^2

Unit 12.1 Activity 11: Using pace length to estimate distance and area

1. Tami estimates that she takes 28 paces to cover 25 metres. She decides to make space in her garden to grow taro. Below is a diagram showing her measurements. Use it to calculate:

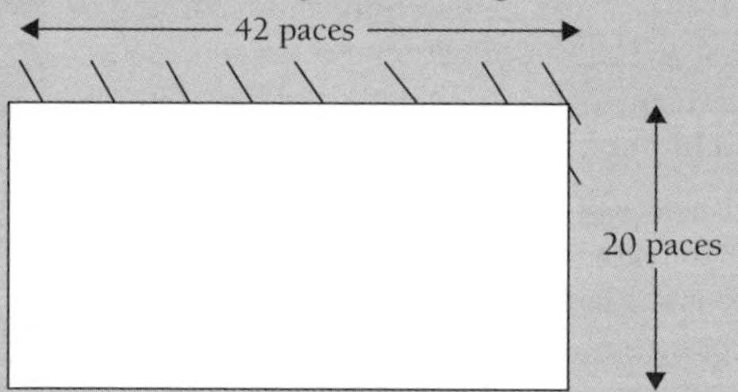

a. The dimensions of the plot.

b. The area of the plot.

2. A student covers 26 metres with 30 paces. He wants to find the area bounded by 5 trees as shown in the diagram below.

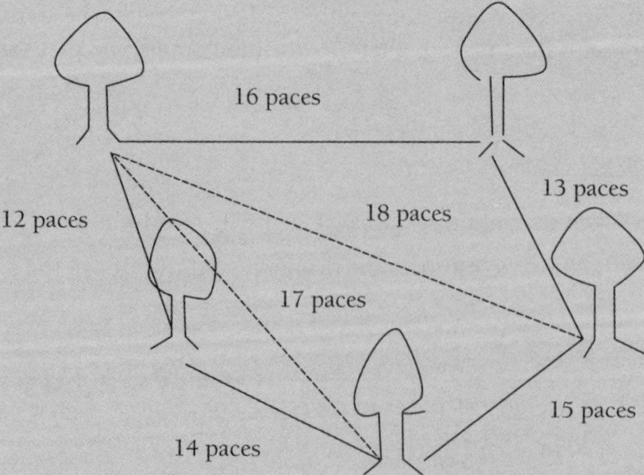

a. Convert the distances between the trees from paces to metres.

b. The region bounded by the trees takes the shape of a pentagon. Use Heron's formula to calculate the area bounded by the trees.

3. Tapeng is the prefect for campus beautification (i.e. planting flowers and keeping the flower beds neat and tidy). One work parade afternoon he decides to plant grass in regions surrounded by the newly constructed footpaths. In order to do that he needs to get grass from regions exactly the same size as the ones surrounded by the footpaths. He paces the sides of the footpaths and discovers that he covers 20 metres with 26 paces. The diagram below illustrates Tapeng's sketch and the measurements of the footpaths.

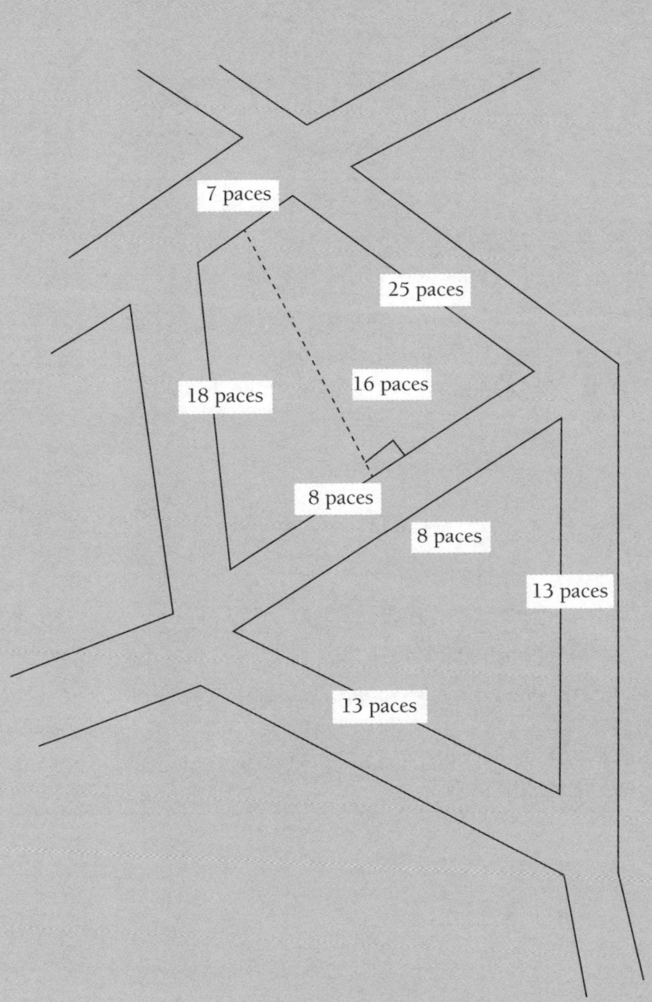

a. Convert the distances in paces to metres.

b. Calculate the area of the trapezoidal region.

c. Calculate the area of the triangular region using Heron's rule.

d. What is the total area needed for grass planting?

4. John wishes to put gravel around his cylindrical water tank so that he can wash or do his laundry there. He has paced out the distances and found that he takes 39 paces to cover 25 m.

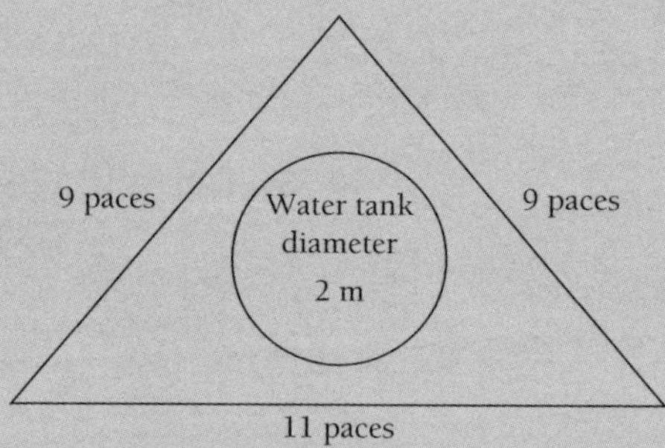

a. Determine the pace length.

b. What are the lengths of the sides of the triangular region in metres?

c. Calculate the area that requires gravel.

5. Dorcas wants to build a fish pond with semicircular ends. She takes 28 paces for each 20 metres, and paces out the area as illustrated in the diagram:

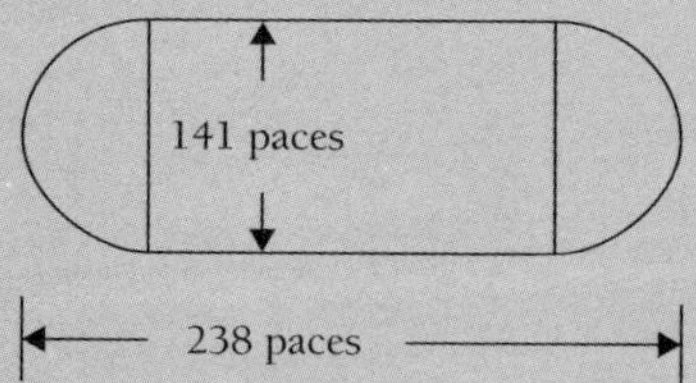

a. Determine the pace length.

b. Convert the distances in paces to metres.

c. Find the area needed for the fish pond. Give your answer in square metres.

6. Select a site or a region on the school grounds and measure the appropriate lengths by pacing. Use your pace length obtained from the Investigation and express the lengths in metres. Then calculate the area.

Unit 12.1 Measurement

Topic 12: Surveying – offset surveys

Continuing our study of surveying on level ground without obstacles, Topic 12 covers survey lines, offset and field book (Syllabus p. 21):

- Defining survey lines, offset and field book.
- Sketching a survey field using field book.
- Interpreting and calculating areas and perimeter using sketch or scale diagrams.

Introduction

Surveyors survey properties for various reasons, including calculating the area of the land enclosed by the boundaries.

The units used to measure the land areas are as follows:

1 square metre (m^2)	$= 1 \text{ m} \times 1 \text{ m} = 1 \text{ m}^2$
1 hectare (ha)	$= 100 \text{ m} \times 100 \text{ m} = 10\,000 \text{ m}^2$
1 square kilometre	$= 1\,000 \text{ m} \times 1\,000 \text{ m} = 1\,000\,000 \text{ m}^2$

Which unit to use depends on the size of the property. For instance, a building block of land is measured in square metres.

There are a number of ways of calculating area, depending on whether it is calculated from a map or from measurements obtained in a survey. Here we will look at ways of calculating areas from survey notes (also called field notes or a field book).

Offset surveys

Offset surveys are used by surveyors to measure irregular blocks of land. The measurements recorded are used to make estimates of the perimeter or area of the block.

To carry out an offset survey of an irregular block of land:

1. The length of the straight line joining the two points furthest apart is measured and recorded. This line, AB in the diagram, is called the **base line**. Its direction is also recorded.
2. Measurements, along AB and perpendicular to AB, are then recorded for various features (P, Q, R and S) of the block of land.

The measurements perpendicular to the base line are called **offset distances**.

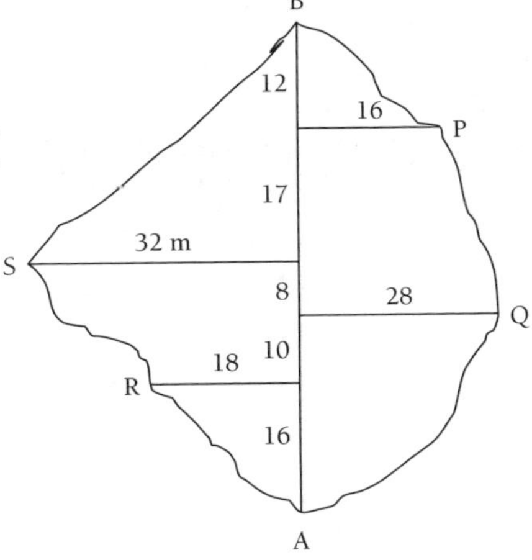

A surveyor would record these measurements as field notes in one of the following ways:

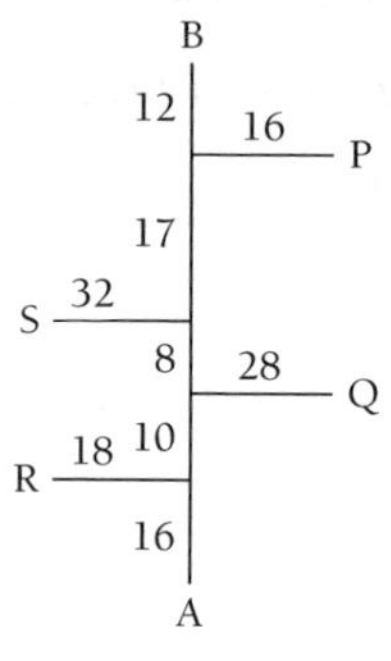

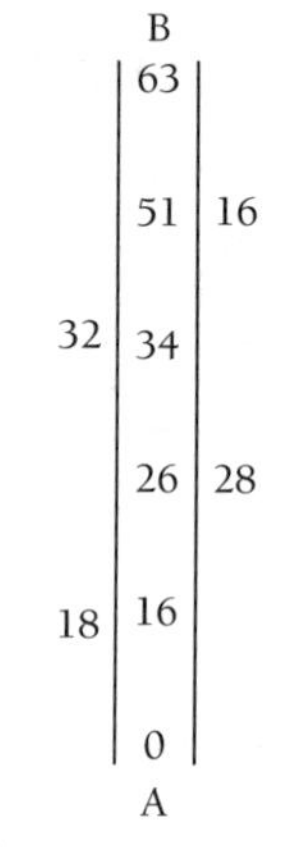

B 12 | 17 16 | 32 8 | 10 28 | 18 16 | 0 A

The measurements along the baseline AB are from point A

Example A

Q. Use the measurements recorded on the field sketch for the offset survey to estimate:

1. The shape of the land.
2. The perimeter.
3. The area.
4. The distance from R to P for the block of land in the diagram.

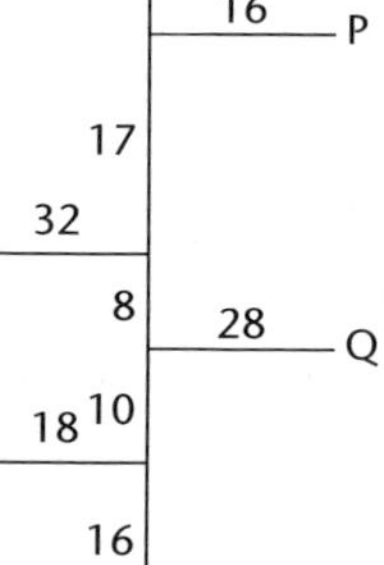

A. **1.** Constructing a scale drawing of the land using the measurements given in the survey gives an idea of the shape of the land.

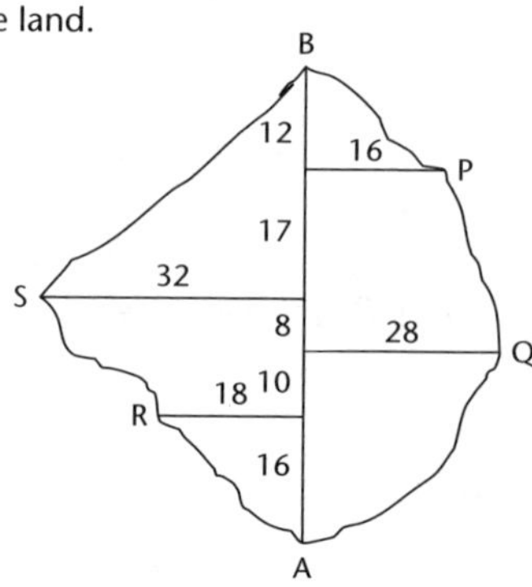

2. Drawing straight lines between adjacent points divides the block of land into six areas that are either right-angled triangles or trapeziums.

Using Pythagoras' rule to find the length BP:

$BP^2 = 12^2 + 16^2$

$= 400$

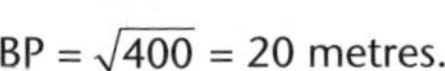

$BP = \sqrt{400} = 20$ metres.

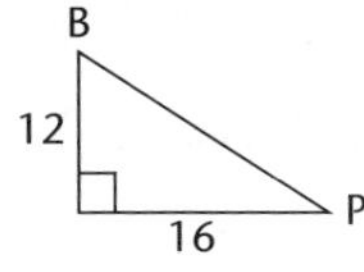

Similarly:

$QA = \sqrt{(16+10)^2 + 28^2} = \sqrt{1\,460} \approx 38.210$ metres

$AR = \sqrt{16^2 + 18^2} = \sqrt{580} \approx 24.083$ metres.

$SB = \sqrt{32^2 + 29^2} = \sqrt{1\,865} \approx 43.186$ metres.

PQ and RS are sides of a trapezium but these lengths can also be found using Pythagoras' rule:

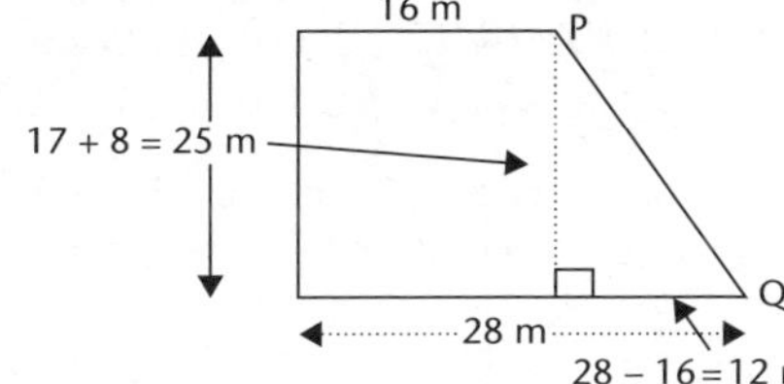

$PQ^2 = 25^2 + 12^2$

$PQ = \sqrt{769} \approx 27.731$ metres

Similarly:

$RS^2 = 14^2 + 18^2$

$RS = \sqrt{520} \approx 22.804$ metres.

Hence an estimate of the perimeter of the block of land is:

$20 + 38.201 + 24.083 + 43.186 + 27.731 + 22.804 = 176.005$ metres.

3. The area of the block can be estimated by finding the area of the four right-angled triangles and the two trapeziums:

The area of the triangle containing side BP:

$= \frac{1}{2} \times 16 \times 12 = 96$ m^2

Similarly:

The area of the triangle containing the side QA $= \frac{1}{2} \times 26 \times 28 = 364$ m^2

The area of the triangle containing the side AR $= \frac{1}{2} \times 16 \times 18 = 144$ m^2

The area of the triangle containing the side SB $= \frac{1}{2} \times 32 \times 29 = 464$ m^2

The area of a trapezium is found using Area $= \frac{1}{2}h(a+b)$ where a and b are the lengths of the parallel sides and h is the perpendicular height.

The area of the trapezium containing the side PQ

$= \frac{1}{2} \times 25 \times (16 + 28) = 550$ m^2

Similarly the area of the trapezium containing the side RS

$= \frac{1}{2} \times 18 \times (18 + 32) = 450$ m^2

Hence an estimate of the area of the block of land is:

$96 + 364 + 144 + 464 + 550 + 450 = 2\,068$ m^2

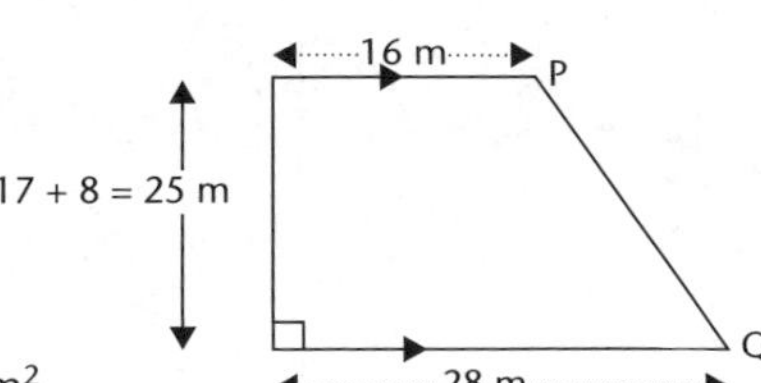

4. To find *RP*, Pythagoras' rule can be used for the triangle RPO:

$RP^2 = 34^2 + 35^2$

$= 2\ 381$

$RP = \sqrt{2\ 381}$

$= 48.80$ metres, correct to two decimal places.

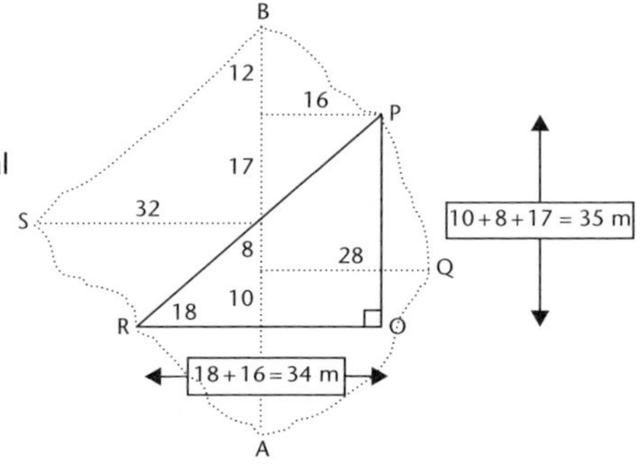

Unit 12.1 Activity 12: Offset surveys

Give answers correct to two decimal places.

1. Draw a scale diagram of the areas recorded by the following field sketches of traverse surveys. (All measurements are in metres.)

a.

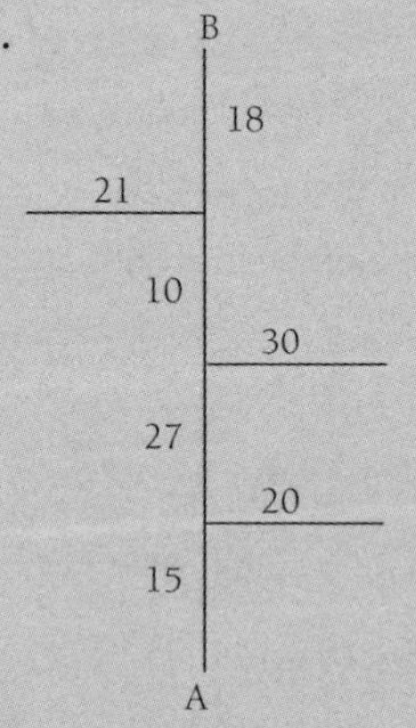

b.

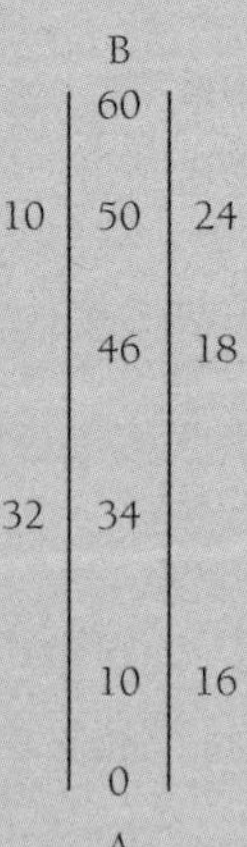

2. For each of the following use the field notes of the survey region to calculate its area and perimeter:

a.

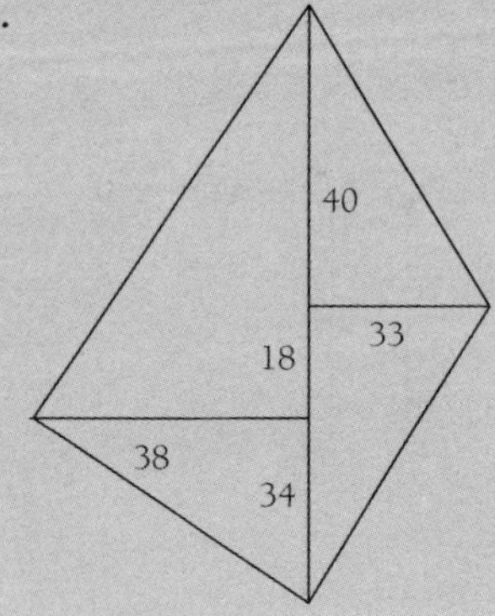

b.

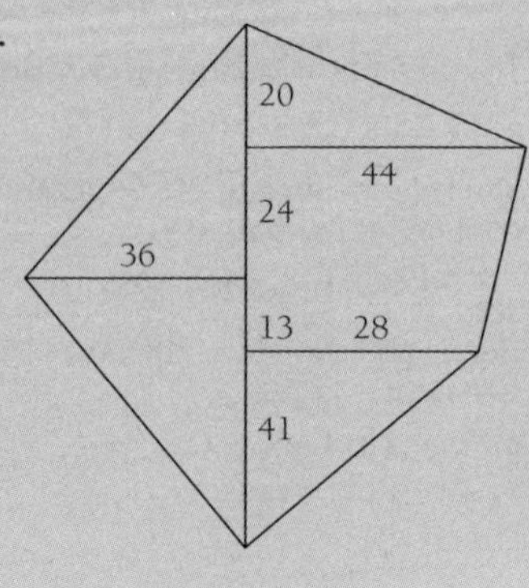

c.

B
18
21
10
30
27
20
15
A

d.

	B	
	60	
10	50	24
	46	18
32	34	
	10	16
	0	
	A	

3. An irregular block of land has been surveyed using offset surveying techniques as shown in the diagram. All measurements are in metres.

a. Find an estimate of the area ABCDE using the measurements recorded in the offset survey.

b. Find the distance between the points:

i. D and B

ii. E and C

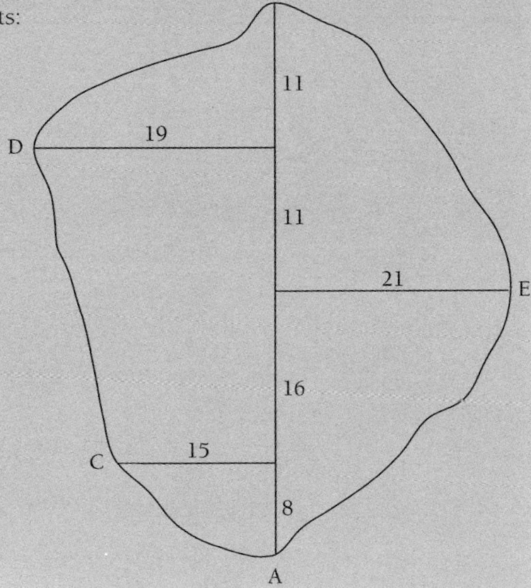

4. The island in the diagram below appears on a map that has a ratio scale of 1:1 000. (You may need to trace this island into your workbook.)

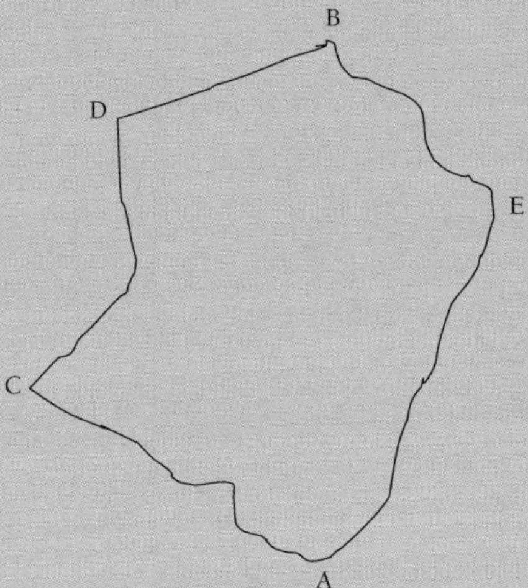

a. Using AB as the base line, draw in the offset lines to the points marked C, D and E. Carefully measure the lines and convert the lengths to metres. Construct a field sketch for your offset survey showing the lengths in metres.

b. Estimate the area and perimeter of the island using your offset survey.

Unit 12.1 Measurement

Topic 13: Surveying – estimating areas with irregular boundaries

Building on the information presented in the last two Topics, we now look at:

- How to estimate areas with irregular boundaries.

Areas with irregular boundaries

Some areas that need to be measured have irregular or curved boundaries or edges. Here is a diagram of an area that has an irregular boundary:

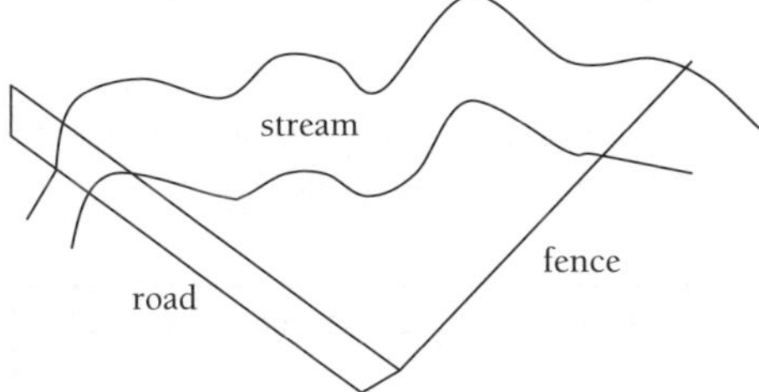

In the field, irregular boundaries are sketched by using a survey line and a number of equally spaced offsets from the survey line to the irregular boundaries.

After the survey line has been drawn, the area of the irregular shape can readily be calculated. The irregular region could be calculated by treating individually each of the regions formed by the offsets.

A survey line is a straight line that divides the survey region into two parts: a regular region and an irregular region.

An offset is a straight line perpendicular to the survey line that meets the irregular edge and divides the irregular regions into calculable regions

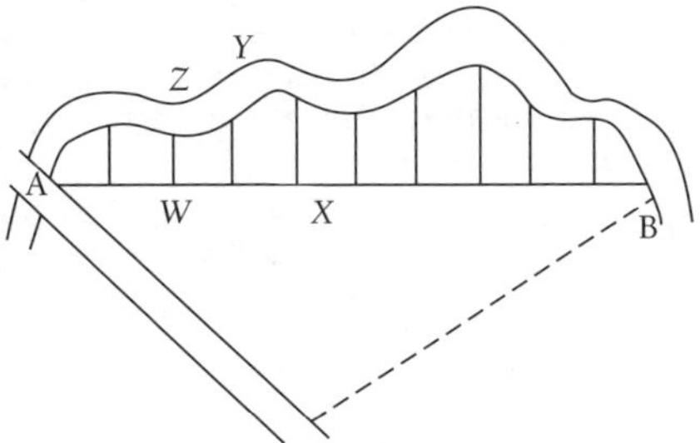

The mid-ordinate rule

If the offsets *WZ* and *XY* are averaged, i.e. $\frac{1}{2}(WZ + XY)$, the resulting line is mid-way between the offsets. The average of the offsets is called the mid-ordinates.

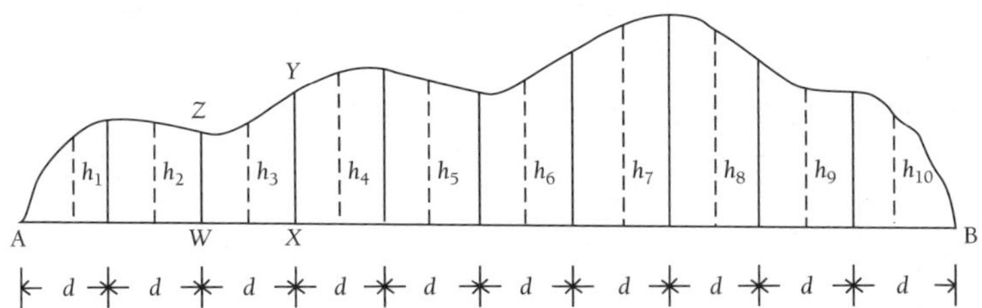

The region *WXYZ* is trapezoid and thus the area of a trapezium is $\frac{1}{2}(a+b)h$.

Substituting the lengths used in the diagram above into the formula for calculating the area for trapezium, $a = WZ$, $b = XY$ and $h = d$, the formula in terms of the letters used in the diagram is:

$$\frac{1}{2}(WZ + XY)d$$

But $\frac{1}{2}(WZ + XY) = h_3$ therefore area of trapezoid $WXYZ = h_3 \times d$ (intervals are **equidistant**)

Area of entire irregular region is calculated as follows:

$$\begin{aligned}\text{Area} &= h_1d + h_2d + h_3d + h_4d + h_5d + h_6d + h_7d + h_8d + h_9d + h_{10}d \\ &= d(h_1 + h_2 + h_3 + h_4 + h_5 + h_6 + h_7 + h_8 + h_9 + h_{10}) \\ &= 10d\left(\frac{h_1 + h_2 + h_3 + h_4 + h_5 + h_6 + h_7 + h_8 + h_9 + h_{10}}{10}\right)\end{aligned}$$

$10d$ as indicated in the diagram above is the length of the entire survey line. This implies that any irregular region similar to the one shown above can be calculated using the mid-ordinate rule, which is stated below:

> Area of irregular region is equal to the **product** of the length of survey line and average of mid-ordinate lengths, ie:
>
> Area = (length of survey line) × (average of mid-ordinate lengths)

First, the mid-ordinates are added and divided by 10 and then the answer is multiplied by the measured length of the survey line.

Example A

Q. An area between the fence line and a patch of banana trees is surveyed. Use the field notes (below), that show the mid-ordinate lengths, to calculate the area of the region. All measurements are in metres.

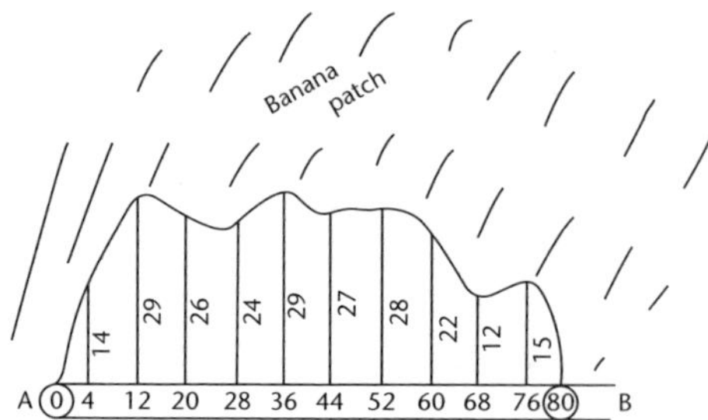

A. Distance between mid-ordinates = 8 m.

Area = length of survey line × average of mid-ordinate lengths

Now, length of survey line = 80 m

Average of mid-ordinate lengths = $\dfrac{14 + 29 + 26 + 24 + 29 + 27 + 28 + 22 + 12 + 15}{10}$

= 22.6 m

Therefore, Area = 80 m × 22.6 m

= 1 008 m²

The trapezoidal rule

The trapezoidal rule is another rule for calculating the approximate area of irregular regions. It is a variation of the mid-ordinate rule.

The region is divided into strips of equal width. The parallel edges of each strip are measured and noted.

The irregular region ABCD, shown below, has been divided into six strips of width w. The edges of each of the strips have been measured and noted as $l_1, l_2 \ldots\ldots l_7$

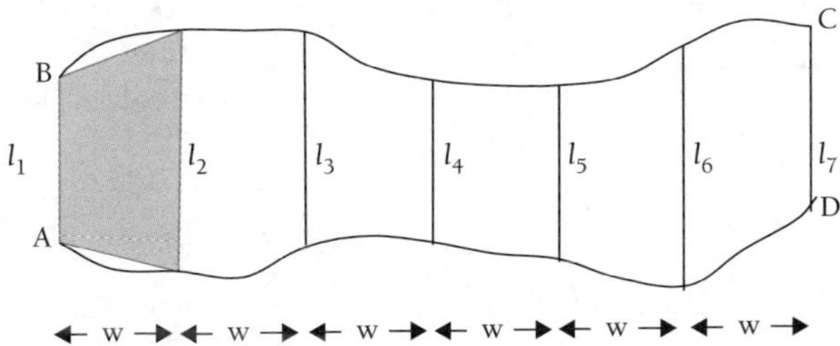

The first trapezium (shaded) has area $\frac{1}{2} \times w \times (l_1 + l_2)$

The second trapezium has area $\frac{1}{2} \times w \times (l_2 + l_3)$ and so on to the last trapezium having the area $\frac{1}{2} \times w \times (l_6 + l_7)$

Adding these six areas gives:

$$\text{Area} = \frac{1}{2} \times w \times (l_1 + l_2) + \frac{1}{2} \times w \times (l_2 + l_3) + \ldots\ldots + \frac{1}{2} \times w \times (l_6 + l_7)$$

$$= \frac{1}{2} \times w \times (l_1 + l_2 + l_2 + l_3 + l_3 + l_4 + l_4 + l_5 + l_5 + l_6 + l_6 + l_7)$$

$$= \frac{1}{2} \times w \times (l_1 + l_7 + 2(l_2 + l_3 + l_4 + l_5 + l_6))$$

From this we can see:

> The approximate area of an irregular region if divided into equidistant strips is calculated using the formula:
>
> $$\text{Area} = \frac{1}{2} \times \textit{width of strip} \times [\textit{sum of end lengths} + 2 \times (\textit{total of middle lengths})]$$

In the diagram above, the end lengths are l_1 and l_7 and the lengths in between are middle lengths l_2 to l_6

Example B

Q. Calculate the area of the region below. All measurements are in metres.

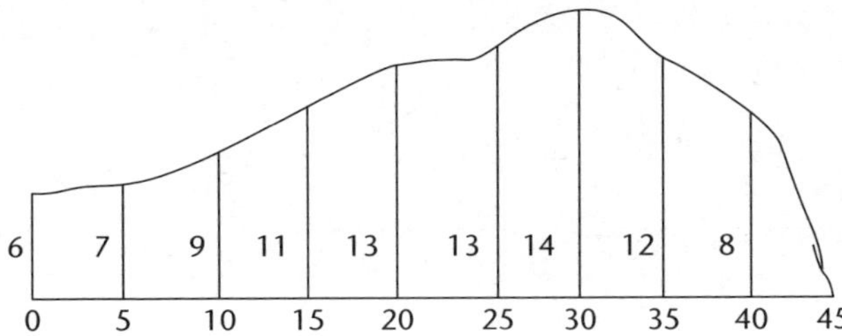

A. Area $= \frac{\text{width}}{2} \times$ (sum of end lengths + 2 × sum of middle lengths)

$$= \frac{5 \times (6 + 0 + 2 \times 87)}{2}$$

$$= \frac{5 \times 180}{2}$$

$$= 450 \text{ m}^2$$

Unit 12.1 Activity 13: Estimating areas with irregular boundaries

1. An area covered with elephant grass next to a main road has been surveyed in order to build a biscuit factory. The following perpendicular offsets have been taken. Calculate the area of the land covered with grass, using the mid-ordinate rule.

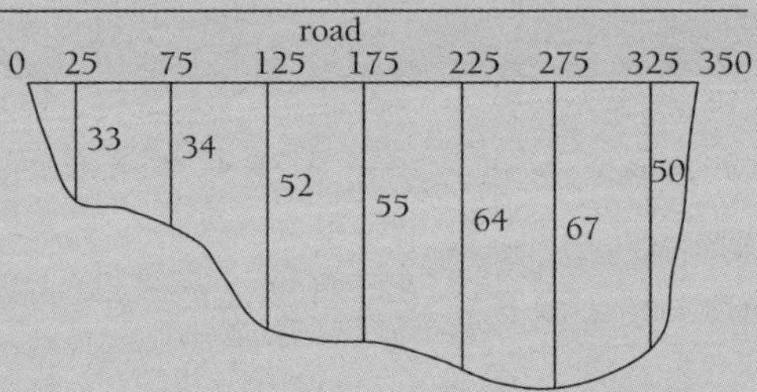

2. An area between the edge of a lawn and a flower bed was surveyed in order to lay gravel. Use the trapezoidal rule and the information provided on the sketch to determine the area where gravel is required. All measurements are in metres.

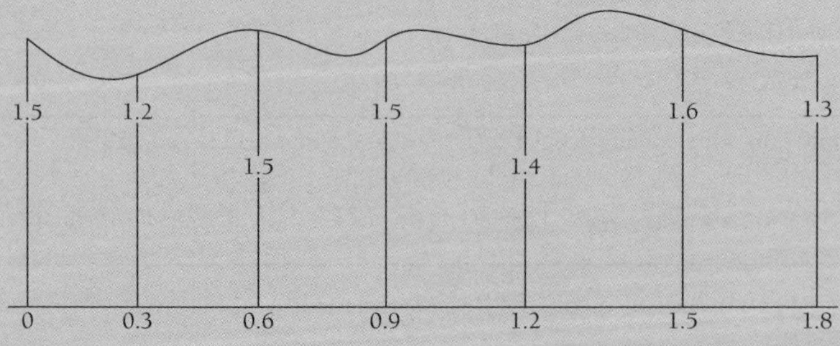

3. Ova dug a plot in his backyard and planted taro seedlings. The representation of the plot is sketched below, with all measurements in metres. Use the mid-ordinate rule to calculate the area of the plot using the survey notes provided in the diagram. The **perpendicular lines** represent the mid-ordinates.

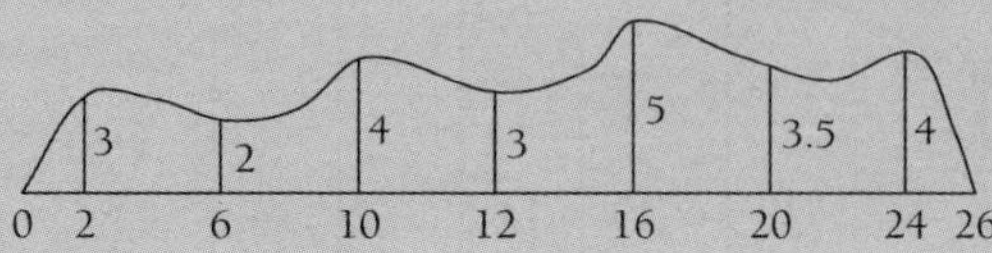

4. Use the trapezoidal rule to calculate the areas of each of the following regions:

a. A small vegetable plot with the lengths 6 m apart:

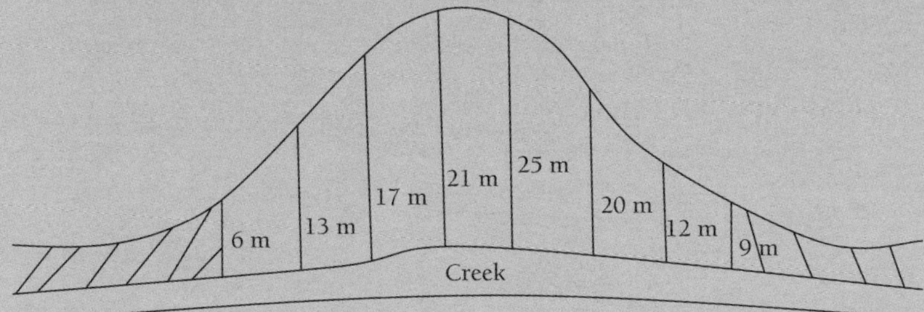

b. Proposed flower garden with 8 m widths:

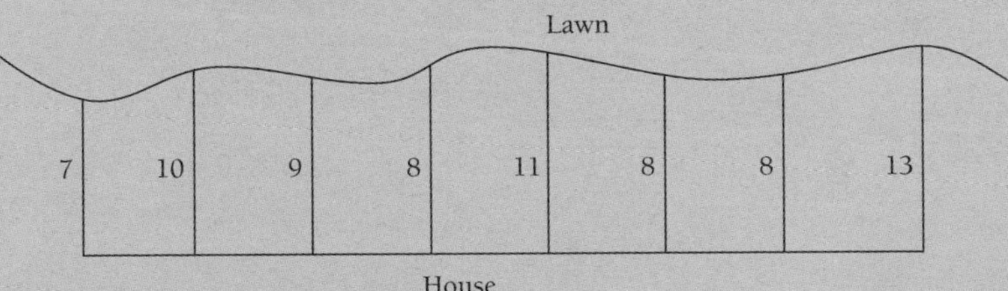

b. Fish pond with 3 m wide strips:

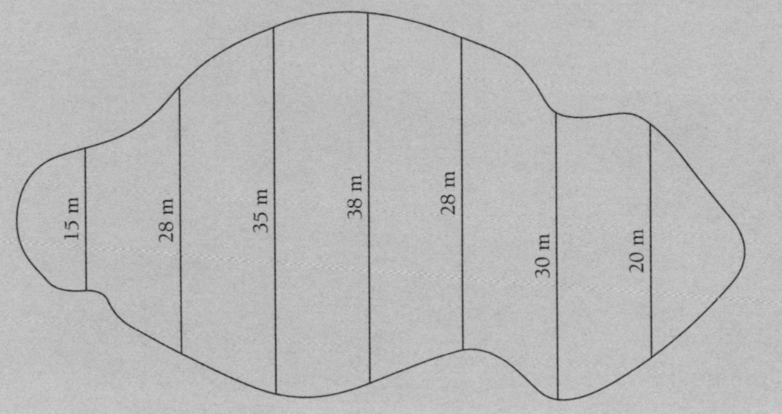

5. The diagram below represents a proposed venue, surveyed for a conference. The region is bounded by a river on one side.

   a. Use the field notes illustrated in the field book entry to calculate the area of the conference venue.

   b. Use the trapezoidal rule to calculate the area to the right of the baseline, AB, and offset survey methods for the area to the left of the baseline.

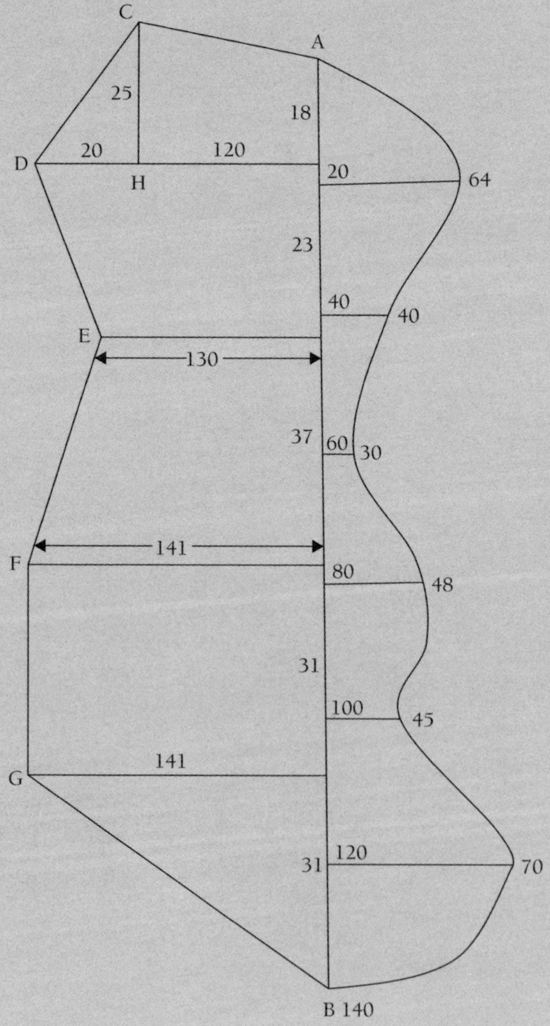

Unit 12.1 Measurement

Topic 14: Surveying – with obstacles

The Syllabus (p. 21) divides the study of surveying into two sections: surveying on ground level without obstacles and surveying with obstacles. Having examined the former in the previous three Topics, we now look at the latter, in particular:

- Offset method and triangulation.
- Methods of applying offset and triangulation to survey around obstacles.

Introduction

Some land areas that are to be surveyed have obstacles or obstructions that make it difficult to measure straight lines. To be able to measure a survey line, the surveyors can either measure around or through the obstacles, depending on the size or nature of the obstacle.

Surveying around obstacles

Three methods are applied to measure a survey line around obstacles.

Method 1

Set an offset near one edge of the obstacle and then measure (away from the survey line) to the edge directly opposite the first offset and thus form a right-angled triangle. Use Pythagoras' theorem to find the distance between the set edges.

Example A

Q. A survey line AB is to be measured around a pond as illustrated in the diagram below:

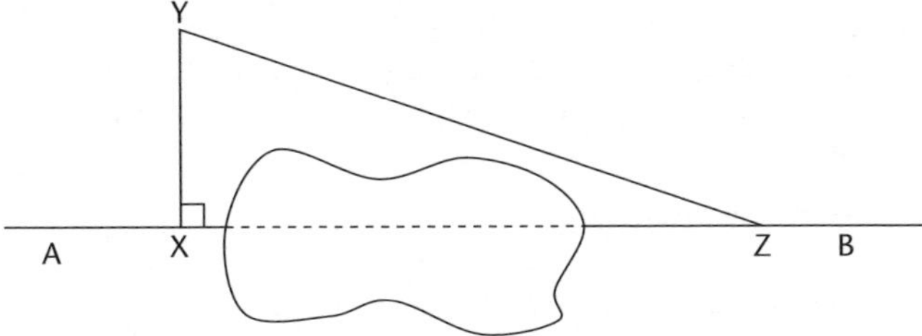

Offset XY is being set and measured and YZ is also measured.

If XY and YZ are measured to be 18.4 m and 65.2 m respectively, XZ can be calculated as follows:

$$\begin{aligned} YZ^2 &= XY^2 + XZ^2 \\ XZ^2 &= YZ^2 - XY^2 \\ &= 65.2^2 - 18.4^2 \\ &= 3\,912.48 \text{ m}^2 \\ &\approx 62.5 \text{ m} \end{aligned}$$

Method 2

Set two offsets of the same length on the opposite edges of the obstacle, then measure the distance between the offsets. The distance between the offsets is equal to the unknown distance.

Example B

Q. Using the same pond as in Example A, the length of the survey line AB is equal to the sum of length AP, PS and SB.

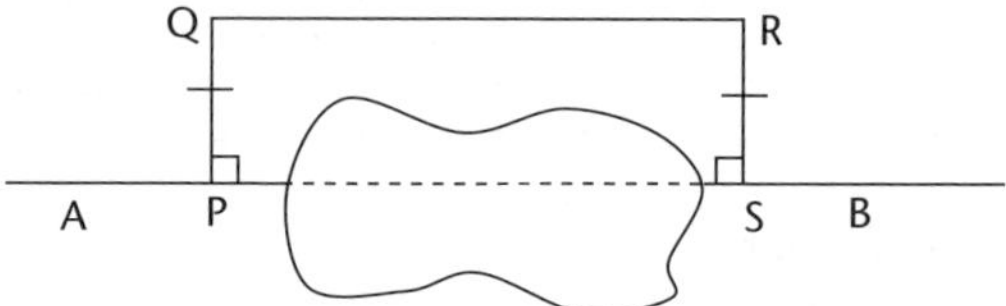

A. To determine the length of PS, offsets PQ and RS are set at opposite edges of the pond. QR is measured and hence is equal to PS (opposite sides of a rectangle are equal in length).

Method 3

Properties of similar triangles can also be applied to determine the unknown part of the survey line. In similar triangles corresponding sides are **proportional** to one another. Let us use the same pond in Example A to illustrate the idea.

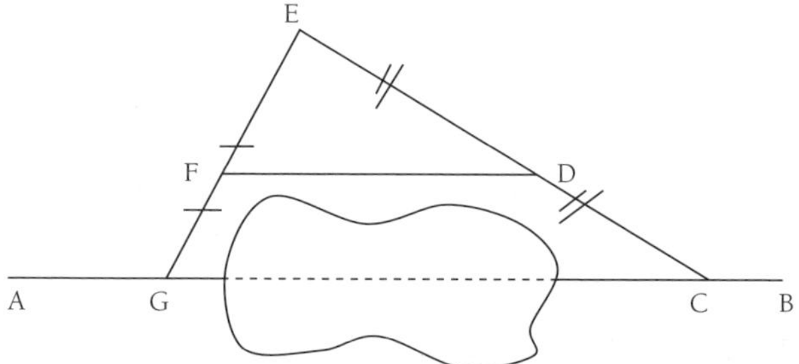

Mark a point, G, at a convenient spot on the survey line. A second point, E, is marked on a location away from the pond. A third point, C, is marked on the survey line on the other side of the pond. Lines CE and GE are measured and bisected at points D and F respectively. The triangles EFD and EGC are similar; the ratio of sides is 1:2. Side FD is measured and hence side GC is twice this length.

It is not possible to measure around some obstacles such as roads and rivers.

In such cases a point P (this could be a tree) is located on one side of the river. On the other side of the river a perpendicular offset is set to a point, Q. Along this offset line a point R is marked with a peg and the offset line is continued to a point S. Another perpendicular offset is set from point S to a point T so that the points P, R and T are in a straight line.

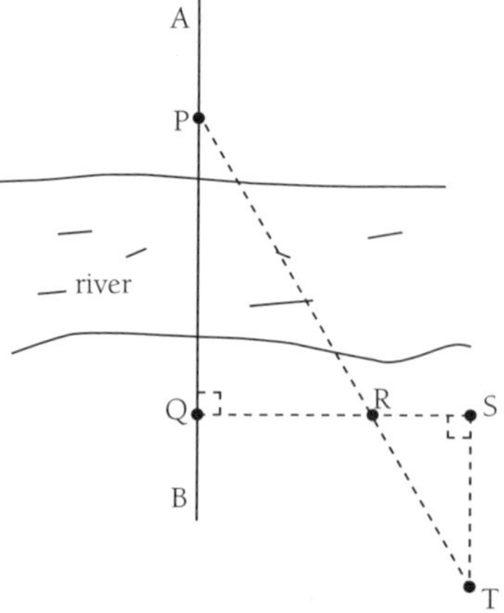

The diagram illustrates the procedure.

The triangles PQR and RST will be similar triangles with the ratio of sides QR:RS (both these lengths are known).

The distances PQ and ST (known) will be in the same ratio.

Example C

Q. If the offset distances QR and RS are 21 metres and 10.5 metres respectively, and the distance ST is measured as 16 metres, find the distance PQ.

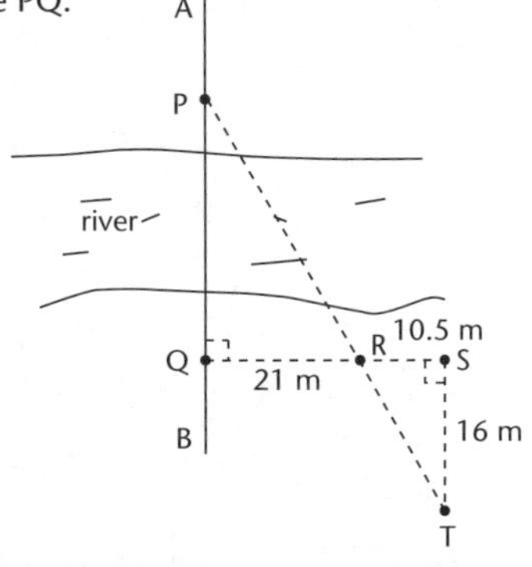

A. PQR and RST are similar triangles with the ratio of sides QR:RS = 21:10.5 = 2:1

The distances PQ and ST will be in the same ratio, so

PQ:ST = 2:1

ST = 16 metres

So

PQ:16 = 2:1 giving $\frac{PQ}{16} = \frac{2}{1}$

PQ = 2 × 16

= 32 metres

Unit 12.1 Activity 14: Surveying around obstacles

Give your answers to two decimal places.

1. An old pineapple garden was surveyed to build a tool shed for a particular school. The garden obstructs the survey line AB. Find the distance CD which is needed to complete the survey.

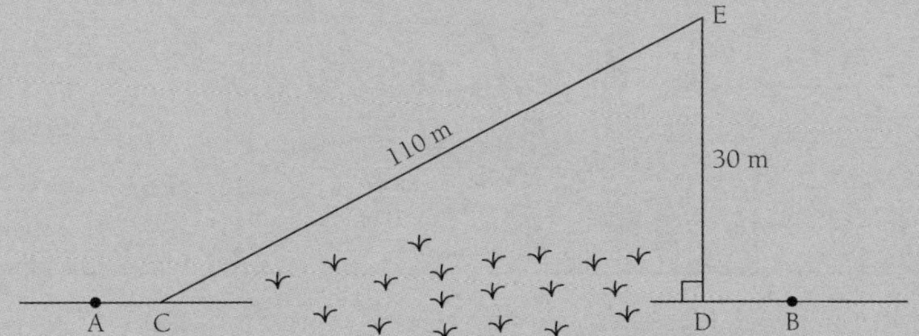

2. A river flows across a survey line PQ. Calculate the distance RS.

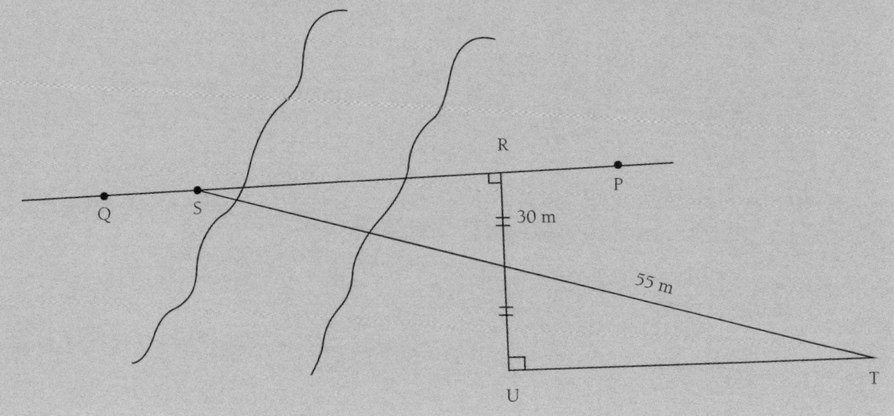

3. A pile of sand obstructs the survey line AB. Calculate the unknown length EF.

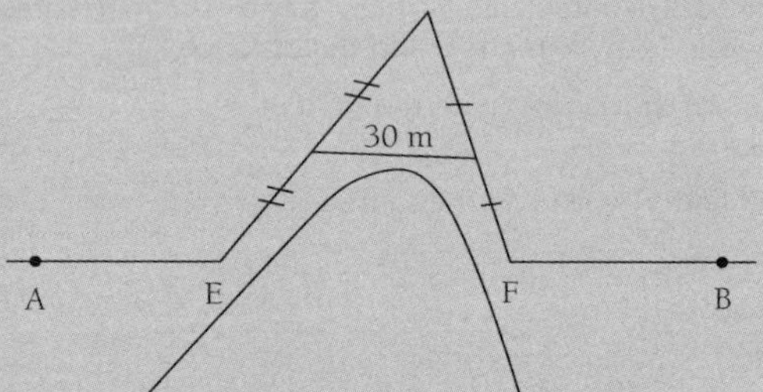

4. A raunwara obstructs the survey line FG. Work out the length of JK.

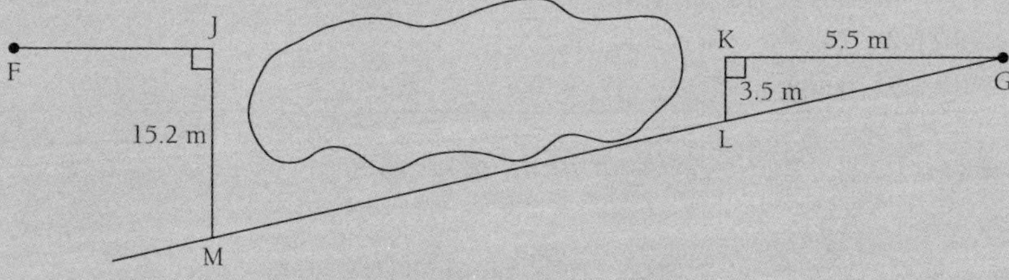

5. A creek is obstructing a survey line. A 50 m offset is marked and another offset 70 m away from the first offset is set and measured to be 35 m. Find the survey line AB.

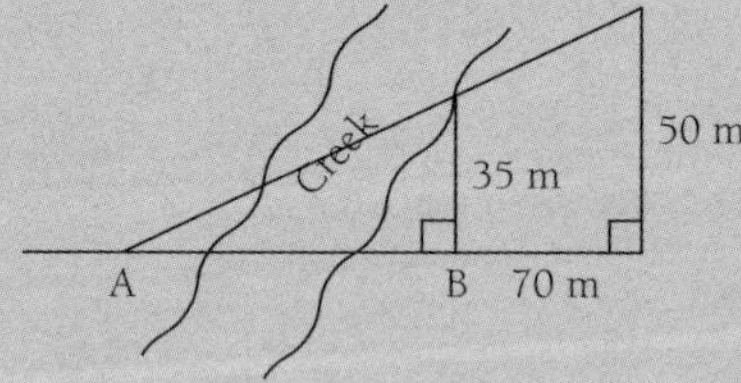

Unit 12.2 Managing Money 2

Topic 1: Interest and inflation – percentages and simple interest

Unit 12.2 focuses on the mathematics relating to money: interest, inflation, loans and investment. The world of business provides a real-life context for developing mathematical knowledge and applying mathematical skills practically, in areas such as compound interest, credit cards and investments. Topic 1 covers:

- An introduction to simple interest.
- A revision of percentages.
- The simple-interest formula.
- Transposing the simple-interest formula to find the principal, interest rate or term.

Interest

When people do not have enough money for personal use or for business use, they can borrow money from a lender and they will pay this lender a fee for using their money. This fee is called **interest**.

Alternatively, when people have more money than they need at the present time, they may **invest** this surplus money. By investing, the value of the amount invested increases over time because the institution that they invest in pays for the use of that money. This payment is also called interest.

- Interest is charged by financial institutions such as banks when a loan is taken out.
- Interest is paid to an investor when she or he 'lends' or invests money in a company or institution.
- Money left in bank accounts attracts a small amount of interest because the bank uses this money (for loans etc.) while it is in their care.

The sum of money invested (or borrowed) is called the **principal**. The amount of interest paid depends on the size of the principal; it also depends on the **term** (length of time) of the investment (or loan) and on the **interest rate**.

The interest rate is usually expressed as a percentage. Most calculations of interest involve **percentages**, so we start by reviewing percentage change.

Percentage change

Per cent (%) means 'part of 100'. So 15% means '15 parts out of 100'.

Percentage increase

If a quantity, Q, increases by $r\%$ then the increase is $\frac{r}{100} \times Q$

The **increased quantity** is $Q + \frac{r}{100} \times Q = Q\left(1 + \frac{r}{100}\right)$

> The **multiplying factor** for a **percentage increase of** $r\%$ is $\left(1 + \frac{r}{100}\right)$
> To find the **increased quantity** we multiply by the multiplying factor.

Example A

Q. The retail price of an item has increased by 8%. If the item cost K7.45 before the increase:

a. How much has the price of the item increased?

b. What is the multiplying factor for this increase and what is its price after the increase?

A. **a.** Increase $= \frac{8}{100} \times 7.45$

$= 0.596$

The increase is 60 toea (rounded to the nearest toea).

b. The multiplying factor is $1 + \frac{8}{100} = 1.08$ and the increased price can be found:

i. By adding the increase : K7.45 + K0.60 = K8.05

ii. By using the multiplying factor: K7.45 × 1.08 = K8.046 ≈ K8.05

Percentage decrease

If a quantity, Q, decreases by $r\%$ then the decrease is $\frac{r}{100} \times Q$

The decreased quantity is $Q - \frac{r}{100} \times Q = Q\left(1 - \frac{r}{100}\right)$

> The **multiplying factor** for a **percentage decrease** of $r\%$ is $\left(1 - \frac{r}{100}\right)$

Example B

Q. The retail prices of all items in a store have been reduced by 25% for the end-of-season sale.

a. If the price of an item before the sale was K69.95, what is its price at the sale?

b. An item is priced at K82.50 at the sale. What was its original retail price?

A. **a.** The price at the sale can be found by:

i. Subtracting the decrease:

Decrease $= \frac{25}{100} \times$ K69.95 = K17.4875 ≈ K17.49

The decrease is K17.49 (rounded to the nearest toea), so the price at the sale is K69.95 – K17.49 = K52.46

ii. Using the multiplying factor:

K69.951 $\times \left(1 - \frac{25}{100}\right)$ = K69.95 × 0.75 = K52.4625 ≈ K52.46

b. K82.50 is the reduced price, so K82.50 is 75% of the original price.

75% of the original price = K82.50

75% × original price = 82.50

$\frac{75}{100} \times$ original price = 82.50

0.75 × original price = 82.50

original price $= \frac{82.50}{0.75}$

$= 110$

The original price of the item was K110.

Example C

Q. A television has increased in price from K595 to K645. What is the percentage increase in price?

A. The increase in price is K645 – K595 = K50.

The original price is K595.

We can substitute these values in:

Increase $= \frac{r}{100} \times$ original price where r is the percentage increase

$$50 = \frac{r}{100} \times 595$$

$$\frac{50 \times 100}{595} = r$$

$$r = 8.4033\ldots$$

The price of the television has increased by 8.4%

Unit 12.2 Activity 1A: Percentages – revision

Give answers correct to two decimal places where necessary.

1. Give the multiplying factor for:

a. An increase of:

i. 20%

ii. 10%

iii. 3%

b. A decrease of:

i. 8%

ii. 12%

iii. 25%

2. State the percentage change that occurs when:

a. K160 increases to K192.

b. K46 decreases by K5.75.

c. K86 is reduced to K79.98

d. K256 increases to K291.84.

3. The retail price of all items in a store has been reduced by 20% for the end-of-season sale.

a. If the price of an item before the sale was K24.95, what is its price at the sale?

b. An item is priced at K67.20 at the sale. What was its original retail price?

4. The prices of products made by a particular manufacturer have increased by 6%.

a. Find the increased price of a product that costs K10.50 before the increase.

b. A product costs K15.37 after the increase. What was its price before the increase?

5. A laptop computer has a selling price of K585. One salesman offers a discount of K45 to customers but another salesman offers a discount of 8%. Which offer is the best for the customer and what is the difference in selling price?

6. At a sale the price of a dress is reduced by 25%.
   a. If the sale price is K249, what was the original price?
   b. If a customer is offered a further 30% off the already reduced price:
      i. What will the customer pay?
      ii. What is the customer's percentage reduction from the original price?
7. A retailer adds 30% to the cost price of all the goods that he sells. He offers a trade discount of 10% to some customers. If a customer receives the trade discount and pays K88.92 for an article:
   a. What was the cost price to the retailer of the article?
   b. What was the retailer's percentage profit for the article?
8. Mary gets a 6% staff discount at the supermarket where she works.
   a. If she buys goods that retail for K54.65, what will she pay?
   b. If Mary buys goods that cost her K72.60, what was their retail value?
9. A car has increased in price by K800. If the new price is K24 600, what was the percentage increase in price?
10. The price of all houses in a particular area has increased by 12% over the last year.
    a. A house was bought for K225 000 a year ago. How much will it be worth now?
    b. A house is for sale at K280 000. How much was it worth a year ago?

Simple interest

Simple interest is interest calculated on the initial amount invested or borrowed (known as the principal) for the entire term of the loan. A simple interest rate is sometimes referred to as a **flat interest rate**.

The following are typical situations where simple interest is charged:

1. *Savings accounts*
   Interest is credited to savings accounts at a fixed rate per annum and can be calculated on either the minimum monthly balance or the minimum daily balance. Money is accessible at any time in a savings account.
2. *Fixed-term deposit accounts*
   The interest is paid at a predetermined rate and paid at the end of the fixed term. The money is not usually accessible during the term of the investment.
3. *Government bonds*
   Money is 'lent' to the government and is bought as government bonds. These have a fixed term and fixed interest rate and the interest is paid at regular intervals. The initial amount invested is returned at the end of the term.
4. *Company debentures*
   If a company wishes to raise capital then it issues debentures to investors. Debentures are issued for a fixed term at a fixed interest rate. Interest is paid to the investor at regular intervals and the initial amount invested is returned at the end of the term.

5. *Hire-purchase agreements (also called time-payment plans)*
   Purchasers of goods agree to 'hire' the goods for a fixed term, paying a fixed, flat rate of interest. During the term of the agreement the purchaser pays a regular instalment, calculated on the original amount owing plus interest, and owns the goods at the end of the term.

6. *Bridging finance*
   Money loaned to cover the time difference between buying an asset and selling an existing asset. Bridging finance is usually used for the short time difference between buying a house and selling an existing house.

The simple interest formula

When **simple interest** is calculated, a set percentage (the **simple interest rate**) of the initial investment (the **principal**) is paid per time period.

The interest rate is usually quoted as a particular percentage per annum (% p.a.), which means percentage each year.

Consider the case where K1 000 is invested in an account paying 8% simple interest p.a.

The amount of interest credited to the account each year would be 8% of K1 000

$$= 1000 \times \frac{8}{100}$$
$$= 80 \text{ kina}$$

After two years the amount of interest accumulated would be $2 \times 80 = 160$ kina.

After three years the amount of interest accumulated would be $3 \times 80 = 240$ kina.

After four years the amount of interest accumulated would be $4 \times 80 = 320$ kina, and so on.

After T years the amount of interest accumulated would be $T \times 80 = 80T$

A **graph** of interest paid (kina) against time (years) shows that there is a linear relationship between interest paid and time:

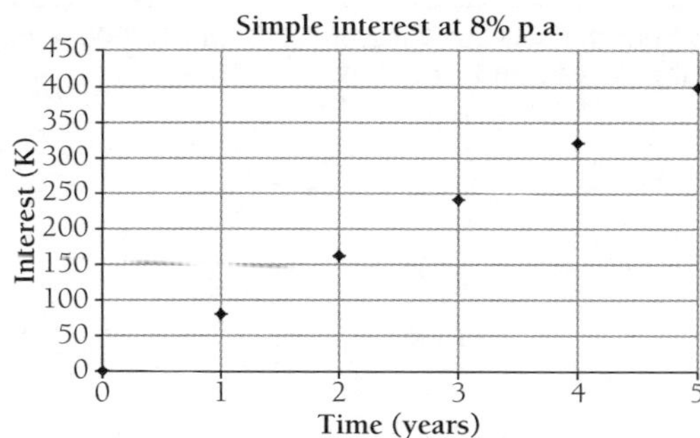

We can see that the points on the graph form a straight line. Simple interest is also referred to as **straight-line interest** or **flat-rate interest**.

To generalise: After T years the amount of interest accumulated would be $T \times 80 = 80T$

$$\text{Interest} = 1000 \times \frac{8}{100} \times T$$
$$= \frac{\text{Principal} \times \textit{rate} \times \textit{Term}}{100}$$

This leads to the **formula for calculating simple interest:**

I kina, on a principal investment of P Kina, for a term of T years at an interest rate of r% per annum:

$$\boxed{I = \frac{PrT}{100}}$$

The total amount accumulated

A kina, on a principal investment of P kina, for T years, at a simple interest rate of r% per annum, is given by:

$$A = P + I$$

$$A = P + \frac{PrT}{100}$$

Taking P out as a common factor gives:

$$\boxed{A = P\left(1 + \frac{rT}{100}\right)}$$

Example D

Q. Calculate the interest paid when K2 460 is invested at 6.5% p.a. simple interest for four years.

A. In this case $P = 2\ 460$, $r = 6.5$ and $T = 4$

Substituting in the formula: $I = \frac{PrT}{100}$ gives

$$I = \frac{2\,460 \times 6.5 \times 4}{100}$$

$$= 639.60$$

The interest paid would be K639.60.

Example E

Q. Find the total amount that will be owing on a loan of K18 000, attracting 8.5% p.a. simple interest, after 3 years and 7 months.

A. $P = 18\ 000$ $\quad$ $r = 8.5$

$T = 3$ years, 7 months $\quad = 3\frac{7}{12}$ years $\quad = \frac{43}{12}$ years

Substituting in the formula $I = \frac{PrT}{100}$ gives

$$= \frac{18\,000 \times 8.5 \times \frac{43}{12}}{100}$$

$$= 5\ 482.50$$

The interest owing after 3 years 7 months is K5 482.50, rounded to the nearest toea.

The total amount that will be owing $= P + I$

$$= 18\ 000 + 5\ 482.50$$

$$= 23\ 482.50$$

The total amount that will be owing is K23 482.50

Example F

Q. Julie has borrowed K2 500 from a bank at 9.25% p.a. simple interest for 185 days. How much will she pay back to the bank?

A. $P = 2\,500;\ r = 9.25;\ T = 185 \text{ days} = \frac{185}{365} \text{ years}$

Substituting in the formula $I = \frac{PrT}{100}$ gives

$$= \frac{2\,500 \times 9.25 \times \frac{185}{365}}{100}$$

$$= 117.2089$$

The interest on the loan is K117.21 to the nearest toea.

Julie will need to pay back K2 500 + K117.21 = K2 617.21

If interest needs to be calculated for a term given as 'from one day of the year to another' then a table numbering the days of the year is helpful:

Day of the year table

Day	January	February	March	April	May	June	July	August	September	October	November	December
1	1	32	60	91	121	152	182	213	244	274	305	335
2	2	33	61	92	122	153	183	214	245	275	306	336
3	3	34	62	93	123	154	184	215	246	276	307	337
4	4	35	63	94	124	155	185	216	247	277	308	338
5	5	36	64	95	125	156	186	217	248	278	309	339
6	6	37	65	96	126	157	187	218	249	279	310	340
7	7	38	66	97	127	158	188	219	250	280	311	341
8	8	39	67	98	128	159	189	220	251	281	312	342
9	9	40	68	99	129	160	190	221	252	282	313	343
10	10	41	69	100	130	161	191	222	253	283	314	344
11	11	42	70	101	131	162	192	223	254	284	315	345
12	12	43	71	102	132	163	193	224	255	285	316	346
13	13	44	72	103	133	164	194	225	256	286	317	347
14	14	45	73	104	134	165	195	226	257	287	318	348
15	15	46	74	105	135	166	196	227	258	288	319	349
16	16	47	75	106	136	167	197	228	259	289	320	350
17	17	48	76	107	137	168	198	229	260	290	321	351
18	18	49	77	108	138	169	199	230	261	291	322	352
19	19	50	78	109	139	170	200	231	262	292	323	353
20	20	51	79	110	140	171	201	232	263	293	324	354
21	21	52	80	111	141	172	202	233	264	294	325	355
22	22	53	81	112	142	173	203	234	265	295	326	356
23	23	54	82	113	143	174	204	235	266	296	327	357
24	24	55	83	114	144	175	205	236	267	297	328	358
25	25	56	84	115	145	176	206	237	268	298	329	359
26	26	57	85	116	146	177	207	238	269	299	330	360
27	27	58	86	117	147	178	208	239	270	300	331	361
28	28	59	87	118	148	179	209	240	271	301	332	362
29	29	*	88	119	149	180	210	241	272	302	333	363
30	30		89	120	150	181	211	242	273	303	334	364
31	31		90		151		212	243		304		365

To calculate the number of days between two dates:

1. If only one of the 'end' days is included:
   Number of days = Number of final date – Number of initial date
2. If both 'end' days are included:
   Number of days = Number of final date – Number of initial date + 1

Note: If the year is a leap year (any year that is a multiple of 4, e.g. 2012, 2016, 2020 etc) then an extra day, 29 February, is included. One is added to the numbers of all the dates starting from 1 March to the end of the year.

Example G

Q. On 19 February 2014 Matilda placed an opening deposit of K12 500 in an account that paid 5.25% p.a. interest. She withdrew the whole amount in the account on 12 August 2014.

- **a.** How many days was the money invested if both of the end days are included?
- **b.** How much interest was paid for this period?
- **c.** How much did she withdraw on 12 August 2014?

A. **a.** Number of days = Number of final date – Number of initial date + 1

$$= 224 - 50 + 1$$

$$= 175$$

b. $P = 12\,500$; $r = 5.25$; $T = \frac{175}{365}$

Substituting in the formula $I = \frac{PrT}{100}$ gives

$$= \frac{12\,500 \times 5.25 \times \frac{175}{365}}{100}$$

$$= 314.6404$$

K314.64 interest was paid for the period of 175 days.

c. Matilda would withdraw the principal plus the interest on 12 August.

A total of K12 500 + K314.64 = K12 814.64

Unit 12.2 Activity 1B: Simple interest

Give answers correct to two decimal places where necessary.

1. Calculate the simple interest payable if:

- **a.** K1 200 is invested for four years at 5% p.a.
- **b.** K2 800 is borrowed for 3.5 years at 6.7% p.a.
- **c.** K5 280 is invested for 8 years at 5.75% p.a.

2. Calculate the simple interest payable if:

- **a.** K800 is borrowed for 30 months at 8.5% p.a.
- **b.** K15 000 is borrowed for 205 days at 6.2% p.a.
- **c.** K56 000 is borrowed for $3\frac{1}{4}$ years at 7.75% p.a.

3. Pila has borrowed K3 800 at 7.25% p.a. for 150 days. How much in total will he owe at the end of this period?

4. Julius has placed K5 000 in a term deposit account paying 6.5% p.a. for three months. How much will his investment be worth at the end of the three months?

5. An investment account pays 6.25% p.a. simple interest. If Bertha places K8 500 in the account:

a. How much interest will she earn if she withdraws the money after 300 days?

b. How much will be in the account after 2 years?

6. Levi buys K84 000 worth of government bonds for a period of five years. The bonds pay a flat rate of 8.2% p.a. interest, payable quarterly.

a. Calculate the total amount of interest paid to Levi over the five years.

b. What is the quarterly interest payment that Levi receives?

7. K56 000 worth of company debentures are purchased for a period of three years. The debentures pay 7.5% p.a., simple interest, paid in monthly instalments.

a. Calculate the total amount of interest paid over the three years.

b. Calculate the monthly interest payment.

8. Bridging finance of K220 000 is arranged for 20 days at an interest rate of 8.35% p.a. What is the interest cost of this bridging finance?

9. On 10 January 2014 Tobias deposited K9 600 in an account paying a flat rate of 6.75% p.a. interest. He withdraws this amount, plus interest, on 21 August 2014.

a. Use the 'Day of the year table' (p. 85) to calculate the number of days that the money attracted interest if both 'end' dates are included.

b. Find the amount of interest Tobias's money earned in this time.

10. Vina took out a personal loan for K5 500 on 11 November 2014 and repaid the loan and the interest owing on 16 May 2015. If the interest charged for the loan was 8.25% p.a., calculate the total amount that was repaid. (Interest is not charged for 16 May 2015.)

11. Francis and Delilah have arranged bridging finance of K254 000 for their new home from 18 July 2014 to 20 August 2014. If the interest charged is 7.45% p.a., calculate the cost of this finance. Interest is charged on 18 July but not on 20 August.

Transposing the simple interest formula to find the principal, interest rate or term

The simple interest formula $I = \frac{PrT}{100}$ can be transposed to make the **subject** of the formula, either:

P (the principal), or

r (the interest rate), or

T (the term).

$$\frac{PrT}{100} = I$$

Multiplying both sides of the equation by 100:

$$PrT = 100 \times \text{I} \qquad \text{... (1)}$$

To make P the subject: Divide both sides of (1) by rT: $P = \frac{100I}{rT}$

To make r the subject: Divide both sides of (1) by PT: $r = \frac{100I}{PT}$

To make T the subject: Divide both sides of (1) by Pr: $T = \frac{100I}{Pr}$

Example H

Q. A total of K108 interest is paid on an amount that has been invested for 3 years at 4% p.a. simple interest. What was the principal amount invested?

A. $I = 108$; $T = 3$; $r = 4$; $P = ?$

Substituting in $P = \frac{100I}{rT}$

$$P = \frac{100 \times 108}{4 \times 3}$$
$$= 900$$

The principal amount invested was K900.

Example I

Q. Over a period of 264 days Steven was charged K252.21 interest on a loan of K4 500. What was the flat rate of interest charged for this period?

A. $I = 252.21$; $P = 4\,500$; $T = 264$ days $= \frac{264}{365}$ years; $r = ?$

Substituting $r = \frac{100I}{PT}$

$$r = \frac{100 \times 252.21}{4\,500 \times \frac{264}{365}}$$
$$= 7.7488$$

Steven was charged a flat rate of 7.75% p.a. for the loan.

Example J

Q. Over a period of time K12 500 has increased to K15 875 in an account paying 6.75% simple interest. How long has the money been invested in this account?

A. The interest, I = Final balance – Principal

$$= 15\,875 - 12\,500$$
$$= 3\,375$$

$P = 12\,500$; $r = 6.75$; $T = ?$

Substituting $T = \frac{100I}{Pr}$

$$= \frac{100 \times 3\,375}{12\,500 \times 6.75}$$
$$= 4$$

The money has been invested in this account for 4 years.

Example K

Q. The growth of an investment is shown on the graph below. Find the rate of interest being paid.

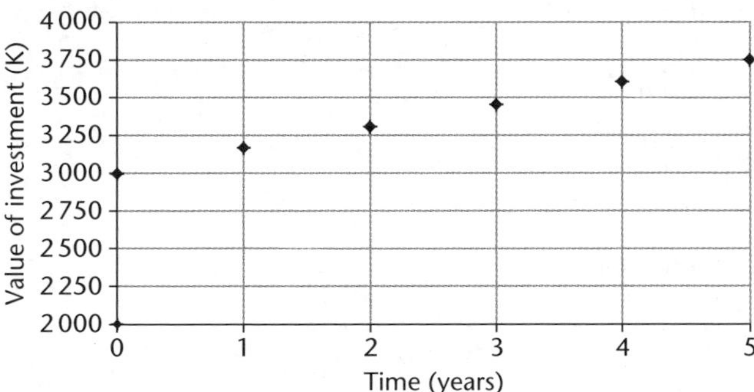

A. The graph is a straight line; this indicates that the type of interest being charged is simple interest.

From the graph it can be seen that the initial investment, P = K3 000.

Another point on the graph that can be read accurately is (5, 3 750) which means that the investment had grown to K3 750 in 5 years.

Hence $I = 3\ 750 - 3\ 000 = 750$;

$T = 5$

Substituting in $r = \frac{100I}{PT}$

$$r = \frac{100 \times 750}{3000 \times 5}$$
$$= 5$$

5% p.a. simple interest is paid on the investment.

Example L

Q. For three years Agnes has received a monthly interest payment of K1 080 from debentures that pay 6.75% p.a. simple interest. What value in debentures did she initially purchase?

A. Interest paid = 1 080 per month

= 1 080 × 12 per year

I = 12 960 per year

$R = 6.75$; $T = 1$; $P = ?$

Substituting $P = \frac{100I}{rT}$

$$P = \frac{100 \times 12\,960}{6.75 \times 1}$$
$$= 192\ 000$$

Agnes initially purchased K192 000 worth of debentures.

Example M

Q. After 2.5 years Emmanuel's debt, which is being charged a flat rate of interest of 7.5% p.a., has grown to K6 080. What amount did he initially borrow?

A. $T = 2.5$; $r = 7.5$; $I = ?$; $P = ?$

Principal + interest = 6 080

$$P + I = 6\,080$$

$$P + \frac{PrT}{100} = 6\,080$$

$$P(1 + \frac{rT}{100}) = 6\,080$$

Substituting the values that we know:

$$P(1 + \frac{7.5 \times 2.5}{100}) = 6\,080$$

$$P(1 + 0.1875) = 6\,080$$

$$1.1875P = 6\,080$$

$$P = \frac{6\,080}{1.1875}$$

$$P = 5\,120$$

Emmanuel initially borrowed K5 120.

Unit 12.2 Activity 1C: Transposing the simple-interest formula

1. Find the principal if:

a. K2 561 is paid in interest over four years on a loan that is charged 6.5% p.a. simple interest.

b. A loan at 7.2% simple interest, that runs for 42 days, accumulates K302.40 in interest.

c. An investment, paying a flat rate of 7.6% p.a., has paid a total of K760 interest over 32 months.

2. Find the interest rate per annum if:

a. K62 500 worth of debentures pays a total of K15 625 in interest over four years.

b. A fixed-term deposit for 3 months pays K81.78 in interest on an initial investment of K5 640.

c. A flat-rate loan of K8 500 over 2.75 years is charged a total of K1 963.50 in interest.

3. Find the term if:

a. K8 000, invested at 6.95% p.a. simple interest, accumulates K1 390 in interest.

b. A loan of K48 000 borrowed at a flat rate of 7.2% p.a. accumulates K4 896 in interest.

c. A term deposit of K10 000, paying simple interest of 5.8% p.a., pays K290 in interest.

4. Kerem's bank account pays interest at 2.5% p.a. and was credited with K8.64 interest for the six months. If the balance didn't change in this time, how much did he have in the account?

5. K66 000 invested in company debentures pays K374 per month in interest. What is the annual rate of interest being paid?

6. Mark took out bridging finance of K245 000 when he bought a new home and was charged K926.80 in interest. If the rate of interest was 9.2% p.a., find the term of the loan in days.

7. After how long will the simple interest on K7 500 at 4.5% p.a. be K731.25?

8. A term deposit pays K118.30 interest for a three-month investment at an interest rate of 5.6%. How much was invested?

9. Ravu's investment has grown from K5 165 to K5 320 over 5 months. What is the flat rate of interest being charged?

10. From 22 April to 31 August Paul placed his savings in an account earning 6% p.a. simple interest. If Paul's savings earned K475.20 interest in this time, what was his initial deposit? Assume interest was paid on both 'end' dates.

11. The interest earned on an investment of K3 000 is graphed below. Find the annual interest rate.

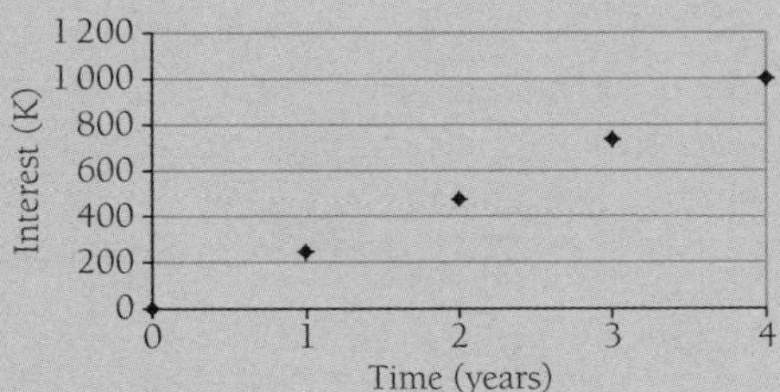

12. The graph below charts the amount owing on a loan over a period of 8 quarters. Find the annual rate of interest.

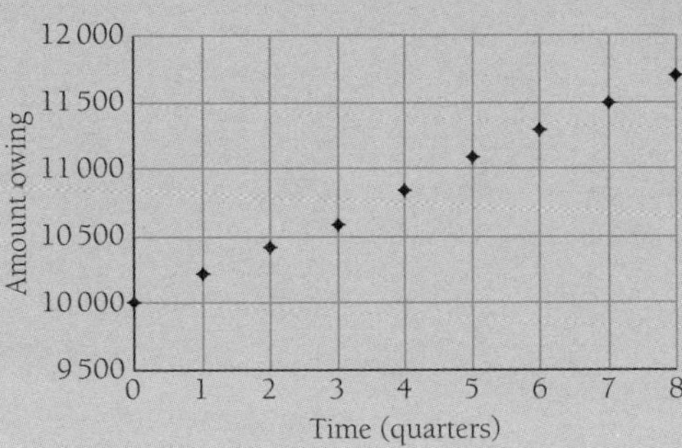

13. The graph below shows the interest earned on an investment of K73 000. Find the rate of interest per annum.

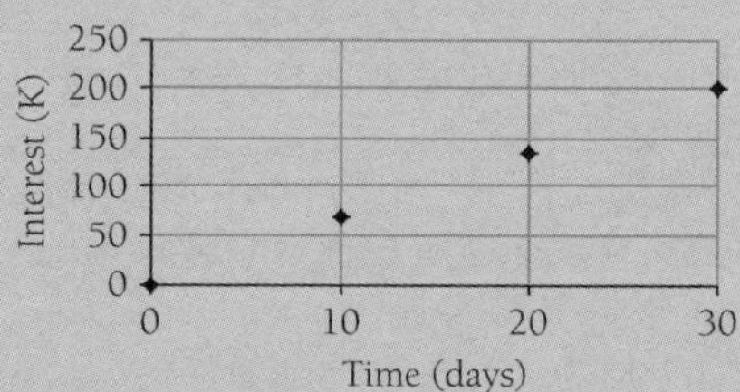

14. After earning interest for 6 months at 5.75% p.a., Alice's investment has grown to K41 150. How much did she initially invest?

15. In 30 months Ekime's loan, which is being charged a flat rate of interest of 9.5% p.a., has grown to K2 574. What amount did he initially borrow?

Unit 12.2 Managing Money 2

Topic 2: Interest and inflation – interest on savings accounts

Unit 12.2 focuses on the mathematics relating to money: interest, inflation, loans and investment. The world of business provides a real-life context for developing mathematical knowledge and applying practical mathematical skills, in areas such as compound interest, credit cards and investments. Having introduced the concept of interest in Topic 1, Topic 2 looks at:

- The practical application of interest in regard to bank accounts.

Balances and interest on savings accounts

Savings accounts are used to manage everyday money transactions and there can be many transactions (deposits and withdrawals) in a month.

Bank charges and government charges apply to savings accounts. Although most banks regard the provision of savings accounts as a service to the customer (for which the customer pays through bank charges), banks still credit a small amount of interest.

The rate of interest, and the way the interest is calculated, vary from bank to bank but there are generally two methods for calculating interest:

- Interest is calculated each month on the minimum monthly balance.
- Interest is calculated daily on the minimum daily balance.

Example A

Q. A savings account was opened on 20 May and the following bank statement shows the transactions in the bank account over two months:

Date	Debit	Credit
20 May		450.00
28 May	80.00	
31 May	0.27	
3 June		364.00
10 June	220.00	
17 June		364.00
30 June	300.00	

a. Calculate the balance in the account at the end of June.

b. If interest is paid at 1.5% p.a. on the minimum monthly balance, find the interest paid on this account for the month of June.

A. **a.** Creating a balance column to show the amount in the account after each transaction:

Date	Debit	Credit	Balance
20 May		450.00	450.00
28 May	80.00		450.00 – 80.00 = 370.00
31 May	0.27		370.00 – 0.27 = 369.73
3 June		364.00	369.73 + 364 = 733.73
10 June	220.00		733.73 – 220 = 513.73
17 June		364.00	513.73 + 364 = 877.73
30 June	300.00		877.73 – 300 = 577.73

The balance at the end of June is K577.73.

b. The minimum balance for June is K369.73, which occurs on 1 and 2 June.

Using $I = \frac{PrT}{100}$ to find the interest on this amount:

Substituting $P = 369.73;\ r = 1.5;\ T = 1/12$

$$I = \frac{369.73 \times 1.5 \times \frac{1}{12}}{100}$$

$$= 0.4621$$

Hence 46 toea in interest is paid on the minimum balance in June.

Example B

Q. Find the interest paid on this account for the month of August if interest is paid at 2.5% p.a., calculated on the minimum daily balance.

Date	Debit	Credit	Balance
1 August			2 540.00
8 August	140.00		2 400.00
12 August		1 060.00	3 460.00
20 August	1 630.00		1 830.00
26 August		1 060.00	2 890.00

A. The table below shows the calculation of days and interest.

Note: On 8 August a withdrawal was made so that the minimum daily balance for that day was the amount *after* the withdrawal. On 12 August a deposit was made but the minimum daily balance for that date was the amount *before* the deposit.

Date	Number of days	Balance	Interest
1 Aug to 7 Aug	7	2 540	$\frac{2\,540 \times 2.5 \times \frac{7}{365}}{100} = 1.2078$
8 Aug to 12 Aug	5	2 400	$\frac{2\,400 \times 2.5 \times \frac{5}{365}}{100} = 0.8219$
13 Aug to 19 Aug	7	3 460	$\frac{3\,460 \times 2.5 \times \frac{7}{365}}{100} = 1.6589$
20 Aug to 26 Aug	7	1 830	$\frac{1830 \times 2.5 \times \frac{7}{365}}{100} = 0.8774$
27 Aug to 31 Aug	5	2 890	$\frac{2\,890 \times 2.5 \times \frac{5}{365}}{100} = 0.9897$
	Total 31		Total 5.5557

K5.56 would be credited to the account as interest for August.

Unit 12.2 Activity 2: Interest and balances

1. Find the balance in this account at the end of February:

Date	Debit	Credit	Balance
1 Feb	0.36		2 856.23
8 Feb		483.48	
14 Feb		20.00	
14 Feb	140.00		
22 Feb	299.95		
28 Feb		1 020.78	

2. Find the balance in the account at the end of the year:

Date	Debit	Credit	Balance
1 Jan			2 856.23
3 Apr		856.05	
8 Jun	550.00		
23 Sept		450.72	
10 Oct		225.00	
28 Dec	420.00		

3. Find the interest payable for a savings account:

a. For September, which had a minimum monthly balance of K858.25 and interest is credited at 3.5% p.a.

b. In April, where the minimum monthly balance was K1 094.28 and the account credited interest at 2% p.a.

c. In January, on a minimum monthly balance of K5 026.38 where the interest was paid at 4.6% p.a.

4. Find the interest payable for a savings account that calculates interest on the minimum daily balance at a rate of 3% p.a.

   a. For June, where the balance for the whole month was K2 560.

   b. For October, where the balance on 1 October was K4 530 and there was a withdrawal of K500 on 1 October.

   c. For April, where the balance on 1 April was K4 782.78 and there was a deposit of K680 on 5 April.

5. A savings account had a balance of K3 000 on 1 January and only one transaction for the year; a deposit of K1 200 on 20 June. The account pays interest at a rate of 4.5% p.a. Find the interest payable for the year if:

   a. It is calculated on the minimum monthly balance.

   b. It is calculated on the minimum daily balance.

6. The statement below gives the transactions in a bank account for March. If interest is credited at 2% p.a., find the interest payable for March if:

   a. Interest is calculated on the minimum monthly balance.

   b. Interest is calculated on the minimum daily balance

Date	Debit	Credit	Balance
26 Feb		300.00	2 563.00
12 Mar	240.00		2 323.00
23 Mar		580.00	2 903.00
3 Apr	180.00		2 723.00

7. The transactions for the financial year of a savings account are given in the table below.

   a. Find the balance for the account at the end of the financial year.

   b. Find the interest payable on the savings account for the year if:

      i. Interest is calculated on the minimum monthly balance.

      ii. Interest is calculated on the minimum daily balance and interest is credited at 3% p.a.

Date	Debit	Credit	Balance
1 July			4 685.00
20 Aug		1 000.00	
10 Dec	1 250.00		
14 Feb	800.00		
23 May		2 450.00	
12 Jun	300.00		

Unit 12.2 Managing Money 2

Topic 3: Interest and inflation – compound interest

The introduction of interest and the application of interest have been covered in the previous two Topics. In Topic 3 we turn to compound interest, covering:

- The calculation of compound interest for various compounding periods.

Compound interest

Simple interest is called 'simple' because it is relatively uncomplicated to calculate: the interest is a percentage of the principal, which does not change, throughout the term.

The calculation of **compound interest** is more complex because after each time period where interest is credited, this interest is added to the principal, thus creating a new principal. The interest for the next time period is calculated from this new principal, and so on.

This creates a compounding effect on the interest – the investor is getting interest on the interest – hence the name compound interest. (In English grammar, a compound word is a word created from two other words, e.g. bathroom; similarly, compound interest is created from two sources: interest on the principal and interest on the interest.)

Calculating and graphing an amount accumulating compound interest

Consider an investment of K1 000 invested in an account that pays interest at 8% p.a., compounded annually.

'Compounded annually' means that the interest is calculated after each year and this interest is then added to the principal, creating a larger principal to accumulate interest in the next year.

Calculating the interest and the amount in the account:

After the first year: Interest paid = 8% of K1 000 = $\frac{8}{100} \times 1\,000$ = **K80**

Amount in the account = K1 000 + K80 = K1 080

After the second year: Interest paid = 8% of K1 080 = $\frac{8}{100} \times 1\,080$ = **K86.40**

Amount in the account = K1 080 + K86.40 = K1 166.40

After the third year: Interest paid = 8% of K1 166.40 = $\frac{8}{100} \times 1\,166.40$ = **K93.31**

Amount in the account = K1 166.40 + K93.31 = K1 259.71

After the fourth year: Interest paid = 8% of K1 259.71 = $\frac{8}{100} \times 1\,259.71$ = **K100.78**

Amount in the account = K1 259.71 + K100.78 = K1 360.49

and so on.

From the calculations above we can see that each year the amount in the account is increasing and hence the interest is increasing.

Graphing the amount (K) in the account against time (years), over 10 years, clearly shows that the points on the graph are not in a straight line but are increasing by an ever-increasing amount each year. This type of graph is called an **exponential graph**.

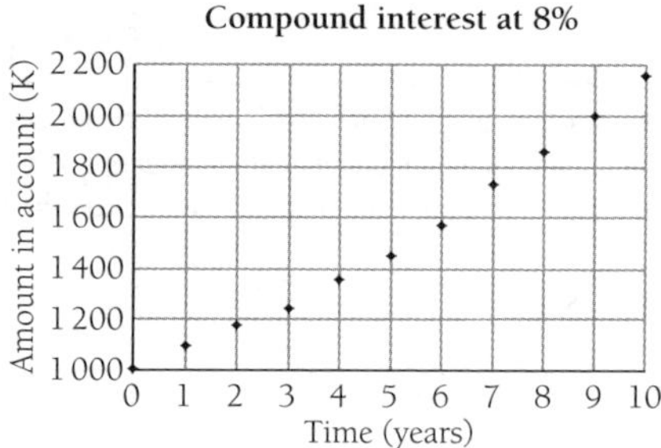

A comparison with simple interest of 8% p.a. over the same time scale shows the difference that compound interest makes to the accumulated amount.

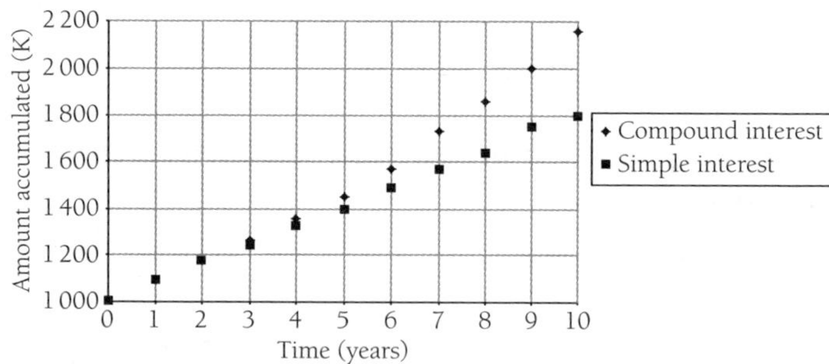

The compound interest formula

We can use a formula to calculate the amount accumulated, A Kina, when compound interest at r% p.a. is applied to a principal amount of P Kina for a term of n years:

If A_n is the amount in the account after n years then:

$A_0 = P$

$A_1 = P + r\% \text{ of } P$

$= P + \frac{r}{100}P$ P is a common factor of both the terms

$= P\left(1 + \frac{r}{100}\right)$

$= PR$ where $R = \left(1 + \frac{r}{100}\right)$

$A_2 = PR + r\% \text{ of } PR$

$= PR + \frac{r}{100}PR$ PR is a common factor of both the terms

$= PR\left(1 + \frac{r}{100}\right)$

$= PR \times R$

$= PR^2$

$A_3 = PR^2 + r\%$ of PR^2

$= PR^2 + \frac{r}{100}PR^2$

$= PR^2\left(1 + \frac{r}{100}\right)$

$= PR^2 \times R$

$= PR^3$

In a similar manner we would find that $A_4 = PR^4$, $A_5 = PR^5$, and so on, which leads to the formula:

If P kina is invested at $r\%$ per annum compound interest for n years, then the amount accumulated (or accrued) is given by:

$$A = PR^n \text{ where } R = 1 + \frac{r}{100}$$

Note: The **interest earned** when P kina is invested at $r\%$ per annum compound interest for n years is $A - P$.

Example A

Q. Calculate **i.** The amount in the account.

ii. The interest earned.

when K10 000 is invested in an account that pays 5.5% p.a., compounded annually.

A. i. $P = 10\ 000$

$r = 5.5\%$ p.a. hence $R = 1 + \frac{5.5}{100} = 1.055$

$n = 5$

Substituting in $A = PR^n$: $A = 10\ 000 \times 1.055^5$

$= 13\ 069.60$

K13 069.60 is in the account after 5 years.

ii. The interest earned = amount accumulated – initial investment

= 13 069.60 – 10 000

= 3 069.60

The interest earned in 5 years is K3 069.60.

Using the compound-interest formula for interest compounded half-yearly, quarterly, monthly, fortnightly, weekly or daily

Compound interest is not always credited annually, although the interest rate is conventionally quoted 'per annum'. If a compound interest rate is 8% per annum then this is:

$\frac{8}{2} = 4\%$ per half-year

$\frac{8}{4} = 2\%$ per quarter

$\frac{8}{12} = \frac{2}{3}\%$ per month

$\frac{8}{26} = 0.30769...\%$ per fortnight

$\frac{8}{52} = 0.1538\%$ per week

$\frac{8}{365} = 0.0219...\%$ per day

If an investment is over 5 years then this is:

$5 \times 2 = 10$ half-years; 10 time periods

$5 \times 4 = 20$ quarters; 20 time periods

$5 \times 12 = 60$ months; 60 time periods

$5 \times 26 = 130$ fortnights; 130 time periods

$5 \times 52 = 260$ weeks; 260 time periods

$5 \times 365 = 1\ 865$ days; 1 865 time periods

The same compound-interest formula can be used to calculate amounts accumulated when the compounding period is less than a year, if appropriate adjustments are made to the interest rate and the time periods:

$A = PR^n$ where A is the amount accrued

P is the initial amount

$R = 1 + \frac{r}{100}$ where r is the interest rate per period

n is the number of time periods

Example B

Q. Calculate the amount in an account after 5 years when K1 000 is invested at 8% p.a. interest compounded:

a. Yearly.

b. Quarterly.

c. Daily.

A. **a.** $P = 1\ 000$

$r = 8\%$ so $R = 1 + \frac{8}{100} = 1.08$

$n = 5$

substituting in $A = PR^n$

$A = 1\ 000 \times 1.08^5$

$= 1\ 469.33$

After 5 years the account contains K1 469.33 if the interest is compounded annually.

b. $P = 1\ 000$

$r = 8\%$ per year

$= \frac{8}{4}\%$ per quarter

$= 2\%$ per quarter

Hence $R = 1 + \frac{2}{100} = 1.02$

$n = 5$ years

$= 5 \times 4$ quarters

$= 20$ time periods

Substituting these values in $A = PR^n$

$A = 1\ 000 \times 1.02^{20}$

$= 1\ 485.94$

After 5 years the account contains K1 485.95 if the interest is compounded quarterly.

Note: This amount is slightly more than the amount calculated with the interest compounding annually.

c. $P = 1\ 000$

$r = 8\%$ p.a.

$= \frac{8}{365}\%$ daily

$= 0.0219...\%$ daily (This figure should not be rounded but used from the calculator)

Hence $R = 1 + 0.0219.../100$

$= 1.000219...$

$n = 5$ years

$= 5 \times 365$ days

$= 1\ 825$ time periods

Substituting these values in $A = PR^n$

$A = 1\ 000 \times (1.000219...)^{1\ 825}$

$= 1\ 491.76$

After 5 years the account contains K1 491.76 if the interest is compounded daily.

Note: This amount is slightly more than the amounts calculated in both **a.** and **b.**

Example C

Q. Interest is paid at 7.2% p.a., compounding monthly, on an investment of K12 000.

a. How much interest has been credited to the account after the first year?

b. How much is in the account after 3 years?

c. What rate of simple interest, per annum, would give the same return after 3 years?

A. **a.** $P = 12\ 000$

$r = 7.2\%$ p.a.

$= \frac{7.2}{12}\%$ per month

$= 0.6\%$ per month

Hence $R = 1 + \frac{0.6}{100} = 1.006$

$n = 1$ year $= 12$ months

Substituting $A = PR^n$

$A = 12\ 000 \times 1.006^{12}$

$= 12\ 893.090...$

Interest = $A - P$

$= 12\,893.090... - 12\,000$

$= 893.090...$

The interest earned in the first year is K893.09

b. $P = 12\,000$

$R = 1.006$

$n = 3 \times 12 = 36$ time periods

Substituting $A = PR^n$

$A = 12\,000 \times 1.006^{36}$

$= 14\,883.619...$

There is K14 883.62 in the account after 3 years.

c. After three years the amount of interest earned is K14 883.62 – K12 000

= K2 883.62

We need to find the interest rate that will earn K2 883.62 in three years with a principal of K12 000.

$P = 12\,000$

$r = ?$

$T = 3$

$I = 2\,883.62$

Substitute in the simple-interest formula to find rate of interest, r:

$$r = \frac{100I}{PT}$$

$$r = \frac{100 \times 2\,883.62}{12\,000 \times 3}$$

$= 8.01$

A simple interest rate of 8.01% p.a. would accumulate the same amount of interest after three years.

Unit 12.2 Activity 3: Compound interest

1. Find the amount accumulated when:

a. K2 000 is invested at 7% p.a., compounded annually, for 5 years.

b. K125 000 is invested at 6.6% p.a., compounded quarterly, for 3 years.

c. K8 000 is invested at 8.4% p.a., compounded monthly, for 4 years.

d. K10 560 is invested at 5.6% p.a., compounded fortnightly, for 2 years.

e. K24 000 is invested at 10.4% p.a., compounded weekly, for 2 years.

2. Find the amount owing on a loan of:

a. K5 500, being charged interest at 6.2% p.a., compounded daily for a year.

b. K836, being charged 6.76% p.a. interest, compounded weekly for 105 days.

c. K6 600, being charged 4.86%p.a. interest, compounded quarterly for 2.5 years.

3. George has borrowed K7 000 and is being charged 6.75% p.a. interest, compounded quarterly. How much will he owe after 4 years if he makes no repayments in that time?
4. Anna wants to invest her K3 000 savings for three years. She has the choice of an account that pays 7.6% p.a. simple interest or another that pays 6.8% p.a., compounded quarterly. Calculate the interest earned in each case and decide which is the better financial option.
5. Find the difference between quarterly and monthly interest paid if K6 000 is borrowed for 2.5 years at 8.2% p.a. compound interest.
6. Titus has invested K7 500 at 5.6% p.a. interest compounding monthly for 3 years. What simple interest rate would produce the same amount of interest after 3 years?
7. Ruth has borrowed K2 500 and is being charged interest at 7.8% interest, compounded weekly.
   a. How much will she owe after 6 months?
   b. If she repays K1 000 after 6 months, how much will she owe after another 6 months?
8. Joseph has a personal loan of K4 800 and is being charged 8.85% interest compounded monthly. He has decided to pay back K1 000 every six months. How much will he owe straight after his:
   a. First payment?
   b. Second payment?
   c. Third payment?
9. After 2 years the difference between an investment earning 8% p.a. interest compounded monthly and 8% p.a. simple interest is K86.32. K9 000 was invested in the compound interest account.
   a. Which investment would earn the most interest in the two years?
   b. How much compound interest was earned in the two years?
   c. How much simple interest was earned in the two years?
   d. Find the principal invested in the simple interest account.

Unit 12.2 Managing Money 2

Topic 4: Interest and inflation – inflation

In Unit 12.2 (Syllabus p. 22) we focus on mathematics that deals with money: interest, inflation, loans and investment of money. Having looked at simple interest and compound interest in the previous two Topics, we now cover:

- Inflation (and appreciation and depreciation).

Loans and investment are covered in later Topics.

Introduction

Inflation is the general upward movement of the prices of goods and services in an economy.

An increase in inflation is often caused by an increase in the supply of money. The annual rate of inflation can fluctuate greatly, ranging from nearly zero to more than 20%.

Governments like to ensure that the rate of inflation does not exceed more than 2–3% so that the spending power of people's money is maintained. If the **inflation rate** is high, prices will increase and people will not be able to purchase the same goods with the same amount of money.

Because inflation is usually quoted as a percentage rate per year, the calculation of inflated prices requires the same theory as growth with compound interest.

Example A

Q. If the inflation rate is expected to remain at 2.5% p.a. for the next 5 years, what will be the cost of an item in five years' time if it costs K1 000 today?

A. The following table shows the calculations over the five years:

Year	Price increase (K)	Cost (K)
0		1 000
1	2.5% of 1 000 = $\frac{2.5}{100}$ × 1 000 = 25	1 000 + 25 = 1 025
2	2.5% of 1 025 = $\frac{2.5}{100}$ × 1 025 = 25.63	1 025 + 25.63 = 1 050.63
3	2.5% of 1 050.63 = $\frac{2.5}{100}$ × 1 050.63 = 26.27	1 050.63 + 26.27 = 1 076.90
4	2.5% of 1 076.90 = $\frac{2.5}{100}$ × 1 076.90 = 26.91	1 076.90 + 26.91 = 1 103.81
5	2.5% of 1 025 = $\frac{2.5}{100}$ × 1 103.81 = 27.60	1 103.81 + 27.60 = 1 131.41

An item that costs K1 000 today will cost K1 131.41 in five years' time if inflation remains at 2.5% p.a.

Note: The increase in value each year is not the same because the increase is increasing. In other words, inflation increases at a compound rate.

A formula can be developed to calculate this inflated value. This involves the same theory as the formula for a compound interest increase.

If P is the original price of the item and $r\%$ is the inflation rate, then the inflated price after n years is given by:

$$\text{Inflated value after } n \text{ years} = PR^n \text{ where } R = 1 + \frac{r}{100}$$

A graph showing the effects of inflation at 2.5% on an item costing K1 000 over ten years is given below.

A straight line (dotted) is drawn to highlight the fact that the graph of the value of the item over ten years is clearly not linear.

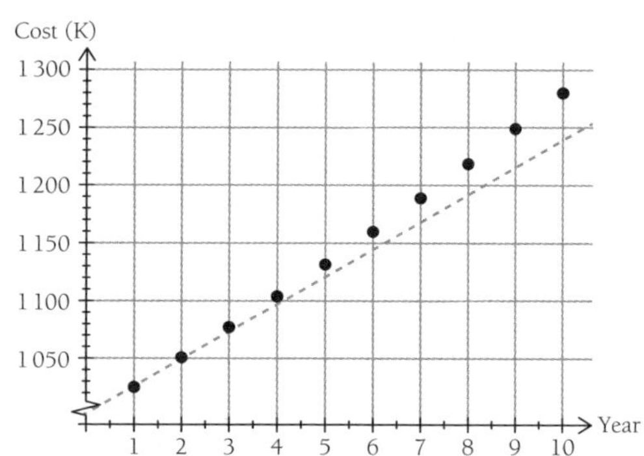

Example B

Q. John bought a house for K156 000 and sold it seven years later for K248 000. If inflation averaged 3.5% over the seven years:

a. Find the inflated value of the house after seven years.

b. Determine the real profit or loss that John made on the sale of the property.

A. **a.** $P = 156\ 000$, $r = 3.5$, $R = 1 + \frac{3.5}{100} = 1.035$

$n = 7$

Inflated value after seven years $= PR^7 = 156\,000 \times 1.035^7 = 198\,475.56$

b. John has made a real profit of K(248 000 – 198 476) = K49 524.

Unit 12.2 Activity 4A: Inflation

Give answers correct to the nearest whole number.

1. If an item costs K100 today, how much will it cost in 5 years' time if inflation remains at 4% per year over the 5 years?
2. What would you expect a car, which costs K19 500 today, to cost in 10 years' time if the rate of inflation is 5% per year over the 10 years?
3. Neil receives increases in his salary that are in line with inflation. If the rate of inflation is 3%, and is expected to stay at this rate, and Neil's salary is currently K48 000 p.a., calculate:

a. His salary in 5 years' time.

b. The number of years before his salary will be at least K60 000.

4. Ari bought a flat 4 years ago for K76 000. If inflation has averaged 2.5% per year and he sells the flat now for K102 000, find the profit or loss that he will make.

5. The inflation rate over the last three years has been:

2008: 2.8%

2009: 2.6%

2010: 3.2%

Find the cost now of an item bought for K440 at the beginning of 2008, assuming that the price has increased with inflation.

6. A household insurance policy automatically increases the amount insured each year by the inflation rate. Eight years ago Flossie took out a household insurance policy with an amount insured of K56 000. If inflation has been averaging 3.7% per year over the eight years, what is the amount insured now?

7. The cost of a car four years ago was K21 000. The cost of the car has now increased due to inflation to K22 910. What has the rate of inflation been over the four years, assuming it is the same for each of the four years?

Appreciation

Appreciation is the increase in value of an item or quantity over time.

Items that appreciate are property (houses, land, shares) or items that are called 'collectables' such as gold, diamonds or artwork. Some items that are highly valued or desirable appreciate at a very high rate each year.

The rate of increase in appreciation is similar to the rate of increase in compound interest.

For an item purchased for P kina and appreciating at a rate of r% per year for n years:

$$\text{Appreciated value} = PR^n \text{ where } R = 1 + \frac{r}{100}$$

Example C

Q. Shares in a mining company have appreciated by 9% each year for the last five years. If the shares were bought for K23 five years ago, what is the value of the shares now?

A. $P = 23$

$r = 9$

$R = 1 + \frac{r}{100} = 1 + \frac{9}{100} = 1.09$

The value of the shares after 5 years $= 23 \times 1.09^5 = 35.388 \approx$ K35.39.

Unit 12.2 Activity 4B: Appreciation

Give answers correct to the nearest whole number.

1. Pink diamonds are predicted to appreciate at 15% per year. A pink diamond is bought for K24 000. Find:
   a. The expected value of the diamond in 3 years' time.
   b. The number of years before the value of the diamond has doubled in value.
2. A company has invested K32 000 in artwork that it predicts will increase in value by 12% per year. Find:
   a. The expected value of the artwork in 3 years' time.
   b. The number of years before the value of the artwork has doubled in value.
3. A block of land is expected to appreciate at 7% each year. If the land was bought for K28 000, how much is it expected to be worth in 6 years' time?
4. If the price of gold increased by an average of 18% over the last 5 years, and the price of an ounce of pure gold was K1 300 five years ago, what is the price of one ounce of gold now?
5. A house in a desirable area has increased at the rate of 8% each year for the last 4 years. If this house is valued at K286 000 now, what was its value four years ago?

Depreciation

Items that are used in businesses, such as machinery, vehicles, computer equipment etc., are called capital items and are classed as assets of the business. These items will wear out with time and hence lose value, or **depreciate in value**.

Depreciation is the loss in value of a capital item over time.

The value of a capital item at a particular time is called the **book value** of the item at that time.

A capital item is considered to have a **useful life** while it is producing revenue for the business. When the item is deemed to be no longer useful to the business it will be **sold off** and its value at this time is called the **scrap value**.

We will consider three types of depreciation:

1. Flat-rate depreciation.
2. Reducing-balance depreciation.
3. Unit-cost depreciation.

Flat-rate depreciation

Flat-rate depreciation is sometimes called **straight-line depreciation**.

Flat-rate depreciation means that an item decreases in value by the same amount each year. This amount can be quoted as a constant percentage of the purchase price or as a fixed value.

Example D

Q. A machine is bought for K22 000 and is considered to have a useful life of 8 years, after which time it will be sold off for a scrap value of K6 000. If flat-rate depreciation is applied to this machine:

a. Find the amount that it depreciates each year.

b. Draw up a table that will show its book value for its useful life.

c. Plot a graph of 'book value' versus 'age' for the useful life of the machine.

A. **a.** Purchase price – scrap value = K22 000 – K6 000

= K16 000

If the machine has a useful life of eight years, it will depreciate a total of K16 000 over the eight years. Since it depreciates the same amount each year this amount will be

$\frac{\text{K16 000}}{8}$ = K2 000

b.

Age of machine (Years)	Depreciation (K)	Book value (K)
0		22 000
1	2 000	20 000
2	2 000	18 000
3	2 000	16 000
4	2 000	14 000
5	2 000	12 000
6	2 000	10 000
7	2 000	8 000
8	2 000	6 000 ← Scrap value

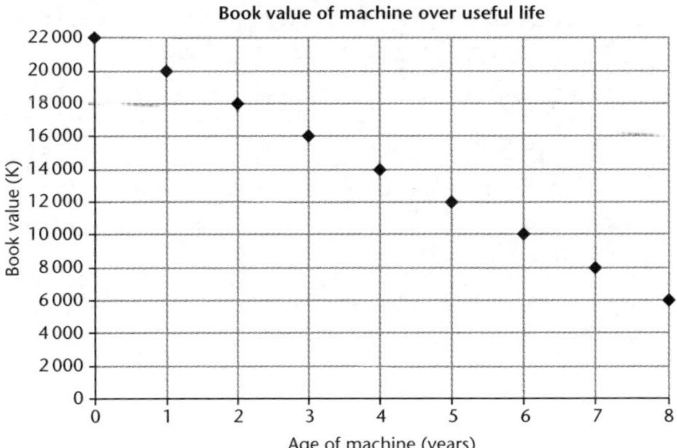

c. Notice that the points on the graph are in a straight line. Hence the other name for flat-rate depreciation: straight-line depreciation.

A formula for the book value of an item with an initial value of P kina, after n years, depreciating at a flat rate of r% p.a.:

$$\text{Book value after } n \text{ years} = P - n \times \frac{Pr}{100}$$

Example E

Q. A business has purchased office equipment for K26 000 that will depreciate by a flat rate of 10% of its purchase price each year. It will be replaced after 6 years.

a. How much will the equipment depreciate each year?

b. What is the scrap value?

A. a. $10\% \text{ of K26 000} = \frac{10}{100} \times \text{K26 000}$

$= \text{K2 600}$

The equipment will depreciate by K2 600 per year.

b. Book value after 6 years = K26 000 – (6 × K2 600)

= K10 400

After 6 years the equipment will be replaced; at that time it will have a book value, and hence a scrap value, of K10 400.

Unit 12.2 Activity 4C: Depreciation

1. A capital item purchased for K2 000 depreciates at a flat rate of 12% each year. Find:

- **a.** The amount the item depreciates each year.
- **b.** The book value of the item after 3 years.
- **c.** The scrap value of the item if it has a useful life of 6 years.

2. A printer, purchased for K6 000, depreciates at a flat rate of 15% per year. It will be replaced when its value falls below K1 000. Find:

- **a.** The amount the printer depreciates each year.
- **b.** The useful life of the printer.

3. A laundromat purchases a washing machine for K4 200. It will be scrapped after 4 years when its value is K300. Assuming flat-rate depreciation, find:

- **a.** The amount the washing machine depreciates each year.
- **b.** The book value of the washing machine after two years.

4. A commercial dishwasher is purchased for K4 800. It is expected to have a useful life of 5 years, after which it will be sold for an estimated scrap value of K600. Assuming straight-line depreciation, find:

- **a.** The depreciation each year as a percentage of the purchase price.
- **b.** The value to the business after 3 years.

5. The following graph shows the depreciation of a capital item over the 6 years of its useful life.

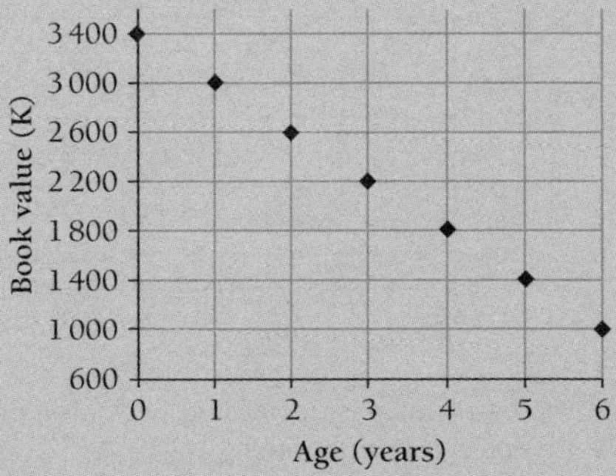

   a. What type of depreciation is being applied in this case?

   b. What was the purchase price?

   c. What was the scrap value?

   d. How much did the item depreciate each year?

   e. Write down an equation connecting 'book value' and 'age'.

Reducing-balance depreciation

Reducing-balance depreciation means that an item depreciates by a constant percentage for each year of its useful life.

Example F

Q. A lathe, purchased for K16 000, will depreciate by 20% for the 6 years of its useful life.

   a. Draw up a table showing the book value of the lathe over its useful life.

   b. Find the scrap value of the lathe.

   c. Plot a graph of the book value of the lathe over the 6 years of its useful life.

A. **a.**

Age (Years)	Depreciation	Book value (K)
0		16 000.00
1	20% of 16 000 = 3 200.00	16 000 – 3 200 = 12 800.00
2	20% of 12 800 = 2 560.00	12 800 – 2 560 = 10 240.00
3	20% of 10 240 = 1 848.00	10 240 – 1 848 = 8 192.00
4	20% of 8 192 = 1 638.40	8 192 – 1 638.4 = 6 553.60
5	20% of 6 553.6 = 1 310.72	6 553.6 – 1 310.72 = 5 242.88
6	20% of 5 242.88 = 1 048.58	5 242.88 – 1 048.58 = 4 194.30

b. The scrap value is the book value after 6 years. The scrap value is K4 194.30.

c.

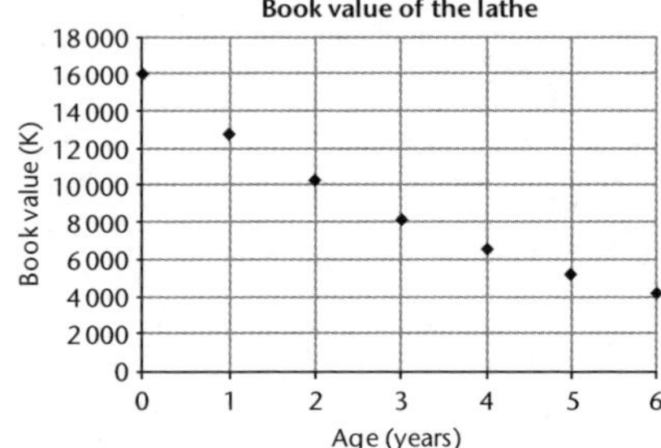

The graph shows that the book value is decreasing with time but that the decrease is less each year. The points are *not* in a straight line.

A formula can be developed to calculate this depreciated value and it is based on the same theory as a compound interest decrease. Hence the formula for the book value after n years of an item purchased for P kina and depreciating at r% per year:

$$\text{Book value after } n \text{ years} = PR^n \text{ where } R = 1 - \frac{r}{100}$$

Example G

Q. A machine is purchased for K12 850 and depreciates by 18% each year.

a. Find the book value after 3 years.

b. If the machine has a useful life of 5 years, what is its scrap value?

A. a. Using the formula to find book value:

$P = 12\,850$

$r = 18\%$ hence $R = 1 - \frac{18}{100} = 0.82$

$n = 3$

Substitute: Book value after n years $= PR^n$

Book value after 3 years $= 12\,850 \times 0.82^3$

$= 7\,085.0788$

The book value after 3 years is K7 085.08

b. Scrap value = Book value after 5 years in this case.

$= PR^5$

$= 12\,850 \times 0.82^5$

$= 4\,764.006\ldots$

The scrap value of the machine is K4 764.01

Example H

Q. A butcher pays K9 560 for a meat saw that will depreciate by 15% each year. It will be scrapped when its value falls below K4 000. After how many years' use will it be scrapped?

A. We want to find the year when the book value falls below K4 000, so we have to solve:

Book value after n years $= PR^n$ for n

$P = 9\,560$

$r = 15$ so $R = 1 - \frac{15}{100} = 0.85$

Book value = 4 000

Substituting these values in the formula:

$4\,000 = 9\,560 \times 0.85^n$

Use trial and error and your calculator to try values for n in the right-hand side of the equation:

$9\,560 \times 0.85^4 = 4\,990.38$ *too high*

$9\,560 \times 0.85^5 = 4\,241.82$ *too high*

$9\,560 \times 0.85^6 = 3\,605.55$ *below* K 4 000

After 6 years the value has fallen below K4 000 so the saw will be scrapped after 6 years.

Unit 12.2 Activity 4D: Reducing-balance depreciation

1. Office furniture purchased for K8 500 depreciates by 10% of its value each year using reducing-balance depreciation. It will be replaced after 7 years. Find:
   - **a.** The amount it depreciates in the first year.
   - **b.** Its book value after 5 years.
   - **c.** Its scrap value.
2. A machine to be used in a factory is purchased for K205 000 and is considered to have a useful life of 15 years. If the value of the machine depreciates by 15% each year, using reducing-balance depreciation, find to the nearest dollar:
   - **a.** The amount of depreciation in the first year.
   - **b.** The book value after the tenth year.
   - **c.** The scrap value of the machine.
3. Plot, on the same set of axes, the book value for the first 6 years of a capital item, purchased for K12 000, which depreciates by 15% per year if the depreciation is:
   - **a.** Flat-rate depreciation.
   - **b.** Reducing-balance depreciation.
4. A car, purchased new for K28 860, is traded in for K13 850 after 4 years. What is the flat-rate depreciation rate applicable in this case? Would the reducing-balance depreciation rate be greater or less than the flat rate in this case?
5. The following graph shows a comparison of flat-rate and reducing-balance depreciation on an item purchased for K5 000 and depreciated over 5 years:

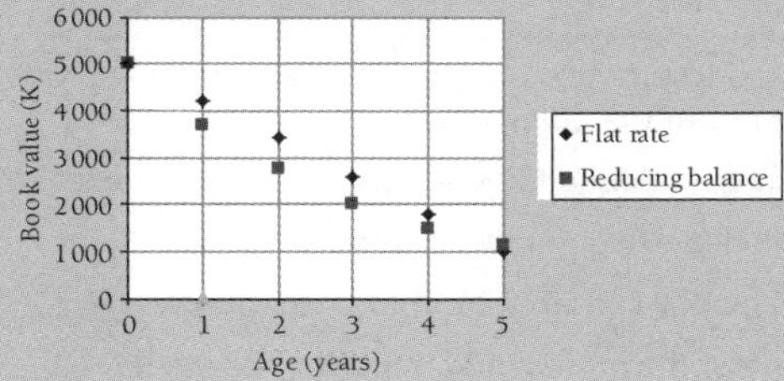

a. If the book value using flat-rate depreciation is K1 000 after 5 years, find the flat rate of depreciation.

b. If the book value using reducing-balance depreciation is K3 700 after the first year, find the rate of reducing-balance depreciation.

c. Find the difference between the book values after 4 years.

6. A business has purchased K48 500 worth of furniture for its restaurant. If the furniture depreciates by 18% each year, using reducing-balance depreciation, find:

   a. The amount the furniture depreciates in its first year.

   b. The amount the furniture depreciates in its second year.

   c. The book value after the fourth year.

   d. The useful life of the furniture if it is to be replaced when its value falls below K10 000.

7. Computer equipment bought for K5 400 depreciates by 30% each year. Find the difference between reducing-balance and flat-rate depreciation for this equipment after 3 years.

8. A capital item purchased for K98 600 depreciates by 22% reducing-balance depreciation each year. The item will be sold off when its value falls below K20 000. What is the useful life of the item?

9. A truck, bought for K58 000, is sold off when its value falls below K22 000. How long will this take if the depreciation is:

   a. Flat rate at 15%?

   b. Reducing balance at 15%?

10. The useful life of a capital item is considered to be the number of years until its book value falls below $\frac{1}{4}$ of its purchase price. If reducing-balance depreciation applies, how long will this be if the depreciation rate is:

   a. 10%?

   b. 25%?

Unit-cost depreciation

The value of some items depends, not on the age, but on the amount of *use* the item has had.

A plaster mould, for example, used to produce porcelain models, may only be able to produce a set number of models before its surface deteriorates. Its value depends on the number of models it has already produced. Or the value of a taxi or PMV could depend on the number of kilometres it has travelled.

Example I

Q. A plaster mould used to produce porcelain models can only produce 100 models before its surface deteriorates to the point where it is no longer useful.

   a. If the mould was purchased for K240 and its scrap value is zero:

      i. What is the depreciation rate per model?

      ii. What is its book value after 36 models are produced?

   b. Graph the value of the mould after *n* models are made.

A. a. i. If the scrap value is zero then the depreciation rate per model is K240/100 = K2.40

ii. The value of the mould to the business decreases by K2.40 for every model produced. If the original value was K240 then:

Book value after 36 models = K240 – 36 × K2.40

= K153.60

b.

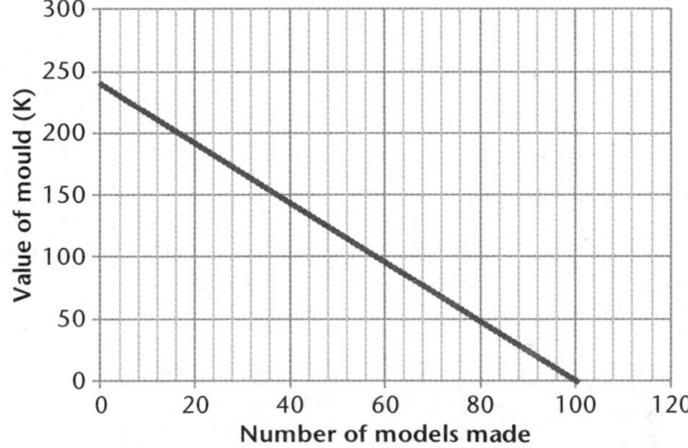

It can be seen that the points on the graph, showing the value of the mould as subsequent models are made, form a straight line.

Unit 12.2 Activity 4E: Unit-cost depreciation

1. A company purchases a delivery van for K28 000. It will be replaced when it has travelled 250 000 kilometres, at which time the scrap value is estimated to be K8 000.

a. Find:

i. The unit-cost depreciation per kilometre travelled.

ii. The unit-cost depreciation per 1 000 kilometres travelled.

iii. The book value of the vehicle after 100 000 kilometres.

b. If the company sells the delivery van after it has travelled only 180 000 kilometres, what price should it expect to get?

2. It is estimated that a die-casting machine can produce 100 000 units in its useful life. If the purchase price of the machine was K86 000 and it will have a scrap value of K3 000, find:

a. The unit-cost depreciation of the machine.

b. The equivalent flat-rate depreciation if it is scrapped after 3 years.

3. A business maintains a company car that it purchased for K34 600. The business will replace this car when it has travelled 150 000 km, at which time the trade-in value will be K16 000.

a. Find the unit-cost depreciation of the car:

i. Per kilometre.

ii. Per 1 000 kilometres.

b. What is the book value of the car after it has travelled:

i. 35 000 km.

ii. 110 000 km.

c. If the car travels an average of 30 000 km per year, find:

i. The useful life of the car.

ii. The equivalent flat-rate depreciation.

4. The unit-cost depreciation of a car is 15 toea per kilometre.

a. If the car was purchased for K43 500, find the book value of the car after it has travelled 30 000 kilometres.

b. If the car is replaced after it has driven 120 000 kilometres, what is its scrap value?

5. The value of a photocopier depreciates by 4 toea for each copy made. The photocopier was originally purchased for K15 000.

a. What is its value after it has made 150 000 copies?

b. How many copies can be made before its value falls to K2 000?

c. How many copies can be made before its value falls to K0?

d. The photocopier is sold for K4 300 when it has made 260 000 copies. Did the seller make a book profit or loss?

6. From graphs **I** to **IV** below, find the graph that best shows the change with time for:

a. Compound interest.

b. Straight-line depreciation.

c. Inflation.

d. Appreciation.

e. Reducing-balance depreciation.

f. Flat-rate depreciation.

g. Simple interest.

h. Unit-cost depreciation.

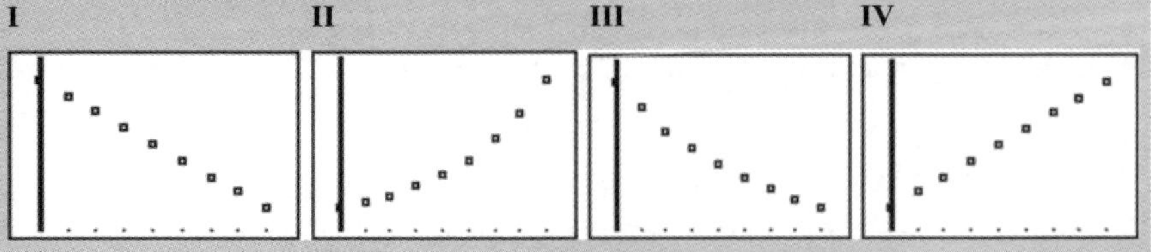

Unit 12.2 Managing Money 2
Topic 5: Consumer credit

The focus of Unit 12.2 is mathematics relating to money, and the application of mathematical skills in practical areas such as interest, credit, inflation, loans and investment. In Topic 5 we look at credit and its different forms:

- Personal loans and credit cards.
- Hire-purchase agreements.
- Reducing-balance loans.
- The annuities formula.

Personal loans

A personal loan is money borrowed for the purchase of goods or services by the borrower.

- **Personal loans** can be obtained from banks or other financial institutions. To obtain a loan from one of these institutions the borrower will need to demonstrate that they can repay the loan. The amount borrowed and the rate of interest charged will depend on whether the loan is secured or unsecured – higher rates of interest are usually payable on unsecured loans. A secured loan is a loan where the borrower has other property that the lender can access if the borrower is unable to pay back the loan.
- When a personal loan is given, the amount borrowed, the interest rate, the term of the loan and the repayment amount will all be specified.
- The financial institution will often charge the borrower for taking out a loan: usually a loan establishment fee and a monthly fee.
- Loans that are for a very **short term** will be charged simple interest and the total amount owing will be repaid at the end of the term. A **bridging loan**, which covers the time between purchasing a new property and selling an existing property, is an example of a short-term loan.
- **Credit card** loans are short-term loans where the interest payable is calculated each month using simple interest principles.
- **Longer-term loans** such as housing loans, are usually given as **reducing-balance loans** where the amount of interest charged reduces as the balance of the loan reduces.
- There are other businesses that lend money for personal use where evidence of the ability to repay the loan is not as strict. Because these loans are at greater risk of not being repaid (called defaulting on the loan) as those given by banks, the lender usually charges a higher rate of interest. These lenders base their loans on a simple interest rate, which is often not specified, over the term of the loan and usually require regular repayments. **Hire-purchase agreements** are examples of this type of loan.

Short-term loans involving some simple interest calculations have been covered elsewhere in this Unit – see Topics 1 and 2. Calculations of the interest paid on credit cards and hire-purchase agreements is covered below

Credit cards

Credit can be provided by a financial institution issuing a credit card, or by a retailer offering a store loyalty card.

Banks and other financial institutions will offer a credit card to a customer if the customer has demonstrated that they have a stable income and that they have a history of being able to manage their money. This is called their credit history.

Credit cards usually have a limit (maximum amount that they can accumulate in credit each month) and this is different for different individuals.

The credit card owner is sent an account each month and all or part of the credit that has been accumulated can be repaid within a specified time. There is a minimum amount that is payable and any amount that is not repaid will accumulate interest. This is usually at a higher rate than the rate for other personal loans.

It is easy to make a purchase with a credit card but it is also easy to overspend and then have to pay large amounts in interest. Interest is charged in various ways with credit cards:

- Cash is usually charged interest from the day the cash is borrowed.
- Some cards have an interest-free period but charge an annual fee for the use of the card. If the amount that is owed is not paid within the interest-free period then the cardholder is charged interest from the date of the purchase. Interest-free periods are different for different credit cards and can be up to 55 days.
- Some cards have no interest-free period but do not charge an annual fee.
- A minimum repayment amount (about 3%) is set each month. If this is the only amount that is repaid then the amount owed can continue to increase and interest rates of 16–20% will be charged on the balance.

Store loyalty cards are organised along the same lines as credit cards but can only be used in the store that issues the card. There are often discounts and other incentives attached to store loyalty cards that are used to encourage customers to shop only in that store. Customers cannot usually access cash on store loyalty cards.

Example A

Q. Michael has a 'no interest-free period' credit card that charges 15.6% interest. He purchased a jacket for K65.00 on 15 April and took out a cash advance of K50.00 on 20 April. His credit card statement arrived on 25 April. How much interest would be included in Michael's monthly statement?

A. Michael will owe interest on K65.00 for 25 – 15 + 1 days = 11 days and interest on the K50.00 for 25 – 20 + 1 days = 6 days. *Note*: The numbers of days include both the purchase date and the end date.

$$\text{Interest} = \frac{65 \times 15.6 \times \frac{11}{365}}{100} + \frac{50 \times 15.6 \times \frac{6}{365}}{100} = 0.43$$

K0.43 will be included in Michael's monthly statement.

Unit 12.2 Activity 5A: Credit cards

1. Esther was charged K12 interest for one month on a K800 credit card balance. What was the monthly interest rate?
2. Tomi obtains a K400 cash advance on his credit card on 8 April. His credit card statement is dated 26 April and he is being charged 18.5% p.a. interest. If the cash advance is the only item on the statement, how much interest is Tomi charged on the credit card statement?

3. On 1 June Helena buys a jacket, costing K148, using her credit card. Her credit card has a 25-day interest-free period and her statement is dated the 15th day of the month. If she does not repay any of the amount until 20 July, and the credit card has an interest rate of 18%, what amount of interest will be charged on the statement on:
   a. 15 June?
   b. 15 July?
4. Malaki has a 'no interest-free period' credit card that charges 16.5% p.a. interest. On 9 January he purchases a phone costing K395.00. His January statement is dated 28 January and the phone is the only item on it.
   a. How much interest is charged on the January credit card statement?
   b. Malaki pays K100.00 off the amount owed on 28 January and makes no more purchases or payments before the next credit card statement arrives on 28 February. How much will this statement show that Malaki owes?
5. Pania has a 'no interest-free period' credit card that charges 16.2% p.a. interest. Her statement is dated the 28th day of the month. In the month of May, Pania makes the following transactions on her credit card:

Date	Transaction	Cost (K)
2 May	Purchase of shoes	45
8 May	Cash advance	100
24 May	Utilities payment	124

   What is the total amount owing that will appear on her statement on 28 May?
6. On 12 October Kenime purchased household furniture, using his credit card, to the value of K2 856. Kenime's credit card has a 'no interest-free period' and charges 17.5% p.a. interest. The statement is dated the 20th day of each month. Kenime plans to pay off the minimum amount, K50, each month until February when he plans to pay off all that is owing. Assuming that Kenimi makes the payment of K50 on the 20th of the month, find the total amount owing on the 20th of:
   a. October.
   b. November.
   c. December.
   d. January.
   e. February.

Debit cards

Because debit cards are much more common than credit cards in PNG, let's look briefly at the difference between them.

A credit card is a short-term loan, where you borrow from the bank, spend now, and pay interest later as well as paying back the money you borrowed. A debit card is *not* a loan, because you can only withdraw money if you already have deposited that money in a savings account or cheque account at your bank. The advantage is that a debit card allows you to spend your own money without carrying cash. That is the main difference and it means that you do not pay interest and the fees charged by the bank are lower than for credit cards.

In PNG there are different types of debit cards such as Kundu Cards, Save Cards and Access Cards. Apart from the differences mentioned above, debit cards operate in a way that is similar to credit cards. You must sign to complete a transaction, or use EFTPOS and enter a PIN (personal identification number); and the bank sends you regular statements that show your transactions and the balance in your account. For those who do not have a credit history and for those who find it difficult to budget, a debit card has advantages.

Hire-purchase agreements (or time-payment plans)

If a purchaser cannot pay the full price of an item, the retailer may offer a hire-purchase agreement.

A hire-purchase agreement usually requires the purchaser to pay a deposit (either a set amount or a percentage of the purchase price), then the remainder of the purchase price, plus interest, is paid in an agreed number of equal amounts (called instalments).

With a hire-purchase agreement the purchaser is hiring the item from the retailer until the final payment is made. If the purchaser defaults (does not pay) on any of the payments then the item will be repossessed by the retailer without any return of the payments.

The interest paid on a hire-purchase plan is quoted as a simple interest rate (flat rate) per annum. Even though the repayment amount can seem affordable, sometimes the amount of interest paid is large.

Example B

Q. Bernard is offered a hire-purchase agreement, charging a flat rate of interest of 15%, to buy a television which has a purchase price of K1 250. If he pays a deposit of K250 and will pay the remainder plus interest in 18 monthly payments, how much will he pay each month?

A. The amount owing after the deposit is paid is K1 250 – K250 = K1 000

Interest at 15% on K1 000 for 18 months = $\dfrac{1\,000 \times 15 \times \frac{18}{12}}{100} = 225$; interest is K225.

Bernard will pay a total of K1 000 + K225 = K1 225 in 18 equal instalments.

Bernard will pay $\dfrac{\text{K1 225}}{18} = \text{K68.06}$ each month.

Example C

Q. Ada is using a hire-purchase agreement to buy a refrigerator that has a purchase price of K990. She has agreed to pay K190 as a deposit and the remainder in six monthly payments of K160.

a. How much interest is Ada paying with this contract?

b. What is the flat rate of interest being charged?

A. **a.** The amount owing after the deposit is paid = K990 – K190

= K800

Six one-monthly payments of K160 = 6 × K160

= K960

Ada is paying K960 – K800 = K160 in interest.

b. Ada is paying K160 in interest for 6 months and this would be 2 × K160 = K320 for 12 months (1 year).

The yearly interest as a percentage of the amount owed = $\frac{320}{800} \times \frac{100}{1} = 40$.

Ada is being charged a flat rate of interest of 40% p.a. This is a very high rate of interest which was not immediately apparent from the amount of interest being charged or the amount of the repayments.

Unit 12.2 Activity 5B: Hire-purchase

1. Find the interest paid on a hire-purchase contract that is used to pay off a principal of K1 500 over 2 years and is charging a flat rate of 16% p.a.
2. Find the amount and the flat rate of interest when a hire-purchase agreement is made to pay off a principal of K2 000, plus interest, with 12 one-monthly payments of K200.
3. Calculate the monthly instalments paid for a hire-purchase agreement that pays off a principal of K720, plus interest, charged at a flat rate of 12.5% in 6 monthly instalments.
4. Tobias has agreed to a hire-purchase contract, charging a flat rate of interest of 9%, to buy a car costing K6 300. He is paying a deposit of K1 300 and he has a choice of the number of monthly payments he can make. Calculate the amount he will be paying per month if he makes:
   - **a.** 12 payments.
   - **b.** 18 payments.
5. Jeannie has taken out a time-payment plan to buy household goods to the value of K2 375. She is paying a deposit of K300 and the contract requires 20 monthly payments of K147. Find the flat rate of interest charged.
6. A hire-purchase plan requires the purchaser to pay a deposit of K40 and then 9 monthly payments of K48.80 on goods that cost a total of K400. Calculate the flat rate of interest per annum being charged.
7. Joseph has bought a car costing K4 500. He has paid a deposit of K500 and has agreed to pay K340 per fortnight for the next 6 months to pay off the balance.
   - **a.** How many payments will he make? (There are 26 fortnights in a year.)
   - **b.** How much interest is he paying?
   - **c.** What is the flat rate of interest per annum?
8. Erin has a choice of hire-purchase contracts to enable her to buy a scooter costing K5 640:

   Option A: A deposit of K1 640 and 12 monthly payments of K439.

   Option B: A deposit of K1 640 and 18 monthly payments of K300.

   Find the flat rate of interest in each case and decide which is the best financial option.

Reducing-balance loans

We have already looked at the effect of compound interest on money borrowed or money invested. When we calculated the amount owing for these loans it was assumed that the interest would be paid at the end of the loan period.

In reality most loans require a commitment to regular repayments throughout the term of the loan. These types of loans are called **reducing-balance loans** because each repayment pays off part of the principal owing and so, when this reduces with time, the amount of interest owed also reduces.

Reducing-balance loans are used for any long-term loan such as a loan to buy a house.

In the process of setting up a reducing-balance loan there are five variables that need to be considered:

- The amount borrowed.
- The compound interest rate.
- The term of the loan.
- The frequency of payments.
- The repayment amount.

When people are negotiating a loan to buy a house, for example, all of these variables are considered to find a loan that will fit their circumstances.

Step-by-step calculation of the amount owing on a reducing-balance loan

The amount owing on a loan after each repayment can be found by working through the addition of interest and repayment for each time period. This is a process done by computers today but the process is worth observing to understand the theory behind reducing-balance loans.

Example D

Q. A loan of K10 000 is being charged interest at the rate of 7% p.a. and regular repayments of K2 000 are made at the end of each year, after the interest is credited.

a. Use step-by-step calculations to find the amount still owing after 5 years.

b. How long does the loan run before it is totally repaid? (This is the term of the loan.)

c. What is the amount of the final payment?

d. How much interest is paid for the term of the loan?

A. A table can be set up to show the calculations:

Year	Interest charged (K)	Balance after interest (K)	Repayment (K)	Amount owing at end of year (K)
0				10 000.00
1	7% of 10 000.00 = 700.00	10 700.00	2 000.00	8 700.00
2	7% of 8 700.00 = 609.00	9 309.00	2 000.00	7 309.00
3	7% of 7 309.00 = 511.63	7 820.63	2 000.00	5 820.63
4	7% of 5 820.63 = 407.44	6 228.07	2 000.00	4 228.07
5	7% of 4 228.07 = 295.97	4 524.04	2 000.00	**2 524.04**

Year	Interest charged (K)	Balance after interest (K)	Repayment (K)	Amount owing at end of year (K)
6	7% of 2 524.04 = 176.68	2 700.72	2 000.00	700.72
7	7% of 700.72 = 49.05	749.77	**749.77**	0.00
Total	**K2 749.77**		**K12 749.77**	

Note: The amount owed at the end of each year (the balance) is reducing because of the payments and so the amount of interest paid also reduces.

a. The amount still owing after 5 years is K2 524.04.

b. The loan runs for 7 years before it is totally repaid.

c. The final repayment is K749.77.

d. K2 749.77 in interest is paid over the term of the loan.

Graphs of reducing-balance loans

The two graphs below relate to Example D above.

Graph 1 shows the balance (amount owing) of the loan after each of the years. It can be seen that it is not a straight line. A larger amount is paid off the balance as the years go by.

Graph 2 shows the interest paid at the end of each year and shows that this is a decreasing amount as the years go by. The graph is also not a straight line.

Graph 1

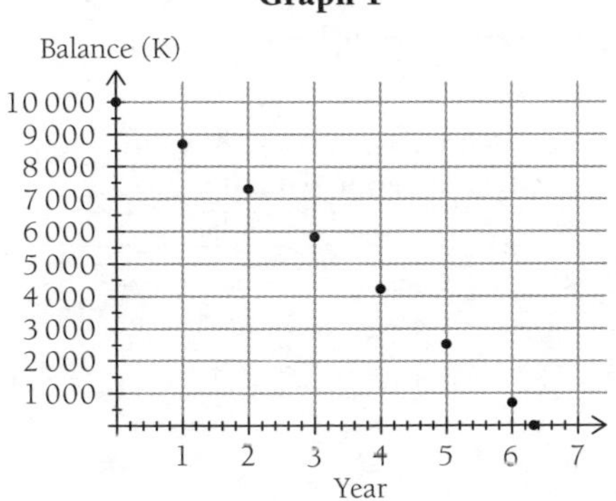

Graph 2

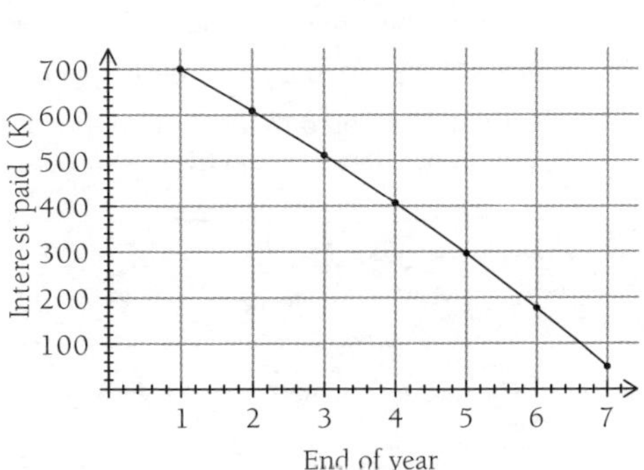

Example E

Q. Nancy has negotiated a loan of K4 000, charging 6.6% p.a. interest calculated on the reducing balance, with monthly repayments of K250.

a. Use step-by-step calculation to find the amount still owing after 5 payments.

b. Find the interest that has been paid over the five payment periods.

c. Compare the interest paid in the reducing-balance calculation (**b.**) with the situation where all the interest is paid at the end of the five months.

A. a. If interest is 6.6% p.a. then this is equivalent to 6.6/12 = 0.55% per month.

Setting up a table for the calculations:

Month	Interest charged (K)	Balance after interest (K)	Repayment (K)	Amount owing at end of month. (K)
0				4 000.00
1	0.55% of 4 000.00 = 22.00	4 022.00	250.00	3 772.00
2	0.55% of 3 772.00 = 20.75	3 792.75	250.00	3 542.75
3	0.55% of 3 542.75 = 19.49	3 562.23	250.00	3 312.23
4	0.55% of 3 312.23 = 18.22	3 330.45	250.00	3 080.45
5	0.55% of 3 080.45 = 16.94	3 097.39	250.00	**2 847.39**
	Total = K97.40			

The amount owing after 5 repayments is K2 847.39

b. The amount of interest paid = 250 × 5 – (amount paid off)

= 1 250 – (4 000 – 2 847.39)

= K97.39

Note: The difference of 1 toea is due to the fact that figures are rounded in the interest charged column.

c. If all the interest is paid at the end of the five months then this amount will be:

$K(4\ 000 \times 1.0055^5 - 4\ 000) = K111.22$

It should be noted that although the interest is more if it is paid at the end of the 5 months, Nancy would have had use of her repayment money in that time.

Unit 12.2 Activity 5C: Reducing-balance loans

1. K5 000 has been borrowed at 9% p.a. interest and regular repayments of K1 200 are made at the end of each year, after the interest is credited. Show step-by-step calculations to find:

a. The amount owing after 4 years.

b. The amount of interest paid in 4 years.

2. A loan of K15 000 is to be repaid in quarterly repayments of K1 000. If interest is charged at 8.4% p.a. on the reducing-balance:

a. What is the quarterly rate of interest?

b. Show step-by-step calculations to find the amount still owing after the first year.

c. Find the interest charged in the first year.

3. Jessie has bought a K780 sound system on her credit card. The credit company charges 18% p.a. interest on the balance outstanding each month and Jessie only repays the minimum required, K50, for each of the first three months.

a. Show step-by-step calculations to find the amount owing after three months.

b. How much interest has she paid in three months?

c. How much more would she owe at the end of three months if she made no repayments?

4. Complete the following tables showing step-by-step calculations:

a. A loan of K7 500 paying 9.6% p.a. interest charged on the reducing balance with repayments of K200 made quarterly.

Quarter	Interest charged (K)	Balance after interest (K)	Repayment (K)	Amount owing at end of quarter.(K)
0				7 500.00
1	....% of 7 500.00 = 180.00	7 680.00	200.00	7 480.00
2			200.00	
3			200.00	
4			200.00	

b. A loan of K200 000 paying 6.9% p.a. interest charged on the reducing balance, with repayments of K2 400 made monthly.

Month	Interest charged (K)	Balance after interest (K)	Repayment (K)	Amount owing at end of month.(K)
0				200 000.00
1	0.575% of ....=		2 400.00	198 750.00
2	0.575% of 19 8750 = ...		2 400.00	
3			2 400.00	
4			2 400.00	
5			2 400.00	

5. Peter has bought an annuity with K300 000 of his superannuation payout, to provide him with income each month. If interest is paid at 8.5% p.a. on the remaining balance, before a payment of K2 400 is paid to Peter, show step-by-step calculations to find the balance of the initial investment remaining after 4 months.

The annuities formula

Annuity technically means 'annual payment' but in the context of 'the annuities formula' it means a regular payment rather than an annual payment.

Finding the amount owing using the annuities formula

The annuities formula is used to find the amount still owing on a loan, which is being charged interest on a reducing-balance basis, where regular payments are being made each time period.

The formula replaces the step-by-step calculation of the amount still owing, and can be established from first principles involving the sum of a **geometric sequence**. This will not be shown here.

The annuities formula is:

$$A = PR^n - \frac{Q(R^n - 1)}{R - 1}$$

where:

- A is the amount owing after n time periods.
- P is the amount borrowed.
- Q is the amount of the regular repayment.
- $R = 1 + \frac{r}{100}$ where r is the rate of interest per time period.
- n is the number of time periods.

Time periods

A time period can be a year, a half-year, a quarter (three months), a month, a fortnight, a week, or a day.

Usually an interest rate is quoted 'per annum' which means for a year, and long-term loans have a term quoted in years. If, however, the repayments and the interest are calculated in a shorter time period then calculations must reflect this shorter time period.

Example F

Q. A loan is charged 9% p.a. interest, compounding monthly over 20 years. How many time periods, n, are involved in this loan and what is the value of r and R?

A. The number of time periods: n = Number of years × number of months in a year.

$= 20 \times 12$

$= 240$

The monthly interest rate, r: $r = \frac{\text{yearly interest rate}}{\text{number of months in a year}} = \frac{9}{12}\% = 0.75\%$

The value of R: $R = 1 + \frac{0.75}{100} = 1.0075$

Use the following to calculate time periods:

1 year = 2 half-years
= 4 quarters
= 12 months
= 26 fortnights
= 52 weeks
= 365 days

Example G

Q. Use the annuities formula to find the amount owing after 5 years for a loan of K10 000, charged 7% p.a. interest on the reducing balance, which has repayments of K2 000 made annually. (Example D is the step-by-step calculation.)

A. For this example:

$P = 10\,000$

$Q = 2\,000$

$r = 7\%$ p.a. so $R = 1 + 7/100 = 1.07$

interest is credited annually

$n = 5$

Substituting in $A = PR^n - \dfrac{Q(R^n - 1)}{R - 1}$

$$A = 10\,000 \times 1.07^5 - \frac{2\,000(1.07^5 - 1)}{1.07 - 1}$$

$$= 2\,548.039$$

The amount owing after 5 years is K2 548.04.

Example H

Q. Calculate the amount still owing after 3 years on a loan of K8 000 that is charging 7.2% p.a. interest on the reducing balance and where payments of:

a. K200 are made each month.

b. K100 are paid each fortnight.

A. a. $P = 8\,000$

$Q = 200$

$r = 7.2\%$ per year; $7.2/12 = 0.6\%$ per month

$R = 1 + 0.6/100 = 1.006$

$n = 3$ years $= 3 \times 12 = 36$ months

Substituting $A = PR^n - \dfrac{Q(R^n - 1)}{R - 1}$

$$A = 8\,000 \times 1.006^{36} - \frac{200(1.006^{36} - 1)}{1.006 - 1}$$

$$= 1\,912.359$$

After monthly payments of K200 for 3 years the amount owing is K1 912.36.

b. $P = 8\,000$

$Q = 100$

$r = 7.2\%$ per year $= 7.2/26\ \% = 0.2769...\%$ per fortnight

$R = 1 + 0.2769.../100 = 1.002769$. (Leave on the calculator.)

$n = 26 \times 3 = 78$ fortnights in 3 years

Substituting in $A = PR^n - \dfrac{Q(R^n - 1)}{R - 1}$

$$A = 8\,000 \times 1.0027...^{78} - \frac{100(1.0027...^{78} - 1)}{1.0027... - 1}$$

$$= 1\,232.757$$

After fortnightly payments of K100 for three years the amount owing is K1 232.76.

Note: The **difference** between the amounts owing for **a.** and **b.** is about K680.This is because the fortnightly payments reduce the balance more frequently and hence less interest is paid. An extra K200 is paid per year in **b.** because there are 26 fortnights in a year but only 12 months.

In general, less interest is paid, and hence the term of the loan is shortened, if payments are made more frequently.

Finding the repayment amount

A question that is often asked when establishing a reducing-balance loan is: 'How much will we have to pay each month to pay off our loan in 5 years?'

In a case like this we are looking for the value of Q in the annuities formula, when $A = 0$, ie. the loan is repaid. (This assumes the values of P, R and are known.)

The annuities formula can be rearranged to make Q the subject:

$$A = PR^n - \frac{Q(R^n - 1)}{R - 1}$$

$$\frac{Q(R^n - 1)}{R-1} = PR^n - A$$

Multiply both sides by $(R - 1)$:

$$Q(R^n - 1) = (PR^n - A)(R - 1)$$

Divide both sides by $(R^n - 1)$

$$Q = \frac{(PR^n - A)(R - 1)}{(R^n - 1)}$$

Example I

Q. Tim wants to pay off his K20 000 loan (for which he is being charged 8.4% p.a. interest on a reducing-balance basis) in 4 years.

a. Find the amount Tim needs to repay each month to pay out his loan in 4 years.

b. How much interest, in total, has Tim paid over the term of this loan?

A. **a.** $P = 20\,000$

$A = 0$ (If a loan is **paid out** then the amount still owing is zero.)

$Q = ?$

$r = 8.4\%$ p.a. which is $8.4/12 = 0.7\%$ per month

$R = 1 + 0.7/100 = 1.007$

$n = 4$ years $= 48$ months

Substituting in $Q = \frac{(PR^n - A)(R - 1)}{(R^n - 1)}$

$$Q = \frac{(20\,000 \times 1.007^{48} - 0)(1.007 - 1)}{(1.007^{48} - 1)}$$

$$= 492.022$$

Tim would need to pay K492.02 per month to pay off his K20 000 loan in 4 years.

b. Tim has paid K492.02 each month for $4 \times 12 = 48$ months.

In total Tim has paid K492.02 × 48 = K23 616.96.

His loan was for K20 000 so he has paid K23 616.96 – K20 000 = K3 616.96 in interest.

To calculate the *total* interest paid for a loan:

> Interest paid = repayment amount (Q) × number of payments (n) – loan amount (P)
>
> $$\text{Interest paid} = Q \times n - P$$

To calculate the amount of interest paid after m payments:

> Interest paid = repayment amount (Q) × m – amount paid off loan
>
> = $Q \times m$ – (loan amount – amount owing after m time periods (A))
>
> = $Q \times m - (P - A)$

Interest-only loans

An interest-only loan is one where the repayments each time period are equal to the interest due after that time period. No amount is paid off the principal in an interest-only loan and so there is no fixed term and the loan does not increase or decrease in value.

Interest-only loans are used to maximise the amount of interest being paid, which is useful if the interest can be a tax deduction, and also to minimise the repayment amount.

If we are to use the annuities formula for an interest-only loan, then the amount owing after n time periods, A, is equal to P, the principal:

$$P = PR^n - \frac{Q(R^n - 1)}{R - 1}$$

$$\frac{Q(R^n - 1)}{R - 1} = PR^n - P$$

$$= P(R^n - 1)\text{: divide both sides by } (R^n - 1)$$

$$\frac{Q}{R - 1} = P$$

$$Q = P(R - 1)$$

$$Q = P\left(1 + \frac{r}{100} - 1\right)$$

$$Q = \frac{Pr}{100}$$

Hence the repayment amount, Q, is equal to the interest calculated for the first time period; a simple-interest calculation.

Example J

Q. Find the fortnightly payment amount for an interest-only loan of K50 000, where the interest rate is 7.8% p.a.

A. The interest rate for fortnightly payments is $\frac{7.8}{26} = 0.3\%$

The payment amount each fortnight = $\frac{50\,000 \times 0.3}{100} = 150$;

K150 needs to be paid each fortnight to pay the interest that is owed.

Unit 12.2 Activity 5D: Using the annuities formula

Give your answers correct to two decimal places where necessary.

1. Find P, Q, R and n values for the loans given in the table below:

	Loan (K)	Interest rate p.a.	Repayment (K)	Compounded	Years
a.	6 000	8.1%	500	yearly	4
b.	15 000	7.2%	250	monthly	3
c.	850	10.8%	120	quarterly	1.5
d.	200 000	8.06%	800	fortnightly	10
e.	82 000	8.32%	180	weekly	10

2. Use the formula $A = PR^n - \frac{Q(R^n - 1)}{R - 1}$ to calculate the amount owing (A) for each of parts **a.** to **e.** in question **1.** above.

3. Ezekiel has an interest-only loan for an investment property. The loan is for K256 000 at an interest rate of 9.8% p.a. with fortnightly repayments. Find the payment that Ezekiel makes each fortnight.

4. Jeffrey has taken out a loan of K12 000 and he is making repayments of K200 per month. He is being charged 8.4% p.a. interest compounded monthly on the reducing balance. Use the annuities formula to calculate the amount he still owes on the loan after:

a. 1 year.

b. 2.5 years.

5. Use the annuities formula to find the amount still owing after 5 years on a reducing-balance loan of:

a. K10 000, being charged 7.75% p.a. interest with repayments of K1 500 per year.

b. K50 000, being charged 8.8% p.a. interest with repayments of K2 500 per quarter.

c. K150 000, being charged 6.96% p.a. interest with repayments of K1 800 per month.

6. Use the formula $Q = \frac{(PR^n - A)(R - 1)}{(R^n - 1)}$ to find the repayment necessary to pay out a loan of K24 000, in 10 years, that is being charged 7.2% p.a. interest on the reducing balance if the repayments are to be made:

a. Quarterly.

b. Monthly.

7. Use the formula $Q = \frac{(PR^n - A)(R - 1)}{(R^n - 1)}$ to find the amount of the repayment required to pay out a loan of:

a. K1 000, being charged 15% p.a. interest, payments being paid monthly over 6 months.

b. K170 000, being charged 7.8% p.a. interest, payments made fortnightly over 20 years.

c. K54 000, being charged 7.2% p.a. payments made quarterly over 10 years.

8. Use the formula $Q = \frac{(PR^n - A)(R-1)}{(R^n - 1)}$ to find the repayment necessary to reduce a loan from K30 000 to K20 000 in 4 years, if the loan is being charged 6% p.a. interest on the reducing balance and the repayments are to be made monthly.

Solving for the principal, *P*

Another commonly asked question is along the lines of:

'How much can I borrow if I can afford to pay K1 250 per month? I want to pay out the loan in 20 years and the interest being charged is 7.2% p.a. calculated on the reducing balance.'

In this case we are looking for P in the annuities formula: $A = PR^n = \frac{Q(R^n - 1)}{R-1}$

To make the subject: $PR^n = A + \frac{Q(R^n - 1)}{R-1}$

adding $\frac{Q(R^n - 1)}{R-1}$ to both sides.

$$\boxed{P = \frac{A + \frac{Q(R^n - 1)}{R-1}}{R^n}}$$

For the example above:

$A = 0$

$Q = 1\,250$

$r = 7.2\%$ p.a. which is $7.2/12 = 0.6\ \%$ per month

$R = 1 + \frac{0.6}{100} = 1.006$

$n = 20$ years $= 240$ months

Careful substitution in the formula $P = \frac{A + \frac{Q(R^n - 1)}{R-1}}{R^n}$

$$P = \frac{0 + \frac{1\,250(1.006^{240} - 1)}{(1.006 - 1)}}{1.006^{240}}$$

$$= 158\,760.54$$

So K159 000 (to the nearest K1 000) could be borrowed under the conditions given.

Unit 12.2 Activity 5E: Using the annuities formula

1. For each of the following:

a. Set up the annuities formula by substituting the appropriate values.

b. Solve for the unknown variable in the annuities formula.

i. The amount owing after 5 years on a loan of K84 000, paying 8.4% p.a. interest calculated on the reducing balance, with monthly repayments of K650.

ii. The amount that can be borrowed when quarterly payments of K5 000 are made so that the loan is paid out over 10 years and where interest is charged at 7.6% p.a. calculated on the reducing balance.

   iii. The payment required to reduce a loan of K100 000 to K50 000 over 10 years with monthly repayments, where interest is charged at 7.2% p.a. calculated on the reducing balance.

2. Match the annuities formula substitutions A to D with the situations **a.** to **d. below:**

   A $0 = 10\,000 \times 1.005^{48} - \dfrac{Q(1.005^{48} - 1)}{1.005 - 1}$

   B $10\,000 = 10\,000 \times 1.005^{48} - \dfrac{Q(1.005^{48} - 1)}{1.005 - 1}$

   C $0 = 10\,000 \times 1.015^{16} - \dfrac{Q(1.015^{16} - 1)}{1.015 - 1}$

   D $20\,000 = P \times 1.025^{16} - \dfrac{250(1.025^{16} - 1)}{1.025 - 1}$

   a. The quarterly payment on a loan of K10 000 that is paid out over 4 years and where interest is charged at 6% p.a.

   b. The amount that is borrowed over 4 years so that K20 000 is still owing and where quarterly payments of K250 are made. Interest of 10% p.a. is being charged on the reducing balance.

   c. The monthly payment on an interest-only loan of K10 000 over 4 years and where interest is charged at 6% p.a.

   d. The monthly payment on a loan of K10 000 that is paid out over 4 years and where interest is charged at 6% p.a.

3. Greg can afford to pay K780 a month as a mortgage payment. If housing loans are being charged 6.9% interest on the reducing balance, how much can Greg borrow, to the nearest K1 000, if the loan is for:

   a. 20 years.

   b. 25 years.

4. Anthony and Maria have taken out a K145 000 interest-only loan for 5 years. If they are paying 7.35% p.a. interest,

   a. How much will they pay each fortnight?

   b. How much interest will they pay over the 5 years?

5. Find the amount a couple can borrow if they can afford to repay K2 400 per month on a loan that is charged 7.15% p.a. on the reducing balance and they would like to repay the loan in:

   a. 15 years.

   b. 20 years.

6. Match the situations **(a)** to **(d)** to the graphs **I** to **IV** of the amount owing over 5 years where interest is charged at 7% p.a. on the changing balance:

   a. K10 000 loan with regular payments of K200 per month.

   b. K10 000 interest-only loan with regular repayments of interest.

c. K10 000 loan with regular repayments of K50 per month.

d. K10 000 loan with regular repayments of K100 per month.

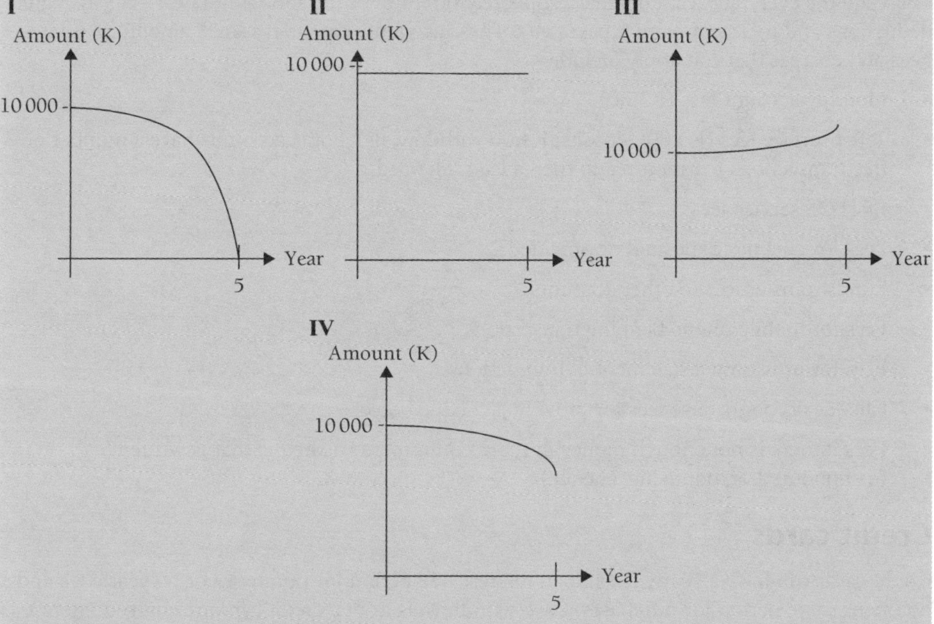

7. Use the appropriate version of the annuities formula to find:

   a. The repayment amount per fortnight for a loan of K35 000 that was reduced to K20 000 over 3 years and where interest of 7.8% p.a. was charged on the reducing balance.

   b. The amount of interest paid per month on an interest-only loan of K90 000 where interest is charged at 6.95% p.a.

   c. The amount that can be borrowed, to the nearest K1 000, if repayments of K850 per fortnight can be made and interest is charged at 9.1% p.a. on the reducing balance over 20 years.

   d. The minimum amount that must be paid each month on a loan of K5 000 so that the loan amount does not increase; interest charged at 9.6% p.a.

   e. The interest charged in the first 10 years of a 25-year loan of K150 000, where interest is charged at 7.28% p.a. on the reducing balance and repayments of K500 per fortnight are made.

Service fees and charges

Banks and other financial institutions have fees and charges for most of the services they provide. These fees and charges differ with every financial institution.

Bank accounts, savings accounts, transaction accounts, cheque accounts

These are the everyday accounts that people rely on for deposits and withdrawals, to pay regular deductions and to accept regular payments. These accounts pay a very small amount of interest. Fees and charges that can apply include:

- Monthly account fees (K3.00).
- Teller service fees (K3.00 deposits; K4.00 withdrawals). Some accounts have a number of free transactions per month and then a fee is charged.
- EFTPOS service fees.
- Fees for cheques deposited or provided.
- Funds transfer fees to other accounts.
- Fees for mobile phone banking transactions.
- Fees for providing a statement of transactions.
- Cheque accounts have fees for providing cheque books.
- Fees if there is not enough money in the account to pay a cheque that is written (dishonoured or 'bouncing' cheque).

Credit cards

Credit cards can have a yearly fee and an interest-free period for purchases or no yearly fee and no interest-free period for purchases. Cash withdrawals from a credit card are charged interest from the day of the withdrawal. Interest charges are quite high for credit cards and owners of credit cards need to have reliable income and a good credit history.

Debit cards

Debit cards allow the owner access to their own funds in an account. Some of the fees that can apply to debit cards:

- Establishment fees.
- Annual fees.
- Card replacement fees.

Term deposits

There are usually no establishment fees but prepayment fees and an interest penalty apply if a term deposit is claimed before the specified term.

Personal loans

In addition to the interest charge a personal loan can have the following charges:

- An application/processing fee.
- Early repayment fees.

- Charges for late repayments.
- 'Bouncing' cheque charges.
- Documentation fees for the verification of your documents.

Home loans

Fees and charges from financial institutions vary significantly between lenders. The following fees and charges may apply to a home loan:

- An application/setup fee that covers the cost of processing the home loan application.
- A fee to establish lenders' mortgage insurance. This is needed if the borrower needs to borrow more than 80% of the value of the property.
- A valuation fee. This is to pay a qualified valuer to assess the market value of a property.
- An exit fee to cover the costs of preparing the documents to finish a loan.
- An early exit fee. Applicable if the loan is repaid before the fixed term.
- Regular account fees, usually monthly.

In addition to these lender fees there are other charges associated with purchasing a home:

- Government stamp duty which is a percentage of the purchase price.
- Legal fees for setting up sale contracts and conducting enquiries and searches of documents relating to the property.

Unit 12.2 Activity 5F: Service fees and charges

This exercise can be done individually or as a group project. It is suggested that only one (or a maximum of two) of the investigations be done, as each requires research out of the classroom. A well-presented report on the findings is essential.

In some instances a hypothetical situation can be used to illustrate the fees and charges; for example when investigating the cost of a loan to buy a house an imaginary property of a certain value can be used.

1. Investigate the service fees and charges for an everyday savings account with a bank. You can do this in one of the following ways:

   a. Ask someone you know who has a bank account for information about the fees and charges they pay.

   b. Approach a local bank and gathering the information on its fees and charges.

   c. Use the internet to access information on a particular bank account.

   Use the list above as a guide to the possible fees and charges (there may be others that are not mentioned on the list).

2. Investigate the service fees and charges for a particular credit card issued by a bank. You should also gather information about the eligibility for owning a credit card and the limits that are placed on the amount of credit. You can do this in one of the following ways:

   a. Ask someone you know who has a credit card for information about the fees and charges they pay. You will probably need to see a monthly statement.

   b. Approach a local bank and gather the information on eligibility, fees and charges for their credit card.

   c. Use the internet to access information on a particular credit card.

   Use the list above as a guide to the possible fees and charges (there may be others that are not mentioned on the list).

3. Investigate the service fees and charges attached to a personal loan with a bank or other financial institution. You should also gather information about eligibility for a personal loan. You can do this in one of the following ways:

   a. Ask someone you know who has a personal loan for information about the fees and charges they pay.

   b. Approach a local bank and gather information on its fees and charges and eligibility criteria.

   c. Use the internet to access information on a personal loan from a particular bank.

   Use the list above as a guide to the possible fees and charges (there may be others that are not mentioned on the list). Often the application forms attached to personal loans are useful sources of information.

4. Investigate the service fees and charges attached to a home loan from a bank or other financial institution. You should also gather information about the eligibility requirements for a home loan. You can do this in one of the following ways:

   a. Ask someone you know who has a home loan for information about the fees and charges they pay.

   b. Approach a local bank and gather information on its fees and charges and eligibility criteria.

   c. Use the internet to access information on a home loan from a particular bank.

   Use the list above as a guide to the possible fees and charges (there may be others that are not mentioned on the list). Often the application forms attached to home loans can be useful sources of information.

Unit 12.2 Managing Money 2

Topic 6: Investments

Investment provides a real-life context for the practical application of mathematical knowledge and skills. In line with the Syllabus (p. 22) Topic 6 covers:

- Types of investment – real estate and the stock market.
- Calculating the profit and value of different types of investment.
- Calculating the dividend of chosen stocks.
- Calculating the dividend on the sale of chosen properties.

Types of investments

Investment with savings institutions that involve regular deposits

If an initial amount, P kina, is invested, and regular payments of Q kina are made in each time period in an investment account compounding r % per time period, then after the first time period the amount accumulated, A kina, will be:

$$A = P\left(1+\frac{r}{100}\right)+Q \qquad \text{where } R = 1+\frac{r}{100}$$
$$= PR + Q$$

After the second time period:
$$A = (PR+Q)+(PR+Q)\times\frac{r}{100}+Q$$
$$= (PR+Q)\left(1+\frac{r}{100}\right)+Q$$
$$= (PR+Q)R+Q$$
$$= PR^2+QR+Q$$

After the third time period:
$$A = \left(PR^2+QR+Q\right)+\left(PR^2+QR+Q\right)\times\frac{r}{100}+Q$$
$$= \left(PR^2+QR+Q\right)\left(1+\frac{r}{100}\right)+Q$$
$$= \left(PR^2+QR+Q\right)R+Q$$
$$= PR^3+QR^2+QR+Q$$

After the fourth time period (using the same method)
$$A = PR^4+[QR^3+QR^2+QR+Q]$$

A pattern is emerging:

After the nth time period:
$$A = PR^n+[QR^{n-1}+QR^{n-2}\ldots\ldots+QR+Q]$$

Using the sum of a geometric **series** to sum the quantities in **brackets** [........] gives *the amount accumulated (A) after n time period as:*

$$\boxed{A = PR^n+\frac{Q(R^n-1)}{R-1}}$$

Note: This is a version of the annuities formula: $A = PR^n - \frac{Q(R^n - 1)}{R-1}$

The amount of interest earned after n time periods is given by:

Interest = Amount accumulated – initial investment – total of regular payments

Example A

Q. Philip has started a savings plan with an initial investment of K1 000 and he is going to add K250 to this amount every month.

a. How much will he have saved after 5 years if interest of 6% p.a. is paid on the increasing balance?

b. How much interest has he earned in the 5 years?

c. Graph the accumulation of Philip's investment over the five years.

A. **a.** $P = 1\ 000$, $Q = 250$ per month, $r = \frac{6}{12} = 0.5$ per month and $n = 5 \times 12 = 60$ months.

Substituting these values in the formula $A = PR^n + \frac{Q(R^n - 1)}{R-1}$ where $R = 1 + \frac{0.5}{100} = 1.005$

$$A = 1000 \times 1.005^{60} + \frac{250(1.005^{60} - 1)}{1.005 - 1} = 18\ 791.36$$

After 5 years Phillip would have saved K18 791.36

b. The amount of interest = Amount accumulated – initial investment – monthly payments

= 18 791.36 – 1 000 – 60 × 250

= 2 791.36

The amount of interest earned is K2 791.36

c. Amount accumulated (K)

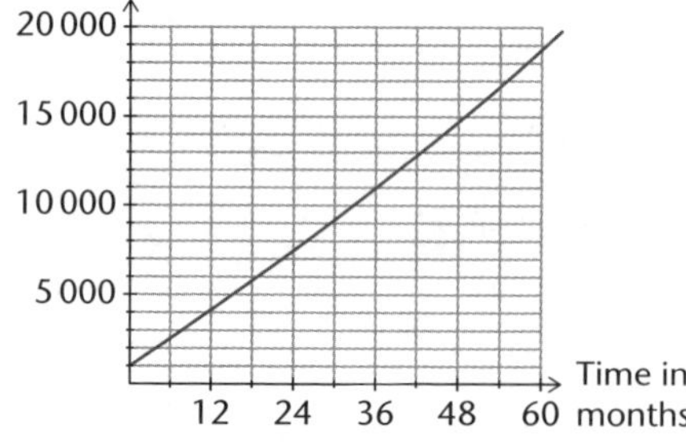

Note: The graph of the accumulated amount is **not linear**.

Calculating the regular payment required to accumulate a set amount

Re-arranging the formula above to make Q the subject means that we can calculate the regular payment required so that a set amount is accumulated.

$$Q = \frac{\left(A - PR^n\right)(R-1)}{(R^n - 1)}$$

Example B

Q. Wendy has saved K500 and wants to have saved a total amount of K3 000 in three years' time. She can earn 5.4% p.a. in a savings account if she makes regular monthly payments. How much will she have to pay into this account each month so that she has at least K3 000 in three years' time?

A. $r = 5.4\%$ p.a. $= \frac{5.4}{12} = 0.45\%$ per month. $R = 1 + \frac{r}{100} = 1 + \frac{0.45}{100} = 1.0045$

$A = 3\,000$, $P = 500$ and $n = 3 \times 12 = 36$ months.

Substituting these values into the formula $Q = \frac{(A - PR^n)(R-1)}{(R^n - 1)}$

Regular monthly payment $= \frac{(3\,000 - 500 \times 1.0045^{36})(1.0045 - 1)}{(1.0045^{36} - 1)} = 61.88$

Wendy will need to make a regular payment of at least K61.88 so that she has at least K3 000 in three years' time.

Unit 12.2 Activity 6A: Types of investments

Give your answers correct to two decimal places.

1. A savings plan is started with an initial investment. Each month an additional K100 is added to the plan and interest is paid at 4.8% p.a. on the increasing balance. Calculate the amount accumulated at the end of four years if the initial investment is:

a. K1 000 **b.** K1 500 **c.** K2 000

2. A savings plan is started with an initial investment of K1 000. Each month an additional amount is added to the plan and interest is paid at 6% p.a. on the increasing balance. Calculate the amount accumulated at the end of four years if the regular monthly payment is:

a. K100 **b.** K150 **c.** K200

3. Courtney wants to have a total of at least K3 000 saved for her daughter's education in three years' time. She has K400 saved and will use this as an initial deposit and then make regular fortnightly payments of K50 into a savings plan earning 5.2% p.a. on the increasing balance. How much will Courtney have in her account after three years?

4. Find the total interest paid, after five years, in a savings account paying 6.24 % p.a. interest on the increasing balance, where:

a. the initial amount invested is K400 and the regular payment is K40 per month.

b. the initial investment is K5 000 and the regular payment is K200 per fortnight.

5. Arnold intends to save a total amount of K10 000 in five years' time. He can earn 6 % p.a. in a savings account if he makes an initial deposit of K1 000 and regular monthly payments.

a. How much will Arnold have to pay into this account each month so that he has at least K10 000 in five years' time?

b. How much will Arnold have to pay into this account each month if he can only earn 4.8% p.a. in the savings account?

6. Andrew and Sonia plan to have a total of K20 000 saved in three years' time. They have already saved K5 000 which they have invested in an account paying 6.24% per annum interest and will regularly add to this amount. What will their regular payments be if they are made:

a. Each month. **b.** Each fortnight (26 fortnights in a year).

7. Mali has started a savings plan with an investment of K1 200 and is adding to this amount each month with an additional K120. If the account is paying 4.8% p.a. interest on the increasing balance, find:

a. The amount saved after 4 years.

b. The time it will take to accumulate K10 000 in the account.

Investing in real estate

When a person invests in real estate they are intending to purchase a property from which they can make a profit either as regular income, from capital growth (increase in value in the future) or from both.

An investment property could be:

- A residential house, unit or apartment that can be rented to provide income.
- A block of land that could be used for residential purposes in the future.
- A business property, such as a shop or a factory.
- Farming land.

Investing in real estate might seem to be a **certain** method to increase wealth, but there are many hidden costs and pitfalls that can mean that this does not always happen.

Initial costs of purchasing real estate

The initial costs of buying any real estate will include:

- The cost of finding a suitable property. If you are buying a property then it is important that it will meet your needs for income and/or capital growth. This can take time.
- The whole purchase price or a deposit.
- The cost of setting up a loan at a lending institution. Most investment properties have a loan attached to them as the interest on the loan is usually tax deductible.
- Conveyancing costs. It is usually necessary to employ a qualified person to check that all contracts and documents are in order and to arrange the transfer of money between the buyer and seller. (Conveyancing is the branch of the law concerned with moving property from one owner to another – *Oxford Advanced Learner's Dictionary*.)
- Stamp duty. Stamp duty is a form of tax payable to the government, by the buyer, at the time of purchase of the property.

Duty amount on the sale of real estate property

Where the value does not exceed K35 000	K5.00 or an amount equal to 2% of the value, whichever is the greater
Where the value exceeds K35 000 but does not exceed K70 000	An amount equal to 3% of the value

Where the value exceeds K70 000 but does not exceed K140 000	An amount equal to 4% of the value
Where the value exceeds 140 000	An amount equal to 5% of the value

Ongoing costs of owning an investment property:

- Interest on the loan.
- Bank fees.
- Rates
- Property manager.
- Insurance.
- Repairs.

Costs on selling the investment property:

- Real estate agent's selling fee.
- Costs of closing the loan. (Banks and financial institutions usually have costs attached to closing a loan before it is repaid in full.)
- Capital gains tax. Profits (after buying/selling costs are deducted) arising from the sale of property acquired for the purpose of resale at a profit are fully taxable as ordinary income.

Things that can go wrong when investing in property:

- Having to sell the property in a hurry and for a reduced price.
- Not being able to meet the loan repayments.
- Not being able to rent the property at the required amount or it being vacant for an extended period.
- Major repairs needed on the property.

Example C

Q. Thom bought an investment property five years ago for K230 000 and has just sold it for K470 000.

a. Calculate the total cost of buying this property if the initial costs of buying the property were K5 400 and Thom paid government stamp duty on the purchase price of the property.

b. Calculate the total cost of selling the property if the selling agent charged a fee of 2.4% of the selling price.

c. If inflation has been averaging 7% per annum in the five years that Thom owned the property, what price would someone buying the property expect to pay now if the price of the property had increased with inflation?

d. Assuming that the ongoing costs of the property were paid for by the rent, how much profit has Thom made by buying and selling this property?

e. Thom has acquired this property for the purpose of making a profit so he will need to pay capital gains tax on the profit that he has made. Thom's other income means that he will be paying 40% on K120 000 of this profit and 42% on the remainder. How much will Thom pay in capital gains tax?

A. **a.** **Buying costs** (cost of property: K230 000)

Initial costs: K5 400

Government stamp duty : 5% of K230 000 $= \text{K}(\frac{5}{100} \times 230\,000) = \text{K}11\,500$

Total buying costs: K16 900

b. **Selling costs:**

Selling agent's fee $= \frac{2.4}{100} \times \text{K}470\,000 = \text{K}11\,280$

Total selling costs = K11 280

c. A 7% increase means that the multiplying factor will be $(1+\frac{7}{100}) = 1.07$ each year.

Someone buying the house today would expect to pay $\text{K}230\,000 \times 1.07^5 = \text{K}322\,587$

d. Profit = Selling price – buying price – costs of buying/selling

= K(470 000 – 230 000 – 28 180)

= K211 820

e. 40% of K120 000 + 42% of (K211 820 – K120 000)

$= \frac{40}{100} \times 120\,000 + \frac{42}{100} \times 91\,820$

= 86 564.40

He will need to pay K86 564 in capital gains tax.

Example D

Q. Bevan has purchased an investment property for K475 000. He has paid a deposit of 30% and is borrowing K450 000 from the bank. The bank has offered him a five-year loan of K340 000 charging 7.8% p.a. interest on the reducing balance, payments being made fortnightly.

a. How much stamp duty will Bevan need to pay on his purchase?

b. How much did Bevan pay as a deposit?

c. Use the annuities formula $Q = \frac{(PR^n - A)(R-1)}{R^n - 1}$ to determine the amount that Bevan will be repaying each fortnight.

A. **a.** Stamp duty will be 5% of the purchase price: 5% of K475 000

$= \frac{5}{100} \times \text{K}475\,000 = \text{K}23\,750$

b. The deposit was 30% of the purchase price: 30% of K475 000

$= \frac{30}{100} \times \text{K}475\,000 = \text{K}142\,500$

c. In the annuities formula, Q is the fortnightly payment, P is the amount borrowed, A is the amount still owing:

$R = 1 + \frac{r}{100}$ where $r = \frac{7.8}{26} = 0.3$ and $n = 5 \times 26 = 130$.

$Q = ?$, $P = 340\,000$, $A = 0$, $R = 1.003$ and $n = 130$

Substituting in the formula:

$$Q = \frac{(340\,000 \times 1.003^{130} - 0)(1.003-1)}{1.003^{130} - 1} = 3\,162.32$$

Bevan will need to repay K3 162.32 each fortnight.

Unit 12.2 Activity 6B: Investing in real estate

1. Using the stamp duty table above, calculate the stamp duty payable on a property whose purchase price is:

a. K29 870.

b. K68 000.

c. K120 000.

d. K360 000.

e. K1 450 000.

2. An estate agent has the following fee scale for the sale of properties: 5% on the first K300 000 + 3.5% on the next K200 000 + 2.5% on the balance. Calculate the selling agent's fees for a property sold for:

a. K290 000.

b. K455 000.

c. K876 000.

d. K1 450 000.

3. Leo has bought an investment property for K560 000. He has paid a 10% deposit and is borrowing the remainder of the purchase price from a bank. The bank has offered him an interest-only loan at 9.1% p.a. interest, paid fortnightly.

a. How much did Leo pay as a deposit on the property?

b. How much interest is Leo paying on the loan each fortnight?

4. Geraldine has purchased an investment property for K386 000. She has paid a 25% deposit and is borrowing K300 000 from a bank that is charging her 8.4% interest on the reducing balance, with monthly payments over five years.

a. How much stamp duty will Geraldine need to pay on her purchase?

b. How much did Geraldine pay as a deposit?

c. Use the annuities formula $Q = \frac{\left(PR^n - A\right)(R-1)}{R^n - 1}$ to determine the amount that Geraldine will be repaying each month.

d. If Geraldine is going to sell the property after five years, and inflation has averaged 6% over the five years, what amount can she expect to get when she sells the property?

e. How much will Geraldine need to pay the selling agent if she sells the property for K520 000 and the following structure is used to calculate the agent's fee:

5% on the first K300 000 + 3.5% on the next K200 000 + 2.5% on the balance

5. Bevan has purchased an investment property for K1 100 000. The initial costs of purchasing the property are:

- Stamp duty.
- Conveyancing costs of K870.
- Bank fees of K1 650 for establishing a loan.

Ongoing costs are:

- Interest on the loan: K6 650 per fortnight.
- Bank fees: K15 per month.
- Property management fees: 7% of the rent paid (which is K7 000 per fortnight).
- Services: K800 per quarter.
- Insurance: K968 per year.
- Repairs: Estimated at K5 000 per year.

Income is: Rent K7 000 per fortnight.

a. Calculate the stamp duty payable on the purchase of this property.

b. Calculate the total cost of purchasing the property.

c. Calculate the ongoing costs per year.

d. What is the difference between the income and ongoing costs per year? How much profit/loss will Bevan make on this investment per year?

6. Dominic is thinking about investing in a housing unit that he has been told will bring in rent of K3 000 per fortnight. He has calculated that other ongoing costs will amount to K250 per fortnight.

a. The bank will allow him to take out a reducing-balance loan at 7.8% p.a., compounding fortnightly, for 20 years. The loan will be paid out in 20 years.

i. Write down the values of A, R and n for this loan in the annuities formula.

ii. Use the formula $P = \frac{1}{R^n}(A + \frac{Q(R^n - 1)}{R-1})$ (a version of the annuities formula) to calculate the amount that Dominic can borrow so that his fortnightly costs are covered by the income for each fortnight.

b. Dominic is also offered an interest-only loan for 10 years with an interest rate of 7.8% p.a. How much could he borrow in this situation where his fortnightly loan payments are covered by the income for each fortnight?

7. Investigate the benefits, or otherwise, of buying an investment property. You could interview someone who has an investment property or you could be an imaginary owner. You will need to gather information from estate agents, conveyancers, lending institutions, insurers and tax experts.

Investing in the stock market

When a company needs to expand its business it can raise money by selling **shares** of its business to the public. A **share** is a part ownership of a company. If you buy shares in a company then you are entitled to a share of the profit of the company; this is called a **dividend**.

If a company wants to sell shares to the public then this is called 'floating' the company and there are strict rules and regulations about the size and management of businesses that are floated. The rules are enforced to protect those buying shares because when a company sells shares it is under no obligation to repay the share price to the buyer.

The process of floating a company, and buying and selling shares in a company, are all overseen by a body called the **stock exchange**. In Papua New Guinea there is only one authorised stock exchange, where shares in both PNG companies and Australian companies can be traded.

When a company is listed on the stock market then its shares can be bought and sold at what is called the **market price**. The market price of a share depends on the **demand** for the share and this can fluctuate according to:

- Profitability of the company.
- Local demand for the product.
- Overseas demand for the product.
- Government legislation.

There is a certain degree of risk associated with buying shares and this is different for different companies. The risk is high when there is a prospect of making a large profit but there is a chance of losing all the invested money.

There are many shares where the risk is quite low and good returns are made over several years.

Why do people invest in shares?

People buy shares in a company so that they can share in the profits of the company, either as a regular income from the dividends and/or from selling the shares at a price greater than the buying price (called **capital growth**).

How do you buy shares?

If you wish to buy shares on the stock market then you need to use the services of a **stockbroker**. Stockbrokers (also called share brokers) are independent financial advisors who are given a licence to trade in shares.

Stockbrokers charge a fee for their service and this is called **brokerage**. Brokerage is usually charged as a percentage of the value of a transaction (buying or selling) but can also be a set amount.

Brokerage on the value of the transaction

Value of transaction (K)	Brokerage
5 000 or below	2.5% (minimum of K65)
5 001 – 20 000	2.0%
20 001 – 50 000	1.5%
50 000 and above	1.0%

Each day the previous day's trading of shares on the stock market is reported in the newspaper: see, for example, p. 148 which reproduces a page from *The National* newspaper for Thursday 26 July 2012.

In the newspaper extract on p. 148, the trading of both local PNG and Australian shares are given in Australian dollars.

The headings on each of the columns mean:

Close The price at which the stock was last traded.

Move The change in price of a share from the previous day's close.

44 The National – Thursday July 26, 2012

BUSINESS

AUSTRALIAN SHARE MARKET

PNG companies/ Associates	Close	Move	Year High	Year Low
AGLEnergy	15.67	-.01	15.80	12.101
AlliedGol	2.14	+.06	4.00	1.42
Bougainvl	.85	-.05	1.22	.60
CueEnergy	.155	+.01	.32	.145
Frontier	.07		.27	.067
GoldAnoma	.007		.048	.006
Goldminex	.065		.155	.06
Highlands	.165	-.005	.345	.125
LNG Ltd	.345	+.015	.55	.225
Marengo	.125	-.005	.30	.12
OilSearch	6.69	+.09	7.58	5.43
ResourceM	.001		.006	.001
Santos	10.30	-.02	14.63	10.04
Steamship	27.00		30.00	24.00

Company name	Close	Move	Year High	Year Low
AAC Ltd	1.09	-.02	1.48	1.02
AAviationServ	2.05		2.36	1.59
Abacus	1.935	-.015	2.18	1.78
Aberdeen	.975		1.17	.975
Academies AG	.49		.60	.44
Acrux	3.82	-.10	4.77	2.59
Adacel	.385		.43	.20
AdelBrtn	3.15	-.04	3.34	2.22
AdvanShar	.67	-.01	.825	.605
AGLEnergy	15.67	-.01	15.80	12.101
AHG Ltd	2.51	-.01	2.72	1.565
Ainsworth	2.14	-.06	2.29	.315
AInvTrust	.395	+.005	.458	.284
Air NZ	.69		.96	.65
AJ Lucas	1.00		1.50	.87
Alchemia	.49	-.005	.72	.25
Alcoa	8.49		13.90	8.49
AleGroup	2.09		2.23	1.72
Alesco	2.01		2.879	.995
Allmine Gp	.125	+.005	.26	.095
Altium	.41	+.02	.41	.096
AMAGroup	.145		.155	.09
AmalgHld	6.55	-.01	6.93	5.064
Amcil	.71	-.01	.76	.60
Amcom Tel	1.11	+.01	1.15	.588
Amcor	7.42	-.07	7.80	5.87
AMP	3.80	-.06	4.79	3.61
AMPCap	.595	-.01	.79	.59
Ansell	12.99	-.02	15.36	11.82
AnteoDiag	.07	-.001	.093	.054
Antisense	.018		.039	.006
ANZ Bank	22.64	-.11	24.05	17.63
AP Eagers	3.52	-.08	3.80	1.84
APAGroup	4.95	-.04	5.34	3.53
APN Prop	.16		.185	.125
APNewsMed	.51	-.025	1.20	.49
ARBCorp	9.28	+.05	9.67	6.65
ArdentLei	1.25	-.03	1.35	1.005
Argo	5.21	-.08	5.70	4.75
Ariadne	.33		.38	.29
Aristocrt	2.54	-.06	3.25	1.88
Arrium Ltd	.765	-.035	1.933	.63
Asciano	4.16	-.07	5.13	3.99
ASF Group	.23		.235	.10
ASG Group	.79	-.01	1.045	.73
AsianMast	.86		1.00	.84
Aus Power & Gas	.48	-.02	.62	.47
Ausdrill	3.22	-.08	4.34	2.52
Ausenco	3.04	-.15	4.72	1.91
AusFound	4.31	-.04	4.57	3.82
AusGovMas	1.49		1.61	1.36
AusMastersCBF5	83.82		84.592	82.031
Austal	1.645	+.015	3.00	1.52
Austbroke	7.00	-.01	7.25	5.75
AustEduc	1.05	+.025	1.06	.76
AustinEng	4.22		5.21	3.14
AustInfra	2.51	+.01	2.57	1.62
Austland	2.55	-.05	2.87	2.17
AustLead	1.20	-.01	1.265	1.005
AustPharm	.345	-.005	.42	.205
AustUInv	5.70	-.03	6.45	5.10
AustVint	.37	+.01	.405	.22
Avita Med	.175		.28	.095
AVJenings	.31		.465	.285
BankQld	7.00	-.03	8.494	6.13
Bega	1.50	-.03	1.91	1.495
BellFinGp	.42		.79	.40
BenAdeBnk	7.97	-.05	9.87	6.82
BeyondInt	.72		.79	.60
BigairGrp	.385		.40	.225
Billabong	1.32	+.005	4.854	.925
Bionomics	.26	-.005	.695	.25
Biota	.69	+.01	1.07	.68
Bisalloy	1.36	-.01	1.70	.75
BKIInvest	1.20		1.25	1.041
Blackmore	28.51	+.01	30.49	24.05
Bluescope	.25	-.01	1.118	.25
Bluglass	.096	+.007	.165	.046
BoartLong	2.44	-.10	4.41	2.34
BoomLog	.25		.33	.19
Boral	3.24		4.49	2.93
BoralBsk				
Bradken	4.64	-.24	8.72	4.62
Brambles	6.17	-.04	7.551	5.76
Bravura	.15		.175	.12
Breville	4.50	+.03	4.66	2.51
Brickwork	10.06	-.02	11.40	8.86
Brierty	.33		.385	.195
BrisBronc	.23	-.01	.35	.22
Brisconn	1.06	+.15	1.06	.69
BrookAust	.05	+.001	.06	.033
Brookfld	3.55		4.25	2.681
BSA Ltd	.21	+.005	.28	.18
BTInvest	1.71	-.015	2.27	1.68
BWP Trust	1.89	-.005	1.978	1.546
Cabcharge	5.26	-.08	6.57	3.95
Cadence	1.27	+.03	1.37	1.11
Calliden	.13		.23	.10
Caltex	14.06	+.08	15.00	8.76
Campbell	49.32	-1.43	69.92	38.24
Capral	.14		.275	.13
Cardno	7.96	-.03	8.20	4.182
Carindale	4.76	-.01	4.77	3.67
CarltonIn	15.82	-.13	16.99	14.40
Carnegie	.045	+.002	.091	.025
CarParkingTech	.26	-.01	.375	.10
Carsales	6.11	-.09	6.32	3.79
CashCnv	.685	-.07	.785	.39
CBA	54.86	+.08	55.90	42.30
Cedarwood	3.70	-.02	4.11	3.07
CentroRetailAu	1.98		2.06	1.655
CenturiaCapita	.395		.61	.37
CenturyAu	.61		.73	.545
CeramFuel	.06	-.004	.19	.06
Cromwell	.70		.74	.60
Crown Ltd	8.30	-.09	9.29	7.45
CS-EMIDT	.915		.96	.915
CSAuTrust	.98		1.015	.96
CSG Ltd	.73	+.01	1.165	.505
CSL Ltd	41.10	+.80	41.67	26.12
CSRLtd	1.18	+.005	2.71	1.145
CTILogist	1.44		1.50	.875
CVC Ltd	.89		.925	.80
CwlthProp	1.045	+.005	1.07	.84
Data3	1.20		1.389	.90
DavidJons	2.31	+.02	3.58	2.10
DecmilGrp	2.29	-.06	3.17	1.707
Devine	.53	-.02	.98	.52
DexusProp	.96		.975	.735
Dicker Data	.42		.49	.23
DivUnited	2.44	-.03	2.72	2.15
DjerriInv	4.00		4.20	3.02
Dominos	9.28	-.02	10.40	5.25
DownerEDI	2.94	-.01	4.11	2.71
Duet	1.885	-.01	2.00	1.48
Dulux Grp	2.97		3.19	2.26
DWS Limited	1.52	-.03	1.57	1.14
DyesolLtd	.115		.58	.105
Echo EG Ltd	4.14	-.02	4.416	3.269
EldersLtd	.225	-.005	.385	.185
ElectroOp	.55		1.09	.505
EmecoHld	.76	-.035	1.198	.745
EmLeaders	.735		.945	.71
Energio	.19	-.01	.29	.145
EnergyAct	1.85	+.01	2.00	1.13
EnergyDev	2.35	+.04	2.73	2.25
Enero Grp	.55		1.224	.468
Engenco	.33	-.025	1.30	.30
Envestra	.85		.855	.555
EnvironCT	.015	-.002	.039	.004
EquityTr	11.80	+.18	14.00	11.00
ERMPower	1.95		2.24	1.245
Eservglob	.18	+.005	.544	.175
EthanePIF	2.23		2.35	1.385
Euroz	.97	+.05	1.70	.89
Fairfax	.53	-.02	.99	.525
Fantastic	2.29	-.01	2.63	1.70
FFI	3.30		3.65	3.20
Fiducian	.94	+.01	1.36	.90
FinbarGrp	1.005		1.165	.795
Fis&PayAp	.435		.495	.25
Fis&PayHc	1.50	-.01	2.10	1.47
FKP Prop	.38		.67	.365
Fleetwood	12.78	-.05	13.46	9.53
FletchBld	4.41	-.04	6.61	4.34
FlexiGrp	2.90	+.01	2.96	1.60
FlightCtr	20.57	-.21	22.50	16.01
Folkestne	.10		.11	.07
ForestPlc	2.30		2.30	1.70
Forge Grp	4.13	-.18	6.86	4.00
FreedomNP	.555		.69	.28
FSAGroup	.35		.385	.23
Funtastic	.155		.22	.026
G8 Educat	.95		.99	.41
GalePac	.275	-.005	.30	.165
Gazal	1.85		1.96	1.568
GBSTHldgs	.73		.97	.665
GeneticT	.115	+.015	.23	.08
GenHealth	.84		.90	.65
Geodynam	.11	+.005	.35	.10
GerardLG	1.02		1.025	.65
GiDynamic	.92		1.16	.75

Year high The highest trading price of that share in the last year.

Year low The lowest trading price of that share in the last year.

Stamp duty is a form of tax payable to the government on the trading (buying/selling) of all shares traded in PNG except for those shares that are listed on the Port Moresby stock exchange. Stamp duty is K0.10 or 1% whichever is the greater.

Unit 12.2 Activity 6C: Investing in the stock market

All prices and answers are to be given in Australian dollars unless otherwise stated. Assume all shares are listed on the Australian Stock Market. Note: Brokerage figures are given in kina. Use the exchange rate AUD 1 = K2.1368.

1. For each of the following buying transactions calculate:

   a. The cost of the shares.

   b. The brokerage paid on the transaction.

      i. 10 000 CueEnergy shares at $0.15 (per share).

      ii. 1 300 OilSearch shares at $6.58.

      iii. 800 Santos shares at $11.56.

2. For each of the following selling transactions calculate:

   a. The value of the shares.

   b. The brokerage paid on the transaction.

   c. The total return in Kina using the exchange rate of AUD1 = K2.1368.

      i. 500 Amcor shares at $7.79.

      ii. 5 450 Highlands shares at $0.165.

      iii. 15 000 LNG Ltd shares at $0.445.

3. Maria wants to sell her 565 Alcoa shares. Use the figures from the newspaper on p. 148 to find:

   a. The price of Alcoa shares at the close of the day.

   b. The value of Mathilda's shares if she sold her shares at the price from part a.

   c. The brokerage Mathilda would pay on the transaction.

   d. The stamp duty Mathilda would pay on the transaction.

   e. The amount Mathilda would receive if she sold her shares at the price quoted at the close of day (include brokerage and stamp duty).

   f. The amount Mathilda would receive if she sold her shares at the highest trading price for the past year (include brokerage and stamp duty).

   g. The percentage decrease in the price of Alcoa shares over the past year.

4. Paul wants to sell his 2 000 Brambles shares. Use the figures in the excerpt from the newspaper above to find:

   a. The price of Brambles shares at the close of the day.

b. The amount that Paul would receive if he sold his shares at the price from part a. Include brokerage and stamp duty in your answer.

c. Paul bought the shares for $7.04. How much profit/ loss did he make from owning these Brambles shares? Include brokerage and stamp duty in your answer.

5. Henri wants to sell his 2 500 Bega shares and buy some Bionomics shares. Use the prices in the newspaper on p. 148 to find the number of Bionomics shares he can buy in this transaction. You must include brokerage in the selling/buying transactions.

Assessing the suitability of companies that sell shares

There are various figures and ratios that help you assess the suitability of buying shares in a company. These figures help someone thinking of buying shares to compare the performance of a company to others in similar industries, and to results in previous years.

1. **The dividend per share (DPS)**

   The dividend that a company pays to a shareholder is calculated as the total amount from the net profit that a company chooses to return to the shareholders, divided by the number of shareholders. A company does not have to return all its net profit to the shareholders in a particular year; it can retain some of this profit to fund growth or pay for projected expenses.

   $$\text{Dividend per share} = \frac{\text{Total dividend paid}}{\text{Number of shares}}$$

   This amount is usually quoted in **cents** per share.

2. **The dividend yield**

   This allows you to compare the income generated by different investments.

   $$\text{Dividend yield} = \frac{\text{DPS}}{\text{Share price}} \text{ expressed as a percentage.}$$

3. **The earnings per share (EPS)**

   The EPS apportions the net profit after tax (NPAT). If a company returns all its profit after tax to its shareholders then EPS = DPS.

   $$\text{Earnings per share} = \frac{\text{Net profit after tax}}{\text{Number of shares}} \quad \text{expressed as cents.}$$

4. **The price–earnings ratio (PE ratio)**

   The PE ratio shows how many years it will take for your purchase of shares to be returned by earnings.

   $$\text{price–earnings ratio} = \frac{\text{Share price}}{\text{EPS}}$$

Example E

Q. A PNG resources company has sold 2 356 400 shares in its company. At present the shares are listed on the stock market at $3.56. The recent annual report showed that the net profit after tax (NPAT) was AUD$56 450 000 and a dividend of 18 cents per share was paid to the shareholders.

a. For this company calculate and **interpret**:

i. The dividend per share.

ii. The dividend yield.

iii. The earnings per share.

iv. The price–earnings ratio.

b. Shares for this company were first listed on the stock market in 2002 at $1.00. Find the average yearly rate of increase in share price if the shares have increased in value from $1.00 to $3.56 in ten years.

A. a. i. The dividend per share (DPS) is 18 cents as stated. This is the share of the profit of the company for each share issued.

ii. The dividend yield = $\frac{\text{DPS}}{\text{Share price}} = \frac{18}{356} \times 100 = 5.06\%$

This is a figure that we can compare with, say, the return on a fixed deposit in a bank.

iii. Earnings per share (EPS) = $\frac{\text{Net profit after tax}}{\text{Number of shares}} = \frac{56\,450\,000}{2\,356\,400} = 23.96 \text{ cents} \approx 24 \text{ cents}$

This is the profit made per share. In this case it is different from the dividend per share which means that not all of the profit has been returned to the shareholders.

iv. Price–earnings ratio = $\frac{\text{Share price}}{\text{EPS}} = \frac{356}{24} = 14.8$

It will take 15 years for the price of shares (at $3.56) to be returned by the earnings.

b. We are looking for the compound interest rate that will cause an increase from 1 to 3.56 over ten years. Substituting in the compound interest formula $A = PR^n$ where $R = 1 + \frac{r}{100}$

$3.56 = 1 \times R^{10}$

Taking the 10th root of 3.56:

$R = 1.135$ and so $1.135 = 1 + \frac{r}{100}$

Solving this equation gives $r = 13.5$

The value of the shares has increased by an average of 13.5% each year.

Unit 12.2 Activity 6D: Assessing the suitability of companies that sell shares

1. Calculate the total dividend paid on:

a. 20 000 Brambles shares that pay a dividend of 26 cents per share.

b. 500 ANZ Bank shares that pay a dividend of 145 cents per share.

c. 4 000 Oilsearch shares that pay a dividend of 3.9 cents per share.

2. A company has made a net profit after tax (NPAT) of $23 500 000 and all of the NPAT was paid as a dividend for the 2 614 000 shares. If the present price of the shares is $1.85, find:

a. The dividend per share.

b. The dividend yield.

c. The earnings per share.

d. The price–earnings ratio.

3. Telstra, a large communications company in Australia, has issued 12 443 074 357 shares. In the last annual report a net profit after tax of $3 424 million was reported and a dividend per share of 28 cents was paid.

a. Was all the NPAT returned to shareholders?

b. What is the dividend yield for Telstra shares if the present price is $4.09?

c. What is the price–earnings ratio for Telstra shares?

d. What does this price–earnings ratio mean for Telstra shares?

4. A large energy company has issued 270 million shares. Its current share price is $17.07 and it has paid a dividend of 45 cents per share.

a. Find the dividend yield based on these figures.

b. Calculate the price–earnings ratio if the EPS is 73.1 cents. What does this ratio tell you?

5. Shares in a company dealing with real estate are currently selling at $8.60. In the last annual report the net profit after tax was $507.9 million and a dividend of 38 cents per share was paid. The earnings per share (EPS) is given as 93 cents.

a. Calculate the dividend yield.

b. Calculate the price–earnings ratio.

c. Compare the performance of this company with that of a similar company with that has a share price of $1.47 and dividend per share of 10.7 cents.

6. Communications company iiNet has paid a dividend of 14 cents, has an EPS of 28.2 cents and an NPAT of $37 million. The current share price is $4.17.

a. Calculate the dividend yield.

b. Calculate the price–earnings ratio.

c. How does the price–earnings ratio compare with the EP ratio of Telstra (question **3**)?

d. Shares in iiNet were issued at $1 in 1999. What is the average yearly increase in capital if the price of the shares has increased from $1.00 to $4.17 over 13 years?

Unit 12.2 Managing Money 2

Topic 7: Insurance

Another real-life context for the practical application of mathematical knowledge and skills is insurance. In line with the Syllabus (p. 23) Topic 7 covers:

- Types of insurances, policies and premium payments.
- Calculating returns over a given time.
- Using simple manipulation of financial formulas.

Introduction

There are things that we all value in life: our health, our family, our house, our car. Although losing any of these things is unlikely, the cost of replacing them can cause great financial hardship to individuals and families.

- Insurance provides some financial protection for the financial hardship caused by unexpected events.
- Insurance is based on risk sharing.

If, for example, the likelihood of your house burning down is about one in one thousand and your house is worth K100 000, then the cost of replacing your house could be covered by 1 000 people each paying K100. The insurance company will work out a cost of insurance, called a **premium**, based on past experience and may set a premium of say, K160, to cover the loss of a house worth K100 000; the extra will allow for running costs of the insurance company and statutory charges.

- Insurance is a contract between two parties, the **insurer** (the insurance company) and the **insured**, where the insurer agrees to pay for certain specified losses if they are suffered by the insured and the insured pays a premium for the insurance.
- The premium is the amount of money that the insured pays to the insurer for the insurance. The insurer issues an **insurance policy** (a contract) when both parties agree on all the details of the insurance.
- The premium will depend on the likelihood of an event happening, and the **term** (the time for which the insurance applies) of the insurance.
- Mathematicians called **actuaries** assess the risk of an event happening and calculate an appropriate premium. Generally, the higher the risk of an event happening, the higher the premium.

Types of insurance

Although it is possible to insure against almost any adverse event, the following are the most common types of insurance.

Life insurance

The loss of a parent can leave the family in financial hardship, so people can insure against their own death, or a relative's death.

The premium is the amount that is paid to the insurance company to insure the life of a person. The amount of the premium usually depends on the age, the health of the person whose life is insured, and the amount for which they are insured. Often the premium will be lower for a young healthy person than for an older person who has some health problems. Insurance companies can refuse to insure people who have major health problems. Health checks are often required before an insurance policy is issued.

When an insurance policy is issued, the life insured and the **beneficiaries** are specified. Beneficiaries are the person(s) who will receive the insurance amount on the death of the person whose life is insured.

Insurance policies can also have declared situations where the insurance amount will not be paid for a particular life insured, for example, where there is a pre-existing medical problem. In these cases the insured has a declared health problem before the insurance policy is issued. When taking out a life insurance policy the person wanting insurance must declare that they are healthy or otherwise. If they make a false declaration the insurance company can refuse to pay the insurance amount upon death. If the person has a health problem there can be an increase in the premium.

Insurance policies can have **exclusions**. This means that if the insured dies because of this cause then the insurer will not pay the insurance amount. Exclusions included in a policy can be situations where the insured dies such as:

- War, civil war and terrorism.
- Insanity, suicide or intentional self-injury.
- Sexually transmitted infection, acquired immune deficiency syndrome (AIDS) or human immunodeficiency virus (HIV).
- Tribal or clan wars.
- Participation in a criminal act.
- Death caused by a beneficiary named under the policy.

There are two types of life insurance:

1. **Term life insurance.** This is short-term life insurance that insures against death for a certain term (one year, five years, ten years, etc.). The premium is calculated for a year but often can be paid fortnightly, monthly etc. Employers can arrange term life insurance for all their employees with the cost of the premium being shared by the employer and the employee. The disadvantage of this type of policy is that if you leave the employment then the insurance no longer applies. Often health checks are not as strict for this type of policy but there can still be exclusions.

2. **Whole-of-life insurance.** The contract is for all of an insured person's life or to an age agreed by the insurer and the insured. The insured amount is payable on death or at the agreed age. Whole-of-life insurance policies have higher premiums because the policy is for a longer term and also because there is a level of savings included in the premium. This savings amount attracts interest each year and so the policy becomes a type of savings account with a cash value. If a person discontinues a whole-of-life policy, the insurance will cease but the policy will have what is called a **surrender value** – a cash value associated with the savings component.

   Loans can be made against the cash value of a whole-of-life policy after a given time and the interest charged is usually lower than that charged by other lending institutions. A reason for the loan does not have to be given but there are usually rules associated with the repayments.

Unit 12.2 Activity 7A: Life insurance

1. An insurance company has the following insured amounts and premiums for a term life insurance policy for an employed person under the age of 60:

Insured amount (K)	6000	11000	22000	30000	40000
Annual premium(K)	72	120	192	276	360

   a. Mary has insured her life for K11 000 but wants the premium deducted from her wages each fortnight. How much will be deducted each fortnight?

   b. James has taken out a term life insurance policy for the amount of K30 000. Unfortunately he dies from a heart attack during the first year but had failed to mention his previous heart problems when he signed the insurance. How much will James's beneficiaries receive?

   c. Calculate the annual cost for each K1 000 of insurance for each category in the table above and use this information to find the amount insured that represents the best value.

   d. Laka's life is insured for K22 000, with the beneficiary being his wife. If he is killed in a car accident while he is insured, how much will the insurance company pay to:

   i. His wife. **ii.** His mother. **iii.** His brother.

   e. Wilson is a valued employee of the company where he works. Wilson's employer has agreed to pay 40% of the premium for a K40 000 insurance policy on Wilson's life. Wilson's wife and his employer are beneficiaries of the policy, where the employer would receive 40% of the insured amount if Wilson died.

   i. How much will the employer pay each year as a premium for Wilson's policy?

   ii. How much will Wilson pay each fortnight for his share of the policy?

   iii. How much will Wilson's wife receive if Wilson dies?

2. It will cost Bernice K380 per year for K30 000 cover on a 20-year life insurance policy. She would like to include K10 of savings per fortnight in this policy.

   a. How much would Bernice's premium be each fortnight?

   b. How much will Bernice have in savings on this policy at the end of one year if the savings part is paid 5% interest? Assume the interest is paid at the end of the year.

   c. If Bernice dies at the end of the fifth year, how much will her beneficiaries receive? Assume that the savings part of her policy is compounding yearly at 5%. You will need to use this formula for the savings calculation:

$$A = PR^n + \frac{Q(R^n - 1)}{R - 1}.$$

   d. Bernice borrowed K300 from her policy and agreed to repay this loan in 26 fortnightly repayments. She is charged 9.1% p.a. interest calculated, and added, fortnightly.

   What amount will Bernice need to repay each fortnight? You will need to use the following formula for the repayment calculation ($A = 0$):

$$Q = \frac{(PR^n - A)(R-1)}{(R^n - 1)}$$

3. Another insurance company has the following table for calculating its basic premium for whole-of-life insurance cover:

Age when the policy is first purchased	Annual premium per K1 000 of life insurance
20	12.65
25	13.25
30	14.58
35	24.90
40	36.64
45	45.60
50	58.24

a. Calculate the annual premium for a 30-year-old person who wants K45 000 of life insurance.

b. Calculate the fortnightly premium for a 40-year-old person who wants K20 000 of life insurance.

c. Marius wants a whole-of-life policy with some savings. He is 25 years old and would like K60 000 of life insurance, and to save K20 per fortnight.

i. Calculate Marius's fortnightly premium.

ii. How much has Marius saved at the end of the first year if the insurance company awards 6.5% p.a. interest compounding yearly?

iii. If Marius dies after 10 years, how much will his beneficiaries receive?

iv. If Marius borrowed K1 000 from his policy and agreed to repay this loan in 52 fortnightly repayments, what amount will Marius need to repay each fortnight? He is charged 10.4% p.a. interest calculated, and added, fortnightly. You will need to use the following formula for the repayment calculation, where $A = 0$:

$$Q = \frac{(PR^n - A)(R-1)}{(R^n - 1)}$$

4. Phillip took out an insurance policy at the age of 50 for K10 000 of life insurance.

a. Use the table from question 3 to calculate his annual premium.

b. If Phillip dies at the age of 63, how much will his beneficiaries receive?

c. How much will Phillip have paid in total premiums before his death?

Health insurance

It is possible for an individual or family to purchase medical insurance that will give financial help in the case of illness or injury. Generally a health insurance plan will cover you (and your immediate family if covered) for medical expenses incurred following an illness or injury. Medical insurance can often be purchased through a person's work and in this case the premium is less than if purchased as an individual. A typical health insurance plan will have:

- Waiting periods for conditions like pregnancy.
- Exclusions for conditions such as HIV/AIDS, infertility, sterilisation, routine physical examinations and health checks, suicide, intentional and self-inflicted injury and work-related injury or illness.

- Expense limits for general illnesses (malaria, flu, fever etc.) for each year.
- Excess payable on each claim.
- Loss-of-life benefit and funeral benefit in the case of death.
- Evacuation costs if you cannot be treated in your area.
- Weekly benefits if you are injured in an accident.
- Some dental and optical benefits.

An example of a health insurance scheme

Yearly premiums and limits for medical insurance that is organised in a workplace:

	Up to 5 members (K)	More than 5 members (K)	Limit of medical and hospital benefits (K)
Single plan	920	480	15000
Couple plan	1190	690	25000
Family plan	1330	870	40000

- A single plan covers a member without dependents.
- A couple plan covers a member with a dependent spouse and no children.
- A family plan covers a member with a dependent spouse and children under 18, or children who are unmarried students under 25 years of age. There is also limited cover, K1000, for the natural parents of the member and spouse.

The insurance policy covers all those insured, up to the given limits, for:

- Medical and hospital expenses incurred following illness or injury.
- Emergency evacuation to a major town, with a limit of K10000 for expenses.
- Funeral expenses: K3000 for an adult spouse, K1000 for a dependent child, K500 for a parent.
- Dental benefits up to K1500 and optical benefits up to K1500.
- Overseas medical expenses.
- Maternity benefit of K3000 for pregnancies after three months of joining the scheme.

Other benefits

- Death of the member: K20000.
- Additional benefits if injured in an accident.

Exclusions on this policy

- Pre-existing conditions: any condition that has been treated in the twelve months prior to joining the scheme.
- Expenses or charges incurred during any waiting period.
- HIV/AIDS, sexually transmitted infection, infertility or sterilisation.
- Routine physical examinations and health checks.
- Suicide, intentional and self-inflicted injury.
- Work-related injury or illness

Excess

- All claims have a 20% excess.

Example A

Q. George is self-employed and has bought a health insurance plan for himself and his dependent wife. He had no children when he first bought the plan.

Use the typical health insurance scheme given above to answer the following questions.

a. What fortnightly amount would George be paying for his health insurance plan?

b. George has an illness and his medical costs amount to a total of K538. How much would he be reimbursed by his health insurance company?

c. One month after starting the health insurance plan George's wife becomes pregnant. How much will she be reimbursed for the costs of the pregnancy?

A. a. George would buy a couple plan in the category 'Up to 5 members' so the cost would be K1 190 a year. This is $\frac{K1190}{26} = K45.77$ per fortnight.

b. All claims have a 20% excess, which means that George will pay 20% of the costs and the insurance company will reimburse him the remaining 80%

80% of K538 $= \frac{80}{100} \times K538 = K430.40$; George will be reimbursed K430.40

c. George's wife would only be reimbursed for her maternity costs if she became pregnant three or more months after joining the scheme. As she became pregnant one month after joining she will not be reimbursed for any of the costs of her pregnancy.

Unit 12.2 Activity 7B: Health insurance

Use the figures and information in the box on the previous page to answer the following questions.

1. A company has 9 employees who have joined a health insurance scheme and the employer pays 30% of the employee's premiums.

a. Nelson is a single man who has health insurance, purchased through this workplace. Calculate the amount that Nelson will pay each fortnight for his medical insurance.

b. The employer is paying 30% of the premiums for his 9 employees: 5 family plans, 2 couple plans and 2 single plans. How much does the employer pay in total each fortnight?

c. Dave is one of the employees and he has an illness for which he incurs K238 in claimable expenses. How much will he receive from his medical insurance company?

d. Abe is an employee of the company and has a family plan for health insurance.

i. How much does he pay each fortnight?

ii. Abe's mother dies while he is insured. How much will he receive from his health insurance for his mother's funeral?

2. Conrad works for a builder and is one of his three employees. Together they have all arranged health insurance. Conrad has a dependent wife and three children, aged 12, 15 and 19. He also looks after his father and his sister.

a. In Conrad's household, who will be covered by his family plan health insurance?

b. How much will Conrad pay each fortnight for his health insurance?

c. Conrad injured his knee in the year before he became insured and has incurred K504 for treatment of this injury since he became insured. How much will he receive from the insurance company for these expenses?

d. Conrad's wife had a routine health check that cost K48. How much will Conrad receive from the insurance company for this check?

e. During the year that he is insured Conrad's household incurred the following medical expenses:

- K346 doctor's consultations for illnesses of his wife and two younger children.
- K62 for his 19-year-old son (illness).
- K154 medical expenses for his father (illness).
- K252 hospital expenses for the 12-year-old.
- K370 dental expenses.
- K422 for spectacles for Conrad and his wife.

How much, in total, will Conrad receive from his insurance company for these medical expenses?

3. Investigate the benefits of health insurance. You may like to investigate the medical, hospital and other expenses that you would have if you had an injury where:

a. You had to be flown to a major hospital for treatment.

b. You spent 10 days in hospital.

c. You were not able to work for 6 weeks.

Investigate the amount that you would be reimbursed by a health insurance company for these expenses. In what circumstances would the insurance company not pay for your medical expenses?

Household insurance

Household insurance protects householders from loss or damage to their house or contents. Items that are covered by policies differ between different companies, but generally households are covered against loss by fire, theft and flood.

When purchasing household insurance it is important to determine the value of the property to be insured.

Co-insurance

If the property is under-insured then the payout from the insurer will be less than is needed to replace the property. This situation is referred to as co-insurance as it is assumed that the insured is taking responsibility for some of the insurance.

For example, if a property is under-insured by 30% then any payout will be reduced by 30%. It is assumed that the property owner is taking responsibility for 30% of the property so if a house, valued at K100 000, is insured for K70 000 (70% of the value) and a claim for damage is for K10 000, then the insurance company may only pay 70% (K7 000) for the damage.

Excess on insurance policies

Excess on insurance policies is a specific amount that the insured is liable to pay before the insurance company pays. Excess is designed to restrict claims for minor damage and to lower the premiums, as the higher the excess, the lower the premium.

Liability

Many household insurance policies include an amount of liability cover. This covers a householder for injury caused to someone while they are on the policy holder's property. This might be for something like a person falling down a faulty set of stairs and breaking their leg.

Unit 12.2 Activity 7C: Household insurance

1. Grace wants to insure her house and contents against fire, theft and flood and has been quoted K3.46 per K1 000 value of insurance for the building and K6.45 per K1 000 for the contents. On top of the premium amounts there are statutory charges of 4% and then goods and services tax (value-added tax) of 10%. There is an excess of K400 for any claims on the contents. Grace's building is worth K120 000 and she is insuring it for K120 000 and her contents are insured for the full amount of K22 000.
   a. Calculate the premium for Grace's building and contents insurance, before charges and tax are applied.
   b. How much will the statutory charges and GST add to the cost of her policy?
   c. How much will Grace pay per fortnight for her house insurance policy?
   d. If Grace's building and contents are destroyed by fire, what amount will she receive from the insurance company?
2. Bernard has insured his house for a total of K60 000 but it is actually worth K90 000. Bernard makes a claim for K21 000 for damage caused by a fire. How much will he receive from the insurance company?
3. Mabel is renting a flat so she insures her contents only for K10 000. She is quoted K6.76 per K1 000 of insurance for the contents plus statutory charges of 5% and then GST of 10%. The policy will have an excess of K400 on any claims.
   a. How much will Mabel pay as a premium per year?
   b. Mabel's flat is broken into . Her music system valued at K460 and iPad valued at K1 240 are stolen. How much will she receive from her insurance company?
4. Winnie has an insurance policy where the amount covered increases each year with inflation. If Winnie purchased her contents insurance five years ago and had K60 000 in cover:
   a. How much cover will she have now if inflation has averaged 5% over the last 5 years?
   b. If the rate of insurance now is K6.82 per K1 000 of cover, plus 5% statutory charges plus 10% GST, how much will Winnie be paying for her insurance now?
5. An insurance company has produced the following table to calculate premiums for different types of policies that they offer for house contents insurance.

   Valuables are items such as jewellery, art works, antiques, collections such as coin collections etc.

A 'no-claim bonus' is a percentage reduction in the premium if the insured has not made a claim in the previous year.

GST of 10% is added to all insurance policies.

Policy type	Cost per K1 000 of cover	Liability	Excess	Valuables cover	Statutory charges	No-claim bonus
Basic	K5.45	K20 000	K400	None	5%	None
Smarta cover	K4.36	K100 000	K1 000	K2 000	5%	5%
Superior	K7.28	K5 000 000	K200	K10 000	5%	10%

a. Give a reason why the cost per K1 000 of cover is less with a Smarta cover policy than with a Basic cover policy.

b. Roland has a Basic insurance policy covering K20 000 worth of household contents.

i. How much does he pay per fortnight for his insurance policy?

ii. If Roland has a break-in at his property and his computer worth K1 250 and a gold necklace worth K480 are stolen, how much will he receive from this insurance company?

c. Rosemary has a Superior insurance policy to cover her house contents.

i. Give two reasons why the premium per K1 000 of cover is greater with the Superior policy.

ii. Rosemary has cover for K56 000 of house contents. How much is her premium per year?

iii. If Rosemary does not make a claim for two years what will be her yearly reduced premium?

iv. Rosemary's property is burgled. K5 200 in household goods and K2 300 in valuables are stolen. How much will Rosemary receive from this insurance company if she claims for these goods?

d. Keith has a Smarta cover policy for his household goods worth K24 000.

i. How much is Keith paying per year for this insurance cover?

ii. An elderly friend of the family trips over a mat in Keith's house and breaks her hip. Her medical expenses amount to K1 865. How much will the friend receive as compensation from the insurance company?

iii. Jewellery worth K3 000 belonging to Keith's wife is stolen. How much will they receive from the insurance company?

iv. If Keith loses all his household contents in a fire, how much will he receive from the insurance company?

6. Investigate the types of building and contents policies offered by an insurance company. A good way to do this is to obtain a policy document. Take particular note of what is covered, what exclusions are declared, the excess, the valuables (and the definition of valuables). Compare your findings with those of other students.

Motor vehicle insurance

Motor vehicle insurance protects you against the liability to pay compensation for the injury or death of persons and/or the damage to property that may occur when you have a vehicle accident. There are two components to motor vehicle insurance:

1. **Compulsory third-party insurance.** This is insurance that provides monetary payment for loss of life or bodily injury for people who are injured by your vehicle. As the name suggests, it is compulsory (under government legislation) for a vehicle owner to have this insurance. The premium is paid yearly with vehicle registration.

Some of the compulsory third-party premiums are shown in the table below:

Vehicle description	Base rate (K)	INS levy 1%	NRSC 5%	Sub-total	VAT (GST) 10%	Total payable (nearest 5t)
Sedan – private use	271.00	2.71	13.55	287.26	28.73	316.00
Station wagon – private use	342.00	3.42	17.10	362.52	36.25	398.80
Van 9 seats or less – private use	421.00	4.21	21.05	446.26	44.63	490.90
Ambulance – hearse	298.00	2.98	14.90	315.88	31.59	347.50
Sedan – business use	301.00	3.01	15.05	319.06	31.91	351.00
Station wagon – business use	364.00	3.64	18.20	385.84	38.58	424.40
Bus – 9 seats or less	359.00	3.59	17.95	380.54	38.05	419.00
Vehicle description	**Base rate (K)**	**INS levy 1%**	**NRSC 5%**	**Sub-total**	**VAT (GST) 10%**	**Total payable (nearest 5t)**
Utility – business use	674.00	6.74	33.70	714.44	71.44	785.90
Van – 9 seats or less	349.00	3.49	17.45	369.94	36.99	406.90
Van – exceeding 9 seats	778.00	7.78	38.90	824.68	82.47	907.15
Truck	950.00	9.50	47.50	1 007.00	100.70	1 107.70
Public Motor Vehicle (PMV)	1 367.00	13.67	68.35	1 449.02	144.90	1 593.90
Taxi – commercial	539.00	5.39	26.95	571.34	57.13	628.50
Motorcycle	171.00	1.71	8.55	181.26	18.13	199.40

INS levy is the insurance levy.

NRSC is the National Road Safety Council.

2. **Insurance cover for loss or damage to your vehicle**, or a vehicle or property that you have damaged due to your fault.

   Third-party damage insurance covers other people's vehicles and property that have been damaged due to your fault, but does not cover your vehicle. This type of insurance is bought when your car is of little value but you are aware that you could be liable in an accident for the damage to a vehicle that is worth a great deal of money.

Comprehensive vehicle cover will cover damage to both your vehicle and another person's vehicle or property that you have damaged in an accident.

The premium for vehicle insurance depends on the type of policy and the value of the vehicle. Each policy is assessed individually as there are many options that can be included in vehicle insurance.

Some insurance companies offer a no-claim bonus: you are rewarded by a reduction in your premium if you have not made a claim on your policy in the previous year.

Excess on insurance policies is a specific amount that the insured must pay before the insurance company pays. Excess is designed to restrict claims for minor damage and to lower the premiums, because the higher the excess, the lower the premium. Excess (and some premiums) is usually higher for young drivers, as historically they are more likely to have accidents.

Unit 12.2 Activity 7D: Motor vehicle insurance

1. Use the table above to answer the following questions about compulsory third-party insurance:

   a. Rosa has a private-use sedan. How much does she pay in compulsory third-party insurance?

   b. David owns three PMVs and a private-use station wagon. How much does he pay each year in compulsory third-party insurance?

   c. What percentage of the compulsory third-party insurance premium involves levies and taxes above the base rate?

   d. If the base rate in compulsory third-party insurance of a trailer is K116, what is the total payable in compulsory third-party insurance?

   e. Why do you think this type of insurance is called 'third-party' insurance? Who are the first two parties?

2. Gavin has third-party damage insurance which has an excess of K500. He has a collision that causes K3 200 damage to the other car and K2 250 damage to his own car. How much will the insurance company pay to Gavin and to the owner of the other car that he damaged?

3. The Nambawan Insurance Company charges a base rate of K32 per K1 000 of value for a comprehensive policy for a private-use motor vehicle. On top of this base premium are charges and levies that total 6% and then GST of 10% is applied to the total. All policies are subject to an excess of K500 for drivers over 25 years of age and K1 000 for those aged 18 to 24. The company also offers a no-claim bonus of 10% of the base premium for each consecutive year that a claim is not made.

   a. Use the information above to calculate the premium (to the nearest kina) for each of the following vehicles:

      i. A sedan bought new for K45 000.

      ii. A station wagon valued at K13 500.

      iii. A utility valued at K32 000.

b. Norma has comprehensive cover for her new car with the Nambawan Insurance Company. Her car is valued at K15 000 and when she is aged 23 she is responsible for a motor-vehicle accident that does K3 460 damage to her car and K4 820 to the other driver's car.

i. How much premium does Norma pay for her car insurance policy?

ii. How much will the insurance company pay out for the accident that Norma has been involved in?

iii. When Norma is 25 her car has decreased in value by 20% and she is eligible for a no-claim bonus. How much will she pay as a premium for comprehensive cover?

c. Jordan is 28 and has comprehensive cover for his car with the Nambawan Insurance Company. His car is valued at K33 500.

i. Calculate the premium that Jordan will pay each year for this comprehensive cover.

ii. Unfortunately Jordan's car is damaged by another driver who has no third-party damage insurance and no money to pay for the repairs. The insurance company agrees to pay for the repairs, which are going to cost K2 840. How much will the insurance company pay towards the repairs?

iii. How much will Jordan be out of pocket if he does not receive his no-claim bonus the next year?

4. Investigate the cost of insuring a car, both third-party damage and comprehensive insurance. You could use a hypothetical situation where you own a particular make of car of a chosen value. Find any restrictions on the policy, the excess and the no-claim bonus. You could compare the premiums and policies offered by two different insurers. Write a report on your findings.

Unit 12.3 Probability and Statistics
Topic 1: Basic probability

Following the Syllabus (p. 24), Unit 12.3 focuses on everyday data and how it is analysed and interpreted. There are many situations where, in order to make informed decisions, people need to gather data and use probability.
In Topic 1 we introduce the basic concepts of probability, covering:

- Probabilities involving multivariate statistical data.
- Probability experiments.
- Theoretical methods.

Probability

An **event** is a happening, or a set of **outcomes** of interest. Events can be described in words, eg 'getting an **even number** when a **die** is rolled', or shown as a list, eg {2,4,6}. The **probability** of an event is a measure of the likelihood or chance of the event occurring.

- If an event is **impossible,** then it has 'no chance' of occurring and its probability is 0.
- If an event is **certain**, then it has a '100% chance' of occurring and its probability is 1.
- All other events have a probability between 0 and 1.

Example A

1. It is impossible that pigs will fly, so the probability that pigs will fly is 0.
2. It is certain that it will rain somewhere in PNG in the next year. So the probability of rain in PNG in the next year is 1.
3. There is a fifty-fifty (or even) chance that a head turns up when an **unbiased** coin is flipped, so the probability of a head is $\frac{1}{2}$ or 0.5.
4. A darts player gets a bull's-eye 10% of the time, so there is a 10% or 0.1 chance that the darts player gets a bull's-eye on his next throw.

The probability of an event is a number between 0 and 1. The probability of an event E is written $P\ (E)$.

- Probabilities can be expressed as fractions, decimals or percentages.
- The closer a probability is to 1, the more likely the event is to occur.
- The closer a probability is to 0, the less likely the event is to occur.

Measuring probability

Probabilities of events can be worked out in two ways:

- *Experimentally*, using the **proportion** of times an event occurs (the **relative frequency** of the event).
- *Theoretically,* by considering what would happen *in theory* in a particular situation.

Experimental probability

The **experimental probability** of an event E is:

$$P(E) = \frac{\text{number of times } E \text{ occurs}}{\text{total number of trials}}$$

The experimental probability of an event is an estimate of the actual probability of the event. The greater the number of **trials**, the more accurate this estimate will be.

In some situations, such as determining the likelihood that a certain drug will cure a disease, the experimental approach is the only possible way of finding the probability. By observing the success rate of the drug over a long period of time, an accurate estimate of the actual probability of its success can be found.

Example B

Over 5 years of field trials, it was found that an antibiotic cream cleared up a skin infection within a week for 7 954 out of 10 000 patients. Since $\frac{7\,954}{10\,000} = 80\%$, the company claims that the probability that the cream is effective within a week is 80% or 0.8.

In some cases the *experimental* probability can be checked *theoretically*.

Example C

Ten students each tossed a coin 50 times and recorded the number of heads. Their results and calculations are shown in the table below.

Student	No. of heads	$\frac{\text{No. of heads}}{50}$	Cumulative no. of heads	Cumulative no. of trials	$\frac{\text{No. of heads}}{\text{No. of trials}}$
1	16	0.32	16	50	0.32
2	24	0.48	40	100	0.40
3	30	0.60	70	150	0.47
4	22	0.44	92	200	0.46
5	28	0.56	120	250	0.48
6	27	0.54	147	300	0.49
7	24	0.48	171	350	0.49
8	19	0.38	190	400	0.48
9	29	0.58	219	450	0.49
10	27	0.54	246	500	0.49

Notes:

1. Column 3 ($\frac{\text{No. of heads}}{50}$) gives the experimental probability of a head for each student.
2. Column 4 (Cumulative number of heads) gives a 'running total' of the number of heads.
3. Column 5 (Cumulative number of trials) gives a cumulative frequency of trials.
4. Column 6 ($\frac{\text{No. of heads}}{\text{No. of trials}}$) is $\frac{\text{Column 4}}{\text{Column 5}}$ and gives the experimental probability as the number of trials increases.

As the total number of trials increases, it can be seen that the relative frequency of a 'head' is getting closer and closer to 0.5 (the theoretical probability), ie the 'long-run relative frequency' is getting closer and closer to the theoretical probability.

Equally likely outcomes

Equally likely outcomes occur when all outcomes have the same chance of occurring, eg, the outcomes 'head' and 'tail' are equally likely when a fair (unbiased) coin is flipped.

For equally likely outcomes the *theoretical* probability of an event E is given by:

$$P(E) = \frac{\text{number of outcomes in } E}{\text{total number of possible outcomes}}$$

Example D

Q. A 6-sided **die** is rolled. Find the probability of the events:

1. An **odd number**
2. A number that is a factor of 6.

A. There are 6 possible outcomes when a die is rolled {1, 2, 3, 4, 5, 6}.

1. There are 3 outcomes that are odd numbers, ie {1, 3, 5}.

   P (odd number)

   $= \frac{3}{6}$ [dividing number of odd numbers by total number of outcomes]

   $= \frac{1}{2}$ [simplifying]

2. There are 4 numbers that are factors of 6, ie {1, 2, 3, 6}.

   P (factor of 6)

   $= \frac{4}{6}$ [dividing number of factors of 6 by total number of outcomes]

   $= \frac{2}{3}$ [simplifying]

Unit 12.3 Activity 1A: Probability

1. The table summarises the results of an experiment by Jason, a Grade 11 student, where a die was rolled 120 times:

Number on die	1	2	3	4	5	6
Frequency	23	18	32	12	19	16

Find the probability that one trial, chosen at random:

a. Resulted in a '6'.

b. Resulted in an even number.

Jason felt that there was something wrong with the die.

c. Explain why Jason may feel there was something wrong with the die.

d. How could Jason 'test' this theory?

2. Petra has been collecting data on a new laundry product, 'DisStain' she is about to market. After testing it on 75 common household stains, she found that 'DisStain' removed 59 of the stains completely. She decides to print the following claim on the packaging for the product.

'DisStain removes over 90% of common household stains from your laundry.'

Comment on Petra's claim.

3. A dartboard is divided into 8 parts and is numbered as shown in the diagram. A blindfolded person throws darts at the board. All the darts hit the board.

a. What is the probability of scoring a 4?

b. What is the probability of scoring a 1?

c. What is the probability of scoring an odd number?

4. A spinner has its face divided into ten equal sectors: two coloured red, three coloured blue and five coloured green. The spinner is spun. What is the probability the spinner stops in a sector which is

a. Red? **b.** Blue? **c.** Green?

5. A spinner has its face labelled 1, 2, 3, 4 as shown. The spinner is spun. What is the probability of getting

a. 1 **b.** 2 **c.** 3 **d.** A number less than 4.

6. Papakura College held its annual fete, where it ran a competition for a prize of a free night's accommodation in a hotel in a nearby town.

For every K1 a person spent at any stall they received a card that had one of the following letters on it (each card is equally likely to be received):

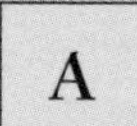

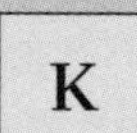

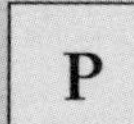

The object of the competition was to collect the cards to spell out the name **PAPAKURA**.

All correct entries were put in a box and the winner was chosen from the box.

a. What is the probability that a person spending K1 at a stall receives a card with a vowel (A, E, I, O, U) on it?

b. Abigail had spent K5 and had the letters **P A – A – – R A**.

What is the probability that the next letter she receives is one that she needs?

c. Simon had spent K12 and got the letters **A A A A K K P P P P U U**.

What is the probability that he spells **PAPAKURA** the next time he spends $1?

7. Three friends (Andrew, Basil and Wesley) played a game which involved each of them flipping a coin. If all three coins land with the same face (heads or tails) up, then that round ends in a draw. If one coin lands with a face up different from the other two, then the one with the different face wins.

a. Andrew and Basil have flipped their coins and both came up with 'Heads'. Wesley now flips his coin. What is the probability that Wesley wins this round?

b. Basil and Wesley have flipped their coins and one came up 'Heads' and the other 'Tails'. Andrew now flips his coin. What is the probability that Andrew loses this round?

c. Andrew and Basil have flipped their coins and one come up 'Heads' and the other 'Tails'. Wesley now flips his coin. What is the probability that Wesley wins this round?

8. Five cards labelled 6, 7, 8, 9, 10 are used in a game. The cards are placed face down and two cards are chosen. The numbers on the cards are added together. If the total of the two numbers is more than 16, the game is won.

a. How many different pairs of cards can be chosen?

b. What is the probability of winning the game?

Probabilities for statistical data

If a selection is made 'at random' from a group, this means that each member of the group is equally likely to be chosen.

Example E

Q. A school has 450 girls and 550 boys on its school roll. A student is chosen at random from the school roll. Find the probability that the student chosen is a girl.

A. There are 450 + 550 = 1 000 students on the school roll.

The probability of choosing a girl is:

$$P\,(\text{girl}) = \frac{450}{1\,000} \quad \text{[dividing the number of girls by the total number of students]}$$

$$= \frac{9}{20}$$

Multivariate statistical data involves more than one variable. When working out probabilities, carefully identify the group from which the **random selection** is to be made. It may not be the entire group under discussion.

Example F

Q. A group of 400 students who travel to school by car, bus or bicycle took part in a survey. The table shows the number of male and female students in each category.

A student is chosen at random from the group.

Find the probability that the student:

	Car	Bus	Bicycle	Total
Female	24	65	75	184
Male	36	95	105	236
Total	60	160	180	400

1. Is female.

2. Travels to school by bus.

3. Is a male who travels to school by car.

A male is chosen at random from the group. Find the probability that he:

4. Rides a bicycle to school

A person who travels by bus is chosen at random. Find the probability the student is:

5. Female

A. **1.** There are 184 females out of 400 students, so $P\,(\text{female}) = \frac{184}{400}$ or 0.46.

2. There are 160 bus travellers out of 400 students, so $P\,(\text{bus}) = \frac{160}{400}$ or 0.4.

3. There are 36 males who travel by car out of 400 students so

   P(male who travels by car) = $\frac{36}{400}$ or 0.09.

4. There are 236 males, of whom 105 ride a bicycle to school, so the probability is $\frac{105}{236}$ or 0.445 (3 sf).

5. There are 160 people who travel by bus to school, of whom 65 are female, so the probability is $\frac{65}{160}$ or 0.406 (3 sf).

Note: 1. In part **4**, the group from which the random selection is made is restricted to 236 males (rather than all 400 students). The event of interest 'bicycle riders' is therefore restricted to only those bicycle riders who are male.

2. Similarly in part **5**, the group from which the random selection is made is restricted to the 160 bus travellers, and the event of interest 'female' is restricted to only those females within the bus travellers group.

Unit 12.3 Activity 1B: Probabilities for statistical data

1. A number is drawn from a hat containing the following 40 numbers:

464	676	558	846	476	968	281	465	696	015
358	245	647	887	409	429	886	390	982	603
882	307	072	656	097	236	440	141	262	032
740	037	862	844	430	905	815	760	497	449

Find the probability the number drawn is:

a. Even. **b.** Above 400. **c.** Less than 100.

2. A class of 30 students recorded the month of their birthdays as follows:

Month	J	F	M	A	M	J	J	A	S	O	N	D
Frequency	1	2	1	3	4	5	3	4	2	0	3	2

a. Find the probability that a student chosen from the class at random will have a birthday:
i. In May. **ii.** In October. **iii.** Before May. **iv.** After May.

b. Find the probability that in a month selected at random, there will be:
i. Four birthdays. **ii.** Fewer than three birthdays.

3. The table summarises the results of 50 students in a test:

Marks	30–34	35–39	40–44	45–49	50–54	55–59	60–64	65–69	70–74	75–79	80–84
Frequency	5	0	4	2	10	10	8	5	2	1	3

Find the probability of a score, chosen at random, of:

a. Between 50 and 54. **b.** Between 45 and 59. **c.** Less than 50.

4. The 205 workers at 'Haus Pik' factory are employed as factory workers or as office workers, as shown in the table.

	Factory	Office	Total
Male	88		115
Female		38	
Total	140		

a. Complete the missing values in the table.

b. The names of all workers are put in a hat each Friday night and a name is randomly selected to receive a prize of a leg of pork. What is the probability that the worker selected is:

i. A male? **ii.** A factory worker? **iii.** A female office worker?

c. The person announcing the prize winner doesn't reveal the winner's name immediately, but delays the result by saying, 'It's a woman'. What is the probability that the woman who has won the prize is a factory worker?

d. A factory worker is randomly selected to draw the prize the following Friday. What is the probability that the factory worker selected is male?

5. Students in Grades 9 and 10 at Hailans Secondary School can study one subject out of Arts, Home Economics, and Practical Skills. The numbers in each option group are shown in the table:

	Arts	Home Economics	Practical Skills	Total
Year 9	125	95	100	320
Year 10	110	80	90	280
Total	235	175	190	600

a. A language student is chosen at random. Find the probability that this student

i. Studies Arts. **ii.** Is in Year 10. **iii.** Is a Year 9 Home Economics student.

b. A student from Year 10 is chosen at random. Find the probability that this student studies Practical Skills.

c. An Arts student is chosen at random. Find the probability that this student is from Grade 9.

d. A student from Grade 9 is chosen at random. Find the probability that this student does not study Arts.

6. The school soccer team keeps a record of its results during a season of 48 games. They also record where the game was played (home or away) and when it took place (during week or at weekend).

Where/when	Win	Draw	Games played
Home on weekday	9	2	12
Away on weekday	7	0	12
Home on weekend	8	3	14
Away on weekend	7	2	
Total			48

The school principal attends one randomly selected game during the season.

a. What is the probability that the game that the principal attended was played at home on a weekday?

b. What is the probability that the game was played away?

c. What is the probability that the game resulted in a loss?

The deputy principal attends one randomly chosen *away* game.

d. What is the probability that the game resulted in a win?

Which of **e** or **f** seems more likely? Justify your answer.

e. A win during a home game or a win during an away game.

f. A loss during a weekday or a loss at the weekend.

7. The table shows the months in which students in five Grade 11 classes at Ailans Secondary School have their birthdays.

	J	F	M	A	M	J	J	A	S	O	N	D	Total
11A	2	1	4	3	0	4	4	3	2	3	2	2	30
11B	0	3	2	3	2	2	5	2	3	2	3	3	
11C	2	1	2	5	1	3	1	3	5	4	1	2	
11D	3	4	3	2	4	0	2	1	3	2	3	2	29
11E	1	2	4	2	3	6	3	0	3	4	2	1	
Total	8		15			15				15		10	

What is the probability that a randomly selected student:

a. In 11B has a birthday in June?

b. Out of the five classes has a birthday in April?

c. In 11E did not have a birthday in the first half of the year?

d. With an October birthday is from 11A?

Unit 12.3 Probability and Statistics

Topic 2: Theoretical probability and simulations

Having introduced the basic concepts of probability in the previous Topic, Topic 2 moves on to discuss:

- Probabilities involving theoretical methods.
- Probabilities involving combinations of events, tree diagrams and conditional probability.
- Simulations.

Probabilities of combinations of events

Events can be combined in various ways to create new events. If A and B are two events, then

- The event **A or B** is the set of outcomes in A or B or both.
- The event **A and B** is the set of outcomes in both A and B.
- The event **A complement** or A′ is the set of outcomes not in A.

To find probabilities for each of these **combinations** of events, list the corresponding outcomes. If outcomes are equally likely, apply the formula for finding the theoretical probability of an event (see page 192).

Example A

Q. A spinner has 10 equal divisions marked with the numbers 1, 2, 3, 4, 5, 6, 7, 8, 9, 10. The spinner is spun and the number noted. The following events are defined.

A: The number is even.
B: The number is greater than 3.

1. List the outcomes corresponding to each of the following events.

a. A **b.** B **c.** A or B **d.** A and B **e.** B′

2. Find the probabilities of each of the events in **1**.

A. 1. From the definitions above, the outcomes for each event are:

a. A = {2, 4, 6, 8, 10} [even numbers]

b. B = {4, 5, 6, 7, 8, 9, 10} [numbers greater than 3]

c. A or B = {2, 4, 5, 6, 7, 8, 9, 10} [numbers even or greater than 3 or both]

d. A and B = {4, 6, 8, 10} [numbers both even and greater than 3]

e. B′ = {1, 2, 3} [numbers which are not greater than 3]

2. There are 10 equally likely numbers possible when the spinner is spun, so the probabilities are:

a. $P\,(\text{A}) = \frac{5}{10}$ or $\frac{1}{2}$ [dividing the number of outcomes in A by total number]

b. $P\,(\text{B}) = \frac{7}{10}$ [dividing the number of outcomes in B by total number]

c. $P\,(\text{A or B}) = \frac{8}{10}$ or $\frac{4}{5}$ [dividing the number of outcomes in 'A or B' by 10]

d. $P\,(\text{A and B}) = \frac{4}{10}$ or $\frac{2}{5}$ [dividing the number of outcomes in 'A and B' by 10]

e. $P\,(\text{B}') = \frac{3}{10}$ [dividing the number of outcomes in B′ by 10]

Note: When the outcomes in B are combined with the outcomes in B′ this makes up the set of all possible outcomes, so $P\,(\text{B}) + P(\text{B}') = 1$, for any event B.

As seen in the example above, there is an alternative way of working out the probability of a **complementary** event.

If A is an event and A′ is the complement of A then $P(A') = 1 - P(A)$

This result can be very useful when the calculation of P (A′) is lengthy but the calculation of P(A) is not.

Example B

Q. Two dice are rolled together and the resulting numbers added. What is the probability the numbers add up to 4 or more?

A. When two dice are rolled there are 36 possible outcomes as shown in the table.

Rather than identifying all the outcomes which add up to 4 or more, it is quicker to count the outcomes corresponding to the complementary event 'numbers add up to less than 4', ie the three outcomes (1,1), (1,2), (2,1).

	1	2	3	4	5	6
1	(1, 1)	(1, 2)	(1, 3)	(1, 4)	(1, 5)	(1, 6)
2	(2, 1)	(2, 2)	(2, 3)	(2, 4)	(2, 5)	(2, 6)
3	(3, 1)	(3, 2)	(3, 3)	(3, 4)	(3, 5)	(3, 6)
4	(4, 1)	(4, 2)	(4, 3)	(4, 4)	(4, 5)	(4, 6)
5	(5, 1)	(5, 2)	(5, 3)	(5, 4)	(5, 5)	(5, 6)
6	(6, 1)	(6, 2)	(6, 3)	(6, 4)	(6, 5)	(6, 6)

Thus P (numbers add up to less than 4) = $\frac{3}{36}$ or $\frac{1}{12}$.

Using the formula for complementary events:

P (numbers add up to 4 or more) = 1 − P(numbers add up to less than 4)

$$= 1 - \frac{1}{12}$$

$$= \frac{11}{12}$$

Unit 12.3 Activity 2A: Probabilities for combinations of events

1. A bag contains 4 blue, 3 green, and 5 yellow balls. A ball is chosen at random. What is the probability that the ball chosen is:

a. Yellow or green? **b.** Blue or yellow? **c.** Not yellow?

2. A number is chosen at random from {1, 2, 3, 4, 5, 6, 7, 8, 9}. What is the probability that it will be:

a. Odd and divisible by 3? **b.** Even or above 5? **c.** Not divisible by 4?

3. A die is rolled and a coin tossed.

a. Copy and complete the table to find the twelve possible outcomes in the sample space. H: head, T: tail.

(H, 1)	(H, 2)	(,)	(,)	(,)	(,)
(,)	(,)	(,)	(,)	(,)	(T, 6)

b. What is the probability of getting an even number on the die *and* a tail on the coin?

4. Two dice, one red and the other white, are rolled together.

Red die \ White die	1	2	3	4	5	6
1	(1, 1)	(1, 2)				
2						
3						
4					(4, 5)	
5						
6						(6, 6)

a. Copy and complete the table to find the sample space of all possible outcomes.

b. How many possible outcomes are there when the two dice are rolled?

c. Find P (total on the two dice is 7).

d. Find P (total on the two dice is less than 5).

e. Find P (total on the two dice is 13).

5. Two dice are rolled together and some events are defined.

A: The two numbers on the dice are the same.
B: The two numbers on the dice are even.

a. Define in words the following events:

i. A and B. **ii.** A or B. **iii.** A′ (the complement of A).

b. Find the probabilities of each of these events.

6. A bag contains 12 marbles, four of each of the colours red, white and blue.

A marble is removed, its colour noted, then returned to the bag. A second marble is then chosen and its colour noted. Find the probability that the two marbles are not the same colour.

Probabilities of sequences of events

To work out the number of ways a sequence of events can occur, the **multiplication principle** is used.

Example C

There are 2 outcomes possible when a coin is flipped – head (H) or tail (T).

There are 6 outcomes possible when a coin is rolled – 1, 2, 3, 4, 5 or 6.

By the multiplication principle, there are $2 \times 6 = 12$ possible outcomes when a coin is flipped then a die is rolled – H1, H2, H3, H4, H5, H6, T1, T2, T3, T4, T5, T6.

This idea is carried over into working out the probabilities of sequences of events.

> If event A occurs with probability p, and event B has probability q of following event A, then the probability of event A followed by event B is pq.

Example D

Q. A spinner has its face marked in quarters, with one quarter coloured black, one quarter coloured white and two quarters coloured red. The spinner is spun twice. Find the probability that the result is black on the first spin and red on the second.

A. The probability of black on the first spin is $\frac{1}{4}$.

The probability of red on the second spin is $\frac{2}{4}$ or $\frac{1}{2}$.

The probability of black on the first *and* red on the second is $\frac{1}{4} \times \frac{1}{2} = \frac{1}{8}$.

Probability trees

Probability trees are branching diagrams that show sequences of events. By listing events in each branch 'pathway', the set of all possible outcomes for the sequence of events can be found.

Example E

A probability tree can be used to find the probability of getting 2 heads and 1 tail when a coin is flipped three times.

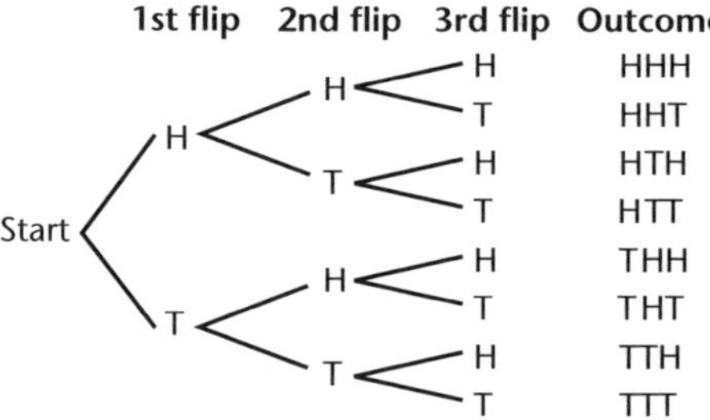

The probability tree above shows the following details:

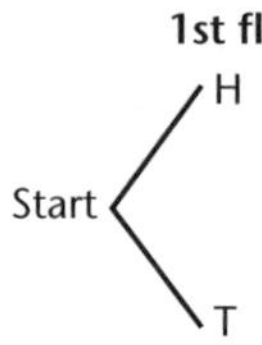

With the first flip, the outcomes are either a head (H) or a tail (T).

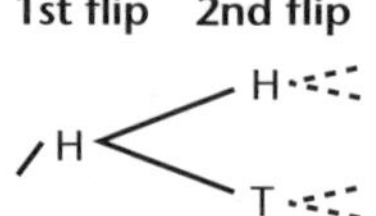

If the first flip is a head, then the second flip could be either a head or a tail, to give HH or HT.

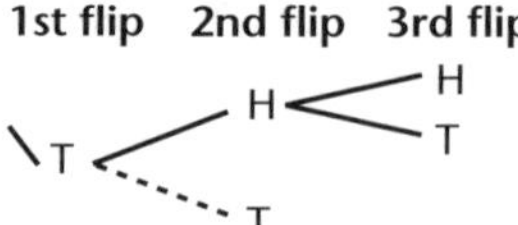

If the first flip gives a tail and the second a head, the outcome for the third flip is either a head or a tail, to give THH or THT.

By listing events in each branch pathway, eg HHT, it can be seen that there are 8 possible outcomes when a coin is tossed 3 times. These outcomes are equally likely.

If E is the event '2 heads and 1 tail',

then E = {HHT, HTH, THH}

and $P(E) = \frac{3}{8}$ [dividing number of outcomes in E (3) by the total number of outcomes]

In the above probability tree, each outcome was equally likely, but this is not usually the case. In general, to calculate probabilities of outcomes at the end of branch pathways, each branch of the probability tree is labelled with the relevant probability. The multiplication principle is then used.

Example F

Q. A jar contains 2 red balls and 3 blue balls. One ball is removed, then a second ball is removed, *without* the first ball being replaced.

1. Find the probability that both the balls removed are red.
2. Find the probability that the two balls are of different colours.

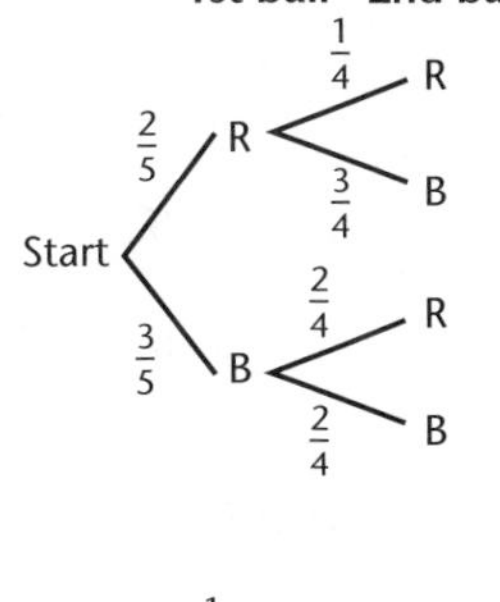

A. The probability tree following shows the probability of each individual outcome.

The probability tree shows the following details:

- For the first ball removed, the chance of selecting a red ball is $\frac{2}{5}$ and the chance of selecting a blue ball is $\frac{3}{5}$.

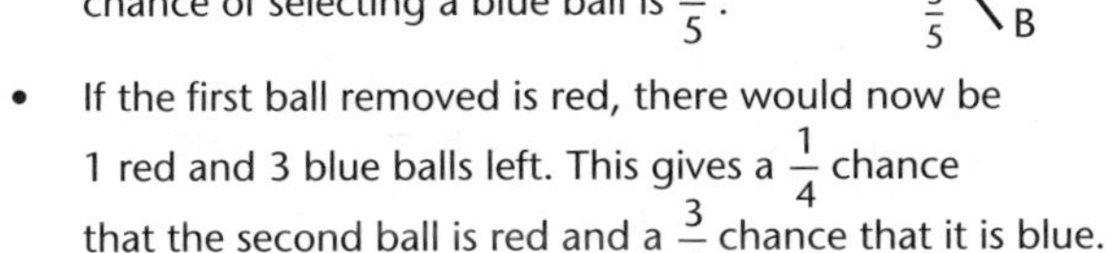

- If the first ball removed is red, there would now be 1 red and 3 blue balls left. This gives a $\frac{1}{4}$ chance that the second ball is red and a $\frac{3}{4}$ chance that it is blue.

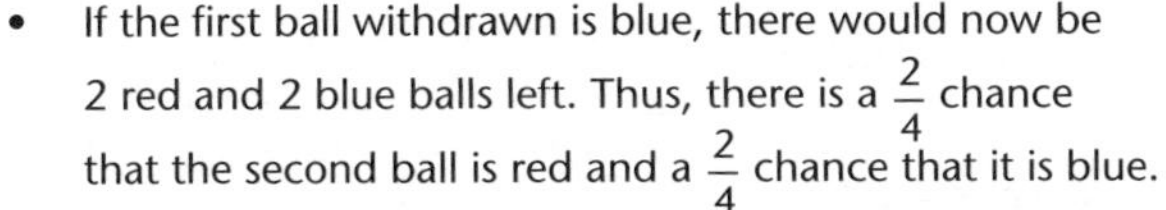

- If the first ball withdrawn is blue, there would now be 2 red and 2 blue balls left. Thus, there is a $\frac{2}{4}$ chance that the second ball is red and a $\frac{2}{4}$ chance that it is blue.

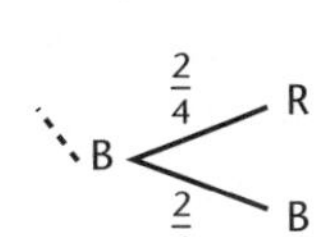

Note that the probabilities in pairs (or more) of branches such as those shown always total 1, eg $\frac{2}{5} + \frac{3}{5} = 1$ or $\frac{1}{4} + \frac{3}{4} = 1$.

The outcomes corresponding to the ends of each pathway of branches are listed and their probabilities are worked out by multiplying along the branches.

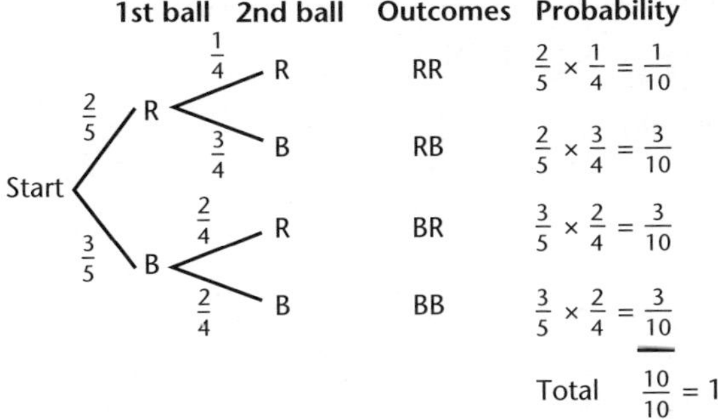

1. The probability of 2 red balls is found by multiplying probabilities along the branch pathway shown:

$$P(\text{RR}) = \frac{2}{5} \times \frac{1}{4} = \frac{1}{10}$$

2. P (2 different colours) = P (RB or BR)

$$= P(\text{RB}) + P(\text{BR})$$

$$= \frac{3}{10} + \frac{3}{10} = \frac{3}{5}$$ [using calculations in probability tree]

The probability tree example above involved both multiplying and adding probabilities. It is important to be sure when to add and when to multiply.

> When calculating probabilities using a tree diagram:
>
> - *Multiply along the branches* – to find the probability of an outcome at the end of a branch pathway.
> - *Add between the outcomes* – when several outcomes (at the end of branch pathways) make up the event whose probability is required.

Note: The sum of all probabilities of outcomes at the ends of branches in a tree diagram is 1 (see Example F).

Example G

Q. Pam is late for school 10% of the time.
Ric is late 15% of the time.

1. What percentage of the time are they both late?
2. What is the probability only one of them is late?

A. Set up a probability tree with probabilities.

Pam		Ric	Outcome
0.1 late	0.15	late	LL
	0.85	not late	LN
0.9 not late	0.15	late	NL
	0.85	not late	NN

1. Percentage both late = P (both late)
   = 0.1 × 0.15
   = 0.015 = 1.5%

2. P(only one late) = P (Pam late, Ric not) + P (Pam not, Ric late)
   = 0.1 × 0.85 + 0.9 × 0.15
   = 0.085 + 0.135 = 0.22

Conditional probability

Conditional probability is the probability of an event assuming that another event has *already* occurred. In this case multiplication of probabilities along the branches of the probability tree will have a different starting point. Part **2** of the following example illustrates this.

Example H

Q. 40% of students at Totara College catch a bus to school. 55% of those students who catch a bus are girls.

1. Find the probability that a student chosen at random from the school roll is a female student who catches a bus to school.
2. Of the girls that catch a bus to school, 35% are senior students. Find the probability that a student who catches a bus to school is a senior female student.

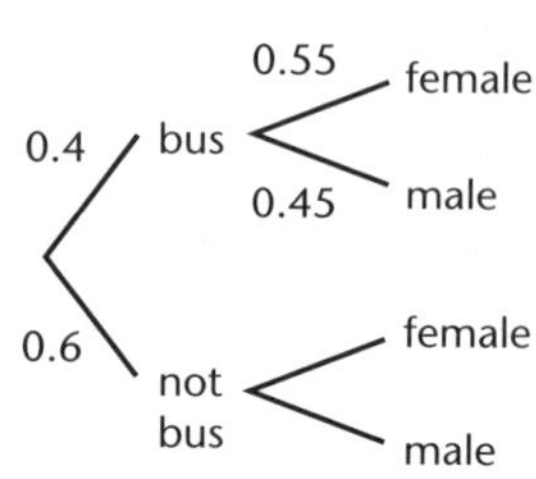

A. 1. P (catches bus and female)
= P (catches bus) × P(female)
= 0.4 × 0.55
= 0.22

2. Given that the student catches the bus:

P (senior female) $= P$ (female) $\times P$ (senior)
$= 0.55 \times 0.35$
$= 0.1925$

0.35 senior
0.55 female
bus
junior
male

Note that the starting point is 'bus' (branches previous to this are ignored).

More complex probability problems may involve reversing processes for working out probabilities.

Example I

Q. Hari's teacher expects him to bring his mathematics textbook to school every day, but she doesn't always check. Hari has worked out that the teacher only checks for textbooks about three times every ten days, on randomly selected days. Hari has to stay in whenever he is caught without his textbook.

The previous term had 50 days and Hari got kept in 6 times. Work out the probability that Hari did not bring his textbook on a day during last term.

A. Let p be the probability Hari does not bring his textbook on a day during last term.

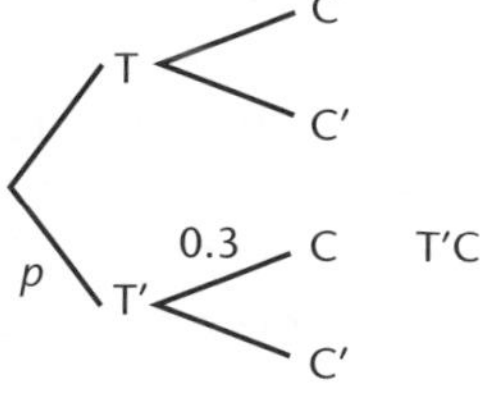

The probability tree diagram shows the events:

T: textbook brought — T′: textbook not brought
C: check done by teacher — C′: check not done

where $P(C) = \frac{3}{10}$ or 0.3 [teacher checks 3 days out of 10]

The probability that Hari was kept in $= \frac{6}{50}$ or 0.12 [he was kept in 6 out of 50 days]

Hari got kept in when he didn't bring his textbook *and* the teacher checked, ie this event corresponds to the outcome T′C.

From the probability tree, $P(T'C) = p \times 0.3$

Setting $p \times 0.3 = 0.12$ [the probability Hari was kept in]

gives $p = 0.4$ [dividing by 0.3]

Thus, Hari did not bring his textbook on 0.4 (or 40%) of the days last term.

Unit 12.3 Activity 2B: Probability trees

1. A dart is thrown at the board shown. A prize is won if the number hit is a multiple of 3 or a multiple of 5. (All darts hit the board.)

29	15	17
39	12	43
18	25	34

a. What is the probability that a prize will be won with one throw?

b. What is the probability that two prizes will be won if two darts are thrown?

c. What is the probability that only one prize will be won if two darts are thrown?

2. Sam estimates that he has a probability of $\frac{3}{5}$ of passing his French exam. Anita thinks that the probability she will pass is $\frac{2}{3}$. *A tree diagram has been started to assist you.*

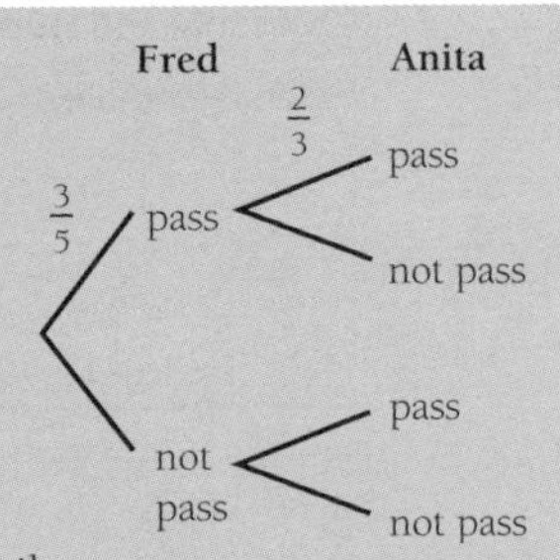

Assuming these probabilities are correct:

   a. Find the probability that both Sam and Anita pass.

   b. Find the probability that both fail.

   c. Find the probability that only one of these two students fails.

3. **a.** Copy and complete the probability tree to show the gender of children in a three-children family.
B: Boy, G: Girl.

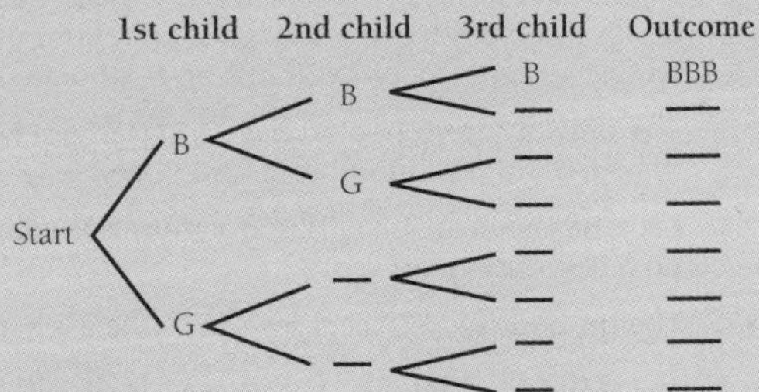

   b. Find the probability that a three-children family has:

      i. 2 boys and 1 girl.

      ii. More girls than boys.

      iii. A middle child who is a girl.

4. A group of students were polled on their leisure activities. 85% of those polled played some type of organised sport. Of those that played organised sport, 42% said that they also read for pleasure. Of those that did not play organised sport, 68% said that they read for pleasure. (A probability tree has been started.)

Find the probability that a student at random from this group:

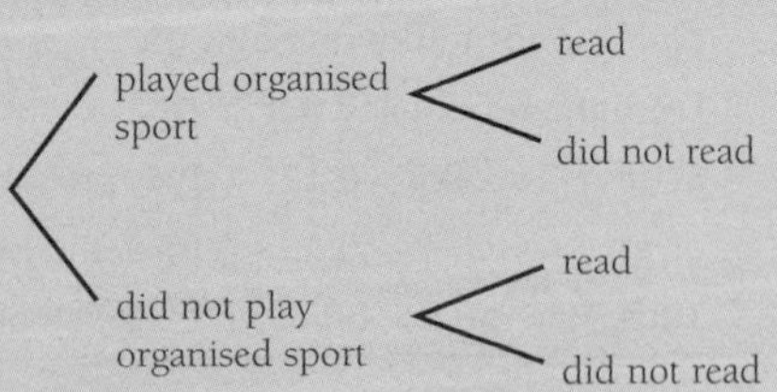

   a. Played some organised sport and read for pleasure.

   b. Did not play an organised sport and did not read for pleasure.

5. Three friends (Anne, Beth and Carol) played a game which involved each of them tossing a coin. If all three coins land the same face (Heads or Tails) up, then that round ends in a draw. If one lands with a face up different from the other two, then the one with the different face up wins the round.

   a. What is the probability Anne wins a round?

   b. What is the probability of a round ending in a draw?

   c. What is the probability of two draws in a row?

6. Out of the Grade 11 students at Gulf College:

- 40% are studying geography, the rest are studying science.
- 65% of those studying geography also study accounting.
- 45% of those studying science also study accounting.

This information can be illustrated on the diagram below:

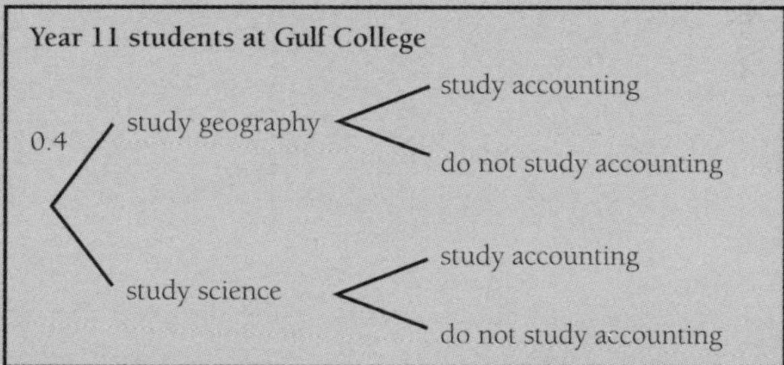

What is the probability that a Grade 11 student at Gulf College :

a. Will study geography but *not* accounting? **b.** Will study accounting?

70% of those who study science and accounting also study economics.

c. What is the probability that a student studying science will also study accounting but will *not* be studying economics?

7. 60% of a class are boys. 40% of the boys in the class take Art, 55% of the girls in the class take Art. Find the probability that a person chosen at random from the class:

a. Is a girl who does *not* take Art. **b.** Is a pupil who takes Art.

32% of the boys in the class who take Art also take Chemistry.

c. What is the probability that a boy in the class will take Art but *not* Chemistry?

8. Mia has a probability of 0.7 of passing her English exam, a probability of 0.8 of passing her mathematics exam and a probability of 0.9 of passing her geography exam. What is the probability Mia passes at least two of these exams?

9. A group of students were polled on their leisure activities. The diagram below shows the results of the poll.

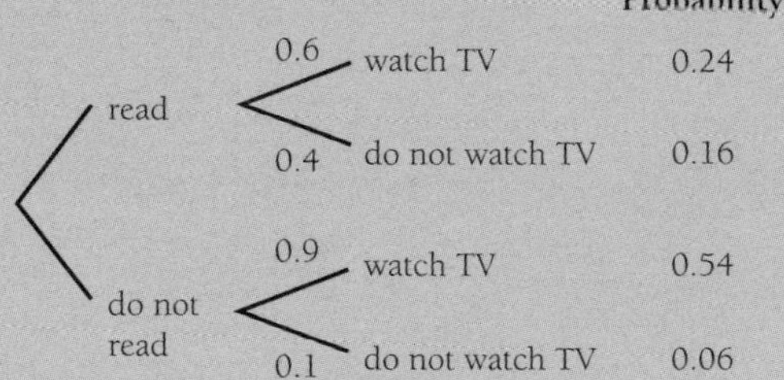

a. Find the probability that a student does not watch TV.

b. Calculate the probability that a student in the group reads.

10. Tami regularly buys her lunch at the school tuckshop. If she buys food she always buys a drink as well, but if she doesn't buy food the probability she buys a drink is 0.2. On 32% of the days Tami buys neither food nor drink. What is the probability Tami doesn't buy food at the tuckshop during a day?

Unit 12.3 Probability and Statistics

Topic 3: Probability of one event and/or another event, and mutually exclusive events

Topic 3 continues the coverage of the content under the heading 'Probability' (Syllabus p. 24), in particular:

- The probability of one event and/or another event.
- Mutually exclusive events.

The probability of one event and/or another event

If A and B are two events, then the event 'A **and** B' is the event that both A and B occur. The **intersection** symbol $\cap$ is frequently used to represent 'and', so $A \cap B$ is the event 'A and B'.

In set terms $A \cap B$ is the set of elements that set A and set B have in common.

On a **Venn diagram** $A \cap B$ is the shaded region at right.

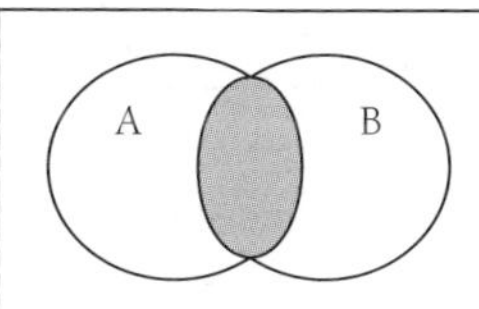

If A and B are two events, then the event 'A **or** B' is the event that either 'A **or** B or **both**' occur. The **union** symbol $\cup$ is frequently used to represent 'or', so $A \cup B$ is the event 'A or B'.

In set terms $(A \cup B)$ is all the elements 'A but not B' plus the elements 'B but not A' plus the elements 'A and B'.

On a Venn diagram $(A \cup B)$ is the shaded region at right.

The following formula can be used to calculate the probability of 'A or B':

$$P(A \cup B) = P(A) + P(B) - P(A \cap B)$$

The derivation of this formula can be seen from Venn diagrams:

$A \cup B \quad = \quad A \quad + \quad B \quad - \quad A \cap B$

By adding sets A and B the set $A \cap B$ has been included twice, hence we need to subtract set $A \cap B$.

Example A

Q. Event A has probability 0.5, event B has probability 0.4 and $P(A \cap B) = 0.3$.
Find the probability that A or B occur.

A. $P(A \text{ or } B) = P(A \cup B)$

$= P(A) + P(B) - P(A \cap B)$ [by formula]

$= 0.5 + 0.4 - 0.3$ [substituting given values]

$= 0.6$

Example B

Q. In a group of people, the probability of having a deep voice is 0.3 and the probability of being tall is 0.6. The probability of being tall or having a deep voice is 0.7. Find the probability a person is tall and has a deep voice.

A. Let T be the event 'being tall' and D be the event 'having a deep voice'.
It is required to find P(D and T) = $P(D \cap T)$. Now

$P(D \cup T) = P(D) + P(T) - P(D \cap T)$	[formula]
$\therefore\ 0.7 = 0.3 + 0.6 - P(D \cap T)$	[substituting given values]
$0.7 = 0.9 - P(D \cap T)$	[simplifying]
$P(D \cap T) = 0.9 - 0.7$	[rearranging]
$\therefore\ P(D \cap T) = 0.2$	

Mutually exclusive events

> Two events are said to be **mutually exclusive** if the fact that one event occurs means that the other event does not (or cannot) occur.

On a Venn diagram two mutually exclusive events, A and B, look like:

The events have no elements in common.

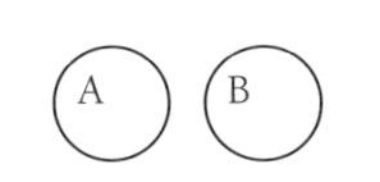

$A \cap B = \varnothing$, the null set.

> For two mutually exclusive events, A and B,
> $P(A \cup B) = P(A) + P(B)$

Example C

Consider rolling a fair die and recording the number on the uppermost side.
Event A is the event 'a number greater than 3' and event B is the event 'a number less than 3'.
A = {4, 5, 6} and B = {1, 2}
Either event A occurs or event B occurs; they cannot occur at the same time. They have no elements in common.

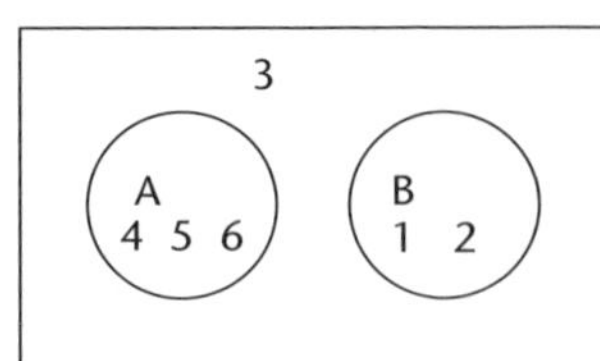

Unit 12.3 Activity 3: Probabilities of one event and/or another event

1. P(A) = 0.8, P(B) = 0.7 and P(A ∩ B) = 0.58. Find P(A ∪ B).
2. P(C) = 0.68, P(D) = 0.35 and P(C ∩ D) = 0.26. Find P(C ∪ D).
3. In a city, the probability a person drives to work is 0.4 and the probability of living in suburb A is 0.3. The probability of driving to work and living in suburb A is 0.2. Find the probablity of living in suburb A or driving to work.
4. The probability John has toast for breakfast is 0.45 and the probability Janet has toast for breakfast is 0.5. The probability of them both having toast for breakfast is 0.4.
   a. Find the probability Janet or John has toast for breakfast.
   b. Find the probability of only one of them having toast for breakfast.
5. P(M) = 0.45, P(N) = 0.65 and P(M ∪ N) = 0.73. Find P(M ∩ N).
6. P(K) = 0.77, P(L) = 0.44 and P(K ∪ L) = 0.66. Find P(K ∩ L).
7. P(A) = 0.5, P(A ∪ B) = 0.75, P(A ∩ B) = 0.35. Find P(B).
8. The tree diagram shows the probabilities of events A, B, C and D. Find:
   a. P(A)
   b. P(B)
   c. P(A ∩ C)
   d. P(B ∩ C)
   e. P(C)

   Hint: C is made up of A ∩ C and B ∩ C.
   f. P(A ∪ C)
   g. P(B ∪ D)

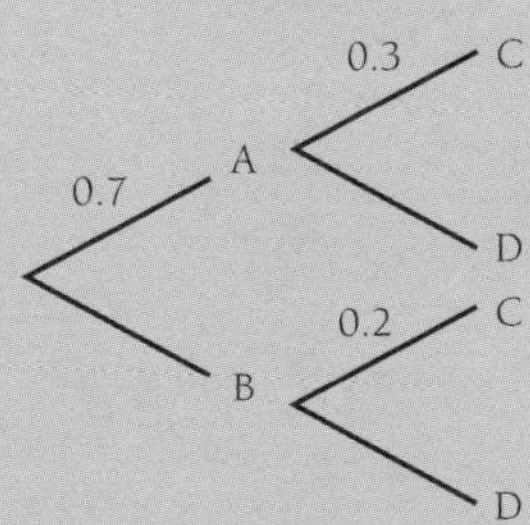

9. A fair die is rolled.

   A is the event 'a number greater than 3'; B is the event 'a number less than 4'.

   C is the event 'the number 5'; D is the event 'an even number'.

   E is the event 'an odd number'; F is the event 'the number 3'.
   a. List the elements for each of the events A to F.
   b. State whether or not the following pairs of events are mutually exclusive:
      i. A and B
      ii. A and C
      iii. B and D
      iv. F and B
      v. A and F
      vi. E and F

c. Find:

i. P(A)

ii. P(B)

iii. P(C)

iv. P(F)

v. $P(A \cup B)$

vi. $P(C \cup E)$

vii. $P(C \cup D)$

viii. $P(E \cup F)$

10. A card is selected at random from a pack of 52 playing cards. S is the event 'the card is a spade'. H is the event 'the card is a heart'. A is the event 'the card is an ace'. R is the event 'the card is red'.

a. State whether the following pairs of events are mutually exclusive:

i. S and H

ii. S and A

iii. H and R

iv. S and R

b. Find:

i. P(S)

ii. P(H)

iii. P(A)

iv. P(R)

c. Find:

i. $P(A \cap R)$

ii. $P(S \cup H)$

iii. $P(R \cap H)$

iv. $P(R \cap S)$

v. $P(A \cup R)$

vi. $P(S \cap A)$

vii. $P(R \cup H)$

Unit 12.3 Probability and Statistics

Topic 4: Independent events and conditional probability

Topic 4 continues the coverage of the content under the heading 'Probability' (Syllabus p. 24), in particular:

- Classify and calculate events as independent and dependent.

In the context of probability, independence has a precise meaning and needs to be carefully defined: two events, A and B, are **independent** if the occurrence of one has absolutely no effect on the occurrence of the other.

Example A

1. Suppose a die is being rolled and a coin is being flipped. If event A is 'a number less than 3 occurs' and event B is 'heads', then A and B are clearly independent.
2. Joe lives in the USA; Joan lives in the UK; neither knows the other.
   E is the event: 'Joe eats cornflakes on July 14',
   F is the event: 'Joan eats muesli on July 14'.
   E and F are obviously independent events except in the most contrived circumstances.

An important property of two independent events is that the probability of events both occurring is the **product** of their probabilities.

$$P(A \cap B) = P(A) \times P(B) \text{ if A and B are independent events}$$

Example B

Q. A is the event that a number less than 3 occurs on the first roll of a die. B is the event that a number greater than 3 occurs on the second roll of a die. Find the probability that both A and B occur if a die is rolled twice.

A. $P(A) = \frac{1}{3}$, $P(B) = \frac{1}{2}$. Require $P(A \cap B)$.

A and B are independent (a number less than 3 occurring on the first roll will have absolutely no effect on whether a number greater than 3 occurs on the second roll).

$\therefore$ probability of both A and B occurring is $\frac{1}{3} \times \frac{1}{2} = \frac{1}{6}$ [using the formula]

ie $P(A \cap B) = \frac{1}{6}$

Unit 12.3 Activity 4A: Independent events

1. A bag contains two blue balls and three red balls. A ball is removed, its colour noted then the ball is replaced in the bag. This is repeated once. Which of the following are independent pairs of events?
   a. (First ball is blue), (second is red)
   b. (First ball is blue), (second is blue)
   c. (First ball is blue), (first ball is red)

   d. (First ball is blue, second is red), (second ball is blue)

   e. (First ball is blue, second is red), (second ball is red)

2. Answer the questions from (1) if the first ball is removed and not replaced.
3. For each of the following pairs of events write down a situation in which they would be independent and a situation when they would not be independent.
   a. James is tall; Dorcas is short.
   b. Sabina is happy; Samuel is sad.
   c. The dog wags his tail; David has a bone.
   d. Sila passes her exam; Sarah passes her exam.
   e. It rains on Saturday; it rains on Sunday.
4. a. If you select a random two-digit number, which of the following are independent pairs of events?
      i. First digit less than 3; second digit greater than 7.
      ii. First digit is prime; second digit is prime.
      iii. First digit is less than 4; sum of digits is 7.
      iv. First digit is a factor of the second; second digit is greater than 5.
   b. For each pair of events which are independent, find the probability of both occurring.
5. The probability of event A is 0.6. The probability of event B is 0.43. The probability of events A *and* B occurring is 0.37. Explain why A and B are not independent.

Conditional probability

A **conditional probability** is a probability worked out using a restricted **sample space**. For example, the probability of randomly selecting a person with grey hair from the general **population** may be as low as 10%. However, suppose the population were restricted to persons over the age of 60. The probability of randomly selecting a person with grey hair from this restricted population would be more likely to be around 90%.

Example C

Q. Assume that 15% of the population is aged 60 or more. Assume that 80% of those over 60 have grey hair. Assume that 5% of those under 60 have grey hair. Represent this on a tree diagram.

A. Let the event 'aged 60 or more' be represented by 60^+. The event 'aged less than 60' is represented by 60^-. Let the event 'has grey hair' be G. The event G′ represents 'does not have grey hair'. The tree appears alongside.

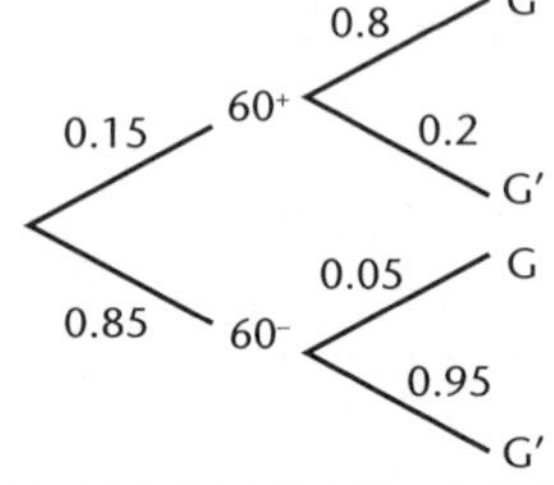

You need to have a good understanding of the meanings of the probabilities on a tree diagram and how to use them. Example D discusses the tree diagram from Example C in more detail.

Example D

Using the tree diagram from Example C, the following aspects should be noted:

1.

0.8 G
0.15 60⁺
0.2 G′
0.05 G
0.85 60⁻
0.95 G′

This says the conditional probability of 'someone having grey hair if they are 60 or more' is 0.8.

2.

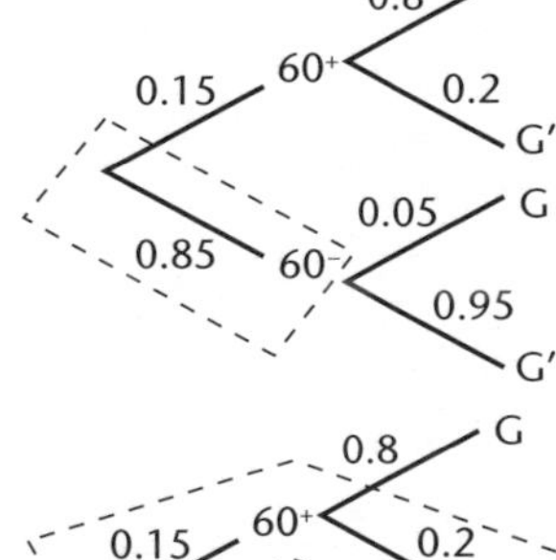

This says the probability of 'someone being under 60' is 0.85.

3.

0.8 G
0.15 60⁺
0.2 G′
0.05 G
0.85 60⁻
0.95 G′

This says the probability of a randomly selected person 'being 60 or more and not having grey hair' is $0.15 \times 0.2 = 0.03$.

Unit 12.3 Activity 4B: Conditional probability

1.

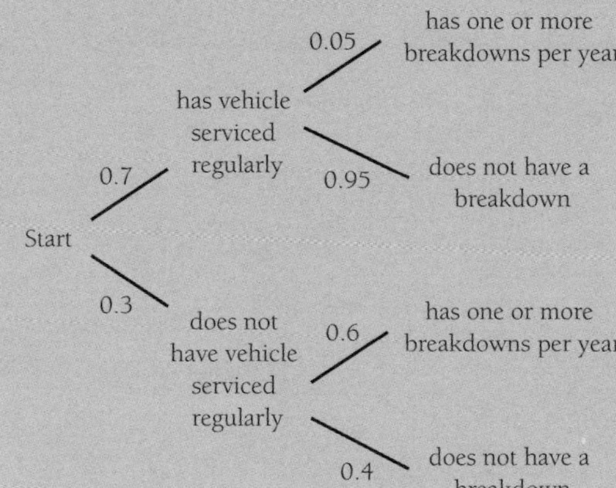

This tree diagram shows the results of a recent survey of users of a certain type of car.

a. What is the probability someone with this sort of car does not have their car serviced regularly?

b. What is the probability someone with this sort of car does not have a breakdown during the year if they have the vehicle serviced regularly?

c. What is the probability someone with this sort of car has one or more breakdowns during the year if they do not have their car serviced regularly?

d. What is the probability someone with this sort of car has their vehicle serviced regularly and has had one or more breakdowns during the year?

2. Copy this tree diagram and put on the six missing pieces of information relating to this diagram. Note A′ is the event 'A does not occur' etc.

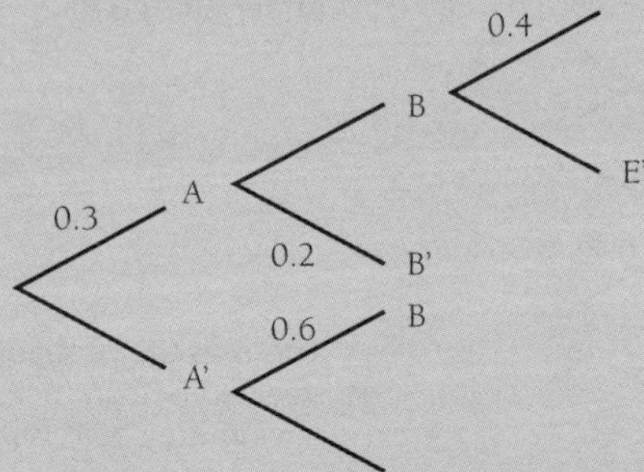

a. What is the probability of B′ occurring if A′ has occurred?

b. What is the probability of E occurring if A and B have occurred?

c. What is the probability of A′ and B both occurring?

d. What is the probability of A, B, and E′ all occurring?

3. In a school all students study maths and physics. The probability of a student passing the maths exam is 0.7. If a student passes the maths exam the probability of passing the physics exam is 0.8. If a student fails the maths exam the probability of passing the physics exam is 0.1.

a. Draw a probability tree showing this information.

b. What is the probability a student does not pass physics if they don't pass maths?

c. What is the probability a randomly selected student passes physics?

4. Use the tree diagram to find the probability of:

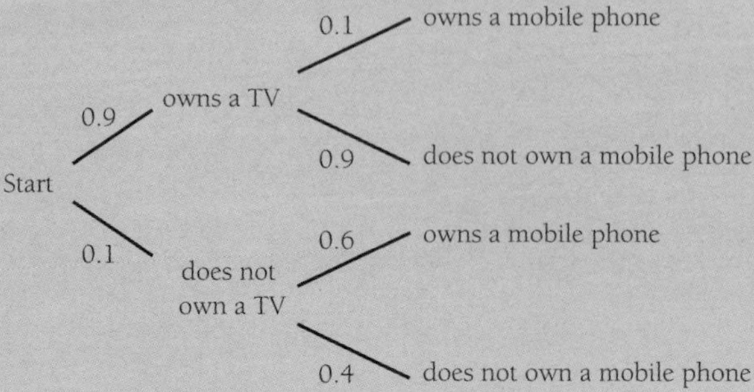

   a. owning a mobile phone if you already own a TV.

   b. not owning a mobile phone if you do not own a TV.

   c. owning a TV.

   d. owning a TV and not owning a mobile phone.

   e. owning a mobile phone and not owning a TV.

5. The probability of studying maths if you study chemistry is 0.85. The probability of studying chemistry is 0.60.

   a. Find the probability of studying chemistry and maths.

   b. If the probability of studying maths is 0.67, find the probability of studying maths if you don't study chemistry.

6. The probability of playing soccer and rugby is 0.32. The probability of playing soccer and not playing rugby is 0.13.

   a. What is the probability of playing soccer?

   b. What is the probability of playing rugby if you play soccer?

   c. If the probability of not playing rugby is 0.31, what is the probability of playing neither soccer nor rugby?

   d. What is the probability of not playing rugby if you don't play soccer?

Expected number or value

If the probability of an event occuring is p and there are N trials, then the **expected number** of occurrences is Np. This expected number is also called the expected value.

Example E

Q. The probability of a gambler winning on any occasion he plays a poker machine is 0.35. If the gambler plays a machine on 180 occasions, what is the expected number of times he will win?

A. Probability of winning, $p = 0.35$

Number of trials, $N = 180$

Expected number of wins $= Np$ [formula]

$= 180 \times 0.35$ [substituting]

$= 63$

Unit 12.3 Activity 4C: Probability and expected values

1. In New Zealand cars driven on the road are required to have a current warrant of fitness and be registered. A recent newspaper article contained the following statement: '80% of cars on the road in New Zealand have a current Warrant of Fitness and 90% are currently registered'.

   A police officer parked on the main street in his town and stopped all passing cars. He issued tickets to motorists found breaching either of these requirements.

   [Assume these statistics are correct, and that a car's having a warrant of fitness is independent of its being registered (an important assumption which needs to be stated).]

   a. What is the probability that a car which is stopped has no warrant of fitness and is not registered?

   b. Calculate the probability that the next driver will be given a ticket.

   c. Of the next 50 cars stopped, what is the expected number of drivers who will be issued a ticket?

2. Mathias is collecting a set of four sports cards from Gutpela cereal packets (there is one card in each packet). His younger sister, Nori, is collecting a set of three plastic dolls from Mobeta cereal packets (one doll per packet).

   Over any three-week period, Nori and Mathias's mother buys Gutpela cereal in two of the weeks and Mobeta cereal in the other week.

   a. Mathias has already collected three different sports cards. What is the probability that when his mother buys the next packet of Gutpela cereal, he obtains the card he needs to complete the set?

   b. Nori also needs one more doll to complete her collection. Find the probability that either Nori or Mathias will complete their set when their mother buys the next cereal packet.

3. The probability of owning an encyclopedia if you own a dictionary is 0.25. The probability of owning a dictionary is 0.93.

   a. Show this information on a tree diagram.

   b. Find the probability of owning a dictionary and an encyclopedia.

   c. If the probability of owning an encyclopedia if you do not own a dictionary is 0.3, find the probability of owning an encyclopedia.

4. The probability of studying history if you study geography is 0.75. The probability of studying geography is 0.5.

   a. Draw a tree diagram for this information.

   b. Find the probability of studying history and geography.

   c. If the probability of not studying history if you do not study geography is 0.6, find the probability of studying neither history nor geography.

Unit 12.3 Probability and Statistics

Topic 5: Statistics – correlation and regression (scatter diagrams)

Following the Syllabus (p. 24), the remaining Topics in Unit 12.3 focus on statistics and the application of statistical knowledge and probability to communicate, justify, predict and critically analyse findings and draw conclusions. Topic 5 looks at:

- Scatter diagrams.

Scatter diagrams

A **scatter diagram**, or **scatterplot**, is a graph obtained from **plotting** numerical, **bivariate** data. It is used to visually demonstrate the relationship between the two variables.

The table below shows the height h (cm) and weight w (kg) of ten Grade 11 students.

height, h (cm)	130	140	142	148	152	155	160	164	170	175
weight, w (kg)	50	55	48	56	65	58	62	68	70	74

This is bivariate numerical data because for each person there are two variables, *height* and *weight*, being measured and recorded.

To construct a scatter diagram for this data a set of Cartesian axes are drawn, each representing one of the variables. The independent variable is plotted on the horizontal **axis** and the dependent variable on the vertical axis.

For the example above: *height* is considered to be the **independent variable** and *weight* the **dependent variable** because it is considered, generally, that *weight* depends on *height*.

The co-ordinate pairs of values (*height, weight*) are plotted as points on the plane.

A scatter diagram for this set of data is shown on the right. The points plotted are (130, 50), (140, 55), ...(175, 74)

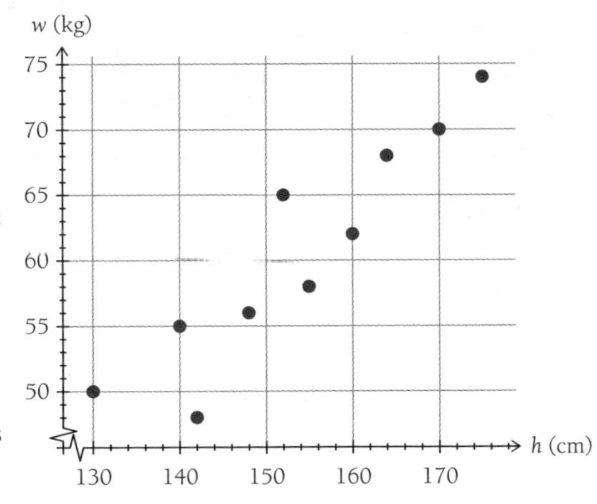

The points on the scatter diagram can be used to indicate whether there is a relationship between the variables. This is referred to as **correlation** between the variables.

The relationship between the variables can be linear or a curve.

The scatter diagram at right displays a linear relationship between *height* and *weight* suggesting that as height increases weight also increases. This is referred to as **positive, linear correlation**. The inference is assumed by observing the dots clustering along an imaginary straight line that has a positive **gradient**.

Interpretation of a scatter diagram

There are four aspects that we consider when commenting on the association between two numerical variables. These are apparent from a scatter diagram.

1. Direction

a. Positive correlation

The points generally go up as x increases, similar to a positive gradient on a straight line. The interpretation is:

'As the *independent variable* (x) increases, the *dependent variable* (y) also increases'.

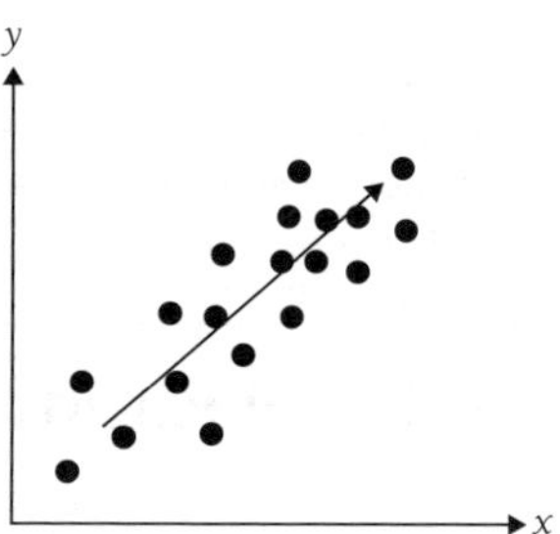

b. Negative correlation

The points generally go down as x increases; similar to a negative gradient on a straight line. The interpretation is:

'As the *independent variable* (x) increases, the *dependent variable* (y) decreases'.

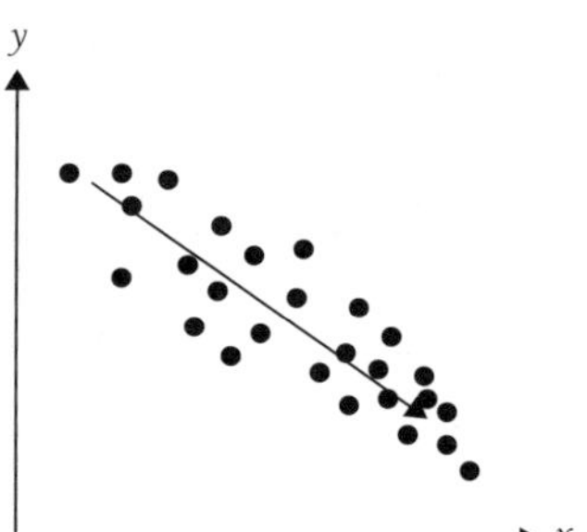

2. Form: linear or not linear

The scatterplots above all appear to be linear, ie the points are generally in a straight line. The scatterplots below would not be of a linear form:

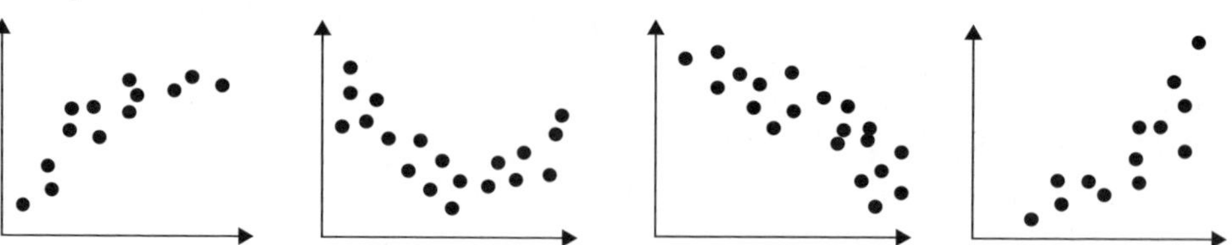

3. Strength

If the points are almost in a straight line then the strength of the association is said to be **strong**. For example:

Strong positive

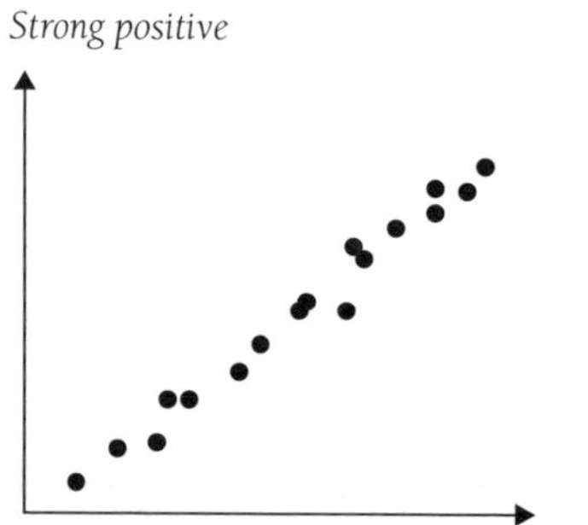

Strong negative

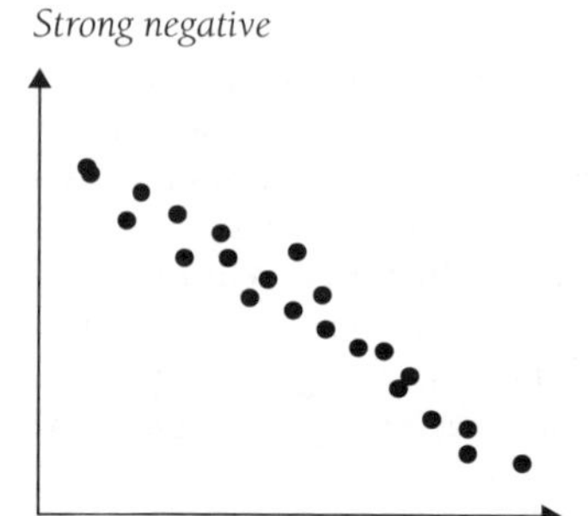

If the points are still obviously in a straight line but more scattered then the strength is said to be **moderate**. For example:

Moderate positive *Moderate negative*

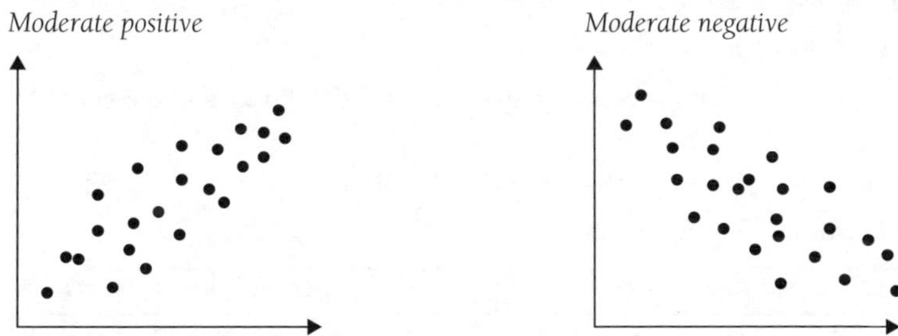

If the points are scattered but a general direction is still observable then the association is said to be **weak**. For example:

Weak positive *Weak negative*

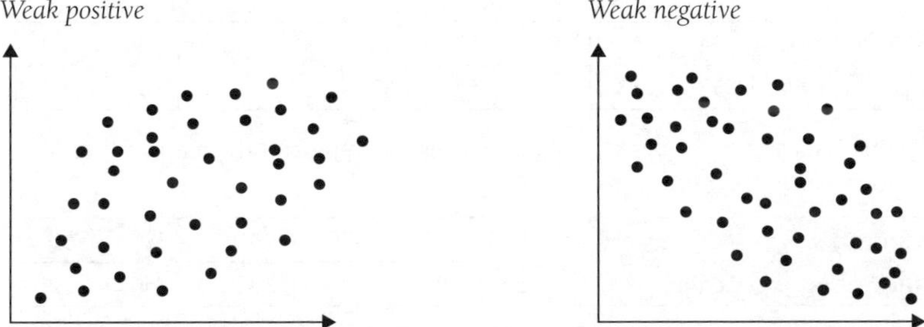

If the points appear to be randomly scattered then there is **no association** between the variables. For example:

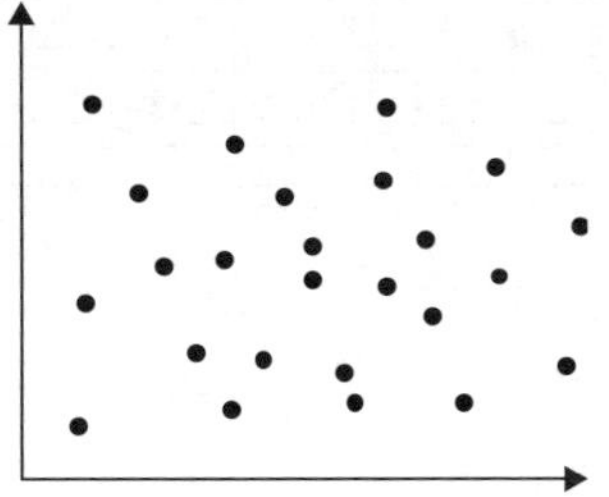

4. Outliers

Outliers stand out from the general body of the data. For example:

Moderate positive correlation with one outlier

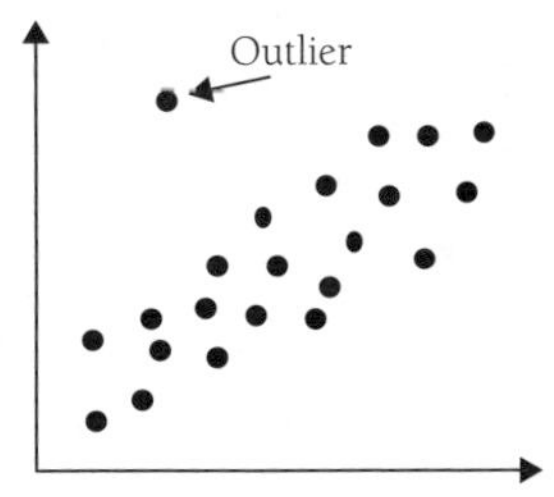

Outliers should be investigated to establish whether they are genuine outstanding data, errors in the data or errors in plotting. A decision can be made to ignore them as they influence correlation measures (correlation **coefficients**) and models (lines) fitted to the data, however, this should only be done after careful consideration.

Unit 12.3 Activity 5: Scatter diagrams

1. For each of the following sets of bivariate numerical data:
   i. Construct a scatter diagram.
   ii. Use the scatter diagrams to comment on the strength, direction, form, and presence of outliers.
   iii. Interpret the direction of data in terms of the variables. Write a sentence in the form:
   'As the (*dependent variable*) increases the (*independent variable*) ....'

a.

x	3	4	4	5	6	7	7	8	8	10
y	14	12	10	10	10	6	8	7	12	6

b. The following data are the marks (%) for ten students in their school exam and their external exam:

School	70	44	43	65	53	65	48	59	86	70
External	65	52	53	63	53	62	51	58	69	60

c. The following data are the percentage of fertile eggs hatching and the temperature of the water for a particular tropical fish:

Temperature (°C)	10	20	30	40	50	60
Percentage of eggs hatching	12	21	26	35	41	52

d. The following data are the measurements of the wrist width (mm) and the forearm length (mm) of a group of young men:

Wrist width (mm)	68	57	60	55	50	51	40	40	58	65
Forearm length (mm)	275	280	275	280	270	249	280	265	273	268

Unit 12.3 Probability and Statistics
Topic 6: Statistics – correlation analysis

Continuing the examination of correlation and regression (Syllabus p. 24), Topic 6 covers:
- Calculating the regression and correlation coefficients.

Correlation

Correlation is a statistical word that means relationship or association. We can talk about the correlation/relationship/association between two variables and mean the same thing. The study of the relationship between two variables is called correlation **analysis**.

The correlation between two numerical variables can be measured by what is called a **correlation coefficient**.

There are several methods for calculating a correlation coefficient. Some calculators will have an inbuilt program for calculating what is called Pearson's correlation coefficient (also called the product-moment correlation coefficient) This is referred to as the 'r' correlation coefficient and is widely used.

Another correlation coefficient, the q correlation coefficient, can be easily found from a scatter diagram.

The *q* correlation coefficient

The steps below can be used to find the q correlation coefficient:

1. Divide the plane into four **quadrants** by finding:
   a. The **median** of the x-values and drawing a vertical line through it.
   b. The median of the y-values and drawing a **horizontal line** through it.
2. Label the quadrants anticlockwise, A, B, C and D.
3. Count the number of points (a, b, c, and d) in each quadrant, excluding the points on the median lines.
4. Calculate q using the formula:
   $$q = \frac{(a+c)-(b+d)}{a+b+c+d}$$
   where a, b, c and d represent the number of points in quadrants A, B, C and D respectively.

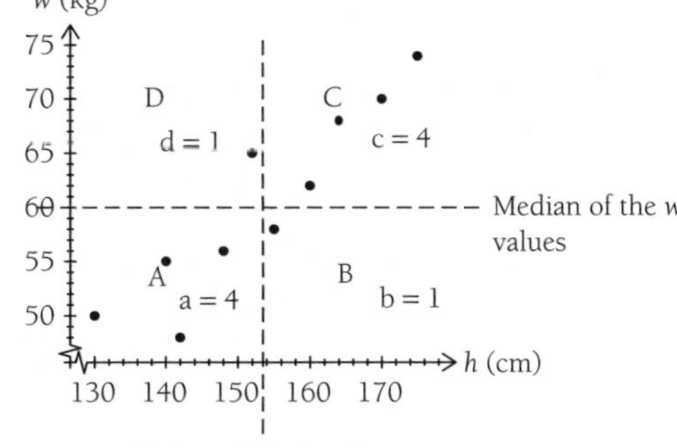

Calculating the q correlation coefficient for the example above:

$$q = \frac{(a+c)-(b+d)}{a+b+c+d} = \frac{(4+4)-(1+1)}{4+1+4+1} = \frac{6}{10} = 0.6$$

The value of q ranges between –1 and +1. If the value of q is $0 < q \leq 1$ then the relationship between the variables is positive; and if the value of q is $-1 \leq q < 0$ then there is a negative relationship between the variables.

The value of q describes the closeness to a linear relationship and gives a numerical value to the **strength of association** between any two variables enabling us to describe the association as weak, moderate or strong.

Strength of relationship	**Value of q (or r)**
Almost no relationship	$-0.25 < q < 0.25$
Weak	$0.25 \leq q < 0.5$ or $-0.5 < q \leq -0.25$
Moderate	$0.5 \leq q < 0.8$ or $-0.8 < q \leq -0.5$
Strong	$0.8 \leq q \leq 1$ or $-1 \leq q \leq -0.8$

For the *Height–Weight* example above, the q correlation coefficient was 0.6 indicating a **moderate**, positive, linear relationship between the variables *height* and *weight*.

Below are the different types of scatter diagram, correlation description with the appropriate q correlation coefficient:

Scatter diagram	Correlation	Value of q
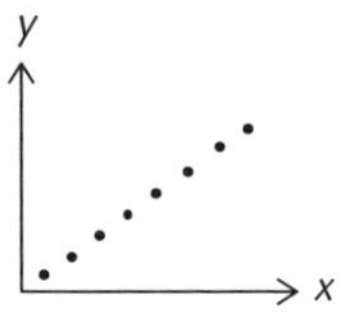 y, x	Perfect positive	1
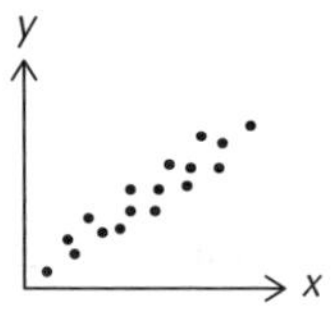 y, x	Strong positive	0.85
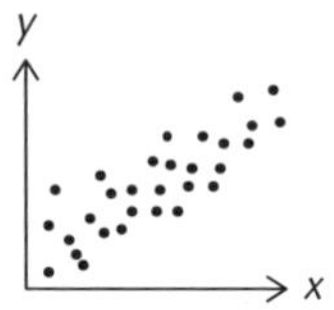 y, x	Moderate positive	0.6
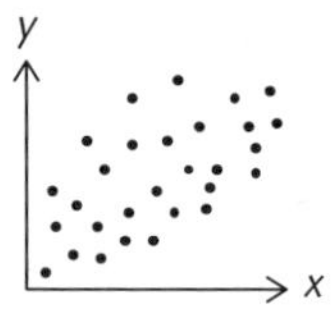 y, x	Weak positive	0.3
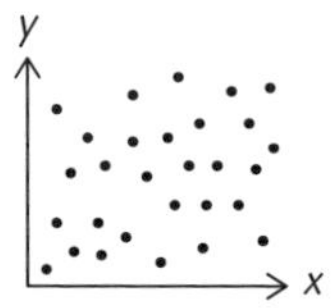 y, x	No correlation	0

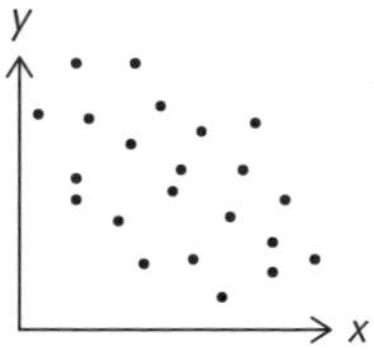	Weak negative	–0.4
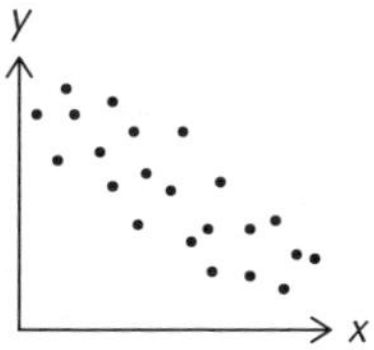	Moderate negative	–0.68
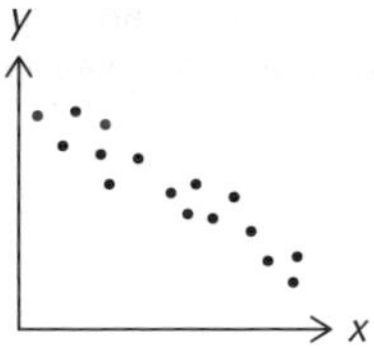	Strong negative	–0.88
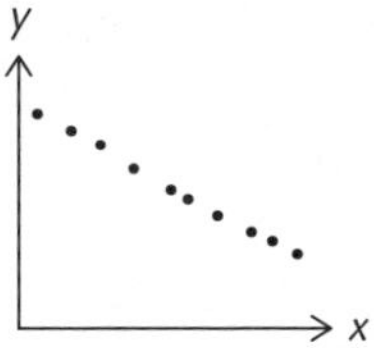	Perfect negative	–1

Example A

Q. The data in the table below were collected from a group of students to investigate the association between pulse rate (in beats per minute) and age (in years):

age	5	6	6	7	8	9	10	12	13
pulse rate	74	70	76	73	71	68	63	64	62

a. Construct a scatter diagram for the data using *age* as the independent variable.

b. Use the scatter diagram to comment on the direction, form and strength of the relationship between the variables *age* and *pulse rate*. Interpret your observation of the direction in terms of the variables.

c. Use the scatter diagram to calculate the *q* correlation coefficient for this data. Interpret this *q* correlation coefficient.

A. a. Pulse rate

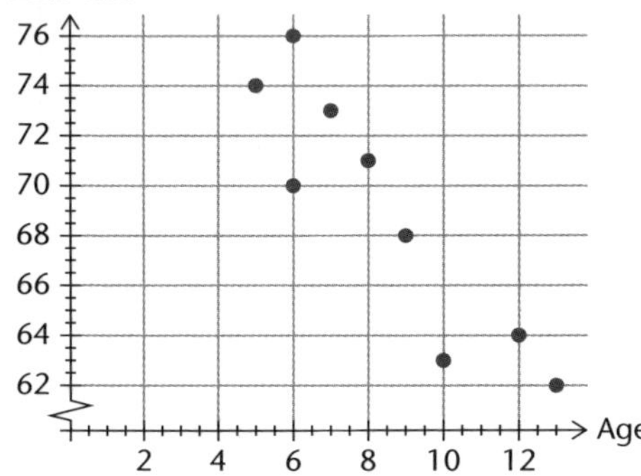

b. The points on the graph show that there is a **strong** (strength), **negative** (direction), **linear** (form) relationship between *age* and *pulse rate.*There are no outliers.The negative direction means that as *age* increases the *pulse rate* decreases.

c. Pulse rate

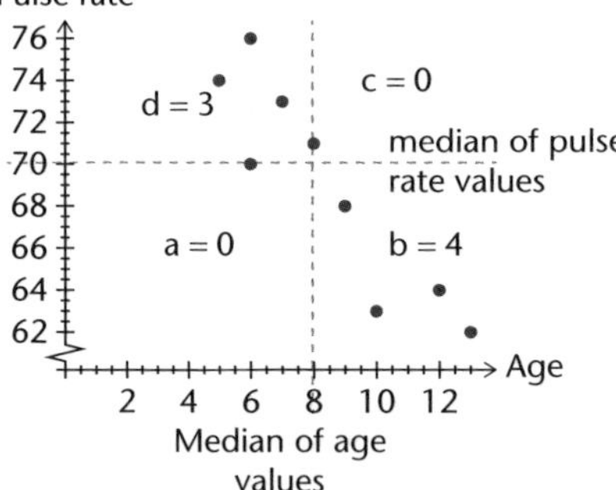

There are nine data points, so the medians will go through the fifth point. Data points on the medians are ignored, so:

$$q = \frac{(a + c) - (b + d)}{a + b + c + d} = \frac{(0 + 0) - (4 + 3)}{0 + 4 + 0 + 3} = -1$$

A q value of –1 indicates that there is perfect negative correlation between the variables. This, however, is not our initial observation from the scatter diagram (strong negative) and this is because there were relatively few data points and the q correlation coefficient is not very accurate in these circumstances.

Note: Pearson's correlation coefficient for this data gives a value of $r = -0.91$, indicating strong negative correlation.

Correlation and causation

It is important to be aware of the difference between correlation and **causation** between variables.

Two variables are said to be **correlated** if an observed change in the level of one variable is accompanied by an observed change in the level of the other variable. There is a **causal relationship** between two variables if a change in the value of one variable causes a change in the other variable.

Correlation does not imply causation. It is possible for two variables to be associated with each other without one of them causing the observed behaviour in the other. When this is the case it is often because there is a third (possibly unknown) causal factor.

Example B

The heights and the reading speeds of a group of children were measured and a strong positive correlation was found. Does this mean that increasing height makes you read faster or that increasing your reading speed will cause you to grow?

These suggestions are obviously not sensible. The strong correlation results because both variables are closely associated with *age*. As *age* increases both the variables *height* and *reading speed* increase.

It is *age* that may influence a change in *height* and *reading speed*.

Unit 12.3 Activity 6: Correlation analysis

1. **i.** Calculate the q correlation coefficient for each of the following sets of data.

   ii. Interpret the q correlation coefficient in terms of direction and strength.

a. 

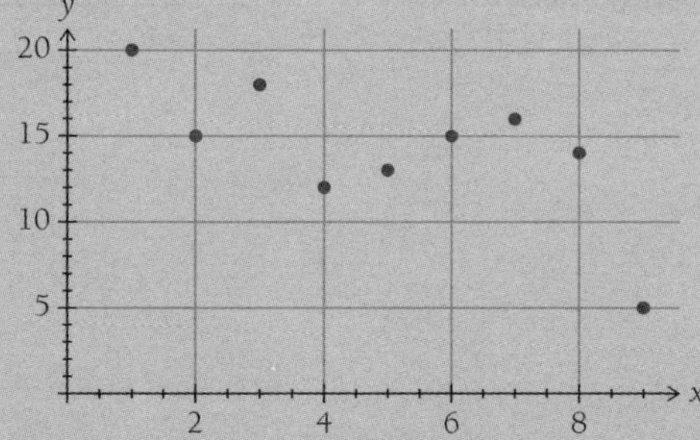

b.

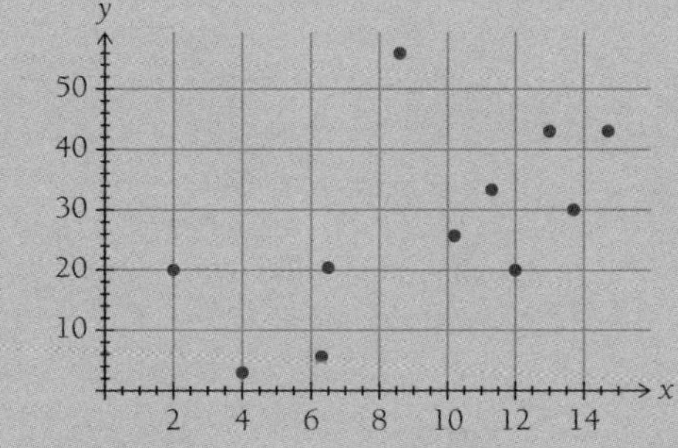

c. 

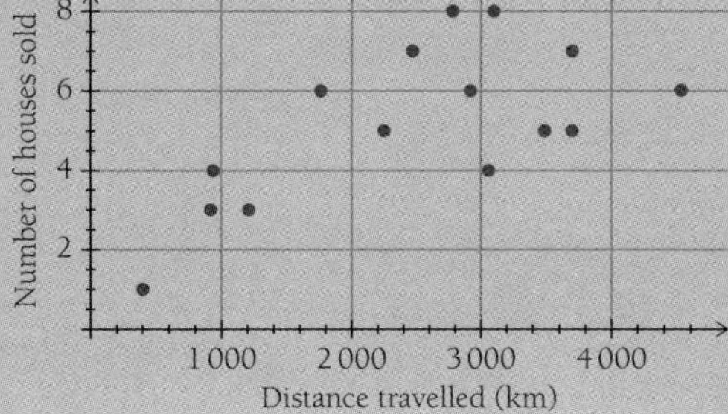

2. For each of the scatter diagrams that you constructed in Unit 12.3 Activity 5, parts **a.** to **d.**:

i. Calculate the q correlation coefficient.

ii. Interpret the q value and compare this to your previous observations.

3. An investigation found that there was a strong positive relationship between the number of dogs in a village and the total number of mobile phones used by the villagers. Which of the following statements could be concluded from this investigation?

a. An increase in the number of dogs in the village causes an increase in the number of mobile phones.

b. As the number of dogs in a village increases the number of mobile phones increases.

c. An increase in the number of mobile phones in a village causes a decrease in the number of dogs.

d. The dogs in the village are buying mobile phones.

e. Both the variables *number of dogs* and *number of mobile phones* are related to the variable number of people in the village.

Unit 12.3 Probability and Statistics
Topic 7: Statistics – regression

In Topic 7 we look at regression:

- Calculating the regression and correlation coefficients.
- Writing the equation of the regression line.
- Interpreting regression and correlation coefficients in the context of a problem

Introduction

Regression is the process of applying a **model** to a set of data points.

For the linear bivariate data that we are looking at in the scatter diagrams, regression means finding a linear model (straight line) that is the best fit for the data. A regression line gives us an algebraic relationship between the variables and from this we can make predictions.

The regression line, or as it is commonly known, **the line of best fit**, is a straight line with an equation of the form $y = mx + c$ where m is the gradient and c is the y-intercept. There are several ways that the equation of a regression line can be found:

1. The line of best fit 'by eye'

The line of best fit by eye is a regression line drawn on a scatterplot in the position that you think best fits the spread of points, and represents the direction, on the scatterplot.

It is not necessary to have access to the actual data to find this regression line. This is not a particularly accurate way of producing a regression line, as everyone has a slightly different view on what is the best line.

Once you have drawn the regression line on the scatterplot then you will need to find the **coordinates** of two points on the line and use the point gradient formula $y - y_1 = m(x - x_1)$ to find the equation where $m = \dfrac{y_2 - y_1}{x_2 - x_1}$

Example A

Q. Construct a line of best fit by eye on the following scatter diagram and find its equation in terms of the variables.

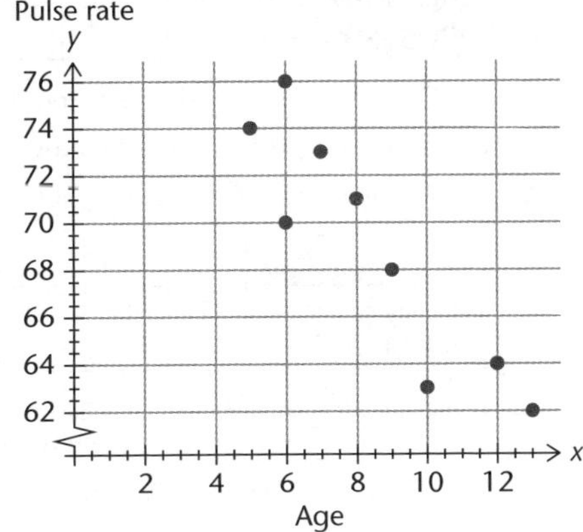

A. This scatter diagram shows strong negative correlation between the variables *age* and *pulse rate*. The strength of the correlation makes it easier to draw a line of best fit by eye on the graph and my line is shown dotted on the graph.

My line goes through the points (6, 73) and (10, 66), so the gradient of the line is $\frac{66-73}{10-6}=\frac{-6}{4}=-1.5$

The equation of the line is $y-73=-1.5(x-6)$

or $y=-1.5x+82$

In terms of the variables the equation is:

$\textit{pulse rate} = -1.5 \times \textit{age} + 82$

Note: It is conventional to write statistical models in the form $y = c + mx$

So the equation would conventionally be written as: *pulse rate* = 82 – 1.5 × *age*

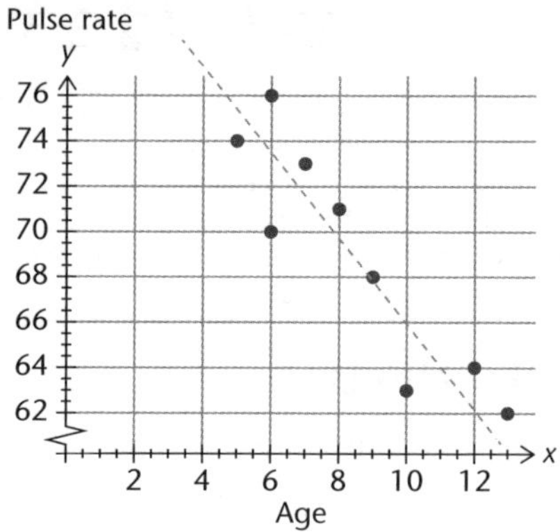

2. The two-mean regression line

To find the equation of the two-mean regression line:

a. Order the data points using the *x*-values then divide the data into two groups, left and right. If there is an odd number of points then use the middle point in both groups.

b. Find the **mean** of the *x*-values and the *y*-values in each group, producing two sets of coordinates: (x_L, y_L) and (x_R, y_R).

c. The two-mean regression line is the line passing through (x_L, y_L) and (x_R, y_R).

d. The equation of the two-mean regression line is found using $y - y_L = m(x - x_L)$, where $m = \frac{y_R - y_L}{x_R - x_L}$

Example B

Q. The fat content, as a percentage, and the energy per 100 g, in kilojoules, of eight foods is measured and the results are given in this table:

Fat content (%)	2	5	12	22	16	10	9	4	10
Energy per 100 g (kJ)	824	1 008	1 626	2 986	2 438	1 456	1 560	988	1 650

a. Construct a scatter diagram and draw the two-mean regression line on this graph.

b. Find the equation of the two-mean regression line in terms of the variables.

A. a. The first step is to order the table using the independent variable values (fat content):

Fat content (%)	2	4	5	9	10	10	12	16	22
Energy per 100 g (kJ)	824	988	1 008	1 560	1 456	1 650	1 626	2 438	2 986

There are nine sets of data, so each group will have five sets in it, the middle set being used in each group. This is shown on the following scatterplot:

The mean of the *x*-values (fat) in the Left group:

$$\frac{2+4+5+9+10}{5}=6$$

The mean of the *x*-values in the Right group:

$$\frac{10 + 10 + 12 + 16 + 22}{5} = 14$$

The mean of the *y*-values(energy) in the Left group:

$$\frac{824 + 988 + 1\,008 + 1\,560 + 1\,456}{5} = 1\,167$$

The mean of the *y*-values in the Right group:

$$\frac{1\,456 + 1\,650 + 1\,626 + 2\,438 + 2\,986}{5} = 2\,031$$

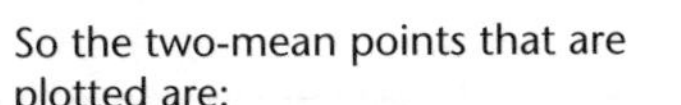

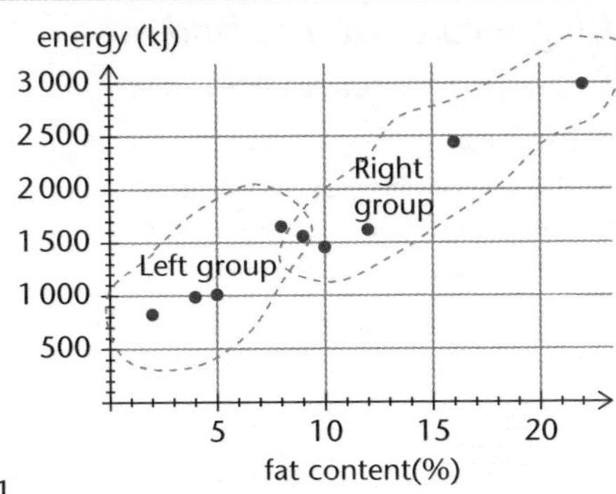

So the two-mean points that are plotted are:

(6, 1 167) and (14, 2 031)

The regression line is drawn through these two points.

b. We will use the coordinates of the two points on the line and use the point gradient formula $y - y_1 = m(x - x_1)$ to find the equation.

$$m = \frac{y_2 - y_1}{x_2 - x_1} = \frac{2031 - 1\,167}{14 - 6} = 108$$

Equation: Substituting the point (6, 1 167) and $m = 108$ gives the equation

$$y - 1167 = 108(x - 6)$$

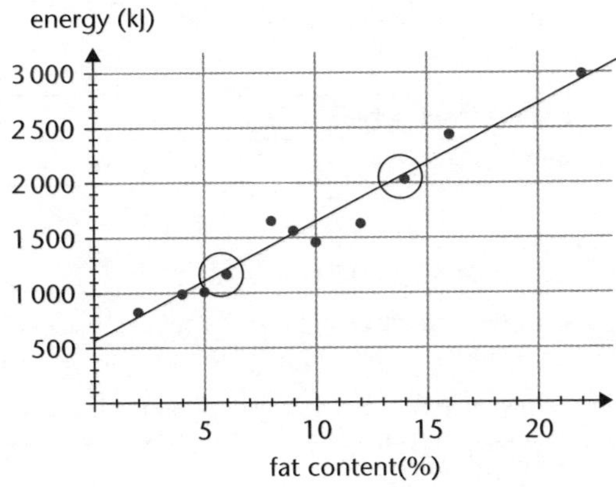

which simplifies to $y = 108x + 519$ or $y = 519 + 108x$

In terms of the variables the equation is: *energy* = 519 + 108 × *fat content.*

3. The least-squares regression line

Some calculators and spreadsheet software have an inbuilt program that will give you the values of *m* and *c* for what is called the 'least-squares regression line'. If your calculator can calculate Pearson's correlation coefficient, *r*, then it should also be able to give you *m* and *c* for the least-squares regression line equation.

The least-squares regression line is a line drawn so that the sum of the squares of the vertical distance from each point on the scatterplot (the dotted lines) is a minimum.

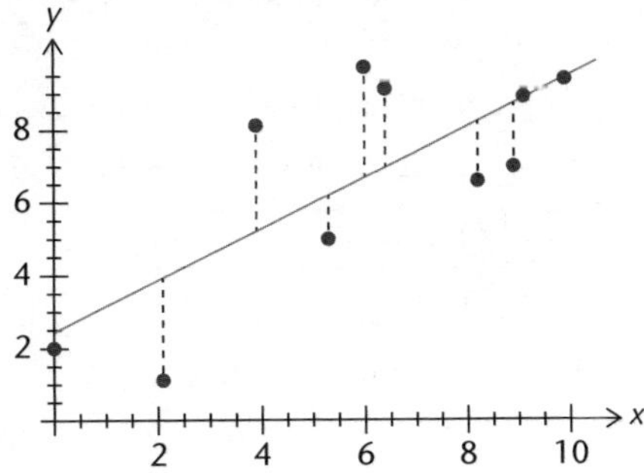

Using a calculator to find the least-squares regression line

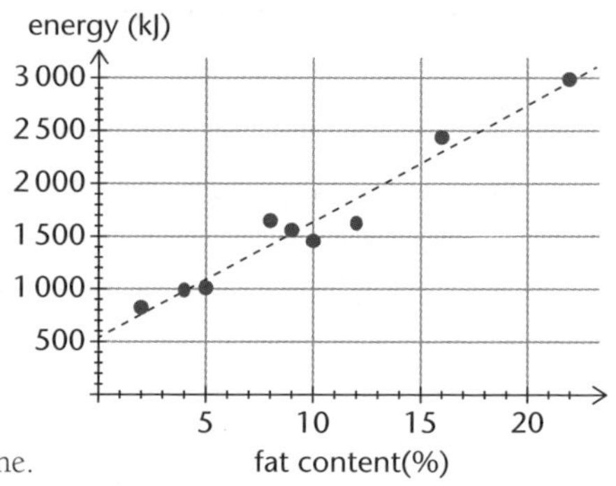

Generally the data are entered into two lists and the appropriate key pressed to give the gradient (**slope**), m, and the intercept, c, of the least-squares regression line.

For the *energy /fat content* data above the least squares regression line is:

$energy = 539 + 110 \times fat\ content$

which is very similar to the two-mean regression line.

The least-squares regression line is shown dotted on the graph above.

Unit 12.3 Activity 7: Regression

1. Marks for assignments obtained by six students in Advanced Mathematics and Physics are as follows:

Advanced Mathematics	3	7	8	8	9	10
Physics	4	6	7	8	10	9

a. Plot a scatter diagram.

b. Find the q correlation.

c. Draw a regression line of best fit by eye on the graph. Find the equation of this regression line and write the equation in terms of the variables.

2. Below are examination marks obtained by nine students in economics and accounting:

Economics	55	20	27	33	73	18	37	51	79
Accounting	72	37	53	74	73	44	59	55	84

a. Plot a scatter diagram.

b. Find the q correlation coefficient and comment on this figure in terms of strength and direction.

c. Apply a two-mean regression line to this data and find the equation of the line. Write the equation in terms of the variables.

d. Draw this regression line on the scatter diagram.

3. Below are the weights and heights of ten Grade 11 students of a particular class:

Weight (kg)	68	68	55	70	57	65	70	41	48	52
Height (cm)	177	180	175	183	175	170	172	161	160	157

a. Plot a scatter diagram of the data.

b. Describe the strength of association of the two variables.

c. Apply a two-mean regression line to this data and find the equation of the line. Write the equation in terms of the variables.

d. Draw this regression line on the scatter diagram.

4. A group of students are interested in the association between the number of buai and the number of cigarettes consumed by six people in a day. They collected the following data:

Number of buai, x	3	6	8	12	14	15
Number of cigarettes, y	2	3	5	6	8	10

a. Plot a scatter diagram for the data.

b. Describe the strength and direction of the association between the two variables.

c. Apply a two-mean regression line to this data and find the equation of the line. Write the equation in terms of the variables.

d. Draw this regression line on the scatter diagram.

Unit 12.3 Probability and Statistics

Topic 8: Statistics – the three-median regression line

Topic 8 extends our study of regression with coverage of:

- The three-median regression line.

Introduction

The three-median regression line is a linear model that is often used to fit bivariate numerical data where there appears to be the presence of **outliers**.

The three-median regression line can be drawn on a scatter diagram or the equation can be found if the data is available.

Example A

Graphical technique of fitting a three-median regression line to a set of data

Q. a. Use a graphical method to fit a three-median regression line to the data graphed here.

b. Find the equation of the line.

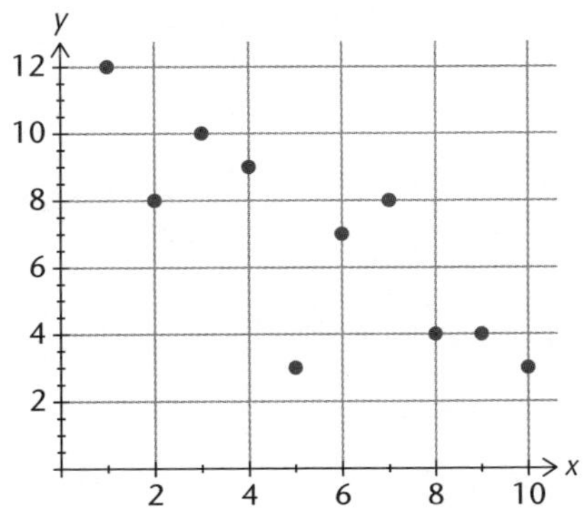

A. a. The data need to be split into three groups. There are ten data points on the graph so we will split them into groups of 3–4–3 points going from left to right.

Note: The points are split into three groups in a 'symmetrical' manner, for example:

If 9 points then the split is 3–3–3

If 10 points then the split is 3–4–3

If 11 points then the split is 4–3–4

If 12 points then the split is 4–4–4 etc.

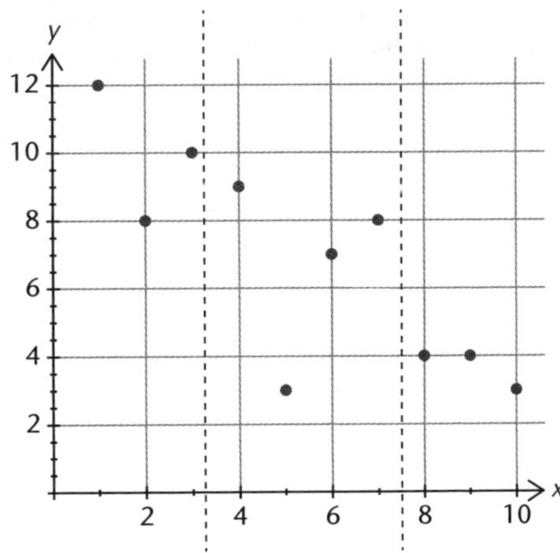

To find the median point of the first group of three points: work across to find the middle point; then work up to find the middle point.

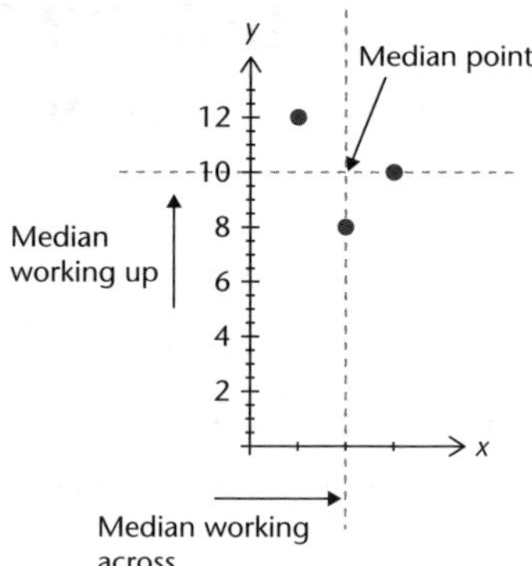

To find the median point of the second group of four points: work across to find the middle point; then work up to find the middle point.

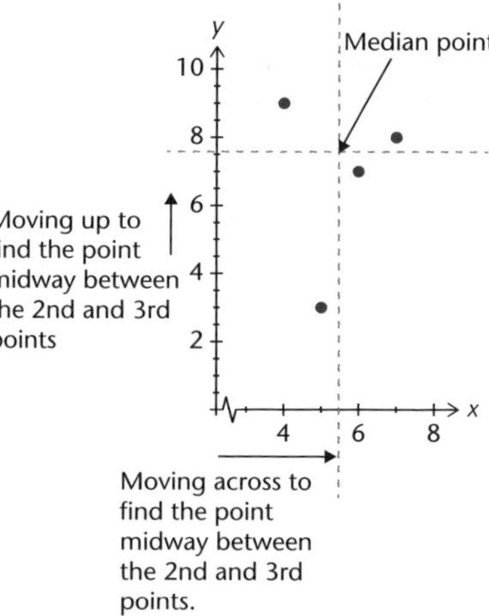

To find the median point of the third group of three points: work across to find the middle point; then work up to find the middle point.

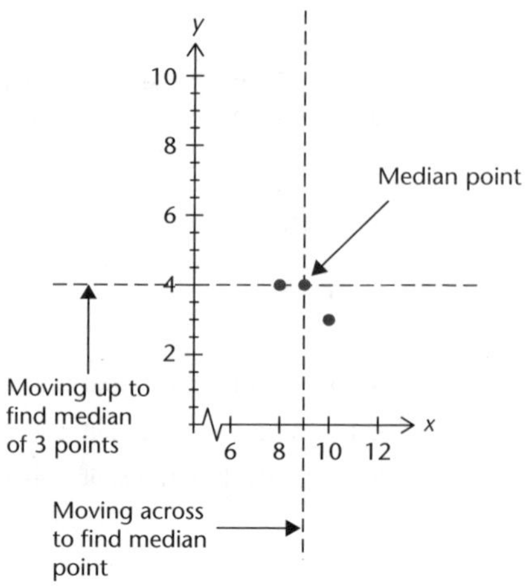

With the three median points clearly marked on the graph, place a ruler on the two outer points and *move one-third of the way towards the middle point.*

Draw a straight line. This is the three-median regression line for the data.

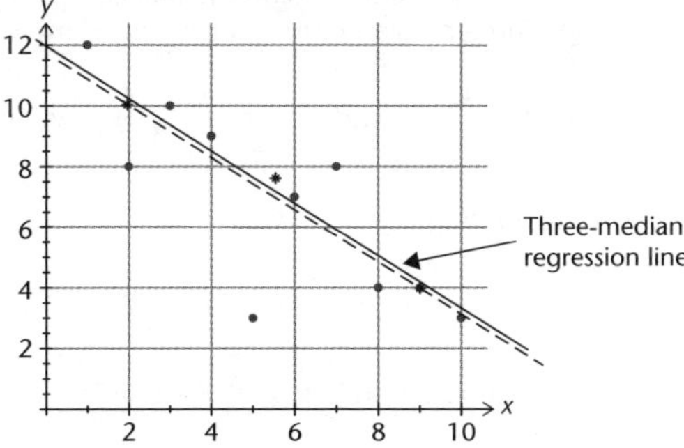

Finding the equation of the three-median regression line from the graph.

i. Choosing two points that are on the line: (0, 12) and (7, 6), use these to find the gradient by substituting in $m = \frac{y_2 - y_1}{x_2 - x_1}$

Gradient $= \frac{6-12}{7-0} = -\frac{6}{7} \approx 0.86$

ii. In this case the point (0, 12) is the y-intercept, so $c = 12$

iii. Substitute these values in the general equation for a straight line

ie. $y = mx + c$

The equation is $y = -\frac{6}{7}x + 12$

Example B

Finding the equation of the three-median regression line if the data is available.

Q. Find the equation of the three-median regression line for the data given in the table below:

x	1	2	3	4	5	6	7	8	9	10
y	12	8	10	9	3	7	8	4	4	3

A. The data need to be sorted by the *x*-values and then split into three groups. The data in the table are already sorted. The median of the *x*-values and the median of the *y*-values need to be found for each group:

The median of the *x*-values is 2	The median of the *x*-values is (5 + 6)/2 = 5.5	The median of the *x*-values is 9

x	1	2	3	4	5	6	7	8	9	10
y	12	8	10	9	3	7	8	4	4	3

The median of the *y*-values is 10	The median of the *y*-values is (7 + 8)/2 = 7.5	The median of the *y*-values is 4

This gives us the three median points (2, 10), (5.5, 7.5) and (9, 4).

If the three median points are labelled (x_L, y_L), (x_M, y_M) *and* (x_R, y_R) then the equation of the three-median regression line is $y = mx + c$ where

$$m = \frac{y_R - y_L}{x_R - x_L} \text{ and } \qquad c = \frac{1}{3}[(y_L + y_M + y_R) - m(x_L + x_M + x_R)]$$

Substituting our point values:

$$m = \frac{4-10}{9-2} = \frac{-6}{7} \text{ and } \qquad c = \frac{1}{3}[(10 + 7.5 + 4) - \frac{-6}{7}(2 + 5.5 + 9)] = 11.881\ldots$$

And the equation is $y = -\frac{6}{7}x + 11.881$

Note: The points in this example are the same as the points graphed in Example A above. The equations are almost the same.

Unit 12.3 Activity 8: The three-median regression line

1. Find the median point of each of the following sets of points graphed below.

a.

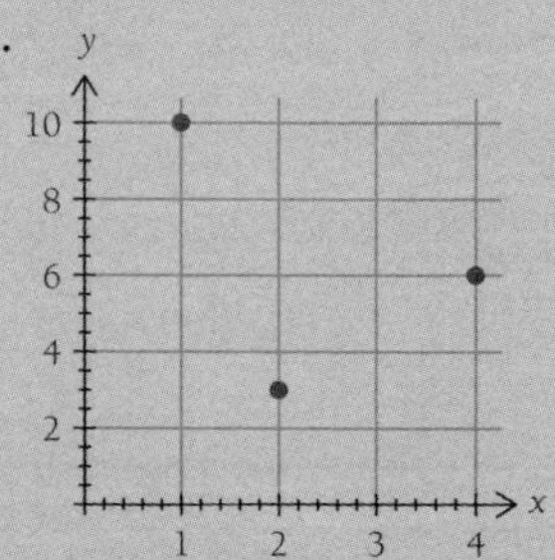

b.

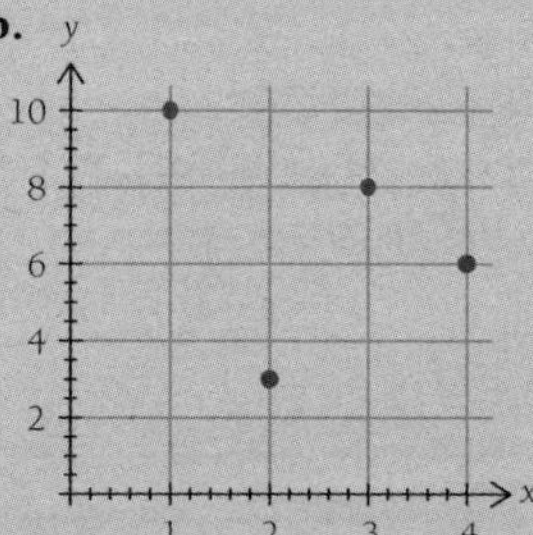

c.

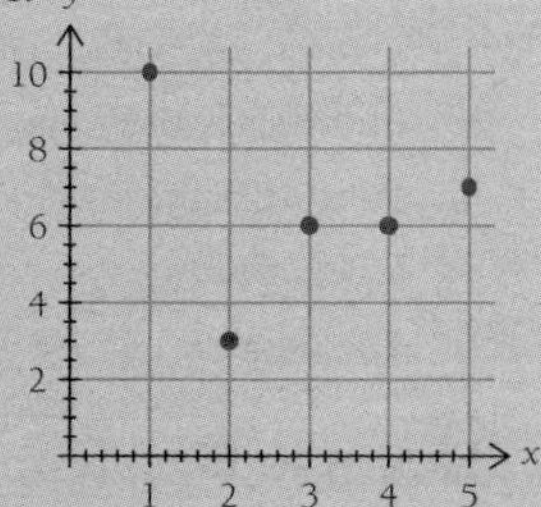

2. For each of the two sets of data graphed below:

a. Find the three median points.

b. Draw the three-median regression line on the graph.

c. Find the equation of the line.

i.

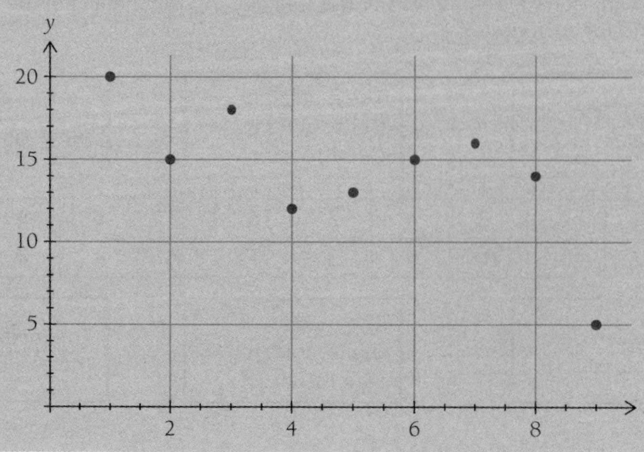

ii.

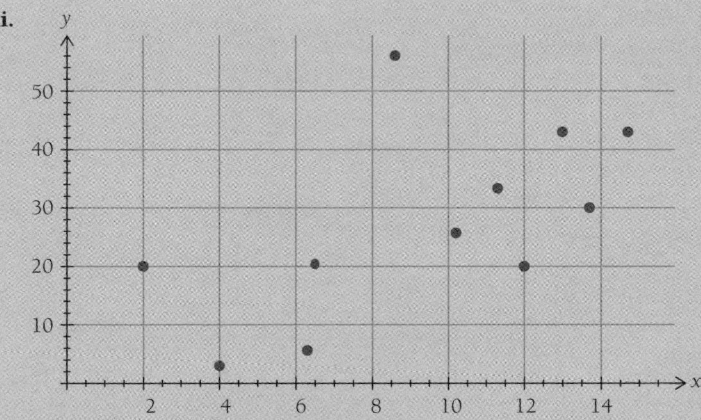

3. Give a reason why you would use a three-median regression line for the set of data graphed in **2. b.** above.
4. For the data graphed below:

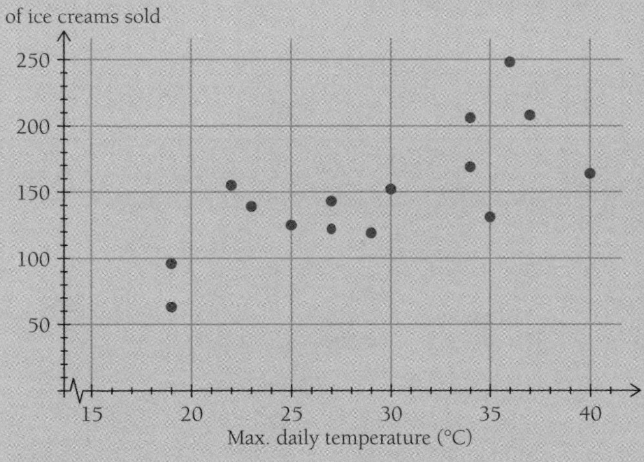

a. Construct a three-median regression line on the scatter diagram.

b. Find the equation of the three-median regression line from the scatter diagram. Write down the equation in terms of the variables.

c. Use the actual data below to find the equation of the three-median regression line. The data is ordered for the independent variable (maximum daily temperature).

Maximum daily temperature °C	19	19	22	23	25	27	27	29	30	34	34	35	36	37	40
Number of ice creams sold	63	96	155	139	125	122	143	119	152	169	206	131	248	208	164

5. The data graphed below and in the table represent the time spent preparing for a test and the score on the test for 15 students.

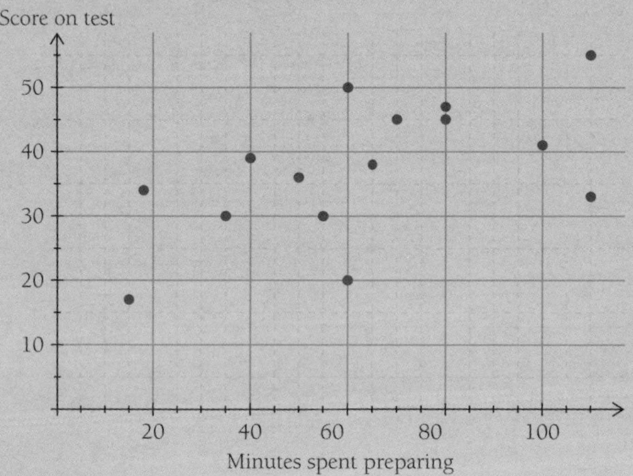

Minutes spent preparing	35	65	110	60	40	80	70	110	18	15	100	60	55	80	50
Score on test	30	38	55	20	39	47	45	33	34	17	41	50	30	45	36

a. Use the data to find the equation of the three-median regression line; you will need to sort the data by the independent variable (minutes spent preparing). Write the equation in terms of the variable.

b. Construct this equation from part **a.** on the scatterplot.

6. The number of houses sold, and the distance traveled in a month, has been recorded for a group of real estate agents. The data is graphed below.

a. Draw a three-median regression line on the scatterplot.

b. Find the equation of your line expressing it in terms of the variables.

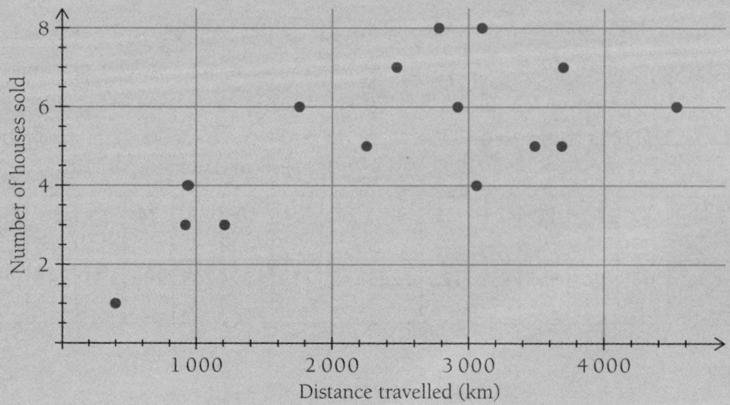

7. A three-median regression line is to be applied to the data graphed below:

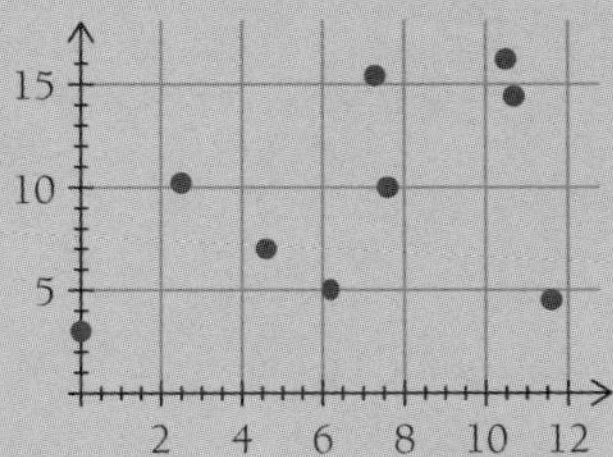

Which one of the following shows the correct line?

A.

B.

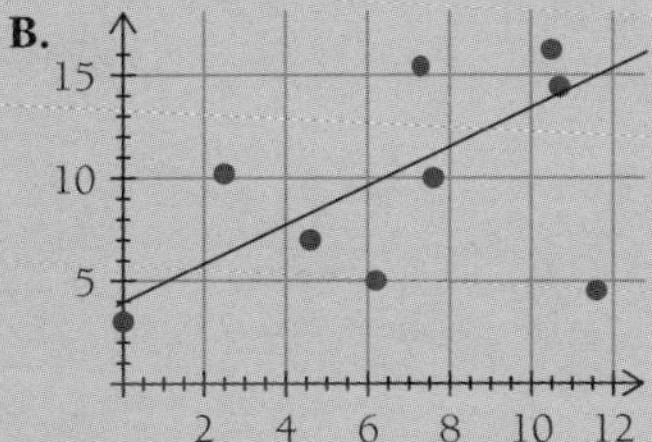

C.

D.

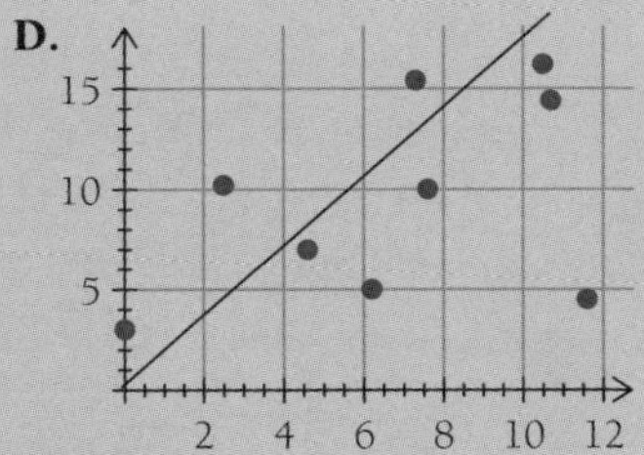

E.

Unit 12.3 Probability and Statistics

Topic 9: Statistics – predicting using a regression line

Topic 9 concludes our study of regression with the focus on interpreting regression in the context of a problem.

Predicting using a regression line

The data below gives the *fat* content (**grams**) and the *energy* (kilojoules) of 17 different foods:

Fat (g)	15	55	18	45	17	24	30	30	30	16	11	9	30	24	24	30	32
Energy (kJ)	1 255	3 555	1 800	1 880	1 670	2 520	2 300	2 300	2 090	1 340	1 130	1 150	2 300	1 670	1 670	2 510	1 460

A scatterplot of the data is drawn below. An inbuilt program has drawn the least-squares regression line on the graph and also given the equation to the line.

The equation of the regression line is: *Energy* = 833.57 + 41.885 × *Fat*

We will round the figures in this equation: *Energy* = 834 + 42 × *Fat*

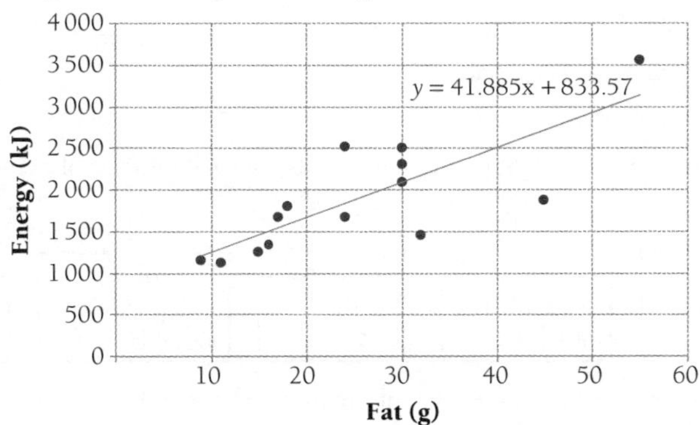

We could use this equation to **predict** the Energy for a given value of Fat. To predict the Energy when Fat = 40 we substitute *Fat* = 40 into the equation:

Energy = 834 + 42 × 10 = 2 514 kilojoules.

Checking this on the graph it looks the correct value.

We could also use the equation to predict the *Fat* content when the *Energy* of a food is 2 000 kJ:

Substituting the value *Energy* = 2 000 into the equation *Energy* = 834 + 42 × *Fat*

2 000 = 834 + 42 × *Fat*

$$Fat = \frac{2\,000 - 834}{42} = 27.8 \text{ g}$$

Interpolation and extrapolation

Interpolation means predicting values from a regression model (equation) for values from *within* the **range** of data from which the regression equation was based. In the example above the *Fat* data ranged from 9 to 55. Using a value *within* this range to predict an *Energy* value would be a case of interpolation.

When we predicted using the equation, that the energy content of a food containing 40g of fat was 2 514 kilojoules this was a case of **interpolation**.

Extrapolation means predicting values from a regression model (equation) for values from outside the range of data from which the regression equation was based. In the example above the *Fat* data ranged from 9 to 55. Using a value outside this range to predict an *Energy* value would be a case of extrapolation.

Predicting, using the equation, the energy content of a food containing, say, 60 g of fat or 5 g of fat would both be cases of extrapolation.

Note: Extrapolation too far beyond the data on which the equation was based may not be reliable.

Interpreting the gradient (slope) and intercept of a regression line

For the example above, the regression equation is:

$$Energy = 834 + 42 \times Fat$$

in which the gradient or slope is the value 42.

The slope can be interpreted as:

'For every increase of one gram of *fat* there is an increase of 42 kilojoules of *energy*'.

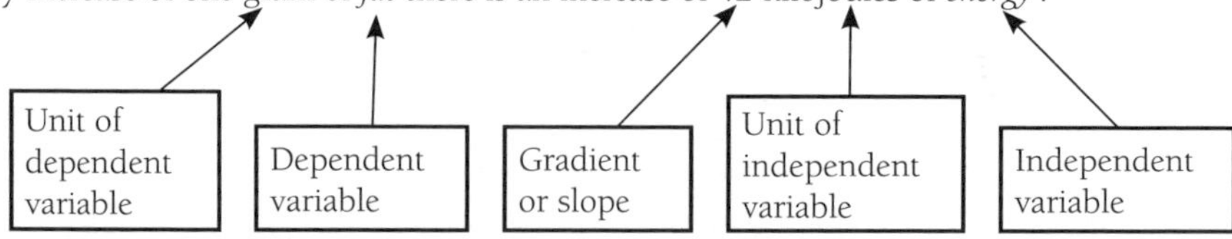

The intercept has the value 834. This can be interpreted as: 'When the fat content of a food is **zero** then the energy provided by the food is 834 kilojoules'.

This interpretation seems sensible as zero fat is a reasonable concept and the point (0, 834) is not too far removed from the data set.

Unit 12.3 Activity 9: Predicting using the regression line

1. A set of data is graphed at right and the equation of a regression line is given.

$y = 5.04 + 0.21x$

a. Use the regression equation to predict:

i. y when $x = 29$.

ii. x when $y = 11$.

b. Are both predictions from part **a.** examples of interpolation or extrapolation?

c. Interpret the slope of the regression line.

d. Interpret the intercept of the regression line.

2. A set of data is graphed at right and the equation of a regression line is given.

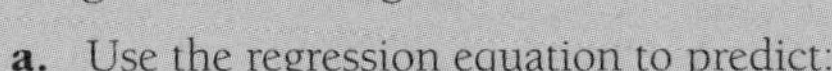

a. Use the regression equation to predict:

i. *Pulse rate* when *age* = 11.

ii. *Age* when *pulse rate* = 70.

b. If we were to predict the *pulse rate* of a 4-year-old, would this be an example of interpolation or extrapolation?

c. Interpret the slope of the regression line.

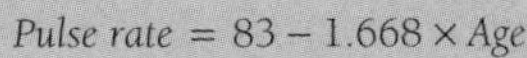

Pulse rate = 83 – 1.668 × *Age*

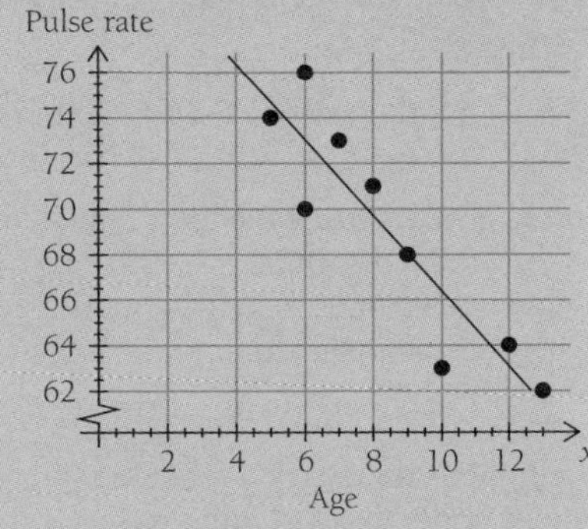

3. Susan is investigating the association between age and body mass index (BMI) and her data is plotted on the scatterplot at right. She has fitted a regression line to the scatterplot.

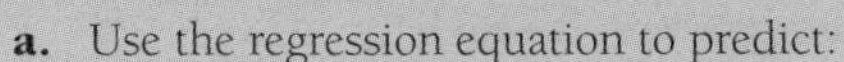

a. Use the regression equation to predict:

i. *BMI* when *Age* = 45.

ii. *Age* when *BMI* = 27.

b. Is a prediction of the *BMI* of a 70-year-old an example of interpolation or extrapolation?

c. Interpret the slope of the regression line.

d. Interpret the intercept of the regression line. Is this a sensible interpretation?

BMI = 22.4 + 0.1045 × *Age*

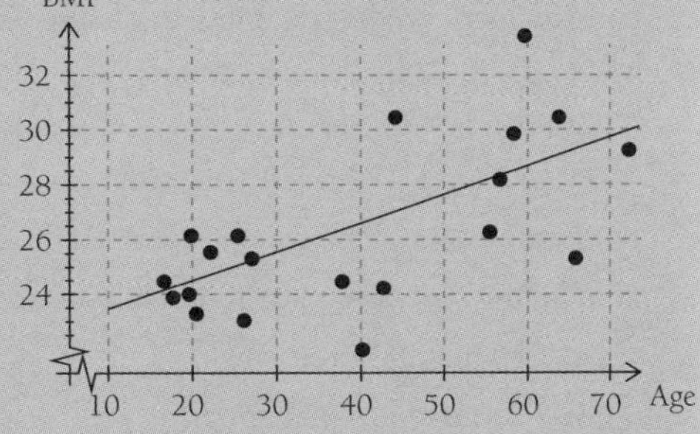

4. The Advanced Mathematics marks and the Physics marks for six students are graphed on the scatter diagram at right. Using this data a regression line with the equation:

Physics mark = 1.23 + 0.814 × *Advanced Maths mark*

is established and drawn on the graph.

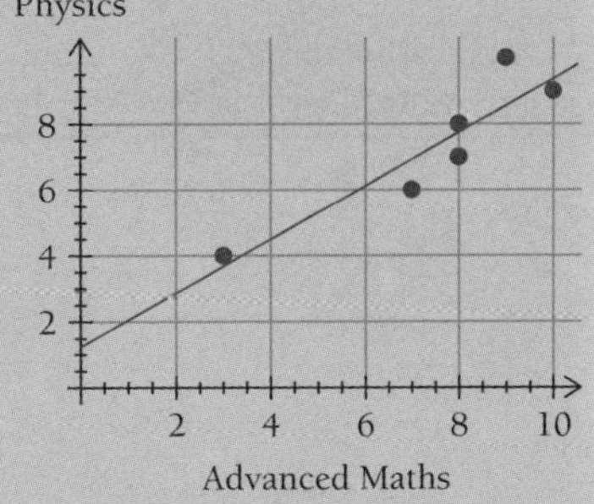

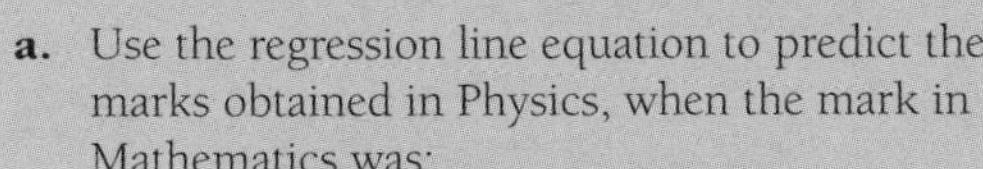

a. Use the regression line equation to predict the marks obtained in Physics, when the mark in Mathematics was:

i. 4 **ii.** 9

b. Use the regression line equation to predict the mark obtained in Maths when the mark in Physics was:

i. 2 **ii.** 6

c. Which of the predictions from parts **a.** and **b.** above are cases of extrapolation?

5. The Economics mark and the Accounting mark for nine students is graphed on the scatter diagram below. A regression line has been fitted to the data and this is also shown on the graph. The equation of the regression line is:

   Accounting mark = 36.24 + 0.57 × *Economics* mark

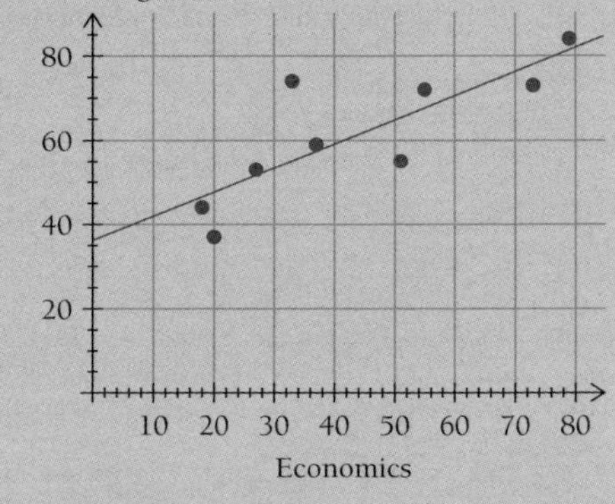

   a. Use the regression line equation to predict the marks obtained in Accounting, when the mark in Economics was:

   i. 40 **ii.** 70

   b. Use the regression line equation to predict the mark obtained in Economics when the mark in Accounting was:

   i. 65 **ii.** 20

   c. Comment on the suitability of the predictions in parts **a.** and **b.**

6. The equation of the least-squares regression line for the data in Unit 12.3 Activity 8, question 5 is:

   Score on test = 24.5 + 0.203 × *Minutes spent preparing*

   a. Interpret the slope and intercept of this equation in terms of the variables.

   b. Use this regression line to predict the score on a test for a student who spends:

   i. 40 minutes preparing.

   ii. 100 minutes preparing.

   c. A student spent 40 minutes preparing for the test and received a score of 39. Find the percentage that this actual score differs from the prediction.

   d. A student spent 100 minutes preparing for the test and received a score of 41. Find the percentage that this actual score differs from the prediction.

Unit 12.4 Algebra and Graphs

Topic 1: Equations – linear equations and inequations

Unit 12.4 (Syllabus p. 25) focuses on interpreting linear equations, quadratic equations and curves, particularly hyperbola, exponential, logarithmic and inequality graphs. Topic 1 involves algebraic methods and solving equations, and covers:

- Solving linear equations such as $5x + 12 = 3 + 2x$.
- Forming and solving linear equations or inequations.
- Modelling by forming and solving appropriate equations.
- Demonstrating algebraic manipulation skills and correct mathematical statements in a well-reasoned, logical presentation.
- Interpretation in context.

Introduction

An **equation** is a **mathematical sentence** containing an equals sign. An **algebraic equation** has at least one **variable**.

In a **linear equation**, the power of the variable is 1.

Example A

The equation $x + 4 = 6$ is linear because the power of the variable x is 1. (x^1 is usually written as x.)

The equation $x^2 + x - 20 = 0$ is *not* a linear equation because it contains x^2 (ie x to the power of 2, not 1).

Solving linear equations

Solving a linear equation means *finding the value of the variable* which makes the equation true. If $x = 2$ is a **solution** of the equation, then $x = 2$ is said to **satisfy** the equation. $x = 2$ is also called the **root** of the equation.

Example B

The solution to the equation $x + 4 = 6$ is $x = 2$, because when $x = 2$ the equation is true, ie $2 + 4 = 6$.

To solve a linear equation, the equation is rearranged so that the *variable is on one side of the equals sign and a number is on the other*. **Rearranging** is done by applying the same **operations** to both sides of the equation in order to isolate the variable.

Example C

Q. Solve the following equations: **1.** $4x = 20$ **2.** $2x + 3 = 5$ **3.** $3x - 2 = 8$

A. **1.** $4x = 20$

$$\frac{4x}{4} = \frac{20}{4} \quad \text{[dividing every term by 4]}$$

$$x = 5$$

2. $2x + 3 = 5$

$2x = 2$ [subtracting 3 from both sides of the equals sign]

$x = 1$ [dividing both sides by 2]

3. $3x - 2 = 8$

$3x = 10$ [adding 2 to both sides of the equals sign]

$x = \frac{10}{3}$ or $3\frac{1}{3}$ [dividing both sides by 3]

Where brackets occur in an equation they should be removed as soon as possible by **expanding**.

Example D

Q. Solve: **1.** $3(x - 4) + 2 = 8$. **2.** $5(x - 3) - 2x = -1$

A. 1. $3(x - 4) + 2 = 8$

$3x - 12 + 2 = 8$ [expanding brackets]

$3x - 10 = 8$ [simplifying]

$3x = 18$ [adding 10 to each side]

$x = 6$ [dividing both sides by 3]

2. $5(x - 3) - 2x = -1$

$5x - 15 - 2x = -1$ [expanding brackets]

$3x - 15 = -1$ [simplifying]

$3x = 14$ [adding 15 to both sides]

$x = \frac{14}{3}$ or $4\frac{2}{3}$ [dividing both sides by 3]

When fractions are involved in an equation they should be removed as soon as possible. This is done by multiplying *every* term in the equation by the **lowest common multiple** of the denominators in the fractions.

Example E

Q. Solve: **1.** $1 + \frac{x}{3} = 5$ **2.** $\frac{2x}{3} + \frac{3}{5} = 4$

A. 1. $1 + \frac{x}{3} = 5$

$1 \times 3 + \frac{x}{3} \times \frac{3}{1} = 5 \times 3$ [multiplying every term by 3 to remove fractions]

$3 + x = 15$ [simplifying]

$x = 12$ [subtracting 3 from both sides of equals sign]

2. $\frac{2x}{3} + \frac{3}{5} = 4$

$\frac{15}{1} \times \frac{2x}{3} + \frac{15}{1} \times \frac{3}{5} = 15 \times 4$ [multiplying every term by 15, the lowest common multiple of 3 and 5]

$10x + 9 = 60$ [simplifying]

$10x = 51$ [subtracting 9 from both sides]

$x = 5.1$ [dividing both sides by 10]

In some linear equations the variable appears on both sides of the equals sign. To solve equations of this type, rearrange the equation so that the terms containing the variable are on one side of the equals sign and all other terms are on the other side.

Example F

Q. Solve the equation $4x - 5 = 2x + 7$.

A.

$4x - 5$	$= 2x + 7$	
$4x - 5 - 2x$	$= 2x - 2x + 7$	[subtracting $2x$ from each side]
$2x - 5$	$= 7$	[simplifying]
$2x$	$= 12$	[adding 5 to each side]
x	$= 6$	[dividing each side by 2]

Note: Always check your solution by **substituting** your answer into the original equation. If the resulting sentence is false, you have made a mistake. In this example,

$x = 6$ gives LHS $= 4 \times 6 - 5 = 19$
and RHS $= 2 \times 6 + 7 = 19$

Since the two sides agree, the answer is correct.

In the following example, the equation is rearranged so that the variable is on the right-hand side of the equals sign. This avoids division by a negative number.

Example G

Q. Solve $5(2 - x) = 7(2x + 4)$.

A.

$5(2 - x)$	$= 7(2x + 4)$	
$10 - 5x$	$= 14x + 28$	[expanding]
10	$= 19x + 28$	[adding $5x$ to both sides]
-18	$= 19x$	[subtracting 28 from both sides]
x	$= \frac{-18}{19}$	[dividing by 19]

Unit 12.4 Activity 1A: Linear equations

Solve each of the following:

1. $x + 3 = 8$ **2.** $x - 8 = 7$ **3.** $x + 9 = 14$

4. $4x = 36$ **5.** $6x = -42$ **6.** $5x = 8.4$

7. $\frac{x}{3} = 7$ **8.** $\frac{x}{8} = -5$ **9.** $\frac{x}{0.4} = 4$

10. $x - 3.4 = 8.7$ **11.** $0.7x = 4.2$ **12.** $\frac{x}{1.5} = 2.4$

13. $2x + 3x = 40$ **14.** $11x - 7x = 30$ **15.** $4x + 1.5x = 110$

16. $7.6x + 11.9x = 351$ **17.** $11.3x - 7.8x = 28$ **18.** $3.6x - 4.7x = 3.3$

19. $3x + 5 = 36$ **20.** $5x + 17 = 7$ **21.** $4x - 11 = 47$

22. $11 - 4x = 24$ **23.** $2x + 7.9 = 12.15$ **24.** $1.5x - 11.5 = 86$

25. $\frac{x}{3} - 5 = 7$ **26.** $\frac{x}{2} + 1 = 13$ **27.** $\frac{x}{3} + 15 = 12$

28. $\frac{3x}{4} + 5 = 17$ **29.** $\frac{5x}{6} - 3 = 9$ **30.** $\frac{2x}{3} - 7 = 8$

31. $4(x + 3) = 32$ **32.** $6(x - 2) = 18$ **33.** $5(2x + 3) = 48$

34. $\frac{3x - 1}{4} = 2$ **35.** $\frac{5x + 1}{4} = 9$ **36.** $\frac{2 - x}{3} = 11$

37. $5(x + 1) + x = 5x + 3$ **38.** $3(x - 5) = 3 - 2x$ **39.** $14x - 8 = 64 - 2x$

40. $1.4x + 2 = 1.5x + 1.8$ **41.** $3(x - 4) = x + 5$ **42.** $\frac{2(x - 4)}{3} = 12$

43. $3(x - 4) = 2(x + 5)$ **44.** $5(x - 3) = 2(x + 7)$ **45.** $4(x + 6) = 5(2x + 11)$

46. $7 - 3(x - 5) = 15$ **47.** $9 + 4(2x - 3) = 77$ **48.** $10 - 2(3x - 7) = 18$

49. $\frac{x - 2}{3} + \frac{2x + 1}{4} = 1$ **50.** $\frac{2x - 3}{2} - \frac{5(x - 1)}{3} = -4$

Note: In all remaining examples, reference to 'both sides' will be omitted.

Example D

Q. Solve the following equations:

1. $\frac{5}{3x + 1} = 2$ **2.** $2\frac{1}{2}(x - 1) - \frac{x + 3}{3} = 4$

A. **1.** $\frac{5}{3x + 1} = 2$

$\therefore\ 5 = 2(3x + 1)$ [multiplying by $(3x + 1)$]

$\therefore\ 5 = 6x + 2$ [expanding]

$\therefore\ 3 = 6x$ [subtracting 2]

$\therefore\ 6x = 3$ [swapping sides]

$\therefore\ x = \frac{3}{6}$ [dividing by 6]

$\therefore\ x = \frac{1}{2}$

2. $2\frac{1}{2}(x - 1) - \frac{x + 3}{3} = 4$

$5(x - 1) - \frac{2x + 6}{3} = 8$ [multiplying by 2]

$\therefore\ 15(x - 1) - (2x + 6) = 24$ [multiplying by 3]

$\therefore\ 15x - 15 - 2x - 6 = 24$ [expanding]

$\therefore\ 13x - 21 = 24$ [simplifying]

$\therefore\ 13x = 45$ [adding 21]

$\therefore\ x = \frac{45}{13}$ [dividing by 13]

$\therefore\ x = 3\frac{6}{13}$

Note: In part **2.** the two steps 'multiplying by 2' and 'multiplying by 3' could be done in one step 'multiplying by 6'.

Unit 12.4 Activity 1B: More advanced linear equations

1. Solve:

a. $5z + 3z + 11 = 12$
b. $3(L + 2) + 5L = 27$
c. $7(2x + 1) + 5(x - 3) = 12$
d. $9(T + 3) - 5(T + 4) = 8$
e. $7 - 2(3 - 2R) = 38$
f. $5(2T + 3) = 2T + 4$
g. $4(2L + 1) = 3(4 - L)$
h. $\frac{1}{2}(3R + 1) = 4$
i. $5(3R + 1) + 3(2R + 7) = 11R + 42$
j. $5T + 3(T + 2) = 2(T + 1)$
k. $4P - 3(P - 4) = 5(P + 11) + 8$
l. $4(P + 1) + 2(2P + 3) = 4(P + 2) - 3(P - 1)$

2. Solve:

a. $\frac{x}{3} + \frac{x}{4} = 2$
b. $\frac{2T}{3} - \frac{T}{4} = \frac{1}{6}$
c. $\frac{3R+1}{4} + \frac{2R-3}{2} = 4$
d. $\frac{3R+1}{2} = 2R$
e. $\frac{P+1}{3} + \frac{P-3}{2} = \frac{P+4}{6}$
f. $\frac{2P+1}{4} - \frac{P-1}{3} = \frac{1}{6}$
g. $\frac{3}{4}(A + 2) = \frac{2}{3}(A - 3)$
h. $\frac{2(A+1)}{3} = \frac{A-1}{5}$
i. $\frac{3}{5}(A + 1) + \frac{2}{3}(A - 1) = 2$
j. $\frac{A+1}{A} = 2$
k. $\frac{3A-1}{2A+5} = \frac{1}{2}$
l. $3 - 2(A - 3) = 4A - 3$

Word problems

Problems given in words (**word problems**) need to be changed to mathematical equations to be solved. This process is called **mathematical modelling**. The general approach is illustrated by the following example.

Example G

Q. Shane is taller than Jason by 2.4 cm. Jason is taller than Ian by 1.3 cm. The three heights total 452 cm. What are their heights?

A. Let Ian's height be h

$\therefore$ Jason's height is $h + 1.3$ [Jason is 1.3 cm taller than Ian]

Shane's height is $h + 1.3 + 2.4 = h + 3.7$ [Shane is 2.4 cm taller than Jason]

Since their combined height is 452 cm

$$h + h + 1.3 + h + 3.7 = 452$$

$$\therefore\ 3h + 5 = 452 \quad \text{[simplifying]}$$

$$\therefore\ h = 149 \quad \text{[subtracting 5 and dividing by 3]}$$

Thus Ian is 149 cm tall; Jason's height is $149 + 1.3 = 150.3$ cm and Shane's height is $149 + 1.3 + 2.4 = 152.7$ cm.

Unit 12.4 Activity 1C: Solving word problems

For each of the following, find: **a.** an equation; **b.** a solution using your equation or any other method.

1. Notebooks cost K1 more than pens. Sandra purchases 5 pens and 6 notebooks. Her total expenditure is K32.40. How much does each pen cost?

2. Tom has some money. Brian and David each have K1 more than Tom, and Paul, Dan and Nicolas each have twice as much as Tom. Together, the six boys have K15.50. How much does each have?

3. A man has three children who each want a calculator for Christmas. He buys the two older children the same model and the youngest child gets a basic model costing K20 less. What is the value of each type of calculator, if the man spends a total of K86.50 on the calculators?

4. The perimeter of a room is 50 m. The length exceeds the width by 3.3 m. Find the length of the room.

5. Four consecutive **natural numbers** have a sum of 110. What is the smallest of these numbers?

6. A girl purchases a number of items each costing K17.35. Her bank account was K1 000.00 before she made the purchase. Afterwards it was K687.70. How many items did she purchase?

7. Tina and Jane purchase raffle tickets and agree to share the prize money in the ratio of how much each paid. They win K117.00 and Jane, who put in K2.00, gets K26.00. What was Tina's share of the cost of the raffle tickets?

8. The distance from B to C is $\frac{7}{11}$ of the distance from A to B. The distance from A to B exceeds that from B to C by 20 km. How far is it from A to B?

9. A man purchases a number of tools at K5 each. Unfortunately, five do not work, but by selling those that do at K12 each he is able to make K45 profit. How many tools did the man purchase initially?

10. All prices are to rise by 15% as a result of a new tax. A shopkeeper had increased the cost of a particular item by K5.00 before the new tax came into effect. As a result of the second price rise, this item now costs K13.80. What did it cost originally?

11. Inflation is running at a certain percentage. After one year, an item which cost K36.00 costs K40.86. What is the inflation rate?

12. The denominator of a fraction is 44 more than the numerator. The fraction, when simplified, is $\frac{13}{17}$. Find the numerator.

13. My son is 31 years younger than I am. In one year's time my son's age will be one quarter of my current age. How old am I?

14. One girl has four times as much money as her friend. She gives her friend K12.00, and as a result they now have the same amount of money. How much did each have originally?

15. A husband and his wife decide to invest in a business. The first time they do this, he invests K3 000.00 and she invests K5 000.00. The next time they invest, they each contribute the same amount. After the two investments are made, the ratio of the husband's investment to his wife's is 13:17. How much did each invest the second time?

Linear inequations

An **inequation** is a mathematical statement that has one of the following four **inequality symbols** in place of an equals sign:

Symbol	Means	Example
$>$	is greater than	$3 > 2$
$\geq$	is greater than or equal to	$-2 \geq -4$
$<$	is less than	$-3 < 3$
$\leq$	is less than or equal to	$5 \leq 5$

A linear inequation has a variable to the power of 1. For example, $3x + 2 \leq 5$ and $x - 7 > 14$ are both linear inequations.

Solving linear inequations

Inequations are solved by finding those values of the variable which make the inequation true.

The procedure for solving linear inequations is the same as that followed for solving a linear equation, with the following exception:

> The inequality sign is *reversed* when multiplying or dividing an inequation by a *negative*.

Example H

Q. Solve each of the following inequations where x is an integer:

1. $3x + 2 \leq 5$ **2.** $4x - 3 > 2x + 7$ **3.** $5 - 3x < 8(x + 3)$

A. 1. $3x + 2 \leq 5$

$3x \leq 3$ [subtracting 2 from both sides]

$x \leq 1$ [dividing both sides by 3]

$\Rightarrow x = \ldots, -2, -1, 0, 1$ since x is an integer

2. $4x - 3 > 2x + 7$

$2x - 3 > 7$ [subtracting $2x$ from both sides]

$2x > 10$ [adding 3 to both sides]

$x > 5$ [dividing both sides by 2]

$\Rightarrow x = 6, 7, 8, \ldots$ since x is an integer.

3. $5 - 3x < 8(x + 3)$

$5 - 3x < 8x + 24$ [expanding brackets]

$5 < 11x + 24$ [adding $3x$ to both sides]

$-19 < 11x$ [subtracting 24 from both sides]

$\frac{-19}{11} < x$ [dividing both sides by 11]

$x > -1\frac{8}{11}$ [re–expressing the inequality]

$\Rightarrow x = -1, 0, 1, 2, \ldots$ since x is an integer.

Note: If the terms in x had been collected on the LHS, the inequation $-11x < 19$ would have resulted. Dividing by -11 (the coefficient of x) would reverse the inequality sign giving $x > \frac{19}{-11}$, the same result.

If an inequation in x is to be solved for real values of x, then the solution is the inequation reduced to its **simplest form** (the values of x cannot be listed). For example, solving $3x + 2 \le 5$ for real values of x gives the solution $x \le 1$. Thus x is any **real number** less than or equal to 1.

Unit 12.4 Activity 1D: Linear inequations

1. Solve each of the following inequations where x is an integer. List the x values.

a. $x + 4 < 9$ **b.** $x - 3 > 10$ **c.** $2x - 5 \le 10$ **d.** $4 - x \ge 5$

e. $2x + 3 > 11$ **f.** $8 - 2x \ge 4$

2. Solve each of the following inequations where x is a real number.

a. $5 - 2x \ge 17$ **b.** $7 - 3x < 1$ **c.** $12 - 5x \le 22$

d. $2x + 3 > x - 4$ **e.** $3x - 2 \le 5x + 6$ **f.** $8x + 2 < 2x - 5$

g. $\frac{2 - 3x}{4} < 2$ **h.** $\frac{x + 3}{2} - \frac{x - 1}{3} \ge 1$ **i.** $\frac{x + 3}{2} < \frac{x - 1}{3} - 1$

j. $1\frac{1}{3}(x - 2) - \frac{1}{2}(x + 3) \ge x + 3$ **k.** $\frac{a}{5} - \frac{2}{3}(1 - a) \ge \frac{a + 1}{2}$ **l.** $\frac{2x + 3}{3} < \frac{3x - 1}{2}$

m. $\frac{4x - 1}{3} \ge \frac{x + 1}{7} - 1$

3. Find the lowest integer such that $5x + 3 > 2x - 11$.

4. Find the highest integer such that $6x - 2 > 8x + 1$.

5. Find all values of x so that $3x < 14$ and $x - 2 > 0$, where x is an integer.

6. Jenni is buying Christmas cards to send her friends. She has K40 to spend and she wants to send out 12 cards. There are two types of card available:

'Good Cheer' costing K3.50 each and 'Tinsel Times' costing K2.00 each.

Jenni wants to buy as many of the 'Good Cheer' cards as she can. Solve the inequality to find the maximum number of 'Good Cheer' cards Jenni can buy, where x is the number of 'Good Cheer' cards:

$3.5x + (12 - x) \times \le 40.$

Unit 12.4 Algebra and Graphs

Topic 2: Equations – choosing formulae

The focus of Unit 12.4 (Syllabus p. 25) is on interpreting linear equations, quadratic equations and curves. The application of mathematical skills to exponential and logarithmic functions is emphasised.

In Topic 2 we manipulate algebraic expressions to solve equations. We cover:

- Choosing algebraic techniques and strategies to solve problems.
- Solving word problems involving linear and quadratic expressions.

Mathematical modelling

Mathematical modelling involves expressing a relationship as a **formula** in which mathematical symbols and letters are used to represent variables. The ability to do this accurately and use the resulting formula is important in many situations.

Example A

Three quantities used in accounting are assets (A), proprietorship (P), and liabilities (L). The relationship between these quantities is that assets are equal to the sum of proprietorship and liabilities. In mathematical form, the relationship can be expressed: $A = P + L$.

Note: A, P and L are variables, and the equation, $A = P + L$ is a formula for A in terms of the two variables, P and L.

Example B

Q. The volume of a cone is one third of the product of its height and the area of its base. Express this relationship in mathematical form.

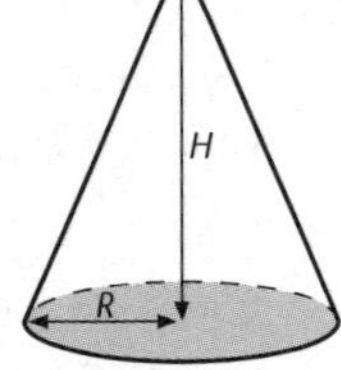

A. Let the volume of the cone be V, the radius of the base be R, and the height be H.

The base area is πR^2 [formula for area of circle]

$\therefore$ the volume is $V = \frac{1}{3}\pi R^2 H$ [from information given]

Example C

Q. A trapezium is a quadrilateral with a pair of parallel sides.

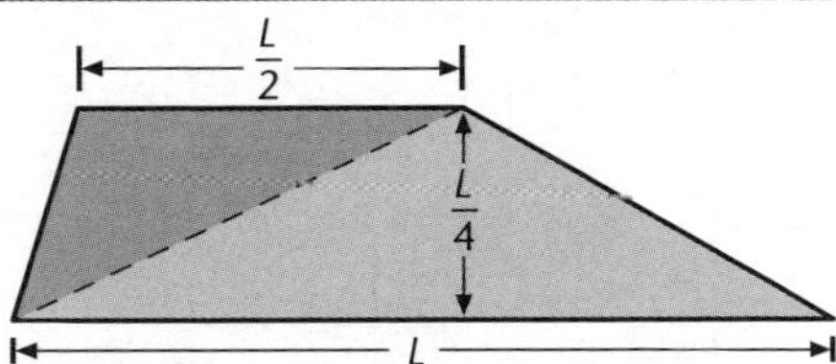

Find an expression for the area of the trapezium shown where one parallel side is one half the length of the other parallel side, and the distance between the two parallel sides is equal to one quarter of the length of the longer parallel side. Express the area in terms of the length of the longer parallel side.

A. Let the length of the longer parallel side be L and the area be A. The area of the trapezium is the sum of the areas of the two shaded triangles (as shown on the diagram, where the shorter parallel side is $\frac{L}{2}$ and the height is $\frac{L}{4}$).

Area of the smaller triangle is $\frac{1}{2} \times \frac{L}{2} \times \frac{L}{4} = \frac{L^2}{16}$ [area is half × base × height]

Area of the larger triangle is $\frac{1}{2} \times L \times \frac{L}{4} = \frac{L^2}{8}$

$\therefore$ area of the trapezium is $A = \frac{L^2}{8} + \frac{L^2}{16} = \frac{3L^2}{16}$ [adding fractions]

Unit 12.4 Activity 2A: Writing formulae

1. A pair of socks costs KD. Write an expression for the cost of the socks in toea.

2. a. Express the time of $3x$ minutes in hours.

b. Express this same time in seconds.

3. Peter weighs x kilograms and y grams.

a. Express Peter's weight in grams. **b.** Express his weight in kilograms.

4. A rectangle has length L metres and width W centimetres. Write an expression for:

a. the area in cm^2. **b.** the area in m^2. **c.** the perimeter in mm.

5. A square has area $4L^2$ in square metres. Write an expression for:

a. the perimeter in cm. **b.** the perimeter in mm.

6. A right-angled triangle has base b and height a. Both have the same unit.

a. What is the area of the triangle?

b. What is the length of the hypotenuse (the third side)?

7. a. Find an expression for the volume of the cylinder shown.

b. Find an expression for the total surface area of the cylinder.

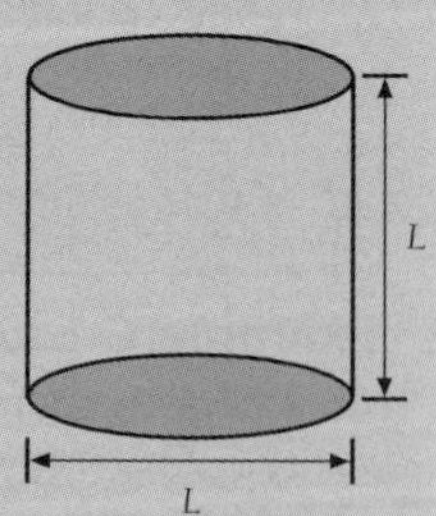

8. **a.** Find an expression for the perimeter of the shape shown.

b. Find a formula for the area of this shape.

9. a. What is the perimeter of the shape shown?

b. What is the volume of the box which forms when the shape is folded along the dotted lines?

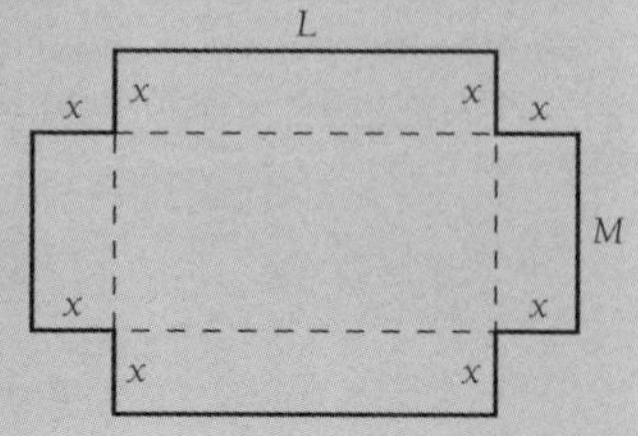

10. The length of the closed box shown is $4L$. The height and width of the box are both half the length.

a. Write down an expression for the volume of the box.

b. Write down an expression for the surface area of the box.

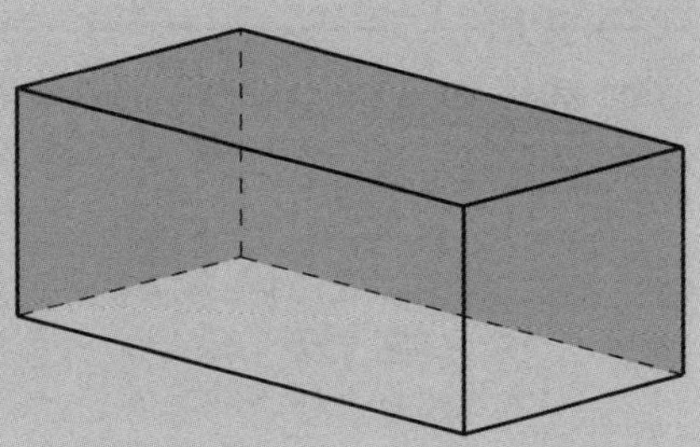

Changing the subject of a formula

The **subject** of a formula is a variable which appears *alone* on one side (usually the left side) of the equals sign. The subject is expressed in terms of the other variables. **Changing the subject** of a formula means **rearranging** the formula (using **inverse operations**) so that another variable becomes the subject.

Example D

Q. Make x the subject of $y = 2x - 6$.

A.

$y = 2x - 6$	
$\therefore y + 6 = 2x$	[adding 6 to both sides]
$\therefore 2x = y + 6$	[swapping sides]
$\therefore x = \dfrac{y+6}{2}$	[dividing both sides by 2]

Sometimes the variable to be made subject appears more than once in the formula. You will need to collect terms in this variable on one side and then **factorise**.

Example E

Q. Make T the subject of $\dfrac{AT + D}{T} = E$.

A.

$\dfrac{AT + D}{T} = E$	
$\therefore AT + D = ET$	[multiplying both sides by T]
$\therefore ET = AT + D$	[swapping sides]
$\therefore ET - AT = D$	[collecting terms in T on one side]
$\therefore T(F - A) = D$	[factorising $ET - AT$]
$\therefore T = \dfrac{D}{E - A}$	[dividing by $E - A$ (providing $E - A \neq 0$)]
Note: $D = 0$ if $E - A = 0$	[substituting $E - A = 0$ in $T(E - A) = D$]

Squaring and taking the **square root** are inverses. Take care with **signs** when taking a square root (if $x^2 = y$ then $x = \pm\sqrt{y}$).

Example F

Q. Make M the subject of $\frac{AM^2 - B}{D} = M^2 + E$ when $A \neq D$.

A. $\frac{AM^2 - B}{D} = M^2 + E$

$\therefore AM^2 - B = DM^2 + ED$ [multiplying by D]

$\therefore AM^2 = DM^2 + ED + B$ [adding B]

$\therefore AM^2 - DM^2 = ED + B$ [subtracting DM^2]

$\therefore M^2(A - D) = ED + B$ [factorising $AM^2 - DM^2$]

$\therefore M^2 = \frac{ED + B}{A - D}$ [dividing by $A - D$ (where $A \neq D$)]

$\therefore M = \pm\sqrt{\frac{ED + B}{A - D}}$ [taking the positive or negative square roots]

Note: Once again these are restrictions on the values the variables can take. Division by zero is undefined, so $A - D \neq 0$. Also taking the square root of a negative quantity is not defined for real numbers, so $\frac{ED + B}{A - D} \geq 0$.

Unit 12.4 Activity 2B: Changing the subject of formulae

Make M the subject of each of the following formulae:

1. $3M = T$ **2.** $AM = T$ **3.** $\frac{M}{P} = T$

4. $\frac{M}{P} = P$ **5.** $\frac{M}{P} = A + B$ **6.** $\frac{2M}{P} = 5$

7. $\frac{5M}{T} = T$ **8.** $M + B = A$ **9.** $M - B = A$

10. $M - B = B$ **11.** $M - B - B^2 = B^2 + 2B$ **12.** $3M - B^2 = 2B^2$

13. $\frac{M}{AB} = BC$ **14.** $\frac{M}{B} - 3 = C$ **15.** $\frac{M}{A} = \frac{C}{B}$

16. $M^2 = A$ **17.** $CM^2 = A + B$ **18.** $\frac{AM - D}{B} = B$

19. $\frac{AM - DM}{B} = B$ **20.** $\frac{M}{A} + \frac{M}{B} = C$ **21.** $\frac{M - A}{M + D} = E$

22. $\sqrt{M} = C$ **23.** $\frac{\sqrt{M}}{C} = D$ **24.** $\frac{2A}{3}\sqrt{\frac{M}{G}} = L$

25. $\frac{A}{C}\sqrt{\frac{G}{M}} = X$ **26.** $\frac{3M^2 - D}{M} = AM$ **27.** $\frac{A\sqrt{M} - D}{B} = \sqrt{M}$

28. $\frac{A}{\sqrt{M}} + \frac{C}{\sqrt{M}} = E$ **29.** $y = \frac{AM - 5}{BM - 4}$ **30.** $y = \frac{3M^3 - E}{4M^3 - F}$

Using formulae

Formulae need to be used accurately. Using formulae often means **substituting** numbers for some of the variables and solving the equation to find the value of an unknown variable.

Example G

Q. From Example B, the volume of a cone is $\frac{1}{3}\pi R^2 H$,

where R is the radius of the base and H is the height. (In this example, π is taken to be 3.142.)

1. Find the volume when $R = 2.152$ cm and $H = 4.365$ cm.
2. Find the radius of the base when the volume is 63.25 cm^2 and the height is 5.230 cm.

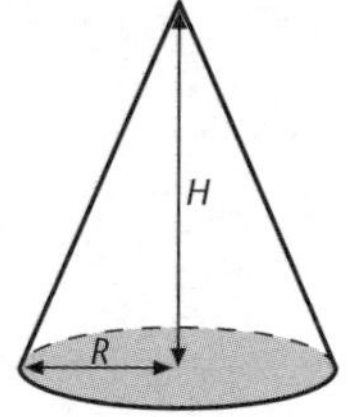

A. 1.

$$V = \frac{1}{3}\pi R^2 H$$

$$= \frac{1}{3} \times 3.142 \times (2.152)^2 \times 4.365 \qquad \text{[substituting into } V = \tfrac{1}{3}\pi R^2 H\text{]}$$

$$= 21.17 \text{ cm}^3 \text{ (2 dp)} \qquad \text{[rounding answer]}$$

2.

$$V = \frac{1}{3}\pi R^2 H$$

$$\therefore 63.25 = \frac{1}{3} \times 3.142 \times R^2 \times 5.230 \qquad \text{[substituting into } V = \tfrac{1}{3}\pi R^2 H\text{]}$$

$$\therefore R^2 = \frac{3 \times 63.25}{3.142 \times 5.230} \qquad \text{[solving for } R^2\text{]}$$

$$R = \sqrt{\frac{3 \times 63.25}{3.142 \times 5.230}} \qquad \text{[taking the positive square root since R is a length]}$$

$$= 3.40 \text{ cm (2 dp)} \qquad \text{[rounding answer]}$$

Note: *Do not* round off calculations until the very end of the calculation. **Rounding** as you progress through a calculation introduces errors at each stage of the calculation, which accumulate over a number of steps producing significant errors in the final answer.

Unit 12.4 Activity 2C: Using formulae

1. The formula for the area of the walls of a room is $A = 2h(l + w)$ where h is the height of the room, l is the length and w the width.

- **a.** Find the area if $h = 2.3$ m, $l = 5.3$ m, $w = 4.5$ m.
- **b.** Find the area if $h = 3.41$ m, $l = 4.6$ m, $w = 5.83$ m.
- **c.** Find the height if the area if 60.42 m^2, the length is 4.86 m and the width is 3.92 m.
- **d.** Find the width if the area is 73.4 m^2, the height is 4.2 m and the length is 6.8 m.
- **e.** Find the length if the area is 165.6 m^2, the height is 5.3 m and the width is 4.8 m.

2. The formula for the volume of a cylinder is $V = \pi R^2 h$, where R is the radius of the cross-section, h is the height and π is taken to be 3.14.

- **a.** Find the volume if $R = 4.6$ cm, $h = 12.3$ cm.
- **b.** Find the volume if $R = 2.67$ cm, $h = 4.76$ cm.
- **c.** Find the height if the volume is 283 cm^3 and the radius is 5.12 cm.
- **d.** Find the radius if the volume is 645 cm^3 and the height is 6.32 cm.
- **e.** Find the radius if the volume is 1367 cm^3 and the height is 15.32 cm.

3. The formula for the simple interest earned by investing KP at R% per year for R years is $I = \frac{PRT}{100}$. Use a graphical calculator or spreadsheet to answer the following questions.

a. Find the principal invested if the rate of simple interest is $12\frac{1}{2}$%, the time is 3 years and the interest earned is K1 031.25.

b. Find the annual rate of interest earned if K4 700 earns K998.75 simple interest after 2.5 years.

c. Patrick borrows K7 500 at $6\frac{1}{4}$ % simple interest per year for a number of years. If he pays a total of K1 875 interest, how long was the loan?

Unit 12.4 Algebra and Graphs
Topic 3: Linear graphs

We now move on to the 'graphs and functions' section of Unit 12.4 (Syllabus p. 25) and this Topic introduces sketching and interpreting graphs. It covers:

- Sketching and interpreting features of linear equations of type $y = mx + c$, $x = a$, $y = b$.
- Sketching and interpreting linear equations written in any form.
- Writing equations for linear graphs.
- Determining and applying an appropriate model for a situation involving graphs.
- Drawing a graph to find the solution to a problem.

Introduction

This Topic involves sketching **linear graphs** (straight line graphs) and **quadratic graphs** (**parabolas**) and interpreting their features, often in a practical context. This Topic focuses on the graph of the straight line.

Plotting points on a graph

Graphs are drawn on a **cartesian plane**, which is formed by a pair of number lines called **axes**, drawn perpendicular to each other and intersecting at the **origin**, O. Points on the plane are described by **coordinates**, which are **ordered pairs** of numbers, such as (2, 3).

- The first coordinate, 2, is called the ***x*-coordinate** of the point. This describes the position of the point relative to the horizontal ***x*-axis**.
- The second coordinate, 3, is called the ***y*-coordinate** of the point. This describes the position of the point relative to the vertical ***y*-axis**.

Points are marked with a small dot, cross or circle and are labelled with a capital letter.

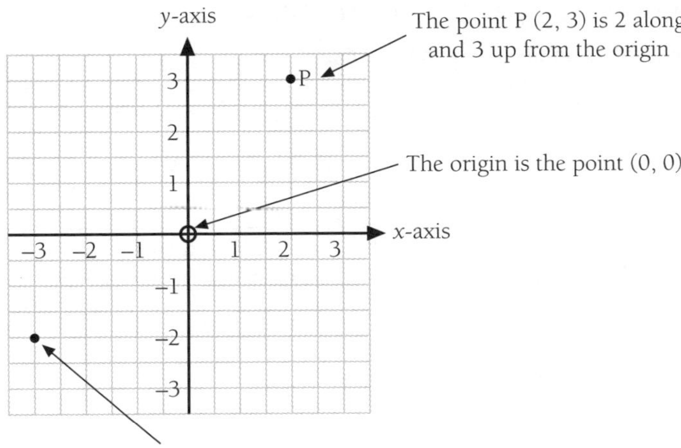

Linear graphs

A **linear graph** is a set of points lying in a straight line. These points obey a rule, which is a **linear equation** relating x and y, in which both x and y are to the power of one, or one variable may be to the power of zero.

> The general equation of a straight line is $ax + by + c = 0$, where a, b, c are numbers

A linear equation can be rearranged into several different forms, eg, $y = 2x + 5$, $x = -4$,

$y = 2$, $2x - 3y = 5$, and so on.

There are three ways of sketching the graph of a straight line – plotting points, the intercept method, and the **gradient-intercept** method.

Drawing graphs by plotting points

This is a basic method for drawing any graph whose equation is known.

1. Draw up a table with some 'sensibly' chosen x-values.
2. Substitute these x-values into the equation to find the corresponding y-values.
3. Plot the points on a graph.
4. Join the points with a straight line or a smooth curve, as appropriate.

Example A

The graph of $y = 3x + 2$ is drawn by setting up a table of values as shown and then plotting the points.

x	–2	–1	0	1	2	3	← chosen x-values
$3x$	–6	–3	0	3	6	9	← 'working'
$y = 3x + 2$	–4	–1	2	5	8	11	← y-values calculated from the x-values

The table shows that the points to plot are:
(–2, –4), (–1, –1), (0, 2), (1, 5), (2, 8) and (3, 11).

These are plotted on the graph on the right.

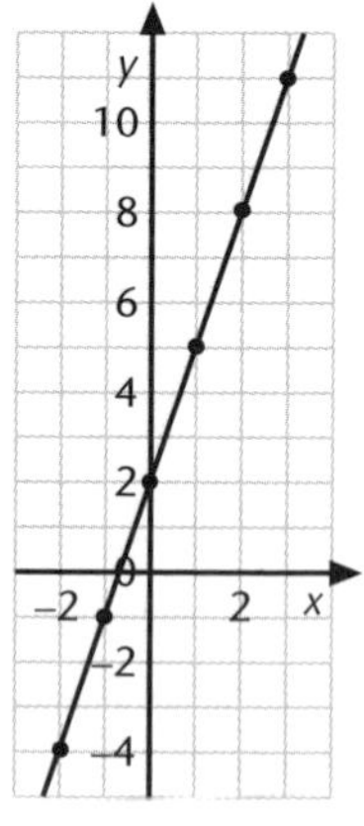

If the graph to be drawn can be recognised as a straight line, then only three points need to be plotted. Two of these points are used to draw the graph and the third point acts as a check.

Example B

Q. Draw the graph of $y + 2x = 4$.

A. Let $x = 1$, then $y + 2 \times 1 = 4$ [substituting in $y + 2x = 4$]

$y + 2 = 4$

$y = 2$ One point on the graph is (1, 2).

Let $x = 3$, then $y + 2 \times 3 = 4$

$y + 6 = 4$

$y = -2$ Another point is (3, –2).

As a check, let $x = 2$, then $y + 2 \times 2 = 4$

$y + 4 = 4$

$y = 0$ Check point is (2, 0)

The point (2, 0) should (and does) lie on the line formed by the points (1, 2) and (3, –2).

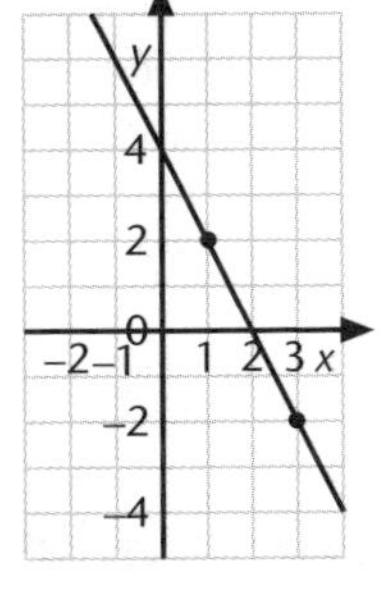

Unit 12.4 Activity 3A: Drawing graphs by plotting points

1. For each of the following, copy and complete the tables:

a. $y = 2x + 1$

x	–2	–1	0	1	2
y	–3				5

b. $y = 3x - 2$

x	–2	–1	0	1	2
y					

c. $y = -x + 2$

x	–2	–1	0	1	2
y					

d. $x + y = 5$

x	–2	–1	0	1	2
y					

e. $y = 4 - x$

x	–2	–1	0	1	2
y					

f. $x + y = 7$

x	–2	–1	0	1	2
y					

2. By finding three points which obey the rules, graph the lines:

a. $y = -x + 2$ **b.** $y = 2x - 4$ **c.** $x + y = 4$

d. $2x + y = 6$ **e.** $y - x = 1$ **f.** $3x - 4y = 12$

Intercept method for drawing linear graphs

The **intercepts** of a graph are the points where the graph cuts each of the axes.

- The ***x*-intercept** is the point where the graph cuts the x-axis. Any point on the x-axis has a y-coordinate of 0.
- The ***y*-intercept** is the point where the graph cuts the y-axis. Any point on the y-axis has an x-coordinate of 0.

By substituting $y = 0$, the x-intercept of a graph can be found, and by substituting $x = 0$, the y-intercept of a graph can be found. When an equation is linear, joining these two intercepts forms the graph of a straight line.

Example C

Q. Draw the graph of $2y + 3x = 6$.

A. Setting $x = 0$ gives $2y + 0 = 6$, $2y = 6$, $y = 3$

Setting $y = 0$ gives $0 + 3x = 6$, $3x = 6$, $x = 2$

The intercepts, (0, 3) and (2, 0) are plotted and joined by a straight line.

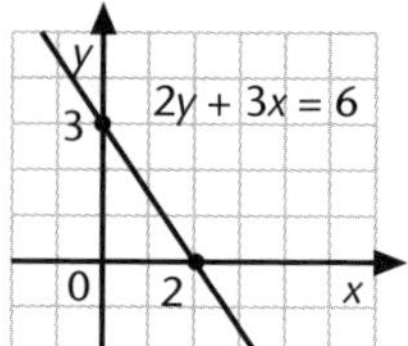

The gradient of a straight line

The **gradient** (symbol ***m***) of a straight line is a measure of the steepness of the line and the direction in which it slopes. The gradient is sometimes called the **slope**. The gradient is calculated by measuring the horizontal and vertical 'steps' between any two points on the line, and using the following formula:

$$\text{Gradient} = m = \frac{\text{vertical step}}{\text{horizontal step}} = \frac{y \text{ step}}{x \text{ step}}$$

Note: Vertically, a step up is positive and a step down is negative. Horizontally, a step to the right is positive, to the left negative.

Example D

Q. Find the gradients of the lines drawn alongside.

1. AB 2. PQ

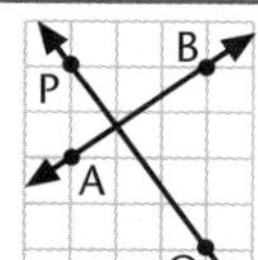

A. 1. Stepping from A to B the vertical step is 2, the horizontal step is 3.

Line AB has gradient $\frac{2}{3}$.

(Alternatively, stepping from B to A, y step = –2, x step = –3, gradient = $\frac{-2}{-3} = \frac{2}{3}$)

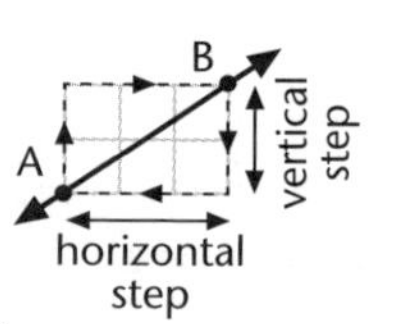

2. Stepping from P to Q, the y step is –4, the x step is 3.

Line PQ has gradient $\frac{-4}{3}$.

(Alternatively, stepping from Q to P, y step is 4, x step is –3. The gradient = $\frac{4}{-3} = -\frac{4}{3}$)

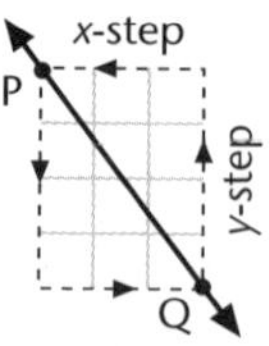

When calculating the gradient of a line, choose points with coordinates that are **integers**, for ease of computation, ie points where the grid lines on the graph intersect.

Example E

The graphs below show six lines with the gradient m of each line calculated.

1.

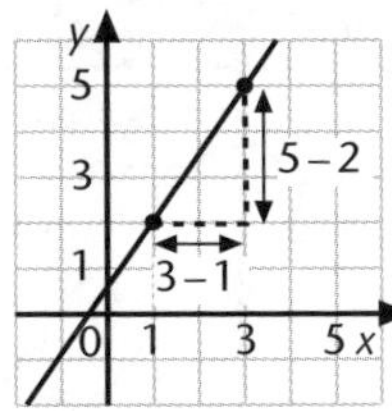

Line passes through (1, 2) and (3, 5)

$$m = \frac{y \text{ step}}{x \text{ step}} = \frac{5 - 2}{3 - 1} = \frac{3}{2}$$

2.

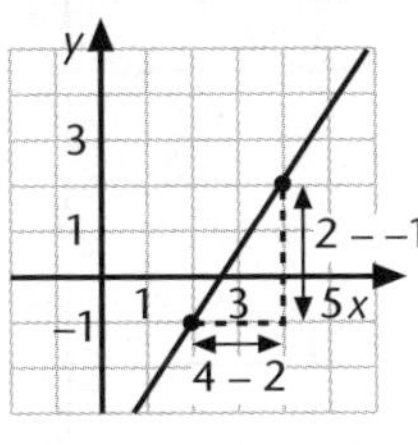

Line passes through (2, –1) and (4, 2)

$$m = \frac{y \text{ step}}{x \text{ step}} = \frac{2 - -1}{4 - 2} = \frac{3}{2}$$

3.

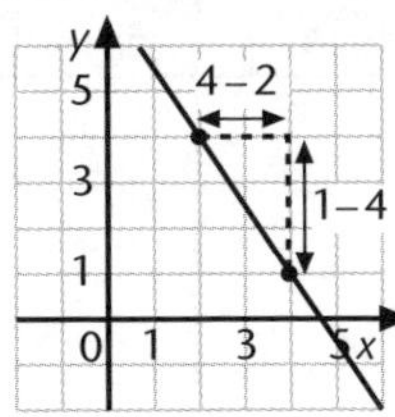

Line passes through (2, 4) and (4, 1)

$$m = \frac{y \text{ step}}{x \text{ step}} = \frac{1 - 4}{4 - 2} = \frac{-3}{2}$$

4.

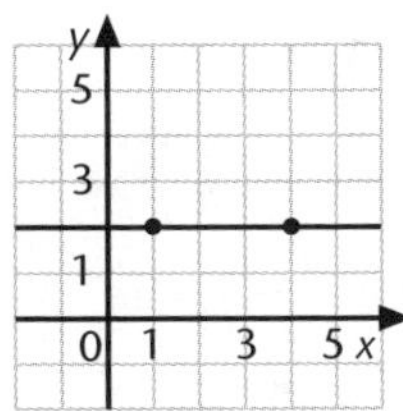

Line passes through (1, 2) and (4, 2)

$$m = \frac{y \text{ step}}{x \text{ step}} = \frac{2 - 2}{4 - 1} = \frac{0}{3} = 0$$

5.

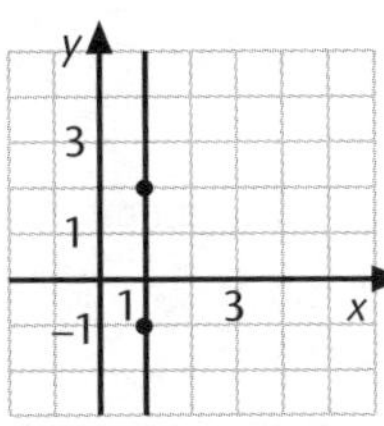

Line passes through (1, –1) and (1, 2)

$$m = \frac{y \text{ step}}{x \text{ step}} = \frac{2 - -1}{1 - 1} = \frac{3}{0} \text{ undefined}$$

6.

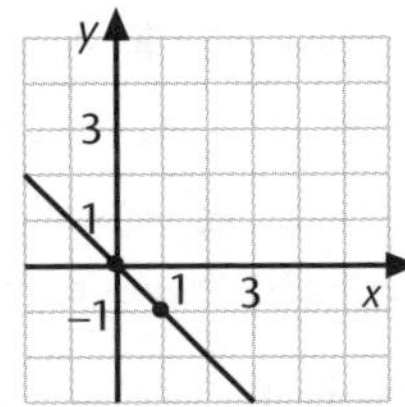

Line passes through (0, 0) and (1, –1)

$$m = \frac{y \text{ step}}{x \text{ step}} = \frac{-1 - 0}{1 - 0} = -1$$

Note:
1. Graphs **1** and **2** are **parallel lines** and have the same gradient.
2. Graph **4** is a **horizontal** line. Its gradient is 0.
3. Graph **5** is a **vertical** line. Its gradient is undefined.

Observing the slopes in Example E shows:

- The first two lines have positive gradients and slope uphill ⁄.
- The third and sixth lines have negative gradients and slope downhill \.

The fact that 'uphill' lines have positive gradients and 'downhill' lines have negative gradients is a useful final check that the sign of a calculated gradient is correct.

Having established the **sign** (+ or –) of the gradient, 'steps' can simply be counted, without regard for the sign. Thus in part **3** of Example E, there are 3 vertical steps and 2 horizontal steps, and the gradient is negative.

Therefore, the gradient $m = \frac{-3}{2}$.

3
y steps
2
x steps

The following diagram summarises the values of the gradient m for various straight line graphs.

Positive Gradient	Negative Gradient	Horizontal Gradient	Vertical Gradient	Equal Gradient
Graph slopes 'uphill', m is positive	Graph slopes 'downhill', m is negative	Graph is horizontal, m is zero	Graph is vertical, m is undefined	Lines are parallel, gradients are equal
The larger the *size* of the gradient, the steeper the line				

Gradient-intercept method for linear graphs

The **gradient-intercept form** of the equation of a straight line is:

$$y = mx + c \text{ where } m \text{ and } c \text{ are constants}$$

In this equation, m is the gradient of the line, and c is the y-intercept of the line.

(Setting $x = 0$ in $y = mx + c$, gives $y = m \times 0 + c$, ie $y = c$. So the y-intercept is c.

For every change of 1 unit in x, the change in y is m units, so the gradient is $\frac{m}{1} = m$.)

By considering the values of m and c, the graph of the line $y = mx + c$ can be drawn.

Example F

Q. Sketch the graph of $y = 2x + 3$ using the gradient-intercept method.

A. $y = 2x + 3$ is compared with $y = mx + c$, giving $m = 2$ and $c = 3$.

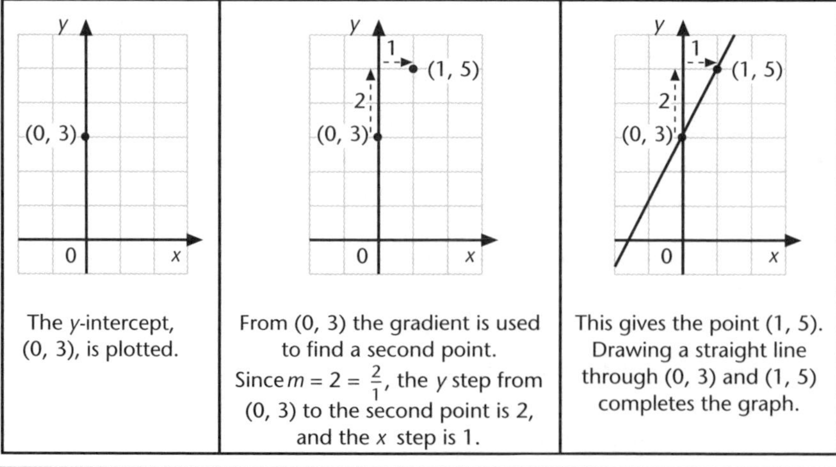

The y-intercept, (0, 3), is plotted.

From (0, 3) the gradient is used to find a second point. Since $m = 2 = \frac{2}{1}$, the y step from (0, 3) to the second point is 2, and the x step is 1.

This gives the point (1, 5). Drawing a straight line through (0, 3) and (1, 5) completes the graph.

The equation of the line may need to be rearranged so that a comparison with $y = mx + c$ can be made. The equation is rearranged so that y is the subject of the equation and appears on the left hand side. The terms on the right-hand side are written with the x term first and the constant second.

Example G

Q. For the line $2y + 3x - 4 = 0$:

1. Find the y-intercept. **2.** Find the gradient. **3.** Draw the graph.

A. Rearranging $2y + 3x - 4 = 0$ gives:

$2y = -3x + 4$ [moving all but the y term to the right-hand side]

$y = \frac{-3}{2}x + 2$ [dividing throughout by 2]

Comparing $y = \frac{-3}{2}x + 2$ with

$y = mx + c$ shows that:

1. $c = 2$, ie the y-intercept is the point (0, 2).

2. $m = \frac{-3}{2}$, ie the gradient is $= \frac{-3}{2}$.

3. The graph is shown opposite.

Note: The gradient is $\frac{-3}{2}$ *not* $\frac{-3x}{2}$.

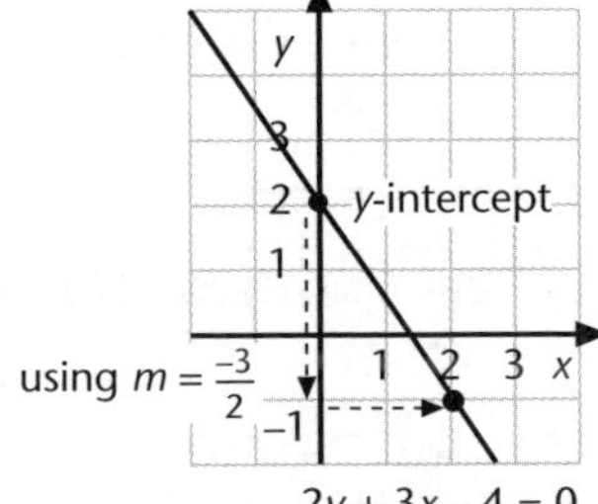

Writing the equation of a linear graph

When the gradient and y-intercept of a line are known, the equation of the line can be written down.

Example H

Q. 1. Find the equation of the line that has a gradient of 3 and a y-intercept of 4.

2. Write down the equation of the line shown.

A. 1. Using $y = mx + c$ with $m = 3$ and $c = 4$, gives $y = 3x + 4$ as the required equation.

2. From the graph, the y-intercept is $c = 1$ and the gradient is $m = \frac{1}{2}$.

The equation of the line is $y = \frac{1}{2}x + 1$

[using the form $y = mx + c$]

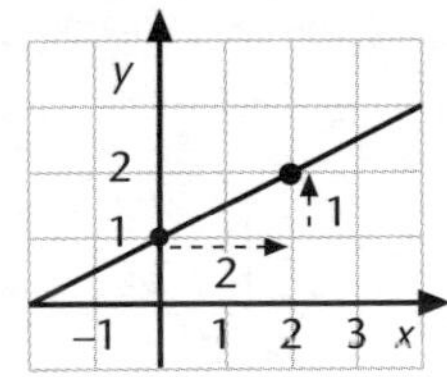

Horizontal and vertical straight-line graphs

Lines parallel to the x-axis (**horizontal lines**) are of the form:

$y = c$, where c is a constant

Lines parallel to the y-axis (**vertical lines**) are of the form:

$x = c$, where c is a constant

Example I

Q. Draw the lines: **1.** $y = 2$ **2.** $x = -3$

A. 1.

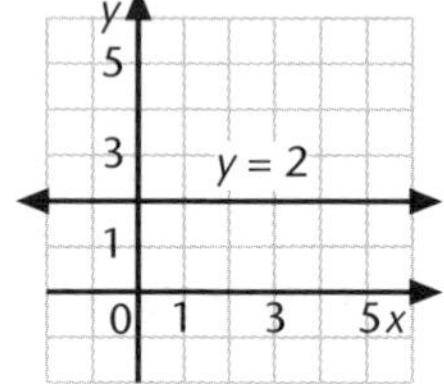

2.

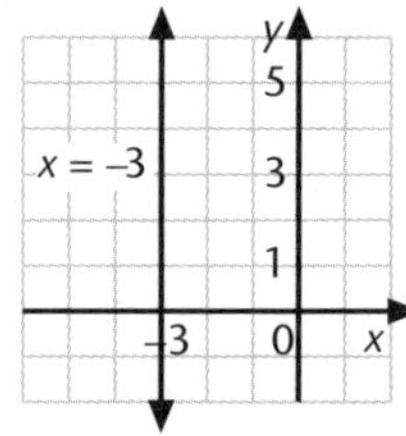

Note: This conforms to the $y = mx + c$ pattern, where gradient $m = 0$ and y-intercept $c = 2$.

Note: Vertical lines have *undefined* gradients, and their equations cannot be expressed in the form $y = mx + c$. (All other lines have equations which can be rearranged into the form $y = mx + c$.)

Unit 12.4 Activity 3B: Linear graphs

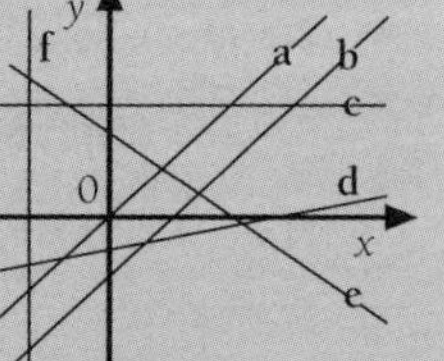

1. Which lines in the diagram have:

a. A positive gradient? **b.** A negative gradient?

c. $m = 0$? **d.** A positive y-intercept?

e. A negative x-intercept? **f.** An undefined gradient?

2. For each of the following equations:

i. Find the gradient. **ii.** Find the y-intercept. **iii.** Draw the graph.

a. $y = x + 2$ **b.** $y = 2x - 3$ **c.** $y = \frac{2}{3}x + 1$ **d.** $x + y = 6$

e. $y = 3x$ **f.** $y = -2x$ **g.** $y + 2x = 5$ **h.** $2y - 3x = 6$

i. $2x + 5y = 10$ **j.** $3x - 4y = 16$ **k.** $y = 3$ **l.** $y = -2$

m. $x = -1$ **n.** $x = 4$ **o.** $y = 2 + 4x$ **p.** $y = 1 - x$

3. For each of the following graphs:

i. Find the gradient. **ii.** Find the y-intercept. **iii.** Write the equation of the line.

a.

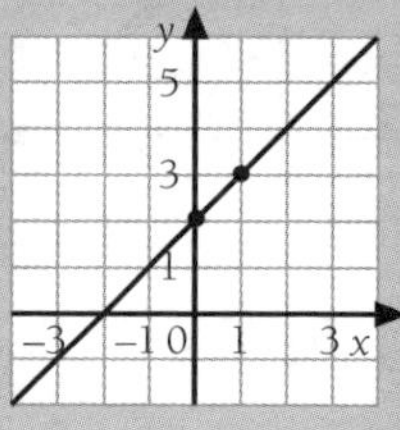

b.

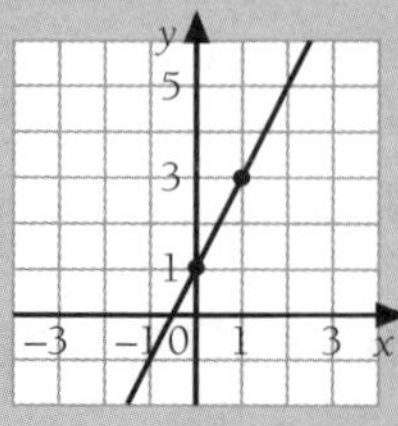

c.

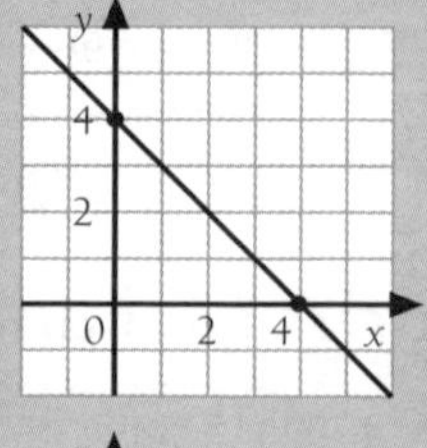

d.

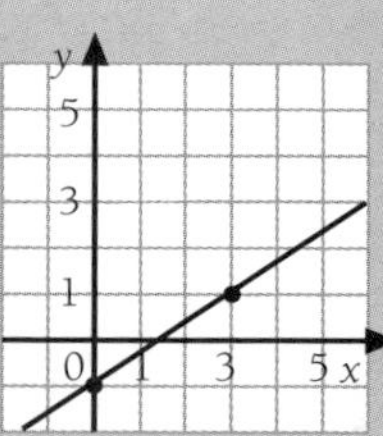

e.

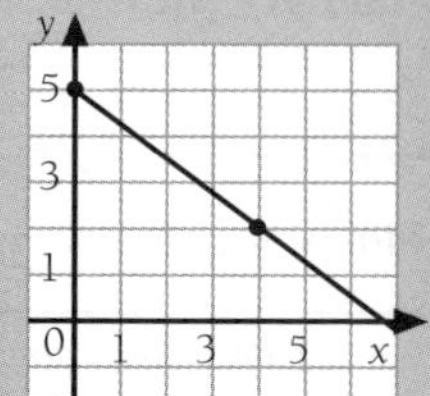

f.

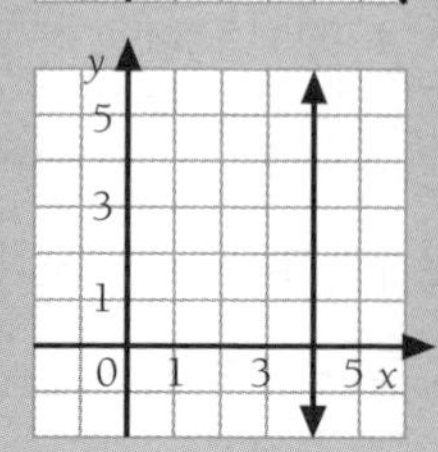

g. 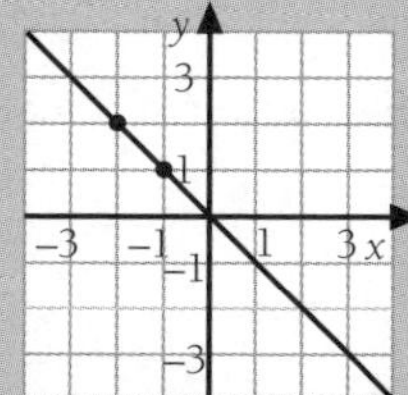

h. 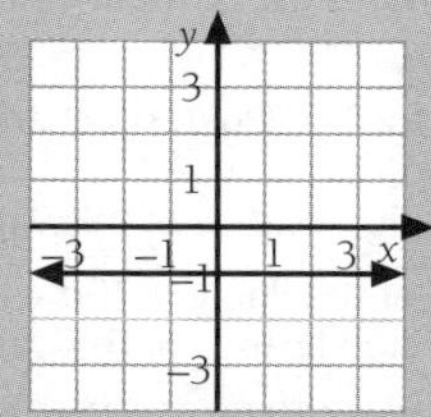

i. 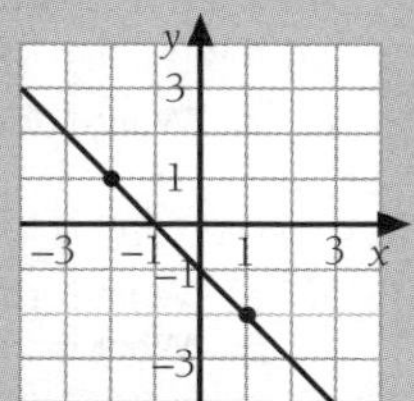

4. Draw the graphs of:

   a. $y = x - 2$ b. $2x + y = 4$ c. $x - 3y = 6$ d. $y = -3$

   e. $2x + 3y = 12$ f. $5x - 2y = 10$ g. $3y + 5x = 15$ h. $x = 2$

5. A line passes through the point (3, 1) and has gradient 2. Draw the graph of this line and use it to find:

   a. The x-intercept of the line. b. The y-intercept of the line.

   c. The equation of the line.

6. A line passes through the points (–2, 3) and (2, –5). Draw the graph of this line and use it to find:

   a. The y-intercept of the line. b. The gradient of the line.

   c. The co-ordinates of the x-intercept. d. The equation of the line.

Linear graphs in practical contexts

Many practical situations can be modelled by linear equations and their graphs. The various features of the straight line may then be interpreted in the context of the application.

Example J

Q. Two venues are considered for a club dinner: The Charlton and The Duxton. On the set of axes below, the graph of the charges of The Charlton are shown.

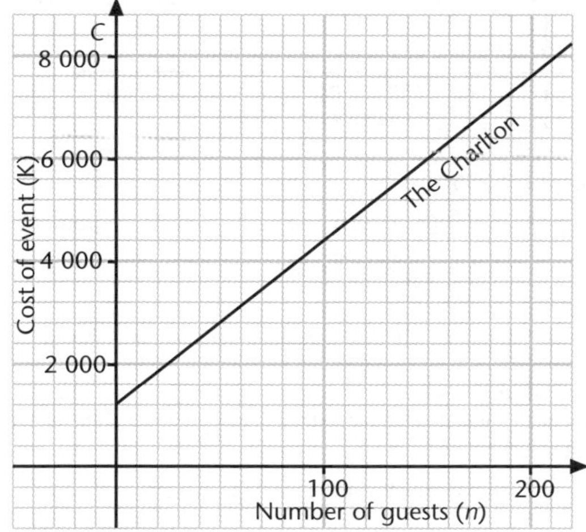

The Duxton charges according to the formula $C = 40n + 400$ where C is the cost (K) and n is the number of guests.

1. a. Where does the line for The Duxton's charges $C = 40n + 400$ intersect the vertical axis?
   b. What is the practical significance of this value?
2. a. What is the gradient of the line $C = 40n + 400$?
   b. What is the practical significance of this value?
3. Draw the equation $C = 40n + 400$ for The Duxton's charges for values of n from 0 to 200.
4. Write an equation in terms of C and n for The Charlton's charges.
5. Given that both venues offer a similar service, which venue would you recommend for:
   a. A group of 80 guests?
   b. A group of 170 guests?
6. For what number of guests is the charge the same at both venues?
7. How is this equal charge shown on the graph?

A. 1. a. K400
   b. This is the fixed fee charged by The Duxton (regardless of the number of guests).
2. a. 40
   b. This is the kina charge per guest (on top of the fixed fee).
3.

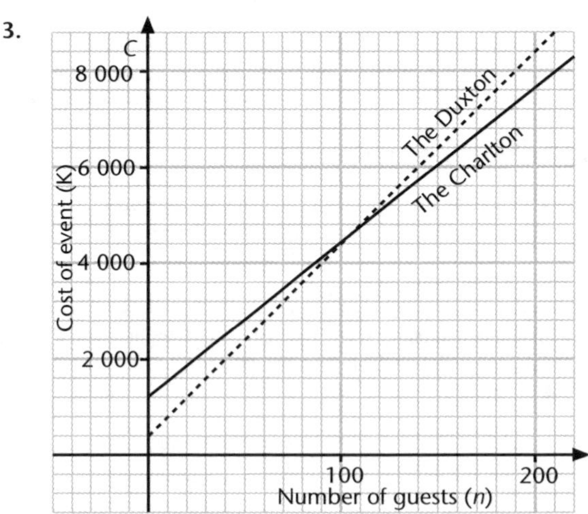

4. C-intercept is 1 200.
   Line passes through (0, 1 200) and (200, 7 600) so gradient is $\frac{7\,600 - 1\,200}{200 - 0} = 32$.
   Equation of line is $C = 32n + 1\,200$.
5. a. The Duxton.
   [charges lower as The Duxton's charge line is below The Charlton's charge line for $n < 100$]
   b. The Charlton
   [charges lower as The Charlton's charge line is below The Duxton's charge line for $n > 100$]
6. The charge is the same (K4 400) for 100 guests.
7. This is the point of intersection of the two lines.

When two different units are connected by a linear relationship, a straight line graph can be used as a **conversion graph**. By plotting two (or more) known conversions, a conversion line can be drawn which can be used to change from one unit to the other.

Example K:

Q. It is known that the boiling point of water is 100 °C or 212 °F, and that the freezing point of water is 0 °C or 32 °F.

1. Use this information to draw a straight line graph that can be used to convert **temperatures** in **degrees Celsius** to degrees Fahrenheit, or vice versa. Put the *x*-axis in degrees Celsius (°C), the *y*-axis in degrees Fahrenheit (°F).
2. Convert 22 °C to degrees Fahrenheit using your graph.
3. Convert 90 °F to degrees Celsius using your graph.
4. Write down the equation of the conversion line.

A. 1. The points (0, 32) and (100, 212) are plotted and joined with a straight line to create the conversion graph.

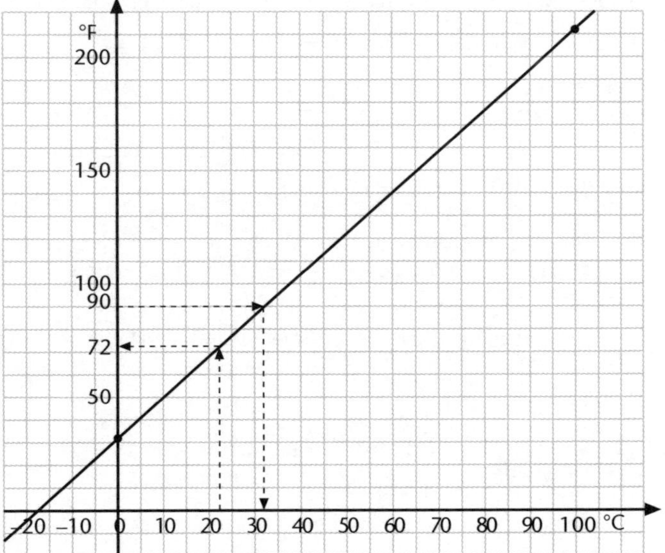

2. To convert 22 °C to degrees Fahrenheit, follow the dotted line up from 22 on the *x*-axis, to cut the straight line graph. Then read off the *y*-coordinate of this point by following the dotted line across to the *y*-axis.

   The conversion is actually 71.6 °F, but this sort of accuracy is difficult to achieve, unless the scale on the graph is large. An approximate value, such as 72 °F in this example, is usually all that a graph affords, ie 22 °C ≈ 72 °F.

3. To convert 90 °F to degrees Celsius, proceed horizontally from 90 on the *y*-axis to cut the straight line graph, then read off the *x*-coordinate (vertically). This gives 90 °F ≈ 32 °C.

4. To get the equation of the conversion line read off the y-intercept, $C = 32$ and

$$\text{gradient } m = \frac{212 - 32}{100 - 0} \quad \left[\frac{y\text{ step}}{x\text{ step}}\right]$$

$$= \frac{9}{5}$$

The equation is $y = \frac{9}{5}x + 32$ or $F = \frac{9}{5}C + 32$. [using $y = mx + c$]

Unit 12.4 Activity 3C: Linear graphs in practical contexts

1. Marie works as a freelance designer. She charges a fee for taking on a job, plus an hourly rate. For her last three jobs her earnings were as shown in the table.

Number of hours worked (h)	12	20	25
Pay in kina (K)	145	225	275

a. Draw a graph with horizontal axis h from 0 to 30 (hours) and vertical axis d from 0 to 300 (kina). Plot the three points from the table on your graph.

b. Draw the straight line which passes through these points.

c. Use your graph to find Marie's:
i. Pay for 18 hours work. **ii.** Number of hours worked for K185.

d. i. What is Marie's fixed fee? **ii.** What is her hourly rate?

e. Write the equation of the straight line.

2. A CNN weather report displays temperatures in degrees Fahrenheit and degrees Celsius. Two cities are shown to have the following temperatures: London 50 °F (10 °C) and Leningrad –4 °F (–20 °C).

a. Use these temperatures to draw up a straight line graph with degrees Fahrenheit on the x-axis (–10 °F to 60 °F) and degrees Celsius on the y-axis (–20 °C to 20 °C).

b. Use your graph to convert: **i.** 41°F to °C. **ii.** –10 °C to °F.

c. Write an equation for degrees Celsius, in terms of degrees Fahrenheit, ie $C = ...F + ...$. Give your y-intercept to the nearest whole number.

3. The Drive-U Taxi Company charges K3 flag fall and K1.50 per kilometre.

a. Copy and complete the table.

Number of kilometres (x km)	Cost (Ky)
2	
4	9
6	12
8	15
	18

b. i. Draw up a grid with the x-axis going from 0 to 12 and the y-axis going from –4 to 20.
ii. Plot the points on the grid.
iii. Join the points to form a straight line.
iv. Find the equation linking x and y.

c. To find the distance you can travel for Kx, the relation becomes $y = \frac{2}{3}x - 2$. Draw this graph on the grid.

4. The recommended cooking time for a roast chicken in a microwave oven is 9 min per 500 g plus 15 min standing time. The line of the graph alongside shows this cooking time.

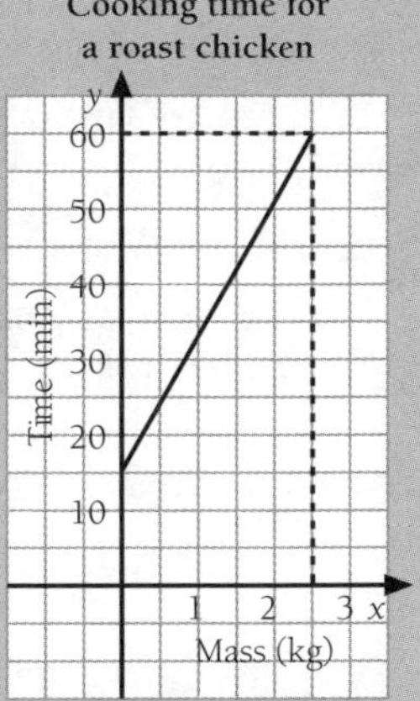

   a. What is the equation of the line?

   The recommended cooking time for a roast chicken in a thermowave oven is 20 min per 500 g.

   b. Copy the graph and draw another line to show the cooking time for a roast chicken in a thermowave oven.

   c. For what approximate mass of chicken are the cooking times the same? How is this shown on the graph?

5. Teri bought a property in a rural area. In 1998, he discovered a small pine tree growing on his section. He measured the height of the pine tree at the same time every year. The graph below shows the height of the tree. A growth **trend line** has been added to the graph.

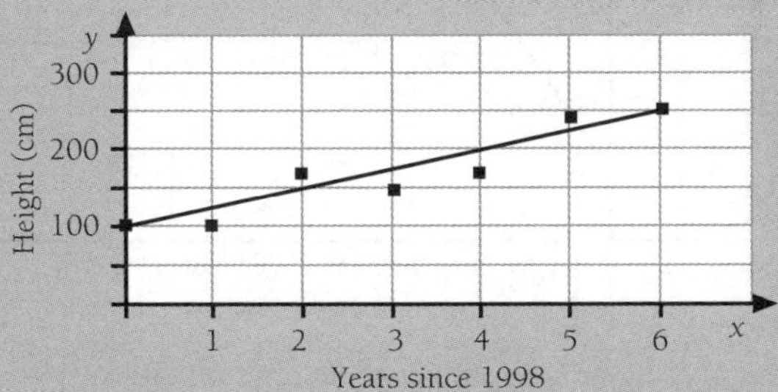

   a. Write down the equation of the growth trend line, where:
   y = height of the tree in cm.
   x = number of years since 1998.

   b. What does the gradient of the growth trend line show about how the tree is growing?

   c. Teri estimates that the tree was about 4 years old when he found it.
   Explain why this is a reasonable estimate.

6. Kara has a farm and on that farm she has a drain that needs cleaning. She gets three different quotes to clean the drain – from *Draincleaners*, *Wetfeet* and *Scrubbers*.

The *Draincleaners* quote is K80 plus 50 cents per metre of drain.

This quote can be described by the equation $C = \frac{1}{2}l + 80$

where C is the total cost to clean the drain in dollars and l is the length of the drain in metres.

a. On the graph below, the graphs of the *Wetfeet* quote and the *Scrubbers* quote have been drawn.

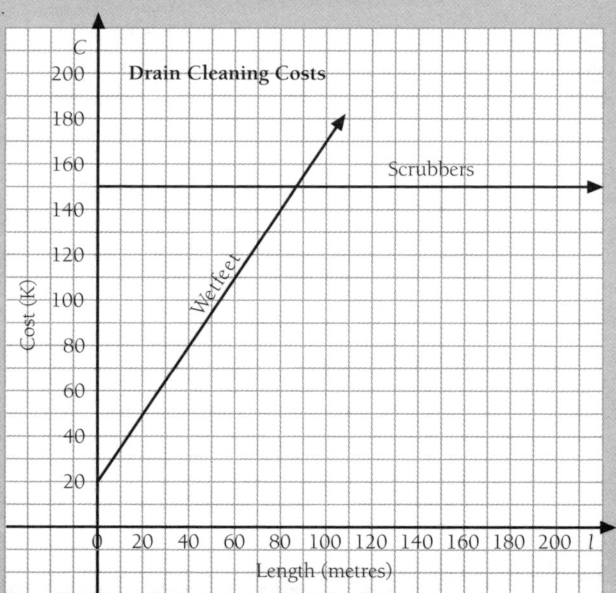

Copy the graphs and on the same set of axes, graph the equation for *Draincleaners* quote: $C = \frac{1}{2}l + 80$. Draw the graph for values of length l from 0 to 200 metres.

b. i. Write the equation ($C = ...$) of the *Wetfeet* quote using the graph.

ii. Write the equation ($C = ...$) of the *Scrubbers* quote using the graph.

c. i. Which quote has the higher fixed charge?

ii. How is this shown by the graph?

d. i. Which quote costs the most per metre to clean a drain?

ii. How is this shown by the graphs?

e. Kara decided she would choose between *Draincleaners* and *Scrubbers*. She says it would not matter which one she chooses, the cost would be the same.

i. How long is the drain she wants cleaned?

ii. How is this shown by the graphs?

7. Lillian is getting connected to the internet. One provider she is looking at offers three different deals: *Flat Rate*, *Middle Road* and *Email Special*.

Middle Road costs K25 plus 20 toea per hour of usage.

The cost for the *Middle Road* deal can be described by the equation $C = \frac{20}{100}t + 25$ where C is the cost each month, in dollars, charged by the provider and t is the time spent connected, in hours, each month.

a. On the graph below, the graphs of *Flat Rate* and *Email Special* deals have been drawn.

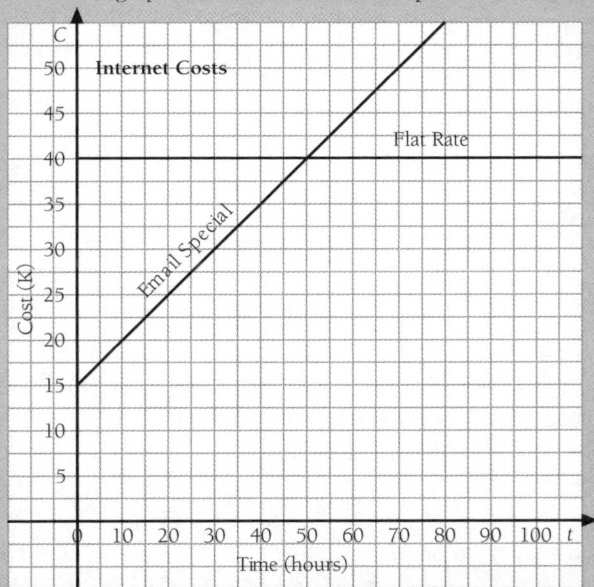

Copy the graphs and on the same set of axes, graph the equation for the *Middle Road* deal: $C = \frac{20}{100}t + 25$. Draw the graph values of time from 0 to 100 hours.

b. **i.** Write the equation ($C = ...$) for the *Email Special* deal using the graph.
ii. Write the equation ($C = ...$) for the *Flat Rate* deal using the graph.

c. **i.** Which deal has the higher fixed charge?
ii. How is this shown by the graph?

d. **i.** Which deal costs the most per hour to be connected?
ii. How is this shown by the graphs?

e. Lillian decided she would choose between the *Email Special* deal and the *Flat Rate* deal.

i. At how many hours of connection would both deals cost the same?
ii. How is this shown by the graphs?

8. The Sotu family were investigating the cost of having some electrical work done in their house. They got information from three different firms: *Sparks*, *Power* and *ACDC*.

The graph below shows the cost for times up to five hours.

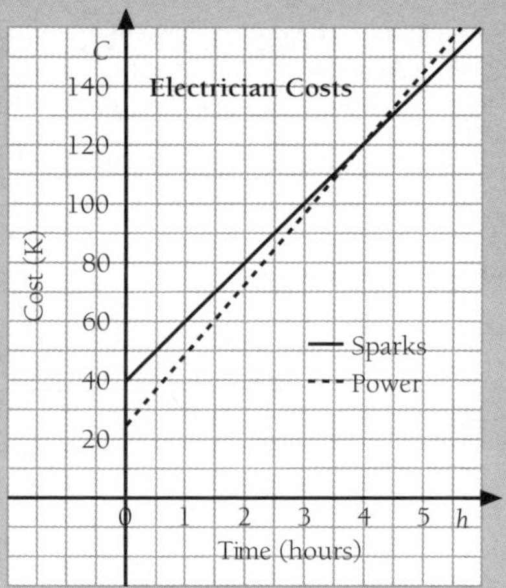

a. The costs for *ACDC* can be given by the equation $C = 20h + 30$ where C is the cost in kina for a given time and h is the time in hours.

Copy the graph above and on it draw the graph for *ACDC*, for t between 0 and 5 hours.

b. **i.** For what time is it the same price for *Sparks* and *Power*?
ii. How is this shown by the graphs?

c. **i.** Which two firms charge the same hourly rate?
ii. How is this shown by the graphs?

d. **i.** What is the hourly rate for *Sparks*?
ii. How is this shown by the graphs?

e. **i.** Use the graph to write the equation for the costs for *Sparks* ($C = ...$).
ii. Use the graph to write the equation for the costs for *Power* ($C = ...$).

Unit 12.4 Algebra and Graphs

Topic 4: Graphs of linear inequations

Continuing our coverage of graphs and functions (Syllabus p. 25),Topic 4 looks at graphs of linear inequations. This Topic covers:

- Optimising an **objective function** for a linear programming problem, where students are expected to form their own constraints and objective function (sensible rounding of answers may be needed).
- Determining the effect of varying the constraints or objective function of a linear programming problem.
- Considering the possibility of multiple solutions to a linear programming problem.

Linear inequations

Linear inequations such as $y < x + 1$ and $2y + 3x \geq 6$ represent **regions** in the Cartesian plane. The region is found by drawing the associated line (obtained by replacing the inequality sign with an equals sign), then testing the coordinates of one point not on the line to determine on which side of the line the region lies.

Example A

Q. Shade the following regions in the plane: **1.** $y < x + 1$ **2.** $2y + 3x \geq 6$

A. 1. First draw the line $y = x + 1$ as shown alongside.

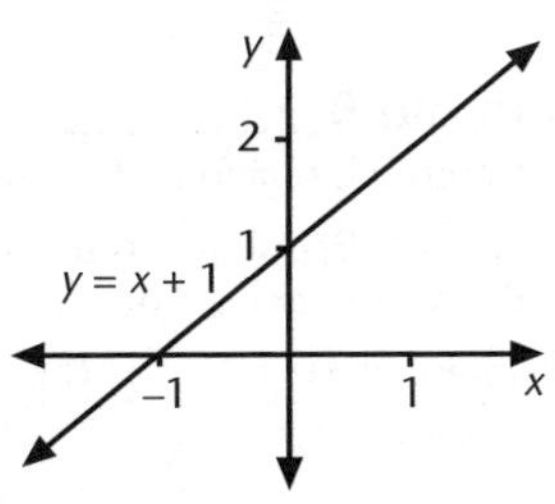

The region is formed by testing the coordinates of one point which does *not* lie on the line. Clearly (0, 0) does not lie on the line. Substituting $x = 0$, $y = 0$ into the inequation gives $0 < 0 + 1$. Since this is true, the point (0, 0) lies in the required region, and hence all points on that side of the line also lie in the required region. This is indicated by shading the region as shown below.

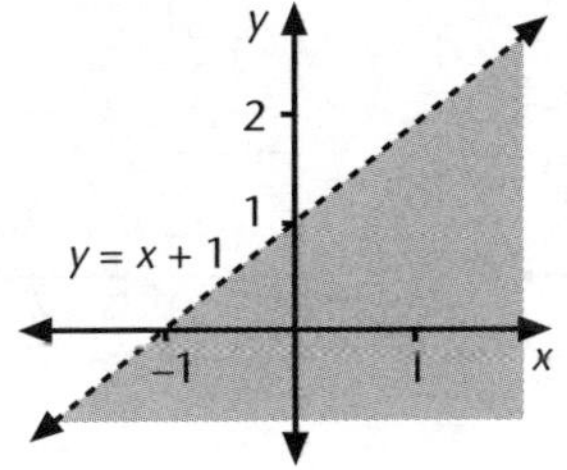

Note: The actual line $y = x + 1$ is dotted, since a dotted line indicates that points on the line are *not* included in the region. (Here y is less than $x + 1$ so points where $y = x + 1$ are excluded.)

2. The line $2y + 3x = 6$ appears as shown alongside. Testing using (0, 0), which does not lie on the line, gives $2 \times 0 + 3 \times 0 \geq 6$. [substituting into the inequation] This is a false statement. Hence the region lies on the other side of the line.

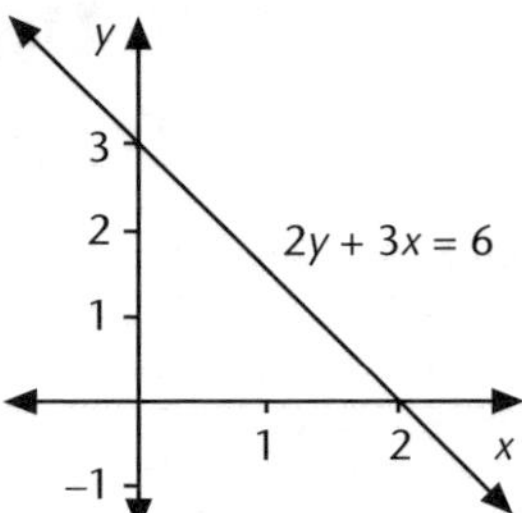

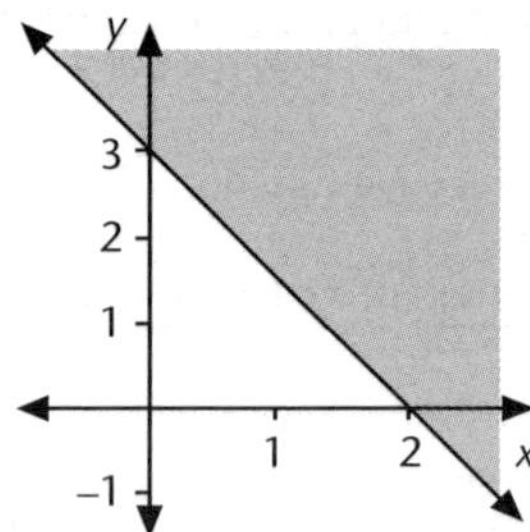

Note: In this case, the line is solid, since points which satisfy $2y + 3x = 6$, ie those points on the line, also satisfy the inequation and should be included in the shaded region.

Combined inequations

When two or more inequalities are combined, the individual inequations are solved first, and then the union or the intersection of the regions is found, as required.

Example B

Q. Graph the region that contains all the points where $x \geq 0$ and $2x + y \leq 6$

A. The required region is the intersection of the points in the region $x \geq 0$ and the points in the region $2x + y \leq 6$:

The region $x \geq 0$ | The region $2x + y \leq 6$ | The intersection of these two regions

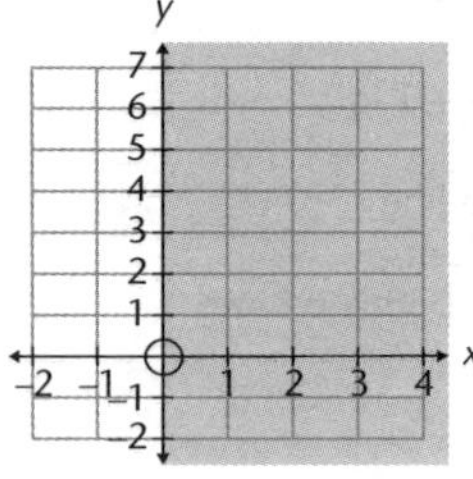

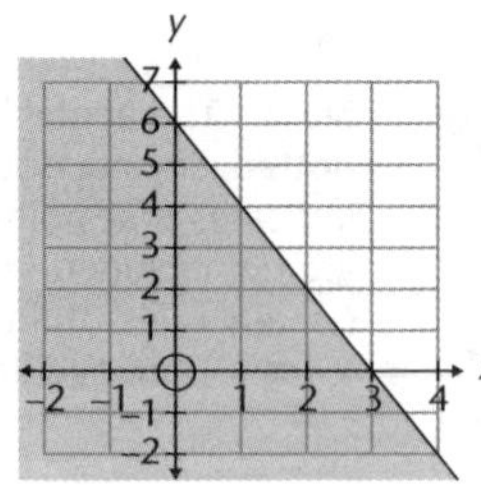

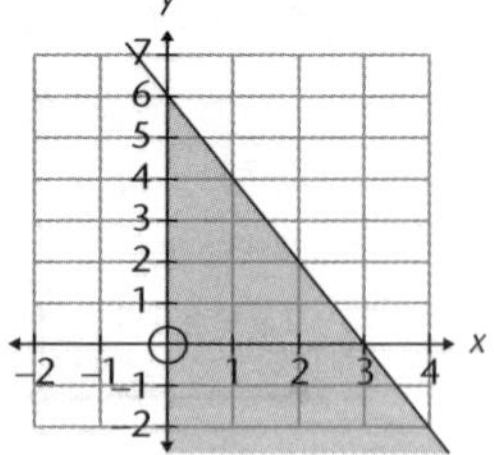

Note: When showing regions for the intersection of two or more inequalities it is easier to use the **shading-out technique**. If you are shading out then the area of intersection will be the unshaded region.

Shading-out for the example above: the required region in each case is the unshaded region.

The region $x \geq 0$ The region $2x + y \leq 6$ The intersection of these two regions

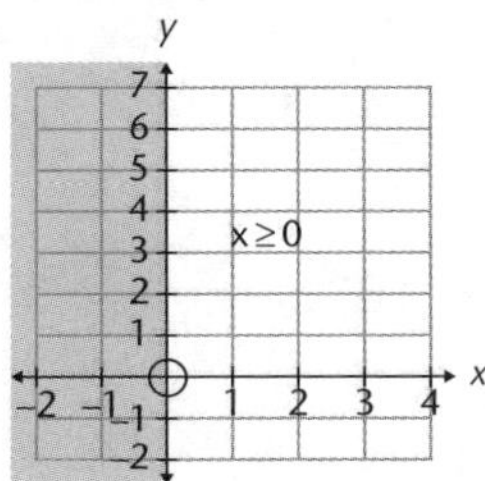

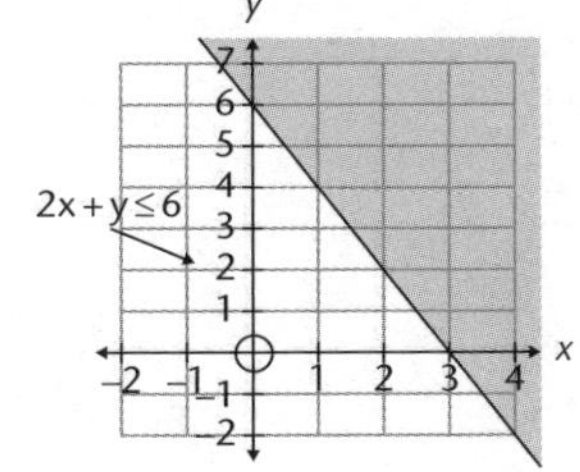

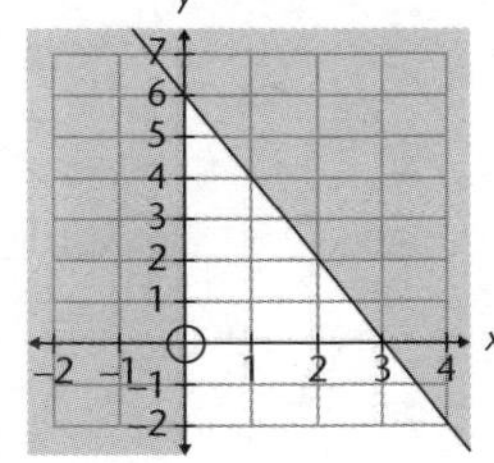

Example C

Q. Shade the region satisfying the inequations

$y \leq x - 2$ *and* $y \leq 2x - 2$.

A. It is necessary to shade the region which satisfies *both* inequalities. [The word 'and' indicates that the shaded region is the *intersection* of the two individual regions.]

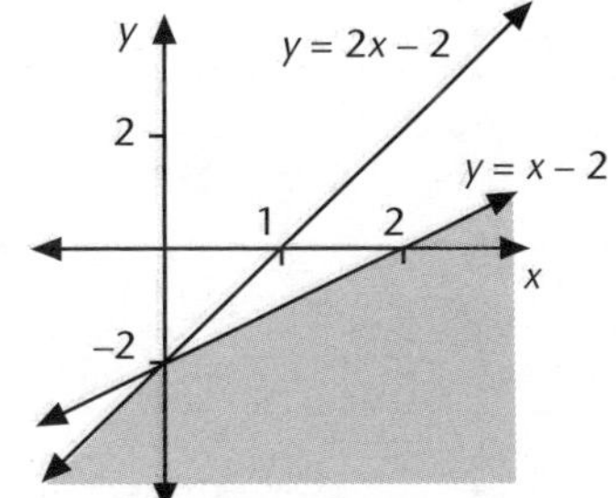

Example D

Q. Shade the region satisfying $y > 3$ *or* $y < x$.

A. In this case, the word 'or' indicates the shaded region should be that which satisfies *either* inequation. [This is the *union* of the two regions.]

Any region shaded is part of the required region.

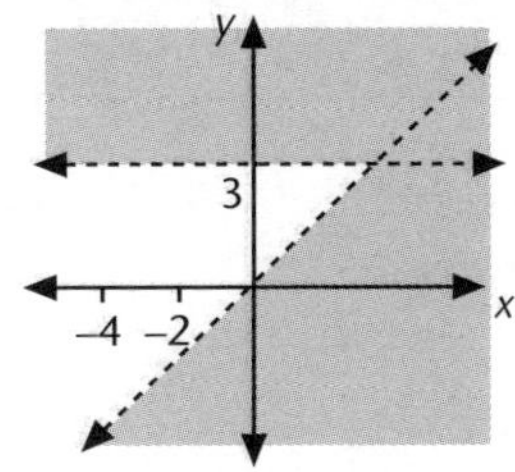

Note: Sometimes the symbol '∩' for intersection is used instead of the word 'and', and the symbol '∪' for union is used instead of the word 'or'.

Unit 12.4 Activity 4: Straight lines and linear inequations

1. Which of the following represent linear relations?

a. $y = 3x + 5$ **b.** $3x - 2y = 6$ **c.** $xy = 11$ **d.** $x^2 - y + 2 = 9$

e. $x^2 + y^2 = 4$ **f.** $x - 7 = 0$ **g.** $y = 3x + 2x$ **h.** $2xy - y = 0$

i. $3x - 7y = 2(x - 3)$ **j.** $x - y = (2x + y)2$

2. Draw the following straight lines:

a. $y = 2x + 1$ **b.** $y = 3x - 6$ **c.** $y = \frac{2}{3}x + 2$ **d.** $y = -2x + 3$

e. $y = 4$ **f.** $x = -2$ **g.** $y = \frac{x - 4}{2}$ **h.** $3x + 2y = 6$

i. $2x - 5y = 10$ **j.** $4y - 3x = 12$

3. Shade the region for each linear inequation:

a. $y < 2x + 1$ **b.** $y \leq 5x + 2$ **c.** $y > 2x + 3$ **d.** $y \geq 3x + 1$

e. $2x < 3y + 6$ **f.** $3x + 2y > 6$ **g.** $2x - 4y < 6$ **h.** $3x + 4y \geq 12$

i. $x < y + 1$ **j.** $2x \geq 5 + 3y$

4. Shade the region:

a. $y < x + 2$ and $y < 2$ **b.** $y < x + 2$ and $x > 2$

c. $y \leq x - 1$ and $y \geq -3$ **d.** $y < 2x + 4$ or $x + y < 2$

e. $x \leq 3$ or $y + x \leq 2$ **f.** $x \geq 2$ and $y \leq 4$ and $y \geq x - 2$

g. $x \geq 0$ and $y \leq 4$ and $y \geq x - 2$ **h.** $x \geq 0$ and $x \leq 4$ and $y \geq 0$ and $y \leq 3$

i. $y \leq x$ and $y \geq x - 2$ and $y \geq 0$ and $y \leq 4$ **j.** $y \leq 3x + 6$ and $y \geq x$ and $y + x \geq 6$

5. Shade the region:

a. $x + y < 1 \cap y > x$ **b.** $2x + 3y < 6 \cap x \leq 2$ **c.** $y \leq x \cap x \geq 4 \cap y < 3$

d. $2x + 3y \geq 12 \cap x \leq 6 \cap y \leq 4$ **e.** $y < x + 1 \cup y > 2x$

Unit 12.4 Algebra and Graphs
Topic 5: Quadratic equations

Our coverage of graphs and functions (Syllabus p. 25) continues with Topic 5, which looks at straightforward algebraic methods and solving equations. It covers:

- Solving factorised equations such as $(x - 1)(x + 3) = 0$.
- Solving simple quadratic equations such as $x^2 + 30x = 400$ and interpreting the results.
- Modelling by forming and solving appropriate equations.
- Interpretation in context.

Introduction

The general form of a quadratic equation in x is

$$ax^2 + bx + c = 0$$

In a **quadratic equation**, the highest power of the variable is 2.

For example: $x^2 - 4x + 1 = 0$, $y^2 = y + 1$, and $3a^2 = 4$ are all quadratic equations.

Solving quadratic equations

The following method for solving quadratic equations depends on an important property of zero:

If two real numbers multiply to make zero, then one or the other (or both) *of the numbers must be zero.*

This result is expressed using symbols as follows:

If $ab = 0$ then either $a = 0$ or $b = 0$ (or both)

This result makes the solving of quadratic equations straightforward if the quadratic is given in factored form (and is equal to zero).

Example A

Q. Solve the quadratic equations: **1.** $(x + 4)(x - 2) = 0$ **2.** $3x(2x + 3) = 0$

A. 1. If $(x + 4)(x - 2) = 0$ then

either $(x + 4) = 0$ or $(x - 2) = 0$ [setting each factor in turn to zero]

$x = -4$ or $x = 2$ [solving both equations for x]

2. $3x(2x + 3) = 0$

$\therefore 3x = 0$ or $(2x + 3) = 0$ [setting each factor in turn to zero]

$\therefore x = 0$ [dividing by 3] or $2x = -3$ [subtracting 3]

$\therefore x = 0$ or $x = -\frac{3}{2}$ [dividing second equation by 2]

If quadratic equations are not given in factored form then the quadratic must be factorised first.

Example B

Q. Solve: **1.** $x^2 + 5x - 6 = 0$ **2.** $2y^2 - 18 = 0$

A. 1.

$x^2 + 5x - 6 = 0$

$(x + 6)(x - 1) = 0$ [factorising]

$\therefore x + 6 = 0$ or $x - 1 = 0$

$\therefore x = -6$ or $x = 1$

2.

$2y^2 - 18 = 0$

$2(y^2 - 9) = 0$ [factorising]

$2(y + 3)(y - 3) = 0$ [factorising]

$y + 3 = 0$ or $y - 3 = 0$

$\therefore y = -3$ or $y = 3$

If necessary, the quadratic equation must be **rearranged** before factorising, so that every term of the equation is on one side of the equals sign, leaving zero on the other side.

Example C

Q. Solve: **1.** $x^2 = 4x + 12$ **2.** $(x + 2)(x - 3) = 50$

A. 1. To solve $x^2 = 4x + 12$, the equation must be rearranged and then factorised.

$x^2 = 4x + 12$

$x^2 - 4x - 12 = 0$ [subtracting 4x and 12 from both sides]

$\therefore (x - 6)(x + 2) = 0$ [factorising]

$x - 6 = 0$ or $x + 2 = 0$

$\therefore x = 6$ or $x = -2$

2.

$(x + 2)(x - 3) = 50$

$x^2 - 3x + 2x - 6 = 50$ [expanding LHS]

$x^2 - x - 6 = 50$ [simplifying]

$x^2 - x - 56 = 0$ [subtracting 50]

$(x - 8)(x + 7) = 0$ [factorising]

$x - 8 = 0$ or $x + 7 = 0$ [solving]

$x = 8$ or $x = -7$

Note: Although the quadratic in part **2** was factorised, the RHS was not zero, so the quadratic could not be solved using the techniques described previously. The quadratic needed to be expanded and the RHS made equal to zero before re-factorising and solving.

The solutions to a quadratic equation can be checked by **substituting** back into the original equation. (This can sometimes be done mentally.)

Example D

Q. Check that the solutions to $x^2 + 5x - 6 = 0$ are $x = -6$ or $x = 1$ (as found in Example B, part **1**).

A. $x = -6 \Rightarrow (-6)^2 + 5 \times -6 - 6 = 0$ [substituting $x = -6$ into $x^2 + 5x - 6$]

$36 - 30 - 6 = 0$

$36 - 36 = 0$ which is true.

$x = 1 \Rightarrow 1^2 + 5 \times 1 - 6 = 0$ [substituting $x = 1$ into $x^2 + 5x - 6$]

$1 + 5 - 6 = 0$

$6 - 6 = 0$ which is true.

$\therefore$ both solutions, $x = -6$ and $x = 1$, are correct.

Once it is recognised that an equation (in x say) is **quadratic** (the equation has terms in x^2 and/or x and/or constants) it is important to remember the correct strategy for solving such equations:

- **Z**ero: the right-hand side of the equation must equal zero.
- **F**actorise: express the quadratic as a product of its factors.
- **S**olve: set each factor in turn to zero and solve for the variable.

A useful reminder of this strategy is the mnemonic **ZFS**.

Example E

Q. Solve $(x - 1)(x + 3) = 0$.

A. A quadratic equation such as $(x - 1)(x + 3) = 0$ has already been made equal to zero (**Z**) and factorised (**F**) so that all that remains is to solve (**S**) the equation by setting each factor to zero.

Thus the solution is $x = 1$ or $x = -3$ [solving $x - 1 = 0$ and $x + 3 = 0$]

Note: A common error in solving equations such as $(x - 1)(x + 3) = 0$ is to expand the left-hand side of the equation, then attempt to isolate the x (as if the equation were linear).

Example F

Solve $3x^2 + x - 4 = 0$

Solution

$3x^2 + x - 4 = 0$

$\therefore (3x + 4)(x - 1) = 0$ [factorising]

$\therefore x = -1\frac{1}{3}$ or $x = 1$ [solving $3x + 4 = 0$ and $x - 1 = 0$]

Unit 12.4 Activity 5A: Solving quadratic equations

1. Solve the following quadratic equations given in factored form:

a. $(x + 2)(x + 3) = 0$ **b.** $(x - 2)(x - 4) = 0$ **c.** $(x + 1)(x - 3) = 0$

d. $(x - 5)(x + 2) = 0$ **e.** $x(x - 4) = 0$ **f.** $x(x + 5) = 0$

g. $(2x - 3)(x + 6) = 0$ **h.** $(2x + 1)(x - 1) = 0$ **i.** $(5x + 3)(2x + 5) = 0$

j. $(3 - 4x)(2x - 9) = 0$ **k.** $(3x - 4)(x - 5) = 0$ **l.** $(3x - 2)(3x + 2) = 0$

2. Factorise and solve the following quadratic equations:

a. $x^2 + 3x = 0$ **b.** $2x^2 - 10x = 0$ **c.** $x^2 + x - 6 = 0$

d. $x^2 + 5x + 4 = 0$ **e.** $x^2 - 3x - 40 = 0$ **f.** $x^2 - 6x + 9 = 0$

g. $x^2 - 8x + 12 = 0$ **h.** $x^2 + 3x - 10 = 0$ **i.** $x^2 - 4 = 0$

3. Solve the following quadratic equations:

a. $x^2 = 2x + 15$ **b.** $x^2 + 3x = 18$ **c.** $x^2 + 3 = 4x$

d. $x^2 = 6x$ **e.** $2x^2 = 5x$ **f.** $x^2 = 49$

g. $x^2 - 2x = 8$ **h.** $3x^2 = 12$ **i.** $2x^2 + x = 6$

j. $3x^2 + 2x - 1 = 0$ **k.** $4x^2 + 8x + 3 = 0$ **l.** $\frac{3}{x+1} = x - 1$

m. $(x + 1)^2 - 2(x + 1) - 3 = 0$

4. Solve the following quadratic equations to find n:

a. $(n + 1)(n - 1) = 63$ **b.** $(n - 8)(n + 2) = 11$ **c.** $(n + 2)(n - 2) = 140$

d. $(2n + 3)(2n - 3) = 91$ **e.** $n(2n + 1) = 36$ **f.** $(3n - 1)(2n + 3) = 35$

Solution of quadratic equations by formula

Sometimes a quadratic equation has a solution, yet cannot be factorised readily. In these cases the **quadratic formula** is used.

The solution to the general quadratic equation, $ax^2 + bx + c = 0$ is:

$$x = \frac{-b \pm \sqrt{b^2 - 4ac}}{2a}$$

Note: The solutions of an equation are often called the **roots** of the equation.

Example G

Q. Find the solutions to the equation $2x^2 + 8x - 3 = 0$, correct to 2 decimal places.

A. Comparing $2x^2 + 8x - 3 = 0$ with $ax^2 + bx + c = 0$ gives $a = 2$, $b = 8$, $c = -3$.

$$x = \frac{-8 \pm \sqrt{8^2 - 4 \times 2 \times (-3)}}{4} \quad \text{[substituting into } x = \frac{-b \pm \sqrt{b^2 - 4ac}}{2a}\text{]}$$

$$\therefore\ x = \frac{-8 \pm \sqrt{88}}{4}$$

$$\therefore\ x = \frac{-8 + \sqrt{88}}{4} \text{ or } \frac{-8 - \sqrt{88}}{4}$$

$$\therefore\ x = \frac{-8 + 9.381}{4} \text{ or } \frac{-8 - 9.381}{4} \quad \text{[working to 3 dp]}$$

$\therefore\ x = 0.35$ or -4.35 (2 dp) [rounding to 2 dp]

A quadratic equation may need to be rearranged into the form $ax^2 + bx + c = 0$ first, as the following example shows.

Example H

Q. Solve $(x + 1)^2 = x + 5$ to 2 decimal places.

A. $(x + 1)^2 = x + 5$

$\therefore\ x^2 + 2x + 1 = x + 5$ [expanding $(x + 1)^2$]

$\therefore\ x^2 + x - 4 = 0$ [collecting terms on one side]

$\therefore\ x = \dfrac{-1 \pm \sqrt{1^2 - 4 \times 1 \times (-4)}}{2}$ [substituting $a = 1$, $b = 1$, $c = -4$ in formula]

$\therefore\ x = \dfrac{-1 \pm \sqrt{17}}{2}$

$\therefore\ x = 1.56, -2.56$ (2 dp)

Unit 12.4 Activity 5B: Solution of quadratic equations by formula

1. Solve each of these equations giving answers correct to two decimal places:

a. $x^2 + 4x + 1 = 0$ **b.** $x^2 + 8x + 3 = 0$ **c.** $2x^2 - 7x + 1 = 0$

d. $3x^2 + 6x - 7 = 0$ **e.** $2R^2 + R - 8 = 0$ **f.** $x^2 + 1.1x - 0.3 = 0$

g. $2.3x^2 - 3.5x - 1.2 = 0$ **h.** $3A^2 + 5A = 2A^2 - 3A + 2$

i. $8U^2 + 3U + 5 = 2U^2 + 11U + 7$ **j.** $\frac{1}{3}x^2 + 1\frac{1}{2}x + \frac{1}{12} = 0$

2. Solve each of these equations to two decimal places:

a. $(x + 3)^2 = 8$ **b.** $(2x - 3)^2 = 7$ **c.** $3(x - 1)^2 + 1 = 6$

d. $x + 1 = \dfrac{3}{x}$ **e.** $(2x + 1)^2 = x^2 - x + 5$ **f.** $\dfrac{x+1}{3} + \dfrac{1}{x} = 2$

g. $\dfrac{2x+1}{x+3} = \dfrac{x+4}{x+1}$ **h.** $3(x + 1)^2 - 6(x + 1) + 2 = 0$ **i.** $\dfrac{x+1}{x^2+x+1} = 0.2$

j. $\dfrac{x^2+x+1}{x^2-x+1} = 3$

Solving word problems with quadratic equations

When solving word problems, the solutions must be interpreted correctly so that they are meaningful within the given context. For example, if the unknown value, x, is a side length, then only positive values for x make sense.

Example I

Q. An aircraft dropped a box of emergency supplies to a group of villagers. The height of the box above the ground, h, at time, t, is given by the formula:

$h = 520 - 5t^2$, where h is in metres and t is in seconds.

To find how long the box takes to fall to the ground, the equation $520 - 5t^2 = 0$ must be solved. Find how long it takes for the box to fall to the ground.

A.

$520 - 5t^2 = 0$	[at ground level, $h = 0$]
$t^2 - 144 = 0$	[dividing by –5 and rearranging]
$(t + 12)(t - 12) = 0$	[factorising]
$t = -12$ or 12	[solving, giving both solutions]

The time taken is 12 seconds [ignore the negative solution, since time taken is positive].

Example J

Q. In an art shop, canvases are sold by the square centimetre. Two canvases in the shop have the same price. One canvas is a rectangle four times as long as it is wide. The other is a square canvas of side length 18 cm longer than the width of the rectangular canvas. Find the dimensions of the square canvas.

A. Let x be the width of the rectangular canvas. Thus, the length of this canvas is $4x$ and the side length of the square canvas is $(x + 18)$.

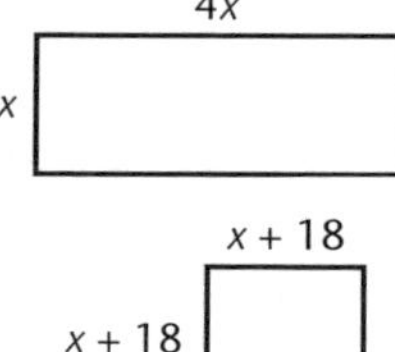

The area of the rectangular canvas is $x.4x = 4x^2$ and the area of the square canvas is $(x + 18)^2$. Equating areas gives the following quadratic to be solved.

$(x + 18)^2 = 4x^2$	
$x^2 + 36x + 324 = 4x^2$	[expanding]
$3x^2 - 36x - 324 = 0$	[collecting terms on one side]
$x^2 - 12x - 108 = 0$	[dividing by 3]
$(x + 6)(x - 18) = 0$	[factorising]
$x = -6$ or $x = 18$	

Only $x = 18$ applies, since x is a length and therefore can't be negative.

The side length of the square is 18 + 18 = 36 cm.

Unit 12.4 Activity 5C: Solving word problems with quadratic equations

1. Sam is the middle child of a 3-child family. His brother is 2 years older than Sam and his sister is 3 years younger than Sam. The product of their (Sam's brother and sister) ages is 176. The equation $(x + 2)(x - 3) = 176$ can be used to find the ages of the 3 children.

a. What does 'x' represent?

b. Solve the equation to find the ages of the 3 children.

2. A pig-farmer found that the equation $L = -n^2 + 12n + 40$ modelled the number of piglets born each day for a period of time on his farm during a year. L = number of piglets born and n = day number.

To find n when 67 piglets were born in a day, the equation $-n^2 + 12n + 40 = 67$ needs to be solved. Solve this equation to find when 67 piglets were born.

3. A field is 20 m longer than it is wide. The area of the field is 384 m^2. In order to find the dimensions of the field, the quadratic equation $x(x + 20) = 384$ must be solved.

x + 20

x

Solve this equation to find the length and width of the field.

4. A square room has side length x m. The room is made 2 m wider and 4 m longer so that its area is now 24 m^2.

   Solve the equation $(x + 2)(x + 4) = 24$ to find the original dimensions of the room.

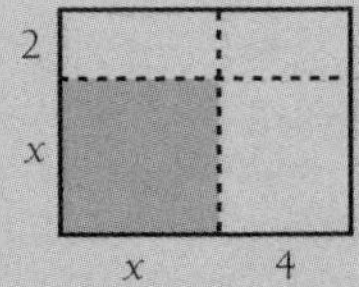

5. William is playing a series of computer games in which he has to capture aliens. William scores one point for each alien captured. So far William has a total score of 45 points. At the end of his next game, William notices that if he squared and doubled the number of points he got for that game then he would get the same number of points as his new total score.

   Solve the equation $2x^2 = x + 45$ to find how many aliens William captured in his latest game.

6. A gift box for soap has a square base and is 6 cm deep. Its top is open. The area of cardboard used for the box is 256 cm^2. The surface area of the outside of the box is $A = a^2 + 24a$, where a is the side length of the base of the box in cm.

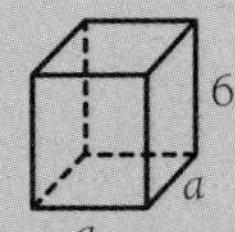

   Solve the equation $a^2 + 24a = 256$ to calculate the side length of the base of the box.

7. Alice and Felix think of two numbers. The numbers are 10 apart and their product is 56. Solve the equation $x(x - 10) = 56$ to work out the numbers Alice and Felix thought of (there are two different possible pairs of numbers).

8. Two square pieces of cardboard are required as backing for a photo. The smaller square has side length 4 cm less than the side length of the larger square.

   If the total area of cardboard required is 400 cm^2, find the dimensions of the larger square.

9. The product of two numbers is 360. If one number is 16 less than double the other number, find the two possible number pairs.

10. Three boys played a game in which they won points for skill. At the endof the game the winner has 6 points less than double the middle player's score. The loser has 4 points less than the middle player's score. The product of the winner's and loser's scores is equal to the square of the middle player's score. Find the three scores.

Unit 12.4 Activity 5D: More word problems

For each of these problems write an equation then solve:

1. A room is 2 m longer than it is wide. It has an area of 100 m^2. Find the width and the length.
2. A room is twice as long as it is wide. It has an area of 200 m^2. Find its width.
3. A triangle has a base which is 2 m longer than its height. It has an area of 10 m^2. Find the height of the triangle.
4. A triangle has a base which is 4 m longer than its height. Its area is 15 m^2. Find the length of the base.
5. John is 29 years older than Peter. The product of their ages is 1 272. Find their ages.

6. The product of two consecutive **natural numbers** is 3 782. Find the numbers.
7. A villager has a garden with an area of $100\,m^2$. The garden is in the shape of a square. He increases all the sides of this garden by a certain amount and now has an area of $150\,m^2$. How much did he increase each side by?
8. The sum of the area and perimeter of a square gives a value of 15. Find the perimeter of the square.
9. A box is constructed so that its length is one metre more than its height which in turn is one metre more than its width. It has a total surface area of $16\,m^2$. Find the lengths of all sides.
10. Find any number which when added to its reciprocal gives 5.
11. The sum of the first n natural numbers is $\frac{1}{2}n(n + 1)$.
    - **a.** Find how many natural numbers you have to add to get a sum of 496.
    - **b.** How many do you have to add to exceed 1 000?
12. The height of a right-angled triangle is one metre more than its base. The hypotenuse of the triangle is 4 m. Find the lengths of the base and height.
13. A girl is asked by her friend how much she gets paid per hour. She answers in the following way: 'The number of hours I work each week is one more than double my hourly pay. I get paid K153 per week. Work it out yourself.' How much does she get paid per hour?
14. A square room is extended so that it is 2.5 m wider and 3.7 m longer. If its new area is $54.4\,m^2$, find its original length and width.

Solving simultaneous linear and non-linear equations

To solve a linear equation and a quadratic equation simultaneously, **substitute** an expression for one of the variables of the linear equation into the quadratic equation.

Example K

Q. Solve $y = x + 1$ and $y^2 - 3x = 13$ simultaneously.

A. Substitute for y (the subject of the linear equation) in $y^2 - 3x = 13$.

$\therefore (x + 1)^2 - 3x = 13$ [substituting $(x + 1)$ for y in the quadratic]

$\therefore x^2 + 2x + 1 - 3x = 13$ [removing brackets]

$\therefore x^2 - x - 12 = 0$ [simplifying]

$\therefore (x - 4)(x + 3) = 0$ [factorising]

$\therefore$ either $(x - 4) = 0$ or $(x + 3) = 0$

$\therefore x = 4$ or $x = -3$

When $x = -3$, $y = -3 + 1 = -2$ [substituting into $y = x + 1$]

When $x = 4$, $y = 4 + 1 = 5$ [substituting into $y = x + 1$]

If necessary, one of the variables will need to be made the subject of the linear equation.

Example L

Q. Give the coordinates of the points where the line $2x - 3y = 6$ cuts the circle $x^2 + y^2 = 9$.

A. The equations $2x - 3y = 6$ and $x^2 + y^2 = 9$ need to be solved simultaneously.

First make one variable (x here) the subject of the linear equation.

$$2x - 3y = 6$$

$$\therefore\ 2x = 3y + 6 \qquad \text{[adding } 3y\text{]}$$

$$\therefore\ x = \frac{3y + 6}{2} \qquad \text{[dividing by 2]}$$

Now substitute this expression for x in the second equation.

$$\therefore\ \left(\frac{3y + 6}{2}\right)^2 + y^2 = 9 \qquad \text{[substituting for } x \text{ in (2)]}$$

$$\therefore\ \frac{9y^2 + 36y + 36}{4} + y^2 = 9 \qquad \text{[expanding]}$$

$$\therefore\ 9y^2 + 36y + 36 + 4y^2 = 36 \qquad \text{[multiplying by 4]}$$

$$\therefore\ 13y^2 + 36y = 0 \qquad \text{[simplifying]}$$

$$\therefore\ y(13y + 36) = 0 \qquad \text{[factorising]}$$

$$\therefore\ y = 0 \text{ or } 13y + 36 = 0$$

$$\therefore\ y = 0 \text{ or } y = \frac{-36}{13}$$

The line cuts the circle at (3, 0) and $\left(\frac{-15}{13}, \frac{-36}{13}\right)$.

When $y = 0$, $x = 3$ and when $y = \frac{-36}{13}$, $x = \frac{-15}{13}$. [substituting into $x = \frac{3y + 6}{2}$]

Unit 12.4 Activity 5E: Simultaneous linear and non-linear equations

Solve each of the following pairs of simultaneous equations:

1. $y = x - 1$, $x^2 + y^2 = 1$

2. $y = 2x + 1$, $x^2 + y^2 = 1$

3. $y = x + 2$, $y^2 + 2x^2 = 4$

4. $y = 2x - 3$, $x^2 + y^2 = 9$

5. $2y - 3x = 6$, $x^2 + y^2 = 9$

6. $y = x + 1$, $x^2 + y^2 = 5$

7. $y = x - 3$, $x^2 + y^2 = 29$

8. $x - 5y = 12$, $y^2 = 2x$

9. $4x + 3y = 25$, $xy = 12$

10. $2x + y = 7$, $xy = 6$

11. $x^2 + y^2 = 13$, $y = x + 1$

12. $x^2 + y^2 = 5$, $y + x = 3$

13. $xy = 8$

$x - y = 2$

14. $y^2 = 3x$

$3x - 2y = 8$

15. $y = \dfrac{1}{x+1}$

$2y + 3x = 4$

16. Give the coordinates of the points where the line $y = x - 1$ cuts the circle $x^2 + y^2 = 25$.

17. Find the x-coordinates of the points where the line $3x - 2y = 6$ cuts the parabola $y = (x - 1)(x - 3)$.

18. Give the coordinates of the points where the line $x + 5y = 2$ cuts the curve $y = 2x^2 - 3x - 26$.

Unit 12.4 Algebra and Graphs
Topic 6: Graphs of quadratic functions

Continuing the focus on graphs and functions (Syllabus p. 25), the material in Topic 6 is about sketching and interpreting graphs. It covers:

- Sketch and interpret features of quadratic graphs.
- Determining and applying an appropriate model for a situation involving graphs.
- Writing equations from a graph to solve a problem (a combination of only two different types of transformation is expected, eg $y = 2x^2 + 3$ or $y = (x - 2)^2 + 1$).
- Drawing a graph to find the solution to a problem.

The parabola

The equation of a **quadratic function** can be written in the form

$$y = ax^2 + bx + c, \text{ where } a, b, c \text{ are constants, } a \neq 0$$

The following are all examples of equations of quadratic functions:

$y = 3x^2 - 2x + 4$ $\quad$ $y = x^2 + 2x$ $\quad$ $y = -x^2 + 3$ $\quad$ $y = -2x^2$ $\quad$ $y = (x - 3)^2$

The graph of a quadratic function is called a **parabola**. The parabola is a smooth curve, with one **axis of symmetry** and a **turning point** called the **vertex**.

The graph of $y = x^2$

The simplest quadratic function has the equation $y = x^2$. The graph is drawn by setting up a table of values and plotting points.

A table of values and graph for $y = x^2$ are shown.

x	−4	−3	−2	−1	0	1	2	3	4
$y = x^2$	16	9	4	1	0	1	4	9	16

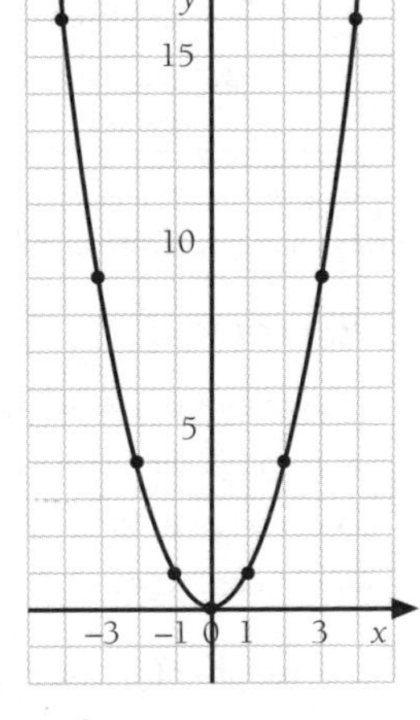

Note: 1. The graph is symmetrical about the y-axis, ie the line $x = 0$.

2. There is a vertex at (0, 0) giving a minimum value for y of 0 when $x = 0$.

3. The points are joined with a smooth curve. Notice how the curve 'flattens out' at the vertex; it is *not* a sharp point.

To see more clearly the behaviour of y between $x = -1$ and $x = 1$ consider the following table of values.

x	−1	$-\frac{3}{4}$	$-\frac{1}{2}$	$-\frac{1}{4}$	0	$\frac{1}{4}$	$\frac{1}{2}$	$\frac{3}{4}$	1
$y = x^2$	1	$\frac{9}{16}$	$\frac{1}{4}$	$\frac{1}{16}$	0	$\frac{1}{16}$	$\frac{1}{4}$	$\frac{9}{16}$	1

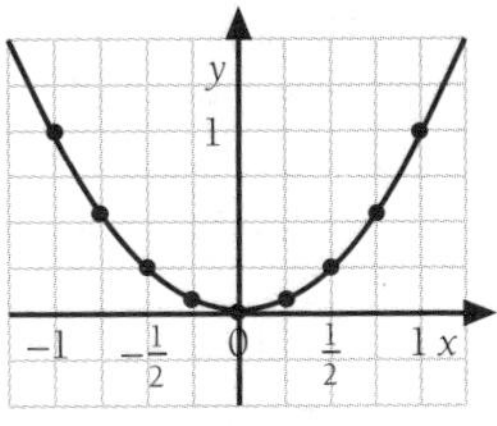

The 'magnification' of the part of the graph near the turning point shows the 'flattening out' of the curve much more clearly.

Transformations of the curve $y = x^2$

Changes to the basic equation of a parabola, $y = x^2$, cause various **transformations** of the curve. The general equation of a transformed parabola is

$$y = a(x - h)^2 + k \quad \text{where } a, h, k \text{ are constants}$$

Four transformations are now discussed. Their graphs are drawn by plotting points.

Graphs of the form $y = ax^2$

By changing the value of a in the equation $y = ax^2$, the steepness of the graph changes. The larger the size of a, the steeper the graph.

Example A

The graphs of $y = 2x^2$ and $y = \frac{1}{2}x^2$ are drawn below and compared with the graph of $y = x^2$.

x	−2	−1	0	1	2
x^2	4	1	0	1	4
$2x^2$	8	2	0	2	8
$\frac{1}{2}x^2$	2	$\frac{1}{2}$	0	$\frac{1}{2}$	2

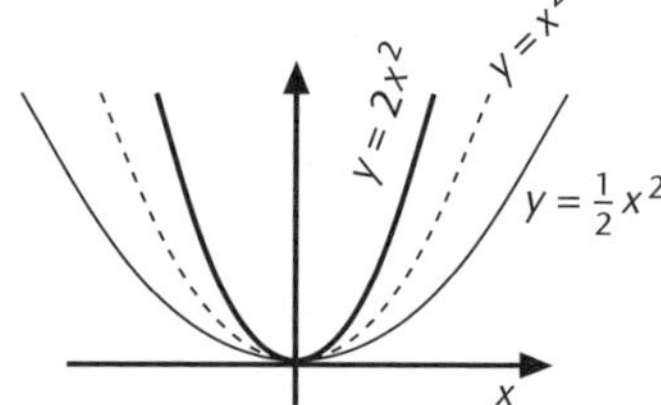

Note: 1. The graph of $y = ax^2$ is steeper than the graph of $y = x^2$ if $a > 1$. For example above, the graph of $y = 2x^2$ lies above the graph of $y = x^2$.

2. The graph of $y = ax^2$ is shallower than the graph of $y = x^2$ if $0 < a < 1$. For example above, the graph of $y = \frac{1}{2}x^2$ lies below the graph of $y = x^2$.

3. For each of these graphs, the vertex is (0, 0) and the axis of symmetry is the y-axis.

If a is negative, the parabola is **inverted** (upside-down). This gives a **maximum** value at the vertex, *not* a **minimum**. (Axis of symmetry remains the y-axis.)

Example B

The graphs of $y = -2x^2$ and $y = -\frac{1}{2}x^2$ are shown and compared with the graph of $y = -x^2$:
[Tabled values for x^2, $2x^2$, $\frac{1}{2}x^2$ (in Example A) are multiplied by −1.]

The graphs of $y = -2x^2$, $y = x^2$ and $y = -\frac{1}{2}x^2$ are *inverted* compared with the graphs of $y = 2x^2$, $y = x^2$ and $y = \frac{1}{2}x^2$.

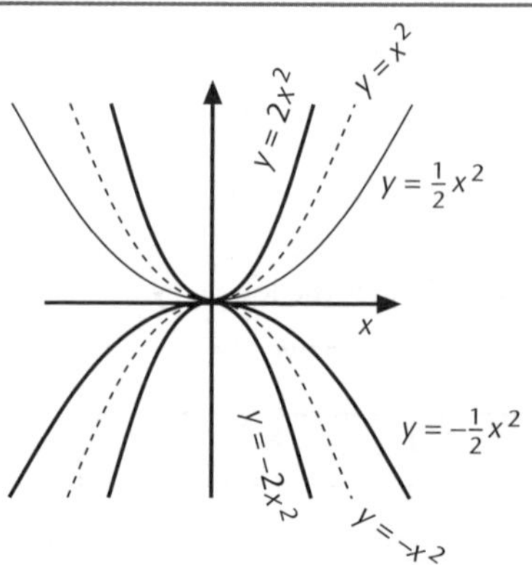

Graphs of the form $y = x^2 + k$

The graph of $y = x^2 + k$ is the graph of $y = x^2$ **transformed** by the **translation** $\begin{pmatrix} 0 \\ k \end{pmatrix}$.

In other words, it is the graph of $y = x^2$ shifted vertically k units.

- If k is positive, the shift is up.
- If k is negative, the shift is down.

Example C

By setting up a table of values, the graph of $y = x^2 + 2$ can be drawn and compared with the graph of $y = x^2$.

x	−3	−2	−1	0	1	2	3
x^2	9	4	1	0	1	4	9
$y = x^2 + 2$	11	6	3	2	3	6	11

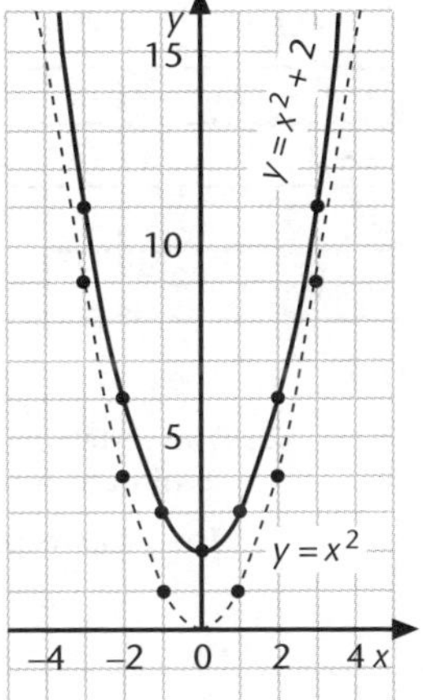

Note: The table and graph show that:

1. The graph of $y = x^2 + 2$ results from moving the graph of $y = x^2$ *up* 2 units.
2. The graph of $y = x^2 + 2$ is symmetrical about the y-axis.
3. The vertex is at (0, 2) and the minimum value for y is 2.

Graphs of the form $y = (x - h)^2$

The graph of $y = (x - h)^2$ is the graph of $y = x^2$ transformed by the translation $\begin{pmatrix} h \\ 0 \end{pmatrix}$ ie the graph of $y = x^2$ moved sideways h units.

- The axis of symmetry is $x = h$ and the vertex is $(h, 0)$.
- The y-intercept is $(0, h^2)$ and the x-intercept is $(h, 0)$.

Example D

Drawing up a table and plotting the graph of $y = (x - 2)^2$, and then comparing it with $y = x^2$, gives:

x	−1	0	1	2	3	4	5
$x - 2$	−3	−2	−1	0	1	2	3
$y = (x - 2)^2$	9	4	1	0	1	4	9

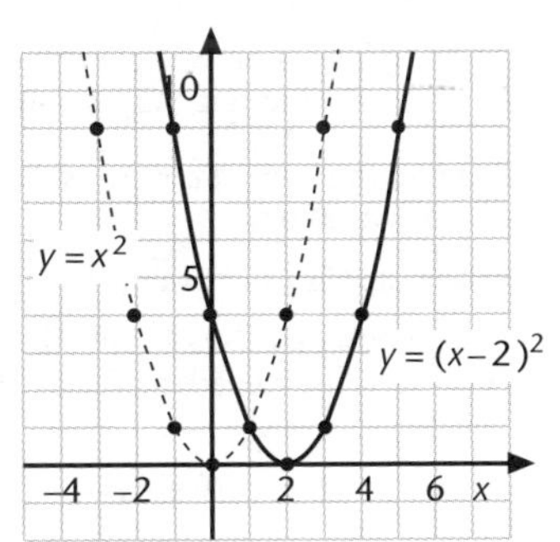

Note:

1. The axis of symmetry is $x = 2$ and the vertex is (2, 0).
2. The y-intercept is (0, 4) and the x-intercept is (2, 0).
3. The graph of $y = x^2$ has been shifted 2 units to the right.

Graphs of the form $y = (x - h)^2 + k$

The graph of $y = (x - h)^2 + k$ is the graph of $y = x^2$ transformed by the translation $\begin{pmatrix} h \\ k \end{pmatrix}$, ie the graph of $y = x^2$ is shifted horizontally h units and vertically k units.

- The axis of symmetry is the line $x = h$.
- The vertex is (h, k), so the minimum value of y is k.
- The y-intercept is $(0, h^2 + k)$.

Example E

Q. Draw the graph of $y = (x + 3)^2 - 2$.

A. Drawing up a table and plotting the points gives:

x	−6	−5	−4	−3	−2	−1	0
$x + 3$	−3	−2	−1	0	1	2	3
$(x + 3)^2$	9	4	1	0	1	4	9
$y = (x + 3)^2 - 2$	7	2	−1	−2	−1	2	7

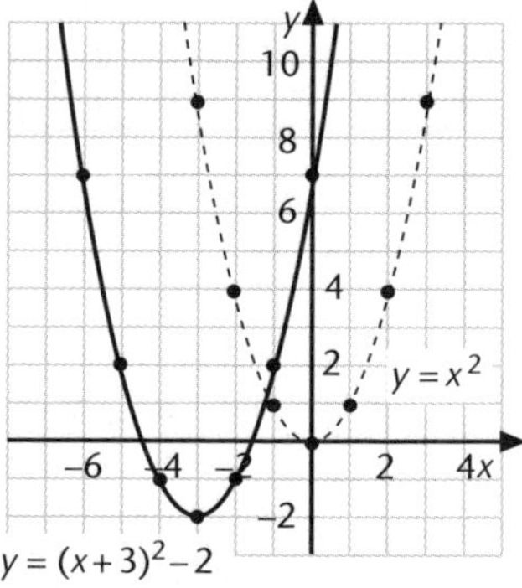

Note: 1. The axis if symmetry is the line $x = -3$.

2. The vertex is $(-3, -2)$, which is a **minimum turning point.**

3. The y-intercept is $(0, 7)$.

Sketching quadratic graphs by transformations

If the equation of a parabola is given in transformation form, $y = a(x - h)^2 + k$, its graph can be sketched by considering the various transformations involved.

Example F

1. The graph of $y = (x + 3)^2 - 2$ was drawn in Example E by plotting points. However, this graph can be drawn directly from the equation. The graph of $y = x^2$ is transformed as follows:

$$y = \underbrace{(x + 3)^2}_{\text{graph is moved 3 units left}} \underbrace{- 2}_{\text{graph is moved 2 units down}}$$

To find the y-intercept, substitute $x = 0$ in the equation.

The graph is as in Example E.

2. The graph of $y = -(x + 3)^2 - 2$ is the graph in part **1** inverted (the negative outside the bracket causes this).

When $x = 0$, $y = -(0 + 3)^2 = -11$

The graph is as shown.

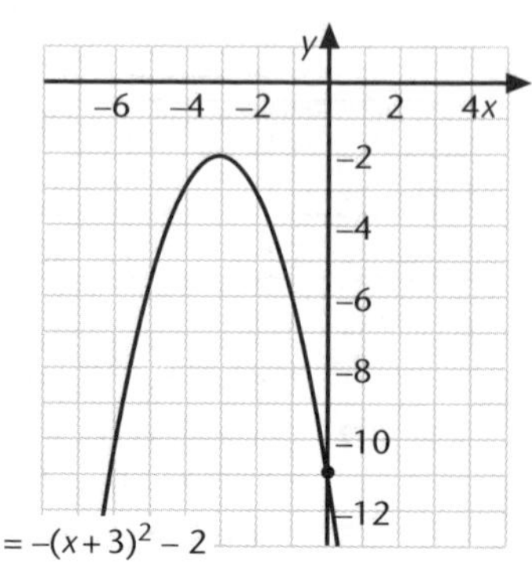

If the coefficient of x^2 is not ± 1, the steepness of the graph changes.

Example G

Q. Sketch the graph of $y = \frac{1}{2}(x - 4)^2 - 3$.

A. The graph of $y = x^2$ is transformed as follows:

$y = \underbrace{\tfrac{1}{2}}_{\text{shallower}} \underbrace{(x-4)^2}_{\text{4 units right}} \underbrace{-3}_{\text{down 3}}$

The vertex is at (4, –3) and the y-intercept is

$y = \frac{1}{2}(0 - 4)^2 - 3$ [substituting $x = 0$]

$= 5$

The graph is as shown.

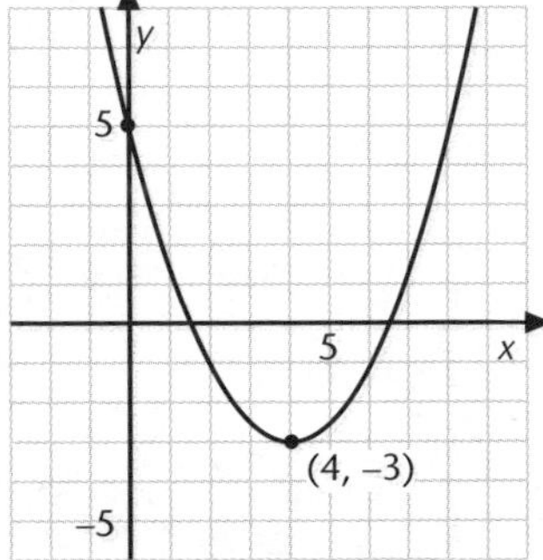

Note: 1. The basic shape of the parabola is the same as that of $y = \frac{1}{2}x^2$, ie from the vertex travel out ± 1 and up $\frac{1}{2}$ to reach a point on the curve. Another point is out ± 2 up 2 from the vertex, etc.

2. The exact positions of the x-intercepts are not given. (If required, substitute $y = 0$ in the equation of the curve and solve the resulting equation.)

Unit 12.4 Activity 6A: Quadratic graphs

1. Complete the tables of corresponding values of x and y which fit the equations:

a. $y = x^2 - 2$

x	–2	–1	0	1	2
x^2	4				
$y = x^2 - 2$	2				

b. $y = (x - 2)^2$

x	0	1	2	3	4
$x - 2$					2
$y = (x - 2)^2$					4

c. $y = (x + 3)^2$

x	–5	–4	–3	–2	–1
$x + 3$	–2				2
$y = (x + 3)^2$	4				

d. $y = 2 - x^2$

x	–2	–1	0	1	2
x^2					
$y = 2 - x^2$					–2

e. $y = x^2 + 3$

x	–2	–1	0	1	2
x^2					4
$y = x^2 + 3$					7

f. $y = 2x^2$

x	–2	–1	0	1	2
x^2					4
$y = 2x^2$					8

2. Match the graphs below with the equations from Question 1.

a.
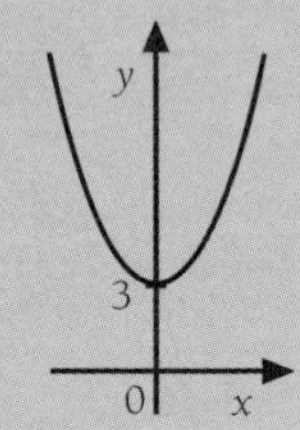

b.
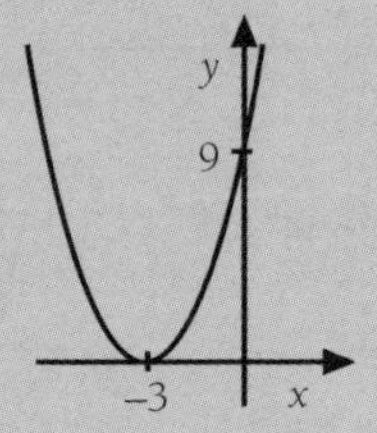

c.
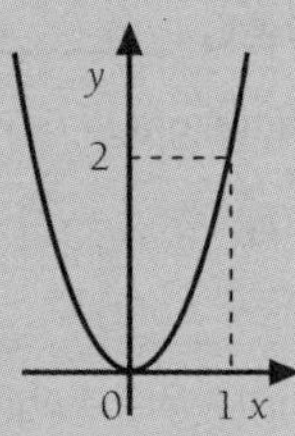

d.
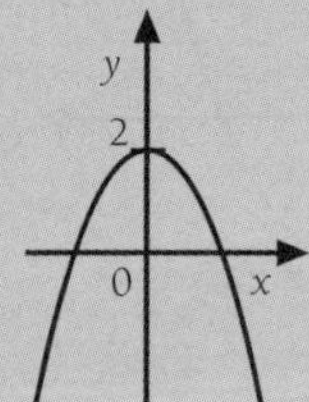

e.
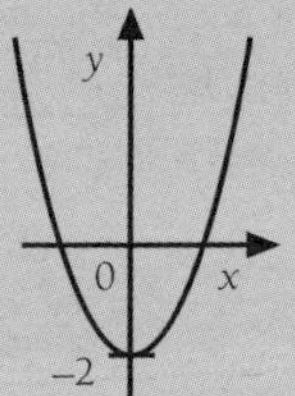

f.
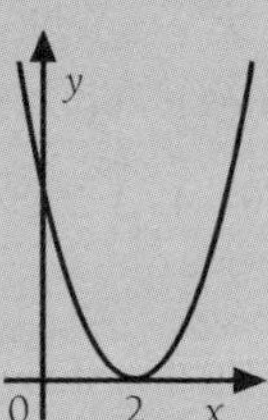

3. Sketch the graphs of the following equations, labelling carefully all intercepts (where possible) and the vertex for each graph.

a. $y = x^2 + 1$ **b.** $y = x^2 - 4$ **c.** $y = (x - 1)^2$

d. $y = (x + 1)^2$ **e.** $y = -2x^2$ **f.** $y = 4 - x^2$

4. Sketch the graphs of the following functions, labelling carefully all intercepts (where possible) and the vertex for each graph.

a. $y = (x + 1)^2 - 4$ **b.** $y = (x - 2)^2 + 3$ **c.** $y = (x + 2)^2 + 1$

d. $y = 2(x - 1)^2 - 3$ **e.** $y = -(x + 4)^2 + 1$ **f.** $y = -\frac{1}{2}x^2 - 2$

Sketching graphs of quadratic functions in factored form

When the equation of a quadratic function is given in factored form, eg $y = (x + 1)(x - 2)$, a different approach is used for drawing the graph. The steps are as follows:

1. Find the *x-intercepts* and the *y-intercept*.
2. Find the *axis of symmetry*.
3. Find the *vertex*.
4. Sketch the parabola.

Example H

Q. Sketch the graph of $y = (x - 4)(x + 2)$.

A. • Intercepts

For the y-intercept substitute $x = 0$

$y = (0 - 4)(0 + 2) = -4 \times 2 = -8$ $\therefore$ y-intercept is $(0, -8)$

For the x-intercepts substitute $y = 0$

$(x - 4)(x + 2) = 0 \Rightarrow x = 4$ or -2 $\therefore$ x-intercepts are $(4, 0)$ and $(-2, 0)$

- The axis of symmetry lies midway between the x-intercepts, so the axis of symmetry is:

  $x = \dfrac{4 + -2}{2}$ [averaging the values of the x-intercepts]

  $x = 1$ [shown by dotted line on graph]

- The vertex lies on the axis of symmetry ie on the line $x = 1$, and will have a y-coordinate of:

  $y = (1 - 4)(1 + 2)$ [substituting $x = 1$ in function]
  $= -3 \times 3$
  $= -9$

  $\therefore$ vertex is $(1, -9)$

The graph is as shown.

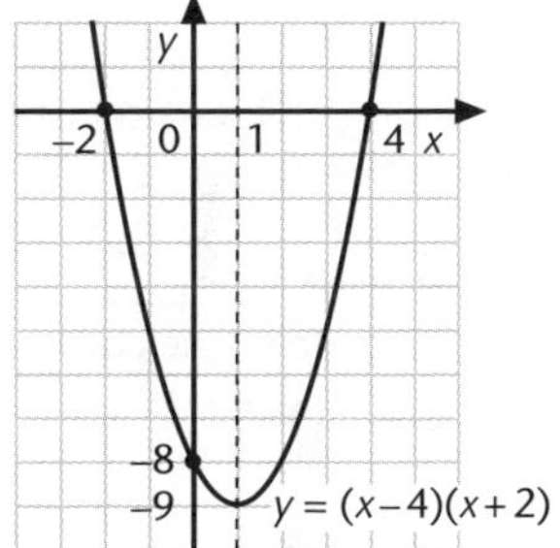

If the equation of the quadratic function is not in factored form, the quadratic will need to be factorised first.

Example I

Q. Sketch the graph of $y = x^2 + x - 6$

A. $y = (x + 3)(x - 2)$ [factorising]

- x-intercepts are -3 and 2

  y-intercept is $3 \times -2 = -6$

- Axis of symmetry is $x = \dfrac{-3 + 2}{2} = -\dfrac{1}{2}$

- Vertex is $(-\frac{1}{2}, -6.25)$

  [substituting $x = -\frac{1}{2}$ in equation]

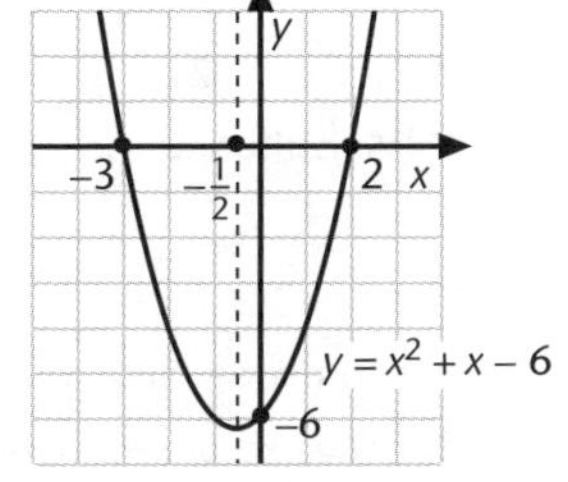

If the term in x^2 is negative then the graph will be inverted (upside down).

Example J

Q. Sketch the graph of $y = (2 - x)(4 + x)$

A. $y = (2 - x)(4 + x)$

- x-intercepts are 2 and -4

  y-intercept is $2 \times 4 = 8$

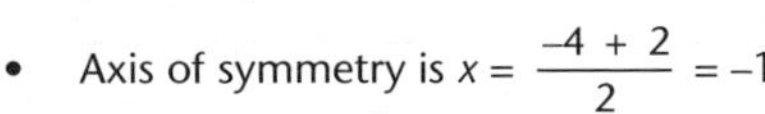

- Axis of symmetry is $x = \dfrac{-4 + 2}{2} = -1$

- Vertex is $(-1, 9)$

  [substituting $x = -1$ in equation]

Note: Expansion of $(2 - x)(4 + x)$ gives $-x^2 - 2x + 8$ so graph is 'upside down.' [coefficient of x^2 is negative]

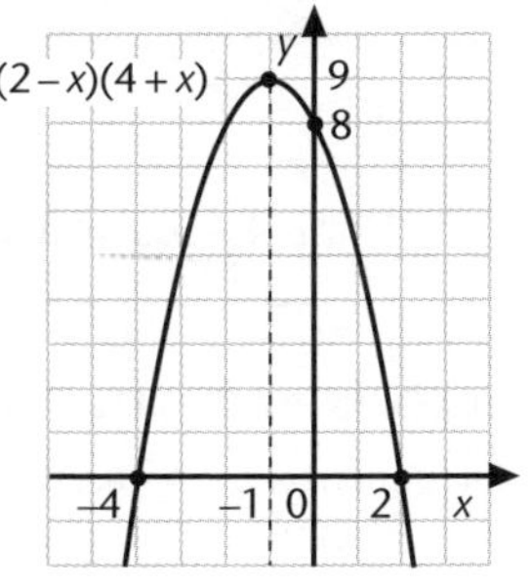

Writing equations of parabolas

By examining features of the graph of a parabola, such as intercepts, vertex, etc, the equation of a parabola can be written down.

Transformation method

If the vertex and one other point are known, the transformation method works well. The equation will be of the form

$$y = a(x - h)^2 + k$$

where h is the sideways translation, k is the vertical translation and a is a factor indicating the change in steepness.

Example K

Q. Write down the equations of the parabolas shown.

1. 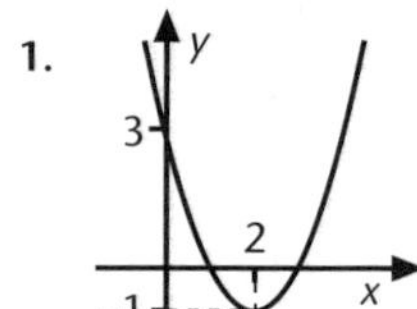

2. 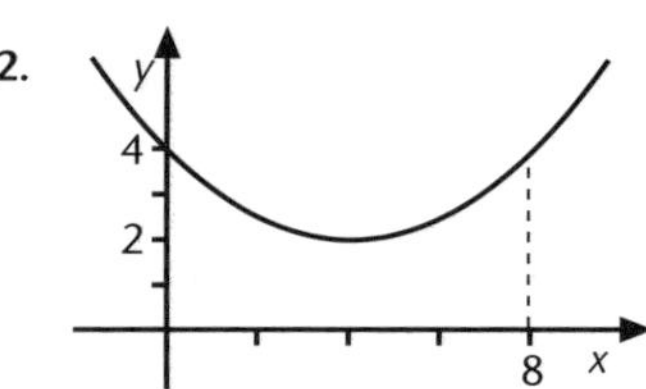

A. 1. The parabola has been translated by the **vector** $\begin{pmatrix}2\\-1\end{pmatrix}$ so its equation is of the form

$y = a(x - 2)^2 - 1$ [general transformation equation with $h = 2$, $k = -1$]

Substituting the point (0, 3) gives

$$\begin{aligned} 3 &= a(0 - 2)^2 - 1 \\ 3 &= 4a - 1 \\ 4 &= 4a \\ a &= 1 \end{aligned}$$

So the equation of the parabola is $y = (x - 2)^2 - 1$.

2. The axis of symmetry of the parabola is $x = 4$. The minimum value is 2.

$\therefore$ the parabola $y = x^2$ has been translated by the vector $\begin{pmatrix}4\\2\end{pmatrix}$ and so its equation is

$y = a(x - 4)^2 + 2$ [general transformation equation with $h = 4$, $k = 2$]

Substituting the point (0, 4) gives

$$\begin{aligned} 4 &= a(0 - 4)^2 + 2 \\ 4 &= 16a + 2 \quad \text{[simplifying]} \\ 16a &= 2 \\ a &= \frac{2}{16} = \frac{1}{8} \end{aligned}$$

So the equation of the parabola is $y = \frac{1}{8}(x - 4)^2 + 2$.

Factorised form method

If the x-intercepts are known, or can be worked out, then the quadratic function can be written down in factored form. The equation of the parabola will be of the form

$$y = a(x - b)(x - c)$$

where b and c are the x-intercepts and a is a factor affecting the steepness of the curve. Another known point is then substituted to evaluate a.

Example L

Q. Write down the equations of the parabolas shown.

1. 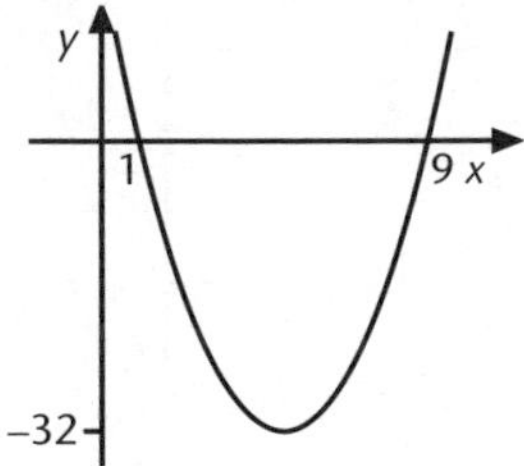

2. 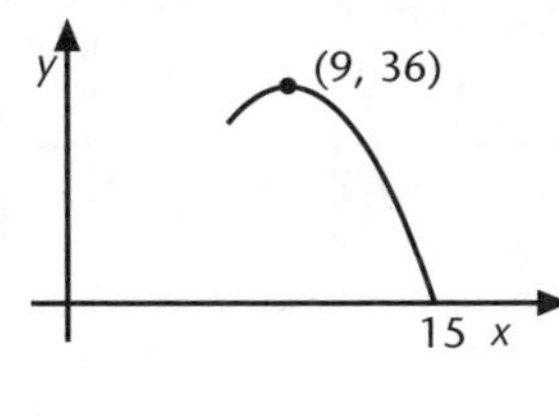

A. 1. The x-intercepts are 1 and 9. Thus, the equation is of the form $y = a(x - 1)(x - 9)$.

The axis of symmetry of the graph is $x = 5$ [since $\frac{1 + 9}{2} = 5$]

From the graph $y = -32$, when $x = 5$:

$-32 = a(5 - 1)(5 - 9)$ [substituting $x = 5$, $y = -32$ in $y = a(x - 1)(x - 9)$]
$-32 = a \times 4 \times -4$
$-32 = -16a$
$a = 2$

$\therefore$ the equation is $y = 2(x - 1)(x - 9)$

The equation could also be written as $y = 2x^2 - 20x + 18$ after expanding.

2. Only one intercept is given, the x-intercept, 15; the maximum point is (9, 36). This means the axis of symmetry is $x = 9$ and the other x-intercept is 3 (by symmetry, $15 - 9 = 6$, $9 - 6 = 3$). So, the equation will be of the form:

$y = -a(x - 3)(x - 15)$ [$-a$ since the parabola is 'upside down']

When $x = 9$, $y = 36$:

$\therefore 36 = -a(9 - 3)(9 - 15)$ [substituting]
$36 = 36a$ [simplifying]
$a = 1$

$\therefore$ the equation will be $y = -(x - 3)(x - 15)$ (or $y = -x^2 + 18x - 45$ in expanded form)

Unit 12.4 Activity 6B: Sketching quadratic graphs

1. Sketch the graphs whose equations are given:

a. $y = (x + 2)(x - 4)$ **b.** $y = (x - 5)(x - 1)$ **c.** $y = x(x - 3)$
d. $y = (x + 3)(x + 1)$ **e.** $y = (x - 4)(x - 2)$ **f.** $y = (3 - x)(1 + x)$
g. $y = x^2 + 6x + 5$ **h.** $y = x^2 + 4x - 12$ **i.** $y = 5x - x^2$

2. The sketch shows the graph of $y = x^2 + 2x - 8 = (x + 4)(x - 2)$.

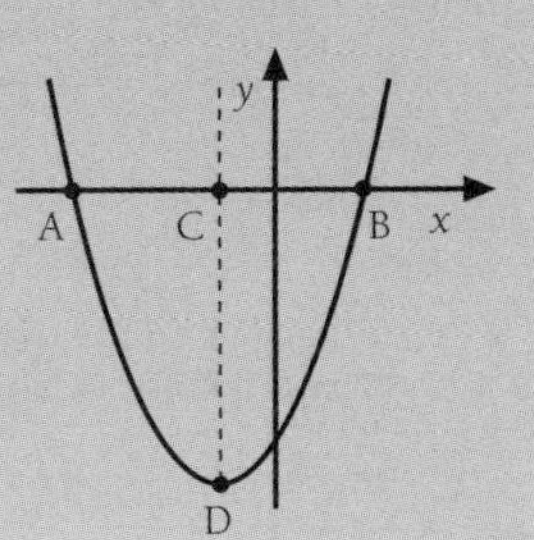

CD is the axis of symmetry.

a. Write the coordinates of:

i. A **ii.** B **iii.** C **iv.** D

b. What is the least possible value of $x^2 + 2x - 8$?

c. Write down the equation of the parabola in the form $y = a(x - h)^2 + k$.

3. Find the equations of the following parabolas (*not* drawn to scale):

a.

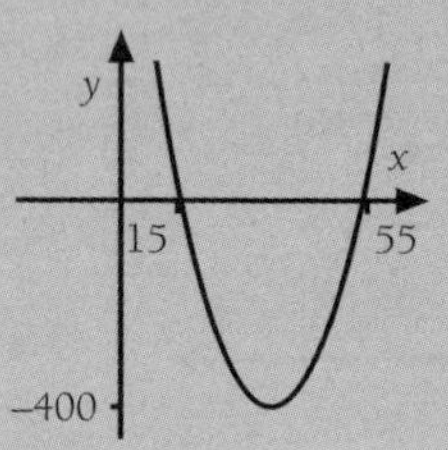

b.

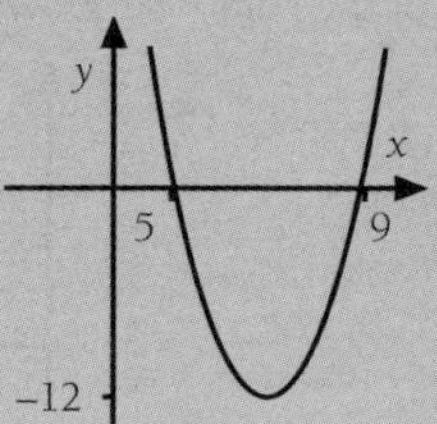

c.

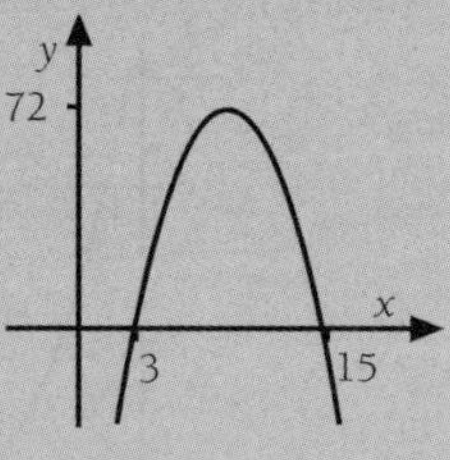

d.

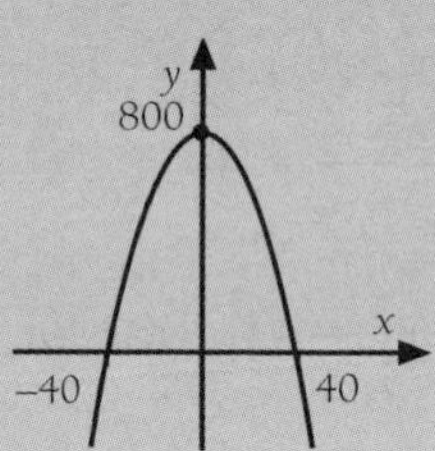

e.

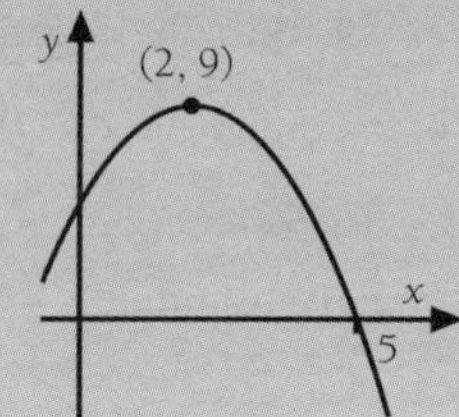

f.

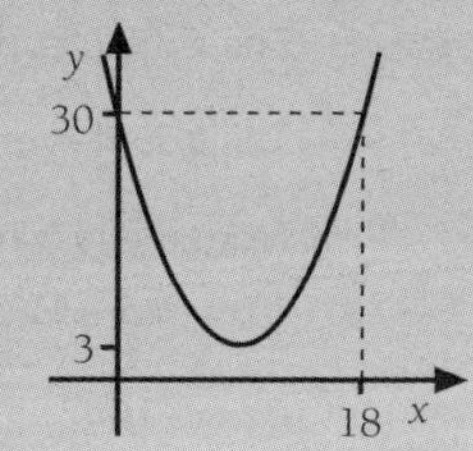

Using quadratic graphs to solve problems

Quadratic functions and their graphs can be used to model many different situations and to solve many problems in a practical context.

Example M

Q. Johnny's birthday present is a bow which fires a small arrow in a parabolic arc.

The equation of the path of the arrow is

$$h = \frac{(4 - d)(d + 2)}{5}$$

where d is the distance in metres from a marker and h is the height in metres of the arrow above the ground.

1. How far away from Johnny does the arrow land?
2. What is the maximum height the arrow reaches?
3. What is the height of the arrow as it passes over the marker?

A. **1.** When the arrow lands, $h = 0$ [height of arrow above ground is 0 metres]

Solving $\frac{(4 - d)(d + 2)}{5} = 0$

gives $(4 - d)(d + 2) = 0$ [multiplying by 5]

$d = 4$ or -2

Relative to the marker, Johnny is at position –2 and the arrow lands 6 m away at position 4.

2. Arrow reaches maximum height halfway between Johnny and the landing point, ie at $d = \dfrac{-2 + 4}{2} = 1$

   When $d = 1$ $\quad h = \dfrac{(4 - 1)(1 + 2)}{5}$ $\quad$ [substituting $d = 1$ in equation of curve]

   $\quad\quad\quad\quad = 1.8$ m

   The maximum height reached is 1.8 m.

3. As the arrow passes over the marker, $d = 0$

   $\therefore h = \dfrac{(4 - 0)(0 + 2)}{5} = 1.6$ m

The properties of the parabola can be used to find maximum or minimum values of quantities modelled by quadratic functions.

Example N

Q. Sue has 100 metres of netting and wishes to build a rectangular pen in the middle of her garden. What are the dimensions of the pen that will give the rectangle the largest area?

A. Let the length of the rectangle be x and the width y.

Hence
$$x + y + x + y = 100 \quad \text{[perimeter = 100 m]}$$
$$2x + 2y = 100$$
$$x + y = 50 \quad \text{[dividing by 2]}$$
$$y = 50 - x \quad \text{[rearranging]}$$

The area, A, of the pen is
$$A = xy$$
$$= x(50 - x) \quad \text{[substituting for } y\text{]}$$
$$= 50x - x^2$$

From the table alongside, the graph of $A = 50x - x^2$ can be drawn.

x	0	5	10	15	20	25	30	35	40	45	50
A	0	225	400	525	600	625	600	525	400	225	0

The graph of $A = 50x - x^2$ shows that the maximum area is 625 m^2 when $x = 25$ m, and hence $y = 50 - x = 25$, ie a square pen of side 25 metres gives the maximum area.

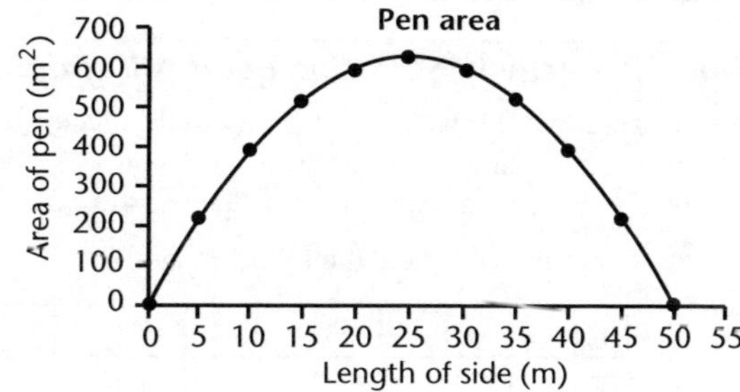

Note: The graph of $A = 50x - x^2$ can also be drawn by factorising first to give $A = x(50 - x)$, then using factored form techniques (x-intercepts at (0, 0) and (50, 0), vertex at $x = 25$, $A = 625$ etc).

The equation of a curve may need to be found in order to solve a problem.

Example O

Q. A large trough is made so that its cross-section is in the shape of a parabola which is 3 m across. A pole inserted in the trough one metre out from one side of the trough shows a depth of 3 m. What is the depth of the trough at its deepest point?

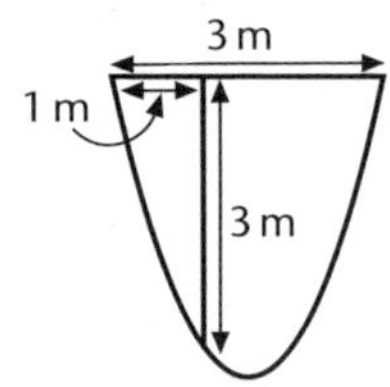

A. The equation of the parabola drawn on the axes shown is

$y = k(x + 1)(x - 2)$
[since $x = -1$ and $x = 2$ are the x-intercepts]

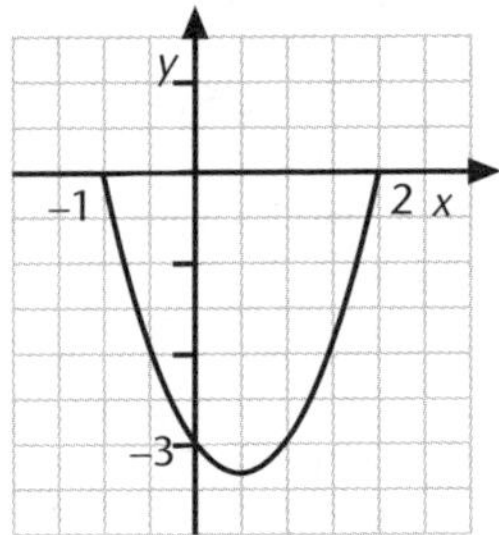

Since the point (0, –3) lies on the curve, it follows that

$-3 = k(0 + 1)(0 - 2)$ [substituting $x = 0$, $y = -3$ in equation of curve]

$-3 = -2k$ [simplifying]

$k = \frac{-3}{-2}$ [dividing by –2]

$= 1.5$

The equation of the parabola is $y = 1.5(x + 1)(x - 2)$.

The trough is deepest halfway between the intercepts, ie at $x = \frac{-1 + 2}{2} = 0.5$

At $x = 0.5$, $y = 1.5(0.5 + 1)(0.5 - 2)$ [substituting $x = 0.5$ in equation]

$= -3.375$

The trough is 3.375 m deep at its deepest point.

Unit 12.4 Activity 6C: Using quadratic graphs to solve problems

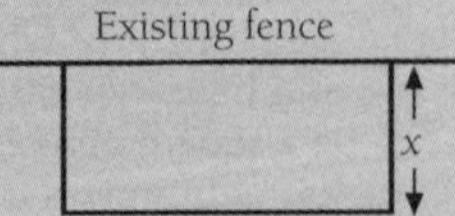

1. A farmer has 24 m of fencing and builds a rectangular pen using an existing fence as one side. The area, A m^2, of the pen is given by $A = 24x - 2x^2$, where x is the width of the pen.

a. Copy and complete the following table:

x	0	1	2	3	4	5	6	7	8	9	10	11	12
A	0	22	40	54									

b. Draw the graph of $A = 24x - 2x^2$, using axes as shown.

c. What is the largest possible area that can be fenced?

d. Why is $x \geq 0$?

e. Use your graph to find the width of the pen when its area is 60 m^2.

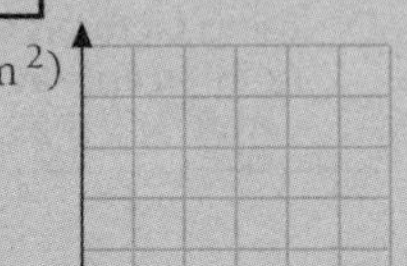
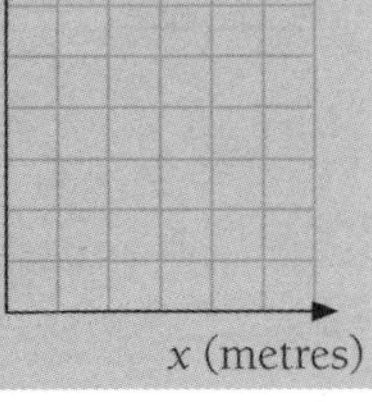

2. David finds that the depth of water in a small drain on his farm is related to the distance across the drain by the quadratic equation

$d = (w - 10)(w - 70)$

where d is the depth, in millimetres, of the water in the drain and w is the distance, in centimetres, measured from a post on one of its banks.

a. Graph the equation for the depth of water in the drain. Draw the graph for values of w from 0 to 80. (Use axes as shown)

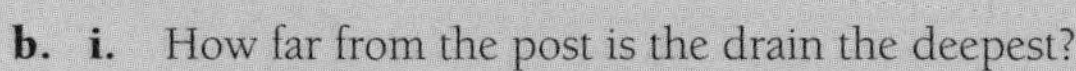

b. i. How far from the post is the drain the deepest?
ii. What is the depth at this distance?

c. i. How wide is the drain at water level?
ii. How is this shown on the graph?

d. What are the two distances from the post that the depth of water is –500 mm?

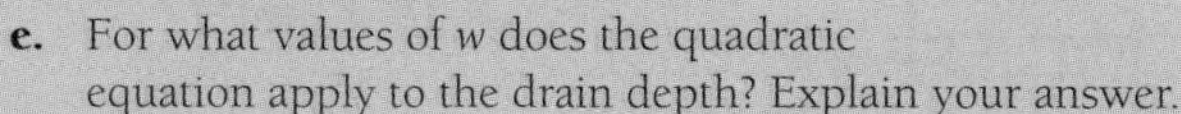

e. For what values of w does the quadratic equation apply to the drain depth? Explain your answer.

f. Give some practical reasons why the *actual* depth of water in the drain may be different from that shown by the graph.

3. On Kara's farm there is an electrical wire suspended between two poles across a gully. The poles are on the edges of the gully and the wire is at the same height on both sides of the gully. The curve of the wire can be modelled by the equation

$h = (d - 6)^2 + 3$

where h is the height, in metres, of the wire above the floor of the gully and d is the distance, in metres, from the left-hand post.

a. Draw the graph of $h = (d - 6)^2 + 3$ for the values of distance, d, from 0 to 12.

b. What is the minimum height that the wire is above the floor of the gully?

c. i. How far apart are the two poles?
ii. How is this shown on the graph?

4. Jodi has a piece of string hung across a wall in her room that has her birthday cards on it. The string takes on a shape which is approximately that of a parabola. The ends of the string are 2 m above the floor of each end, and the string is 1.5 metres above the floor at the lowest point. The ends of the string are 4 metres apart.

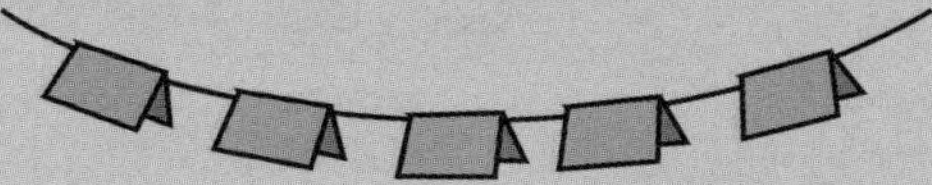

a. Find an equation that represents the shape of the string and draw its graph.

b. Jodi is 1.65 m tall. She stands under the string and 1.2 m to the right of the lowest point of the string. Will her head touch the string?

5. Maratha has a silage stack on her farm built into a small pit. She measures the height, *h* cm, from the floor of the pit, every metre from the left-hand end.

Maratha's data are shown in the table below. A graph of these data is shown

Distance *d* (m)	0	1	2	3	4	5	6	7	8	9	10
Height *h* (cm)	10	60	110	160	210	260	310	360	320	200	0

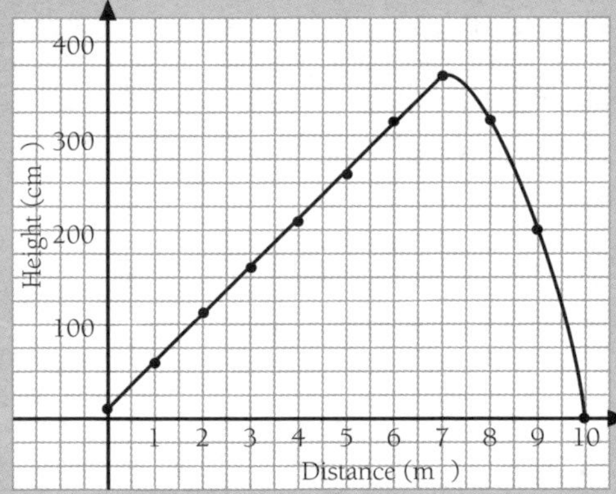

a. Write the equation for *h* in terms of *d* that models the **first seven metres** of the stack.

b. Write the equation for *h* in terms of *d* that models the **last three metres** of the stack. (Assume (7, 360) is the turning point of this quadratic graph).

6. The path of a bouncing ball was recorded by multiple flash photography.

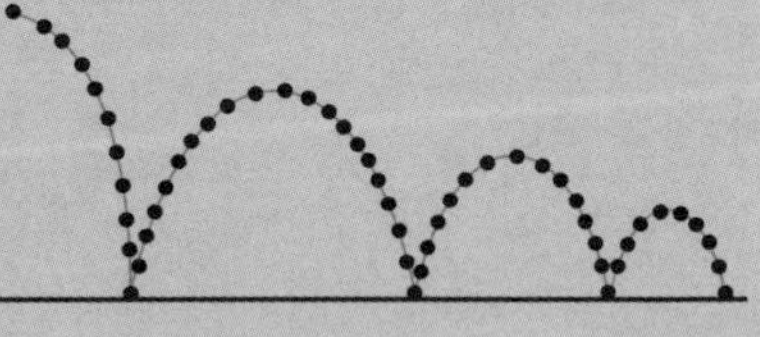

The flash operated every 0.1 seconds and the height of the ball from the floor was measured. The measurements of the path of the first bounce are shown in the table below.

Time (secs)	1.0	1.1	1.2	1.3	1.4	1.5	1.6	1.7	1.8	1.9	2.0
Height (m)	0	0.45	0.80	1.05	1.20	1.25	1.20	1.05	0.80	0.45	0

a. Graph the results, joining the points with a smooth curve.

b. What was the maximum height reached by the ball on its first set of bounces?

c. Write an equation for the curve.

d. How high would the ball have been above the floor after 1.25 seconds?

7. A gutter for the roof of a shed is to have a rectangular cross-section and to be made from a long strip of sheet metal 30 cm wide. The sides are to be turned up at right angles to the base and the sides are to be of equal height.

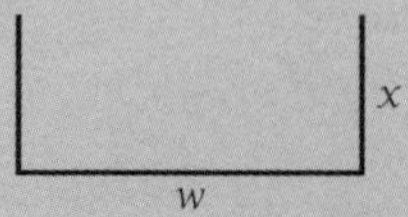

For what value of *x* is the cross-sectional area the greatest?

Hint: Express *w* in terms of *x* then sketch the graph of the quadratic function which models the area of the cross-section.

Unit 12.4 Algebra and Graphs

Topic 7: Cubic graphs

In Topic 7 the focus is on drawing straightforward non-linear graphs:
- Draw factorised polynomials.
- Identify and interpret features of graphs, including intercepts, symmetry, and behaviour of graphs at large values of x and y.

Cubic graphs

Cubic graphs are graphs of **cubic functions** which are **polynomials** of **degree** 3.

Cubic graphs can be written in the form:

$y = ax^3 + bx^2 + cx + d$ where a, b, c and d are numbers

The graph of the simplest cubic, $y = x^3$, is drawn by plotting points.

Example A

The graph of $y = x^3$ can be drawn by filling in a table then plotting the points:

x	−10	−2	−1	0	1	2	10
x^3	−1 000	−8	−1	0	1	8	1 000

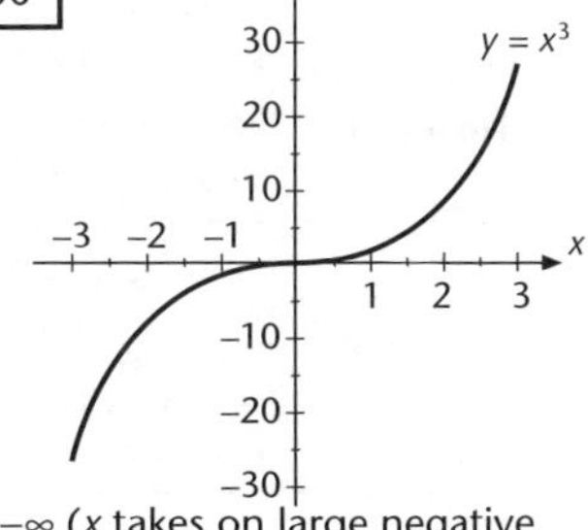

Notice that the graph has **point symmetry** about the point (0, 0).

The x-intercept is $x = 0$ and the y-intercept is $y = 0$.

The graph of $y = x^3$ can be transformed in a similar way to the transformations of

$y = x^2$ in the previous chapter.

As $x \to \infty$ (x takes on large positive values), $y \to \infty$ also. As $x \to -\infty$ (x takes on large negative values), $y \to -\infty$ also.

Cubic graphs in factored form are sketched by finding the x- and y-intercepts and sketching the curve through them.

Example B

The graph of $y = (x - 1)(x - 3)(x - 4)$ can be sketched by finding the intercepts:

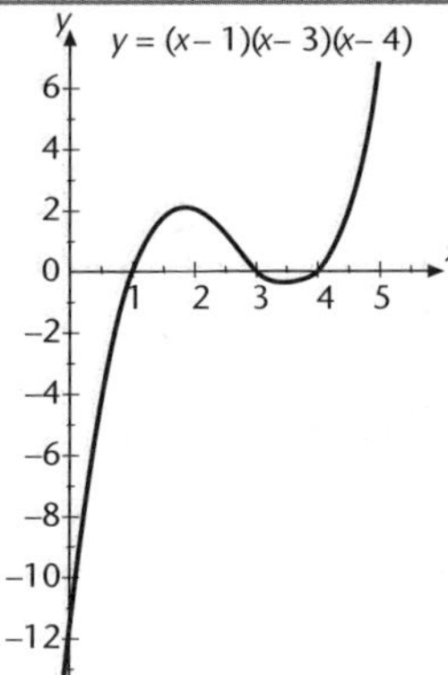

Substituting $x = 0$ in $y = (x - 1)(x - 3)(x - 4)$, gives:

$$y = (0 - 1)(0 - 3)(0 - 4)$$
$$= (-1) \times (-3) \times (-4)$$
$$= -12$$

The x-intercepts are found by substituting $y = 0$, giving:

$$(x - 1)(x - 3)(x - 4) = 0$$
$$\therefore \; x = 1, 3 \text{ or } 4$$

As $x \to \infty$, $y \to \infty$ and as $x \to -\infty$, $y \to -\infty$.

Example C

For the graph of $y = (x + 1)^2(2 - x)$,

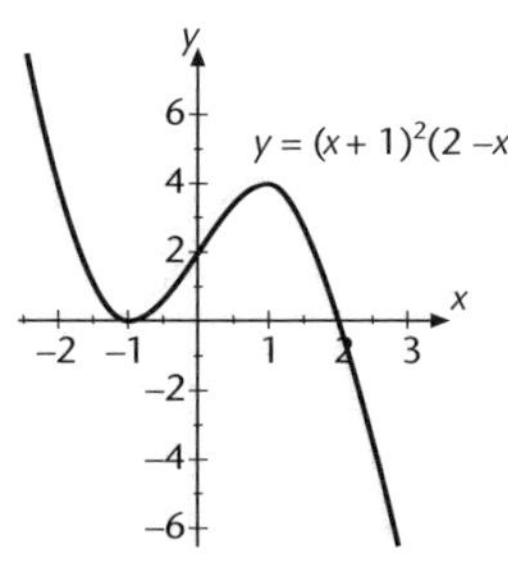

the y-intercept is 2. [substituting $x = 0$]

The x-intercepts are found by setting $y = 0$, giving

$(x + 1)^2 (2 - x) = 0$, so the x-intercepts are -1 and 2.

As $x \to \infty$, $y \to \infty$ and as $x \to -\infty$, $y \to -\infty$.

Note: **1.** The graph *touches* the x-axis at -1. In cubics like $y = (x + 1)^2(2 - x)$ where one of the factors of the polynomial is a **perfect square**, the graph will *touch* the x-axis rather than cut it.

2. In the second bracket the coefficient of x is negative which means that when $y = (x + 1)^2(2 - x)$ is expanded, the coefficient of x^3 is negative. Thus for large positive x, y is negative.

By considering intercepts, the equation of a cubic can be written down.

Example D

Q. What is the equation of the graph shown?

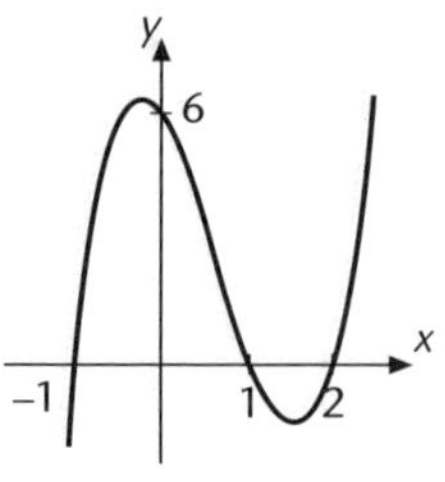

A. The graph has x-intercepts at -1, 1 and 2 so the equation of the graph is $y = k(x + 1)(x - 1)(x - 2)$, where k is a constant.

The graph has a y-intercept of 6.

Substituting $x = 0$, $y = 6$ in $y = k(x + 1)(x - 1)(x - 2)$, gives:

$$6 = k(0 + 1)(0 - 1)(0 - 2) \qquad \text{[substituting]}$$
$$\therefore \; 6 = 2k$$
$$\therefore \; k = 3$$

Hence the equation of the graph is $y = 3(x + 1)(x - 1)(x - 2)$.

Unit 12.4 Activity 7: Cubic graphs

1. Sketch the graphs of the following cubic functions:
   a. $y = (x - 1)(x - 2)(x + 4)$
   b. $y = (x + 3)(x + 2)(x - 1)$
   c. $y = (2 - x)(2 + x)(x - 1)$
   d. $y = x(x - 1)(x + 3)$
   e. $y = x(x - 1)^2$
   f. $y = (x - 2)^2(x - 3)$
   g. $y = (x - 1)^2(2 - x)$
   h. $y = -(2 - x)(x + 1)(x - 3)$

2. Write down the equations of the cubic functions whose graphs are shown below:

a.

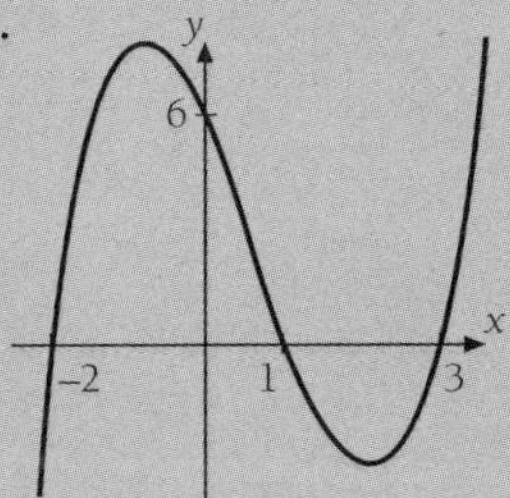

b.

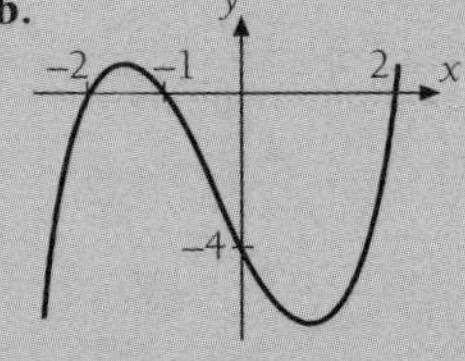

c.

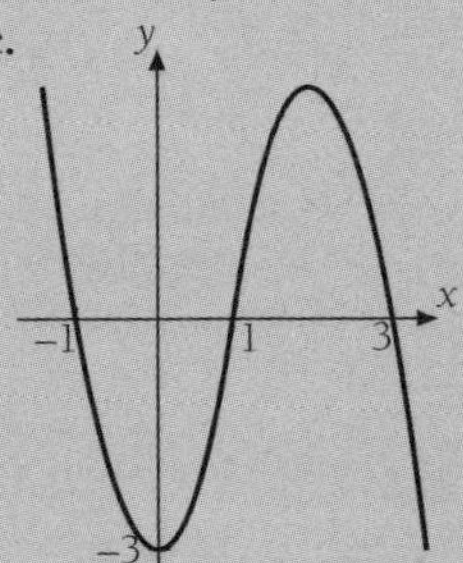

d.

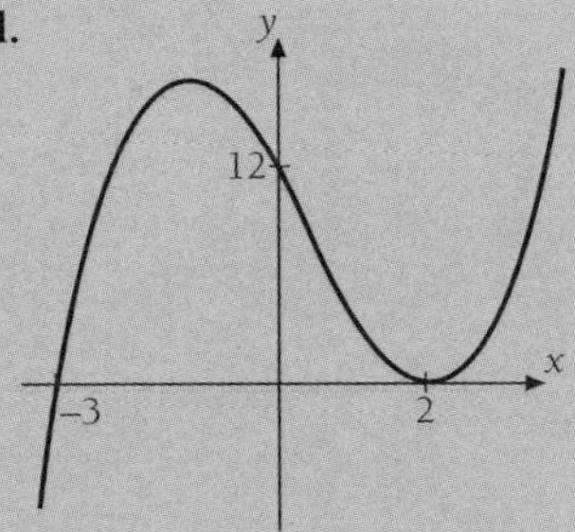

3. Draw the graph of the volume of a box against its length. Its height is two metres less than the length and its width is one metre less than the length.

4. Draw a graph of the volume of a square-based box against side length of the base. Its height exceeds the base length by one metre.

5. Records were kept through the first few months of this year of the kina value above K1.50 of one kilogram of a commodity. The results of this survey are shown in the graph.

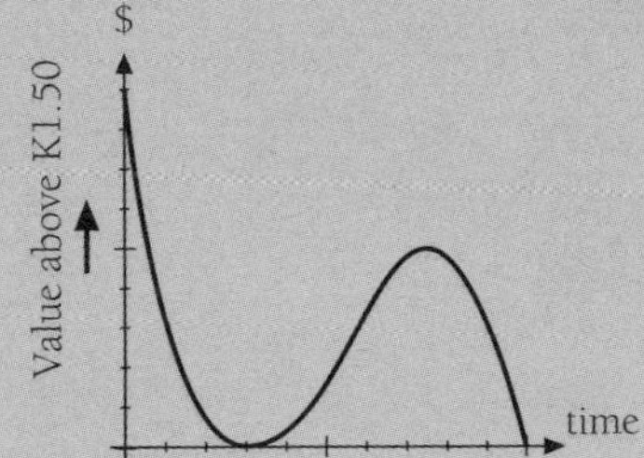

   a. If the horizontal marks represent the passage of one month then what were the dates when the value of one kg was K1.50?
   b. What was the value at the beginning of the year? (You may assume that each vertical mark represents one kina.)
   c. When was the value of one kilogram in excess of K5.50?
   d. What is an equation which would describe such a graph?

Unit 12.4 Algebra and Graphs

Topic 8: Rational expressions

The material in Topic 8 covers the manipulation of algebraic expressions and solving equations.

- Simplify rational expressions.
- Factorise expressions including quadratics.

Rational expressions are sometimes called **algebraic fractions**. They are fractions involving variables (letters) as well as numbers. The procedures used to **simplify** rational expressions are similar to those used with fractions involving only numbers.

Multiplication

Numerators are multiplied together and **denominators** are multiplied together.

Example A

Q. Express $\frac{x}{y} \times \frac{a}{c}$ as a single fraction.

A. $\frac{x}{y} \times \frac{a}{c} = \frac{xa}{yc}$ [numerators then denominators are multiplied together]

Answers are simplified by **cancelling** any common factors.

Example B

Q. Simplify $\frac{x}{y} \times \frac{y}{z} \times \frac{z^2}{x^3}$.

A. $\frac{x}{y} \times \frac{y}{z} \times \frac{z^2}{x^3} = \frac{xyz^2}{x^3yz}$

$= \frac{x.y.z.z}{x.x^2.y.z}$ [writing $z^2 = z \,.\, z$, $x^3 = x \,.\, x^2$]

$= \frac{z}{x^2}$ [cancelling common factors x, y and z]

Some expressions are best factorised *before* any multiplication is done, so that common factors can be identified.

Example C

$\frac{x^2 + 3x + 2}{(x+2)^2} \times \frac{5}{x+1} = \frac{(x+1)(x+2)}{(x+2)(x+2)} \times \frac{5}{(x+1)}$ [factorising]

$= \frac{5}{x+2}$ [cancelling $(x + 1)$ and $(x + 2)$]

Unit 12.4 Activity 8A: Multiplication of rational expressions

1. Simplify each of the following:

a. $\frac{3}{4} \times \frac{a}{b}$

b. $\frac{5a}{b} \times \frac{c}{d}$

c. $\left(\frac{3}{4} \times \frac{a}{d}\right) \times \frac{c}{b}$

d. $\frac{a}{b} \times \frac{a}{c}$

e. $\frac{3}{a} \times \frac{ab}{6}$

f. $\frac{pq}{r} \times \frac{3r}{q}$

g. $\frac{5a^2}{3} \times \frac{9}{a}$

h. $\left(\frac{5}{a} \times \frac{a^2}{10}\right) \times \frac{1}{3}$

i. $\frac{6k}{3q} \times \frac{qk}{3}$

j. $3\frac{1}{4} \times \frac{b}{2}$

k. $\frac{x^2}{y} \times \frac{y^3}{x^4} \times \frac{y}{x}$

l. $\frac{2x^2}{y} \times \frac{xy^2}{6x^4}$

2. Simplify each of the following (factorise first):

a. $\frac{x+1}{x+3} \times \frac{x^2+3x}{x^2+2x+1}$

b. $\frac{x}{x^2-1} \times \frac{x^2+4x+3}{x^2+3x}$

c. $\frac{(x+3)^2}{x^2+5x+6} \times \frac{x^2-9x+14}{x^2-4x-21}$

d. $\frac{4x+8}{x^2+11x-28} \times \frac{x^2+9x+14}{x^2+4x+4}$

Division

Division means multiplication by the **reciprocal** of the **divisor** (the reciprocal of $\frac{a}{b}$ is $\frac{b}{a}$).

Example D

Q. Simplify $\frac{x^2}{3} \div \frac{x}{9}$

A. $\frac{x^2}{3} \div \frac{x}{9} = \frac{x^2}{3} \times \frac{9}{x}$ [multiplying by $\frac{9}{x}$ which is the reciprocal of $\frac{x}{9}$]

$= \frac{9x^2}{3x}$ [multiplying numerators then denominators]

$= 3x$ [cancelling common factors 3, x]

Remember that expressions must be in factored form before common factors can be cancelled.

Example E

$\frac{x^2-y^2}{x^2+2xy+y^2} \div \frac{ax-ay}{2x+2y} = \frac{x^2-y^2}{x^2+2xy+y^2} \times \frac{2x+2y}{ax-ay}$ [multiplication by reciprocal]

$= \frac{(x-y)(x+y)}{(x+y)(x+y)} \times \frac{2(x+y)}{a(x-y)}$ [factorising]

$= \frac{2}{a}$ [cancelling]

Unit 12.4 Activity 8B: Division of rational expressions

1. Simplify each of the following:

a. $\frac{2}{3} \div \frac{a}{c}$ **b.** $\frac{ab}{c} \div \frac{d}{e}$ **c.** $\frac{x^2}{y} \div \frac{y}{3}$

d. $\left(\frac{2}{5} \times \frac{a}{3}\right) \div \frac{2}{3}$ **e.** $\left(\frac{a}{c} \times \frac{5}{3}\right) \div \left(\frac{a}{b} \times \frac{5}{3}\right)$ **f.** $\frac{ab}{3} \div \frac{a^2}{9}$

g. $\frac{4a^2b}{3} \div \frac{2ab}{3}$

2. Simplify each expression to a single fraction.

a. $\frac{2}{3} \div \frac{5}{a+1}$ **b.** $\frac{5}{7} \div \frac{x}{x^2-1}$ **c.** $\frac{b-1}{3} \div \frac{3}{4}$

d. $\frac{2}{3} \div \left(\frac{y}{x+y}\right)$ **e.** $\frac{3x-1}{2} \div \frac{2}{3}$ **f.** $\frac{3x}{x+1} \div 2\frac{1}{2}$

3. Simplify each expression (factorise first where possible).

a. $\frac{5x-10}{x+3} \div \frac{x-2}{x+3}$ **b.** $\frac{x+4}{x-1} \div \frac{x+4}{x+2}$ **c.** $\frac{x^2+x}{x+2} \div \frac{x}{x^2-4}$

d. $\frac{x+3}{x+7} \div \frac{x^2+4x+3}{x^2+10x+21}$ **e.** $\frac{x^2+2x+1}{x^2+3x+2} \div \frac{x^2+6x+5}{x^2-4}$

Addition and subtraction

Addition and subtraction of fractions involves changing the fractions to **equivalent fractions** with a **common denominator** (choose the lowest common multiple of the denominators).

Example F

$\frac{x}{3} + \frac{y}{4} = \frac{4x}{12} + \frac{3y}{12}$ [changing both fractions to equivalent fractions with common denominator 12]

$= \frac{4x + 3y}{12}$ [expressing as a single fraction]

Example G

Q. Express $\frac{a}{x^2} - \frac{b}{ax^3} - \frac{1}{a^2}$ as a single fraction.

A. $\frac{a}{x^2} - \frac{b}{ax^3} - \frac{1}{a^2} = \frac{a^3x}{a^2x^3} - \frac{ab}{a^2x^3} - \frac{x^3}{a^2x^3}$ [changing to equivalent fractions with common denominator a^2x^3]

$= \frac{a^3x - ab - x^3}{a^2x^3}$ [expressing as a single fraction]

Use brackets to avoid errors with negative signs.

Example H

$$\frac{a+3}{2}-\frac{a-3}{3}=\frac{3a+9}{6}-\frac{2a-6}{6}$$ [changing to a common denominator of 6]

$$=\frac{(3a+9)-(2a-6)}{6}$$ [using brackets to avoid errors]

$$=\frac{3a+9-2a+6}{6}$$ [removing brackets (note 6 is positive)]

$$=\frac{a+15}{6}$$ [simplifying]

Factorise denominators first (where possible) so that the lowest common multiple of the denominators can be seen more easily.

Example I

$$\frac{3}{x+y}-\frac{2}{x-y}+\frac{1}{x^2-y^2}$$

$$=\frac{3}{x+y}-\frac{2}{x-y}+\frac{1}{(x-y)(x+y)}$$ [factorising the denominator]

$$=\frac{3(x-y)}{(x+y)(x-y)}-\frac{2(x+y)}{(x+y)(x-y)}+\frac{1}{(x+y)(x-y)}$$ [changing to a common denominator]

$$=\frac{3(x-y)-2(x+y)+1}{(x+y)(x-y)}$$

$$=\frac{3x-3y-2x-2y+1}{(x+y)(x-y)}$$ [removing brackets]

$$=\frac{x-5y+1}{(x+y)(x-y)}$$ [simplifying]

Unit 12.4 Activity 8C: Addition and subtraction of rational expressions

1. Express each of the following as a single fraction in simplest form:

a. $\frac{a}{b}+\frac{b}{3}$ **b.** $\frac{a}{8}+\frac{a}{4}$ **c.** $\frac{a}{b}+\frac{a}{d}$ **d.** $\frac{a}{b}+\frac{1}{4}$

e. $\frac{a}{3}-\frac{b}{4}$ **f.** $\frac{a}{c}-\frac{b}{a}$ **g.** $\frac{1}{a}+\frac{2}{a^2}$ **h.** $\frac{2}{3}\times\frac{a}{b}+\frac{1}{b}$

i. $2\frac{1}{2}+\frac{b}{4}$ **j.** $3p+\frac{2q}{7}$

2. Simplify each of these expressions to a single fraction in simplest form:

a. $\frac{a+3}{2}+\frac{a}{2}$ **b.** $\frac{2b+3a}{3}+\frac{a+b}{3}$ **c.** $\frac{a+b}{2}+\frac{2a+7b}{3}$

d. $a+\frac{a+4}{3}$ **e.** $\frac{5b+11}{2}-\frac{1}{3}$ **f.** $\frac{4a+1}{a}+\frac{1}{2}$

3. Simplify these expressions to a single fraction in simplest form:

a. $\frac{a+1}{3}-\frac{a-1}{4}$ **b.** $\frac{2b+3}{3}-\frac{3-b}{5}$ **c.** $\frac{p-q}{4}-\frac{p+q}{3}$

d. $\frac{3p}{4}-\frac{p-2q}{5}$ **e.** $\frac{a}{3}-\frac{4-3a}{5}$ **f.** $\frac{7-p}{2}-\frac{p}{3}$

g. $\frac{18}{5}-\frac{2-p}{2}$ **h.** $\frac{3+r}{4}-\frac{2-r}{3}$ **i.** $\frac{6p-q}{3}-q$

4. Simplify each of these expressions to a single fraction in simplest form:

a. $\frac{1}{x+1}+\frac{1}{x+2}$ **b.** $\frac{2}{p+3}+\frac{1}{p+1}$ **c.** $\frac{2}{p+1}+\frac{1}{p-3}$

d. $\frac{3}{R+4}+\frac{2}{R-2}$ **e.** $\frac{3}{R}-\frac{1}{R+3}$ **f.** $\frac{1}{A+3}-\frac{1}{R+2}$

g. $\frac{4}{A-3}-\frac{1}{A-1}$ **h.** $\frac{A+2}{A+1}+\frac{A}{A-1}$ **i.** $\frac{x}{2(x+1)}-\frac{1}{3(x+1)}$

j. $\frac{y}{3(x+y)}-\frac{x}{4(x-y)}$ **k.** $\frac{2}{(x+1)(x+2)}+\frac{3}{(x+2)(x+3)}$

l. $\frac{4}{x(x+1)}-\frac{3}{(x+1)(x+2)}$ **m.** $\frac{1}{x-y+2}+\frac{2}{x+y-3}$

n. $\frac{3}{x(x+y-2)}-\frac{2}{x(x-y+2)}$ **o.** $\frac{4}{x(x-y)}-\frac{3}{y(x-y)}$

5. Express each of the following as single fractions, simplifying where possible:

a. $\frac{2ab}{10a^2b^3}\times\frac{25a^3b}{15ab^2}$ **b.** $\frac{6xyz^2}{15x^2z}\times\frac{10y^3z^3}{9xy^2}$

c. $\frac{9a^2c}{a-c}\times\frac{2a^2-2c^2}{6ac^3}$ **d.** $\frac{x^2+4x+4}{x^2-9}\times\frac{x^2+5x+6}{x(x+2)^2}$

e. $\frac{u^2-36}{x^2+5x+4}\div\frac{u^2+5u-6}{x^2-1}$ **f.** $\frac{2a^2-98}{a^2-25}\times\frac{a^2-3a-10}{3a-21}\div\frac{a^2+5a-14}{2a+10}$

g. $\frac{2}{x-y}+\frac{1}{2x-2y}$ **h.** $\frac{2}{x+1}+\frac{3}{x-1}$

i. $\frac{2}{x-2y}-\frac{1}{x+2y}$ **j.** $\frac{1}{x+y}+\frac{y}{x^2-y^2}$

k. $\frac{1}{x+1}+\frac{1}{x+2}-\frac{3}{x^2+3x+2}$ **l.** $\frac{1}{a-4}-\frac{2}{a-5}+\frac{1}{a-6}$

m. $\frac{1}{x^2-1}+\frac{1}{x^2+2x-3}+\frac{2}{x^2+4x+3}$ **n.** $\frac{a(a-1)(a-2)}{4}+\frac{a(a-1)(a-2)(a-3)}{3}$

Unit 12.4 Algebra and Graphs

Topic 9: Formulae

The material in Topic 9 is about using straightforward algebraic methods and solving equations. The Topic covers:

- Substituting values into formulae
- Rearranging formulae.

Introduction

A **formula** is an equation which describes the relationship between two or more variables eg $A = \pi r^2$ or $s = ut + \frac{1}{2}at^2$. The variable appearing on its own on the left-hand side of the formula is called the **subject** of the formula, eg A is the subject of the formula $A = \pi r^2$.

Substituting into formulae

The process of replacing variables with numbers is called **substitution**. To find the numerical value of an unknown variable, the numerical values of the remaining variables are substituted into the formula.

Example A

Q. The surface temperature of an aircraft is given by the formula $S = A + \left(\frac{v}{160}\right)^2$, where A is the air temperature (°C) and v is the air speed (kmh^{-1}). Using the formula, find the surface temperature if the air temperature of an aircraft flying at 640 kmh^{-1} is –11°C.

A. $S = A + \left(\frac{v}{160}\right)^2$ [formula given]

Substituting $v = 640$ and $A = -11$ [values given]

$S = -11 + \left(\frac{640}{160}\right)^2$ [replacing v with 640 and A with –11 in formula]

$= -11 + 4^2$ [simplifying]

$= 5$°C

Note: Units do not need to be used in the working of the problem. Ensure your final answer includes the correct units.

If the unknown variable is not the subject of the formula, substitution results in an equation to be solved.

Example B

Q. The variables a, b, and c are related by the formula $a = 3b + 4c$. Find b when $a = 32.4$ and $c = 7.5$.

A. $a = 3b + 4c$ [given formula]

Substituting $a = 32.4$ and $c = 7.5$ gives

$32.4 = 3b + 4 \times 7.5$ [replacing a with 32.4 and c with 7.5 in the formula]

$32.4 = 3b + 30$ [simplifying]

$2.4 = 3b$ [subtracting 30 from both sides]

$b = 0.8$ [dividing by 3 and swapping sides]

Unit 12.4 Activity 9A: Using formulae

1. Find the values of each of the following expressions if $x = 2$, $y = 3$, and $z = -2$.

   a. $x + y + z$ b. $xy - z$ c. $3x + 5y$ d. $5(4x - y)$

   e. $\frac{x - y}{z}$ f. $(x + y)(x + z)$ g. $\sqrt{x^2 + y^2 - z^2}$ h. $\frac{x + y}{x - y}$

   i. $y(x - z)$

2. The formula for changing a temperature in degrees Fahrenheit (°F) to degrees Celsius (°C) is $C = \frac{5}{9}(F - 32)$. Change each of the following temperatures to °C.

   a. 32 °F b. 212 °F c. 300 °F

3. The formula for changing a temperature in degrees Celsius (°C) to degrees Fahrenheit (°F) is $F = \frac{9C}{5} + 32$. Change each of the following temperatures to °F.

   a. 20 °C b. 100 °C c. −15 °C

4. The volume, V, of a sphere is given by $V = \frac{4}{3}\pi r^3$ where r is the radius. Find the volume of a sphere with a radius of 6 centimetres.

5. The time, T, in seconds, that a pendulum takes for one complete swing is given by $T = 2\pi\sqrt{\frac{l}{g}}$, where g is the constant due to gravity and l is the length of the pendulum. Find T if $l = 2$ m, $g = 9.8$ m s^{-2}, and $\pi = 3.14$.

6. Tessa is a TV repair person. She charges a flat fee of K15 per visit plus K40 per hour. The total cost, C, of a visit of t hours is given by the formula $C = 40t + 15$.

   a. Find the cost of a repair that took 2.5 hours.

   b. How long did Tessa take for a repair that cost a total of K135?

7. The perimeter, P, of a semicircle of radius r can be found using the formula $P = r(\pi + 2)$.

   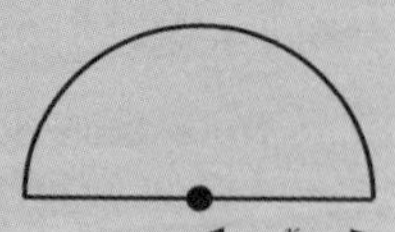

   Using $\pi = 3.14$

   a. Find the perimeter of a semicircle with a radius of 20 cm.

   b. The radius of a semicircle whose perimeter is 65.8 m.

8. A courier company charges KC to send a parcel of a weight w kg a distance d km. The cost is calculated using the formula $C = 3.5w + 0.15d$.

   Use the formula to find out

   a. The cost of sending a 3.5 kg parcel a distance of 236 km.

   b. The weight of a parcel which costs K59.70 to send 300 km.

   c. The distance a 2.5 kg parcel was sent if it cost K86 to send.

9. The total surface area, S, of a cylinder is given by the formula $S = 2\pi r(r + h)$, where h is the height and r is the radius.

a. Find the total surface area of a cylinder whose radius is 5.2 cm and whose height is 12.7 cm.

b. Find the height, to the nearest millimetre, of a cylinder with a radius of 8 cm and a total surface area of 960 cm^2.

10. The formula for calculating the interest, I, earned by a principal of KP at an interest rate of R% per annum for T years is $I = \frac{PRT}{100}$. Find the principal required to earn K900 after 3 years at a rate of 15% per annum.

11. The formula for changing temperature in degrees Fahrenheit to a temperature in degrees Celsius is given in question **2**. For what value of F is the temperature in degrees Fahrenheit the same as the temperature in degrees Celsius?

12. The formula for changing temperature in degrees Celsius to degrees Fahrenheit is given in question **3**. Sharon claimed that she doubled the temperature in degrees Celsius and got the correct temperature in degrees Fahrenheit. What temperature in degrees Celsius does this work for?

Changing the subject of a formula

Changing the subject of a formula means **rearranging** or **transforming** the formula to make some other variable the subject. The process is similar to **solving equations** (see Topic 1, p. 221) in that inverse operations are used to rearrange the formula (+ and – are inverse operations as are × and ÷).

Example C

Q. Rearrange the formula $A = \frac{bh}{2}$ so that b is the subject.

A. $A = \frac{bh}{2}$ [given formula]

$2A = bh$ [multiplying both sides by 2 (to 'undo' division by 2)]

$\frac{2A}{h} = b$ [dividing both sides by h (to 'undo' multiplication by h)]

$\therefore b = \frac{2A}{h}$ is the required formula [swapping sides]

When an unknown variable is not the subject of the formula, it is often useful to rearrange the formula first to make that variable subject, before substituting.

Example D

Q. The variables a, b, and c are related by the formula $a = 3b + 4c$. Rearrange the formula to make b the subject, then find the value of b when $a = 32.4$ and $c = 7.5$.

A. $a = 3b + 4c$

$a - 4c = 3b$ [subtracting $4c$ from both sides]

$b = \dfrac{a - 4c}{3}$ [dividing both sides by 3 and swapping sides]

Substituting $a = 32.4$ and $c = 7.5$ gives $b = \dfrac{32.4 - 4 \times 7.5}{3} = 0.8$ [as in Example B]

Note: Making b the subject first, before substituting for a and c, is particularly useful when there is a series of substitutions to be made to find multiple values of b.

Another pair of inverse operations to remember is squaring and taking the square root.

It is usually best to isolate the square root first before squaring.

Example E

Q. Make l the subject of the formula $T = 2\pi\sqrt{\dfrac{l}{g}}$

A. $T = 2\pi\sqrt{\dfrac{l}{g}}$

$\dfrac{T}{2\pi} = \sqrt{\dfrac{l}{g}}$ [dividing both sides by 2π (to isolate the square root)]

$\dfrac{T^2}{4\pi^2} = \dfrac{l}{g}$ [squaring both sides to remove the square root operation]

$l = \dfrac{T^2 g}{4\pi^2}$ [multiplying both sides by g and swapping sides]

Unit 12.4 Activity 9B: Rearranging formulae

1. For each of the following formulae, change the subject of the formula to the variable shown in brackets:

a. $v = u + at$ (u)

b. $v = u + at$ (t)

c. $C = 2\pi r$ (r)

d. $I = \dfrac{PRT}{100}$ (R)

e. $P = 2(l + w)$ (w)

f. $a^2 = b^2 + c^2$ (b)

g. $s = ut + \dfrac{1}{2}at^2$ (u)

h. $R = \dfrac{V}{I}$ (I)

i. $s = \dfrac{n}{2}(a + l)$ (l)

j. $b = \sqrt{ac}$ (a)

k. $S = ut + \dfrac{1}{2}at^2$ (a)

l. $v^2 = u^2 + 2ad$ (a)

m. $F = \dfrac{9C}{5} + 32$ (C)

n. $v^2 = u^2 + 2ad$ (u)

o. $A = \dfrac{h}{2}(a + b)$ (a)

2. The variables u, v and w are linked by the formula $u = \sqrt{\frac{v + 1}{3w}}$ where v and w are both positive.

a. Make v the subject of the formula.

b. Make w the subject of the formula.

3. Tony wants to concrete a square patio of side length x m at his house. He also wants to concrete a path that is one quarter the width of the patio. He will concrete as much of the length of the path as he can afford.

Tony works out that if he concretes k m of the path then the total cost,

T, of the concrete will be given by the formula:

$T = ax^2 + ax\,\frac{k}{4} + c$, where a is the cost per square metre of the concrete and c is a fixed cost for delivery.

Tony wants to see how much of the path he can concrete for various values of x, a, c and T that he is considering. Find a formula Tony can use which would allow him to work out k for various values of a, c, T and x (ie make k the subject of the formula).

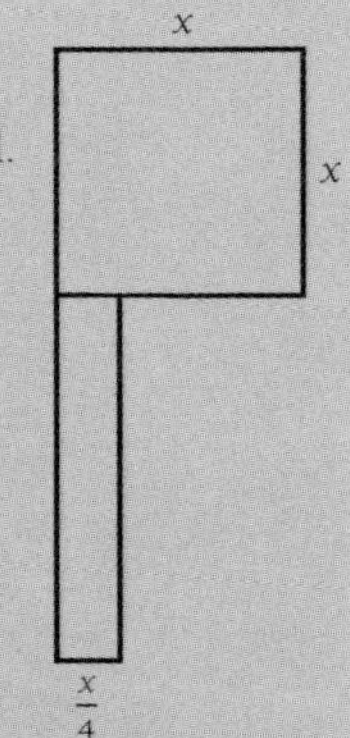

4. The formula for the area, A, of a circle of radius r is given by $A = \pi r^2$. Seth has a series of circular ponds to construct of given areas, so he needs to work out the corresponding radius for each pond. Give a formula that Seth could use to find the radius of a pond when its area is known.

5. The formula for changing temperatures in degrees Fahrenheit to degrees Celsius is $C = \frac{9}{5}(F - 32)$. Rearrange this formula to make F the subject.

Unit 12.4 Algebra and Graphs

Topic 10: Hyperbolae

The material in Topic 10 extends the drawing of straightforward graphs to cover:

- Draw rectangular hyperbolae of the form $y = \frac{a}{bx}$, where a and b are integers.
- Draw rectangular hyperbolae of the form $y = \frac{a}{x-c} + b$.
- Identify and interpret features of graphs including asymptotes and behaviour of graphs at large values of x and y.
- Write equations of graphs.

Hyperbolae

Hyperbolae are graphs of functions of the type $y = \frac{ax+b}{cx+d}$ where a, b, c and d are constants.

Hyperbolae of the form $y = \frac{b}{cx}$

In the case where a, $d = 0$ the equation of the hyperbola is of the form $y = \frac{b}{cx}$ or $xy = \frac{b}{c}$, where b and c are constants (real numbers).

In the simplest case, $b = 1$ and $c = 1$; ie the hyperbola has equation $xy = 1$ or $y = \frac{1}{x}$. This graph is drawn in the following example.

Example A

Q. Draw the graph of $xy = 1$.

A. Completing a table and plotting the corresponding points gives:

x	-3	-2	-1	$-\frac{1}{2}$	$-\frac{1}{3}$	0	$\frac{1}{3}$	$\frac{1}{2}$	1	2	3
$y = \frac{1}{x}$	$-\frac{1}{3}$	$-\frac{1}{2}$	-1	-2	-3	E	3	2	1	$\frac{1}{2}$	$\frac{1}{3}$

[rearranging $xy = 1$ to get $y = \frac{1}{x}$]

Notes:

1. There is no value for y when $x = 0$ because $y = \frac{1}{0}$ is undefined (shown by Ma Error on a calculator display). Thus the domain does not include the value 0, and the branches of the graph do not cut the y-axis.
2. The y-values never reach 0 (because there is no number which will divide into 1 to give 0) and so the **range** does not include the value 0.

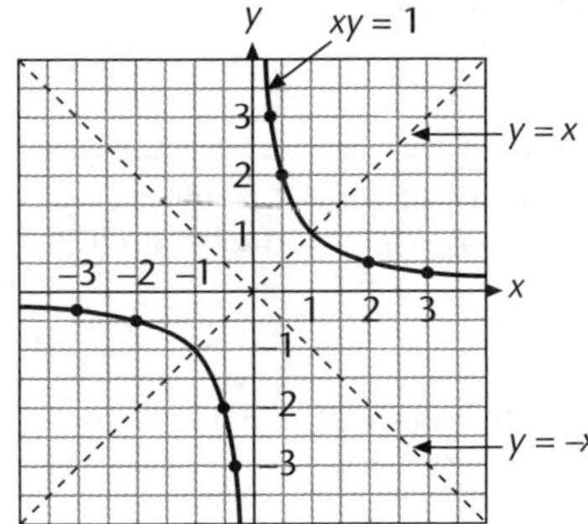

3. The x-axis and the y-axis are **asymptotes** for the curve (for large x values the curve approaches the x-axis, for large y values the curve approaches the y-axis).
4. The two axes of symmetry are shown by dotted lines on the graph. The equations of these lines of symmetry are $y = x$ and $y = -x$. The graph has point symmetry about (0, 0).
5. There is a **discontinuity** (a hole or break in a graph) for $x = 0$ in the graph of $xy = 1$.

When $\frac{b}{c}$ is *positive* in the equation $xy = \frac{b}{c}$, then the branches of the graph are in quadrants 1 and 3 (as in Example A).
If $\frac{b}{c}$ is *negative* in the equation $xy = \frac{b}{c}$, then the branches of the graph are in **quadrants** 2 and 4.

quadrant 2	quadrant 1
quadrant 3	quadrant 4

Example B

The graph of $xy = -2$ has branches in quadrants 2 and 4, because $\frac{b}{c} = -2$.

This can be verified by multiplying by −2 the y-values in the table in Example A.

Alternatively, pairs of numbers whose product is −2 can be readily seen to include:
$(-4, \frac{1}{2})$, (−2, 1), (−1, 2), (1, −2), (2, −1), $(4, -\frac{1}{2})$ etc.

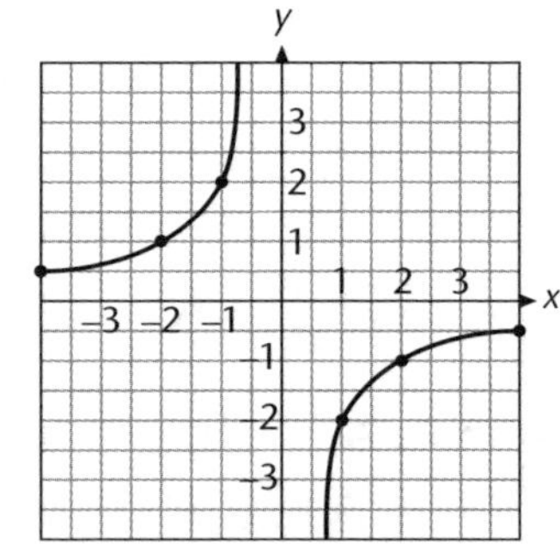

A hyperbola whose asymptotes intersect at the origin has the equation $xy = k$, for some constant k. By substituting a known point into this rule, the equation of the hyperbola can be found.

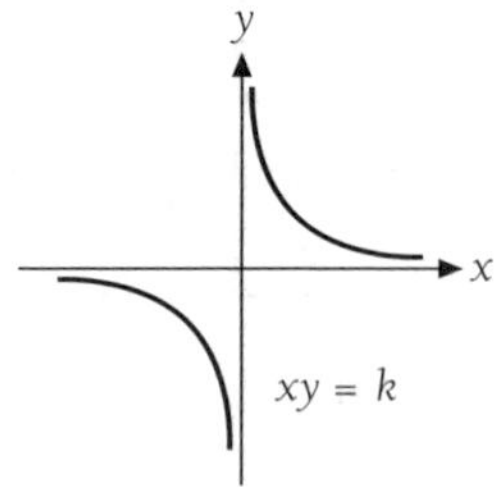

Example C

The graph of a hyperbola is shown alongside. Its equation is $xy = k$ for some constant, k.

Substituting the point (3, 1) into the rule gives $3 \times 1 = k$, so $k = 3$.

The equation of the hyperbola is, therefore, $xy = 3$ or $y = \frac{3}{x}$.

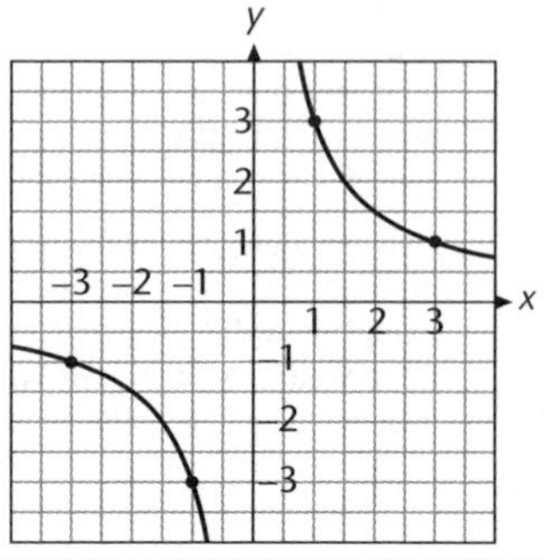

Unit 12.4 Activity 10A: Graphs of $y = \dfrac{b}{cx}$

1. **a.** Fill in a copy of the table below where $A = (6, 1)$, $B = (4, 1.5)$, $C = (3, 2)$, $D = (2, 3)$ and $E = (-1, -6)$. These points obey the rule xy = constant. What is the value of the constant?

Point	x	y	xy
A			
B			
C			
D			
E			

   b. Plot the points (6, 1), (4, 1.5), (3, 2), (2, 3), (1.5, 4), (1, 6), (0.75, 8), (–6, –1), (–4, –1.5), (–3, –2), (–2, –3), (–1.5, –4), (–1, –6) on a set of axes. These points lie on the graph of $xy = 6$.

   c. Join the points with appropriate smooth curves.

   d. What type of curve have you drawn?

2. Draw the graphs of: **a.** $xy = 4$. **b.** $y = \dfrac{-4}{x}$.

3. Draw the graphs of: **a.** $y = \dfrac{5x}{2}$. **b.** $y = \dfrac{-2}{3x}$.

4. Write down the equation of the hyperbolae:

   a.

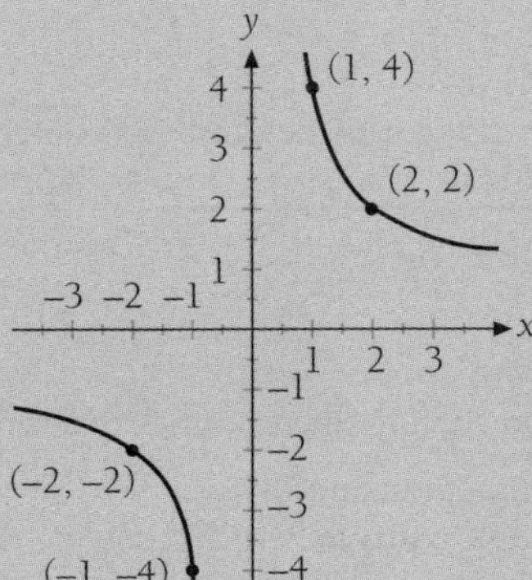

   b.

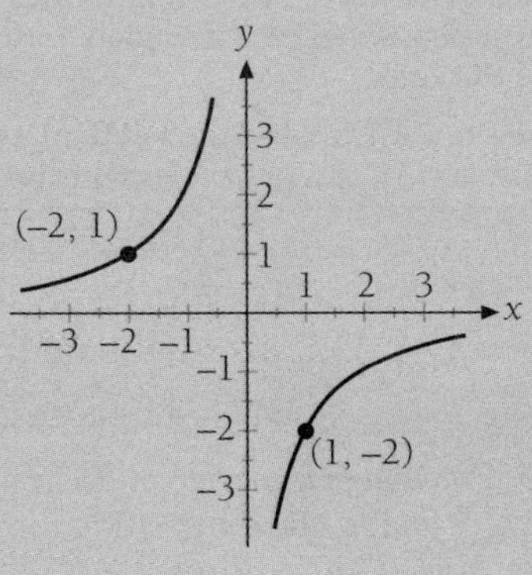

5. Write down the equations of each of the hyperbolae shown below:

   a.

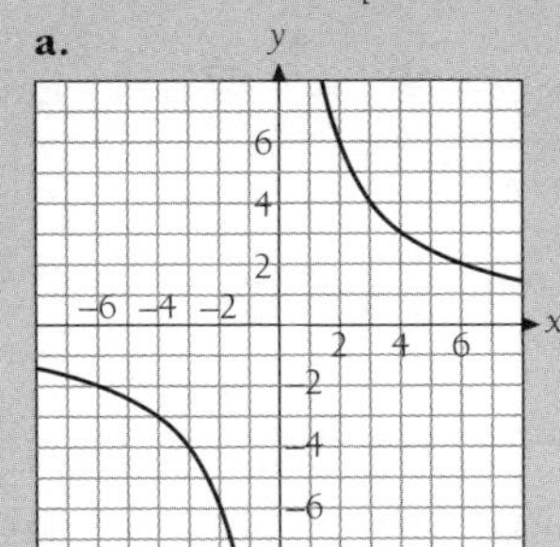

   b.

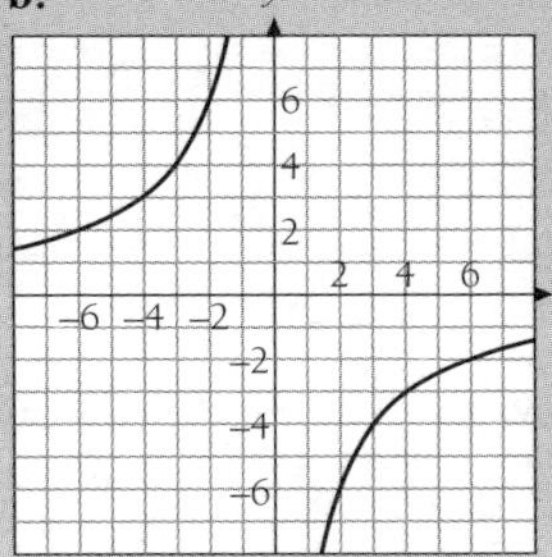

   c.

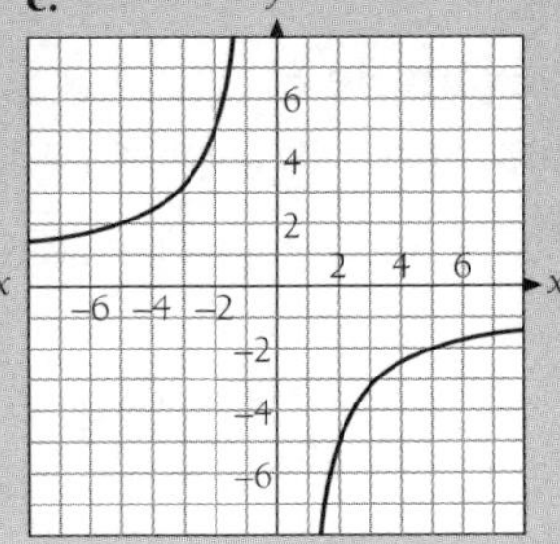

Hyperbolae of the form $y = \frac{a}{x-c} + b$

Important features of hyperbolae of the form $y = \frac{a}{x-c} + b$ are illustrated in the following examples.

Example D

The hyperbola $y = 3 - \frac{3}{x-1}$ is drawn from a table of values:

x	−100	−10	−3	−2	−1	0	$\frac{1}{2}$	0.9	1	1.1	1.5	2	5	10	100
y	3.03	3.3	3.75	4	4.5	6	9	33	E	−27	−3	0	2.25	2.7	2.97

The graph that results is shown below.

Notes:

1. There is no point on the graph with a y-value of 3, ie, 3 is excluded from the **range**.

The line $y = 3$ is called the **horizontal asymptote** of $y = 3 - \frac{3}{x-1}$ (as x gets larger and larger in magnitude, y gets closer and closer to 3).

2. There is no point on the graph with an x-value of 1, ie, 1 is excluded from the **domain**.

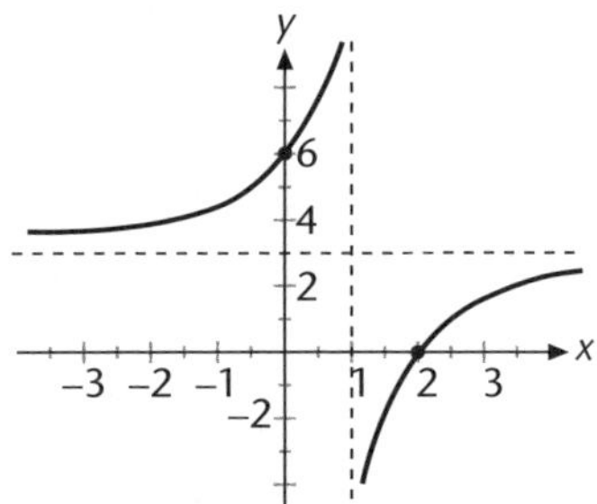

The line $x = 1$ is called the **vertical asymptote** of the hyperbola (as x gets closer and closer to 1, y gets greater and greater in magnitude).

Although it is always possible to sketch hyperbolae by plotting points, they are easily sketched by using the following procedure:

- The *y-intercept* is found by substituting $x = 0$.
- The *x-intercept* is found by substituting $y = 0$, and solving the resulting equation.
- The *vertical asymptote* occurs at the restriction in the domain (division by 0 is undefined) and is found by setting the denominator equal to zero and solving the resulting equation.
- The *horizontal asymptote* is found by finding the **limit** of the function as x gets very large. In the general case the horizontal asymptote is $y = b$.

Example E

Q. Sketch the graph of $y = 3 + \frac{12}{x-2}$.

A. • The y-intercept is given by: $y = 3 + \frac{12}{-2}$ [substituting $x = 0$ into $y = 3 + \frac{12}{x-2}$]

$\therefore\ y = -3$

• The x-intercept is given by: $3 + \frac{12}{x-2} = 0$ [substituting $y = 0$ into $y = 3 + \frac{12}{x-2}$]

$\therefore\ 3x + 6 = 0$ [multiplying by $(x - 2)$ and simplifying]

$\therefore\ x = -2$

- The vertical asymptote is given by: $x - 2 = 0$

  $\therefore \; x = 2$

- The horizontal asymptote is $y = 3$.

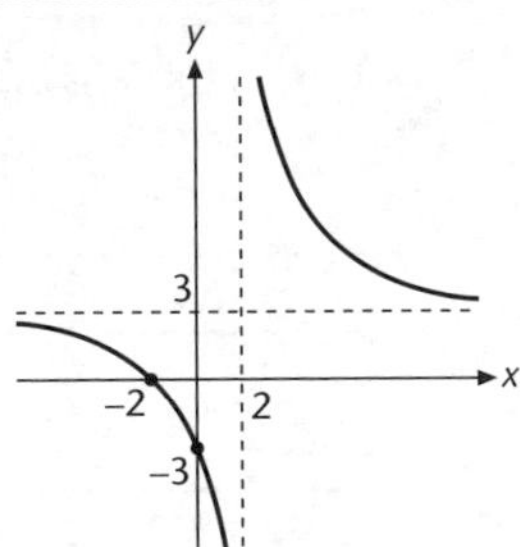

Note: The graph has point symmetry about (2, 3) which is the point of intersection of the two asymptotes.

By considering the features (such as intercepts and asymptotes) of a hyperbola, its equation can be written down. At the 'Merit' level, either b or c is zero.

Example F

Q. Find the equation of the hyperbola shown.

A. The equation is of the form $y = \dfrac{a}{x - c} + b$.

By considering asymptotes $c = 0$ and $b = 2$, so equation is $y = \dfrac{a}{x} + 2$.

Substitute (–1, 0) to get

$0 = \dfrac{a}{-1} + 2$

$a = 2$ [solving]

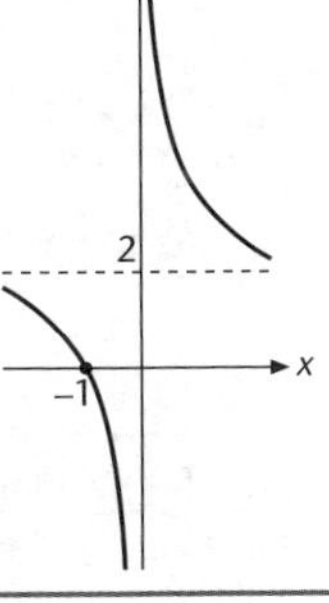

So equation is $y = \dfrac{2}{x} + 2$.

In harder examples, both b and c may be non-zero.

Example G

Q. Find the equation of the hyperbola shown.

A. The equation is of the form:

$y = \dfrac{a}{x - 2} + b$ [since vertical asymptote is $x = 2$]

As the horizontal asymptote is $y = -1$, it follows that $b = -1$.

Substituting the y-intercept (0, –2), gives:

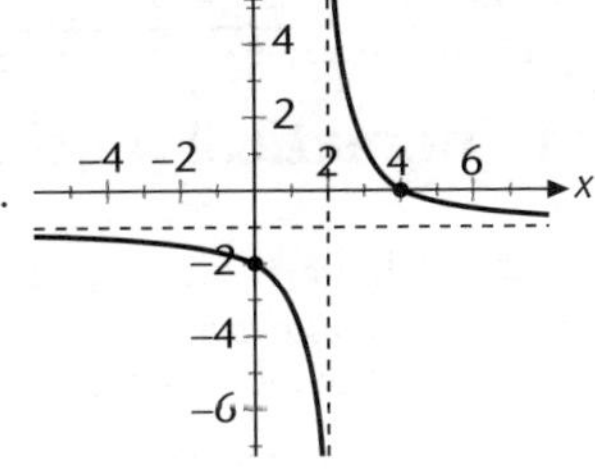

$-2 = \dfrac{a}{-2} - 1$

$4 = a + 2$ [multiplying by –2]

$a = 2$ [solving for a]

Therefore, the equation is $y = \dfrac{2}{x - 2} - 1$.

Unit 12.4 Activity 10B: Graphs of $y = \frac{a}{x-c} + b$

1. Sketch the hyperbolae with the following characteristics:

	Vertical asymptote	Horizontal asymptote	x-intercept	y-intercept
a.	$x = 2$	$y = 2$	3	3
b.	$x = 3$	$y = 6$	1	2
c.	$x = 4$	$y = -3$	2	−1.5
d.	$x = 2$	$y = 3$	6	9
e.	$x = -1$	$y = 0$	none	4

2. Sketch the graphs of the following functions:

a. $y = \frac{3}{x-3} + 3$ **b.** $y = \frac{8}{x-4} + 4$ **c.** $y = -1 + \frac{3}{x+1}$

d. $y = \frac{2}{x-1}$ **e.** $y = \frac{1}{x} + 1$ **f.** $y = \frac{2}{x-1} - 1$

3. Write down the domain and range of each of the hyperbolae in Question 2.

4. Find the equations of the hyperbolae with the following properties:

	Vertical asymptote	Horizontal asymptote	x-intercept	y-intercept
a.	$x = -3$	$y = 0$	none	4
b.	$x = 0$	$y = 3$	2	none
c.	$x = 0$	$y = -1$	3	none
d.	$x = -2$	$y = 2$	−3	3
e.	$x = 5$	$y = -2$	−2	0.8

Mathematical modelling (problems in context)

Example H

Q. A ship sinks and the survivors escape to an island where the local people give each person 5 kg of food to last them until their rescuers arrive.

In addition the survivors save 100 kg of food from the ship which they divide equally among the survivors. The captain refuses any of the food brought ashore.

a. If x is the number of survivors, write an equation for the total amount of food (F) available to each survivor except the captain until the rescue ship arrives.

b. Draw a graph of this amount of food F against x.

A. **a.** The amount of ship's food per survivor is $\frac{100}{x-1}$. [total food ÷ number of people which is $(x - 1)$ as the captain is excluded]

$\therefore$ total food per survivor is $F = 5 + \frac{100}{x-1}$. [add 5 kg given by locals]

c. The graph below is obtained by plotting F (food available per survivor) against x (the number of survivors). The table gives some values.

x	2	3	5	11	21	26	51
F	105	55	30	15	10	9	7

The graph is as below:

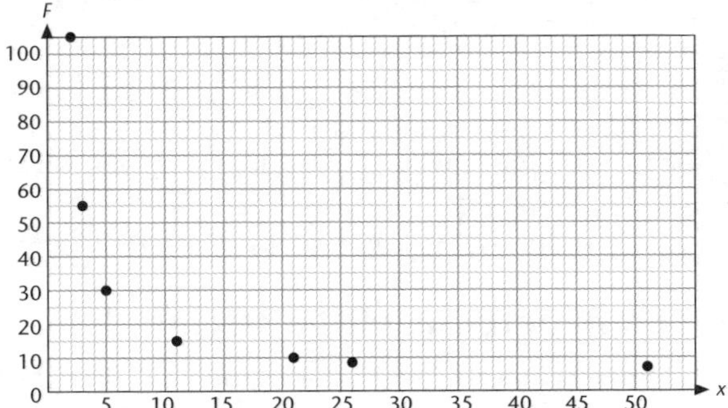

Notes:

1. The points are not joined as you can only have whole numbers for the numbers of survivors.
2. The points all lie on a hyperbola. The reader might like to consider what would happen if $x = 1$.

Unit 12.4 Activity 10C: Modelling problems

1. It costs a ski resort K4 200 daily to operate its chairlifts.

To recover these costs, skiers pay a daily charge to use the chairlifts.

a. If the daily charge is increased, then fewer skiers are needed to recover costs.

Some examples are shown in the table below. Copy the table and complete the three empty boxes:

Daily charge (K)	Number of skiers needed to recover costs	Total costs recovered (K)
10	420	4 200
20	210	4 200
30		4 200
40		4 200
n		4 200

b. The examples in the table are shown on the graph below. Copy the graph then continue it across to the right-hand edge of the grid.

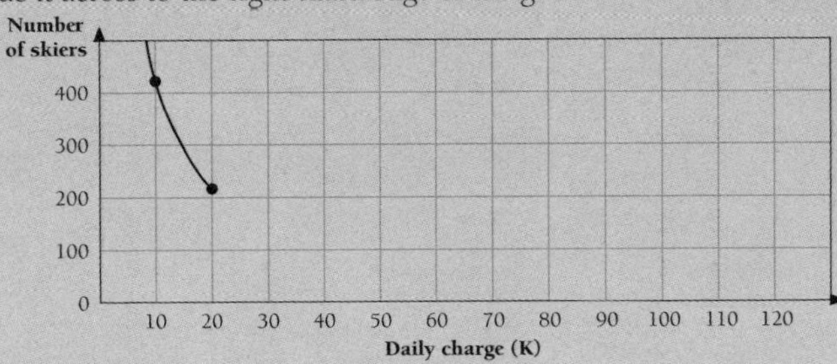

c. What type of graph do the points in **a.** and **b.** lie on?

2. An amusement company runs games. On average per day 10 free games are played on each machine. The cost to the company per game consists of two parts: 5t of fixed costs, and a variable cost. The total cost to the company of a machine per day is K5.
   a. Let x be the number of games that the customers pay for. What is the total number of games played per day?
   b. Let the variable cost be y toea per game. What is the total cost per day?
   c. Write an equation relating y and x.
   d. Draw the graph relating y and x.

3. The cost of production of a product consists of K2 000 fixed costs, plus K10 per kilogram.
   a. If A is the total amount (kg) produced, write down a formula which relates d, the amount produced per kina, to A.
   b. Explain why a graph of d against A would not be a perfect hyperbola.

Intersection of a hyperbola with a straight line

To find the point(s) of **intersection** algebraically, the equations of the hyperbola and the straight line are solved simultaneously.

Example I

Q. Find the points of intersection of the hyperbola $y = \frac{3}{x}$ and the straight line $y = x + 2$.

A. Solving simultaneously (equating the right sides of each equation to eliminate y), gives:

$$x + 2 = \frac{3}{x}$$

$x^2 + 2x = 3$ [multiplying by x]

$x^2 + 2x - 3 = 0$ [rearranging]

$(x + 3)(x - 1) = 0$ [factorising]

$\therefore\ x = -3$ or $x = 1$

$x = -3$ gives $y = -3 + 2 = -1$ [substituting in $y = x + 2$]

$x = 1$ gives $y = 1 + 2 = 3$ [substituting in $y = x + 2$]

$\therefore$ The points of intersection are $(-3, -1)$ and $(1, 3)$.

Note: This can be checked by graph.

Unit 12.4 Activity 10D: Intersection of a hyperbola and a line

Find the coordinates of the points of intersection, if any, of the hyperbola $y = \frac{1}{x}$ with the following straight lines:

1. $y = 3$ **2.** $y = 5$ **3.** $x = 2$ **4.** $x = 4$ **5.** $y = x$

6. $y = 2x$ **7.** $y = 2 - x$ **8.** $y = 2x + 1$ **9.** $y = x - 1$ **10.** $x + y = 0$

Find the coordinates of the points of intersection, if any, of the line $y = x$ with the following hyperbolae:

11. $y = \frac{x}{x-1}$ **12.** $y = \frac{-2}{x}$ **13.** $y = 1 + \frac{1}{x+1}$ **14.** $y = 1 - \frac{1}{x+1}$ **15.** $y = -2 + \frac{9}{x+2}$

16. Find the points of intersection of the hyperbola $xy = 5$ and the straight line $y = x + 4$.

17. Draw graphs to show the three different ways that a hyperbola can intersect with a straight line.

a. With two points of intersection.
b. With one point of intersection.
c. With no points of intersection.

Unit 12.4 Algebra and Graphs

Topic 11: Exponential and logarithmic graphs

In Topic 11 the focus is on the drawing of strightforward non-linear graphs. It covers:

- Draw exponential functions of the form $y = a^x$, $a \in N$.
- Draw logarithmic functions of the form $y = \log_a x$, $a \in N$.
- Draw exponential functions of the form $y = a^{x-b} + c$ and either b or c equal to 0.
- Identify and interpret features of graphs including intercepts, asymptotes and behaviour of graphs at large values of x and y.
- Write equations of graphs.

Exponential (growth) curves

The general equation of an **exponential** (or growth) curve is $y = a^x$ where $a > 0$.

Example A

A table of values of $y = 1.6^x$ is shown.

x	–5	–3	–1	0	12	3
y	0.10	0.24	0.63	1	1.6	4.10

The graph of $y = 1.6^x$ is drawn from these values.

Notes:

1. All exponential curves of the form $y = a^x$ pass through (0, 1) since $a^0 = 1$ for all values of a (except $a = 0$).
2. If $a > 1$ then y gets larger as x gets larger.
3. The *greater* the value of a, the *steeper* the curve.
4. The negative x-axis is an asymptote as $x \to -\infty$, which means that the curve does not cut the x-axis at all.

The graph of $y = a^x$ becomes a decay curve when $0 < a < 1$. This means that as $x \to \infty$, $y \to 0$.

Example B

Q. An insecticide firm claims that its product will kill half of the existing insect population every hour.

Draw the graph of $y = \left(\frac{1}{2}\right)^x$, ie $y = \frac{1}{2^x}$, to show the proportion of the population (y) of insects remaining after x hours.

A. Setting up a table of values for $y = \frac{1}{2^x}$:

x	–3	–2	–1	0	1	2	3
$y = \left(\frac{1}{2}\right)^x$	8	4	2	1	$\frac{1}{2}$	$\frac{1}{4}$	$\frac{1}{8}$

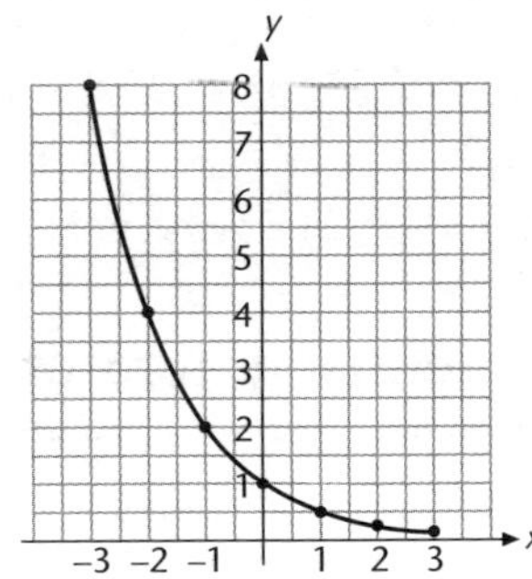

Plotting points and joining them with a smooth curve gives the graph shown:

Note: Only $x \geq 0$ values are relevant to the practical problem described.

Unit 12.4 Activity 11A: Basic exponential graphs

1. a. The points (–1, 0.5), (0, 1), (1, 2), (2, 4), (3, 8), (4, 16) lie on the graph of $y = 2^x$. Plot them on a set of axes.

   b. Join the points with a smooth curve.

2. a. Copy and fill in the table below where $y = 1.5^x$.

x	–2	–1	0	1	2	3
y						

   b. Use the table to sketch the graph of $y = (1.5)^x$.

3. Set up a table of points (as in question 2) and use it to draw the graph of:

   a. $y = (2.5)^x$ b. $y = (0.8)^x$

4. Write down the equations of the following graphs:

a.

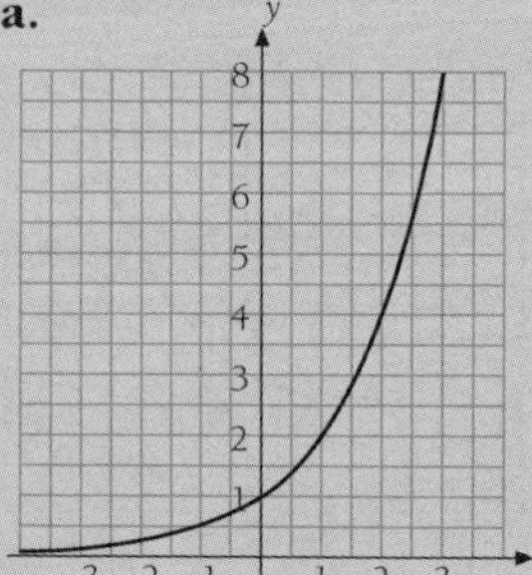

b.

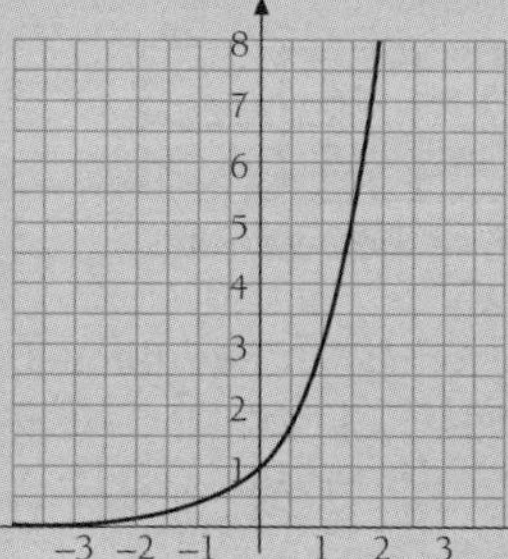

c.

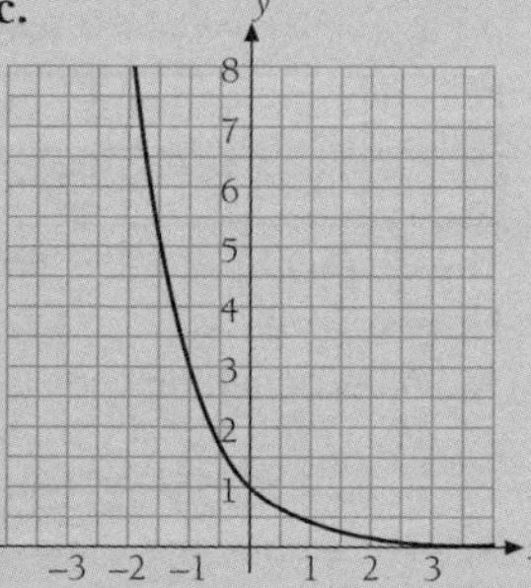

5. Write down the value of a which makes $y = a^x$ the correct equation of the curve.

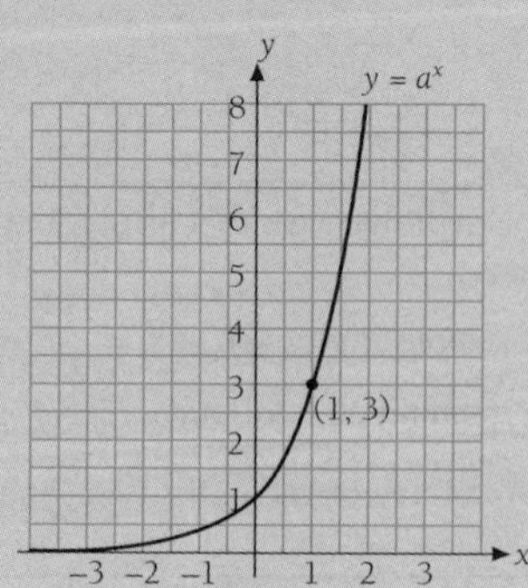

6. For what value of a does the point (2, 25) lie on the curve $y = a^x$?

Exponential graphs of the form $y = a^{x-b} + c$

The graph of $y = a^{x-b} + c$ transforms the graph of $y = a^x$ as follows:

- Replacing x with $x - b$ in the equation $y = a^x$ gives the equation $y = a^{x-b}$.

  The graph of $y = a^{x-b}$ is a horizontal translation of the graph of $y = a^x$ by b units right if $b > 0$, or b units left if $b < 0$.

- Replacing y with $y - c$ in the equation of $y = a^x$ gives $y - c = a^x$ or $y = a^x + c$.

  The graph of $y = a^x + c$ is a vertical translation of the graph of $y = a^x$ by c units up if $c > 0$, or c units down if $c < 0$.

- Replacing x with $x - b$ and y with $y - c$ gives the equation $y - c = a^{x-b}$ or $y = a^{x-b} + c$.

  The graph is a horizontal translation b units and a vertical translation c units of the graph of $y = a^x$.

Some of these graphs are now discussed.

Example C

Q. Draw the graph of $y = 2^{x-1}$.

A. A table of points obeying the rule $y = 2^{x-1}$ is shown below:

x	−2	−1	0	1	2	3	4
y	$\frac{1}{8}$	$\frac{1}{4}$	$\frac{1}{2}$	1	2	4	8

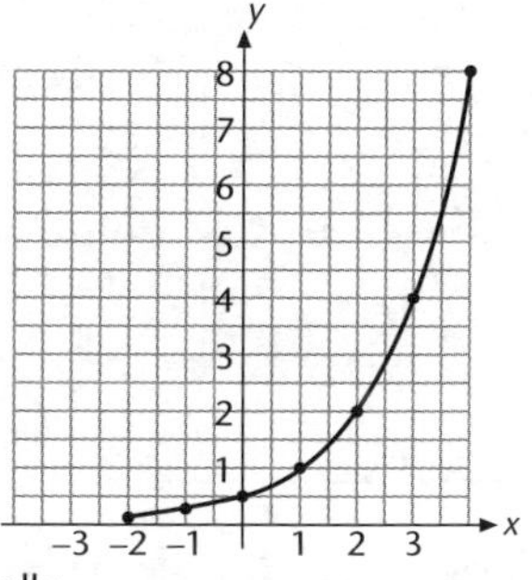

The graph is shown alongside.
This is the graph of $y = 2^x$ translated one unit right horizontally.

Example D

Q. Draw the graph of $y = 3^x - 1$.

A. A table of points obeying the rule $y = 3^x - 1$ is shown below:

x	−2	−1	0	1	2
y	$\frac{-8}{9}$	$\frac{-2}{3}$	0	2	8

The graph is shown alongside.
This is the graph of $y = 3^x$ translated one unit down vertically.

By considering features of the graph, such as points on the curve, intercepts and asymptotes, the equation of an exponential graph can be found.

Example E

Q. Find the equation of the graph shown whose equation is of the form $y = a^{x-b} + c$.

A. The graph shown is that of $y = \left(\frac{1}{2}\right)^x$ translated 1 unit to the left and 2 units up.

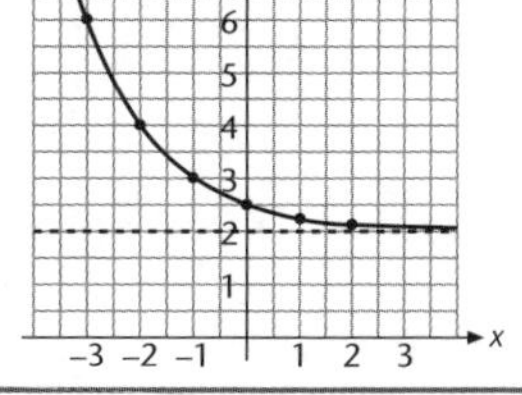

Thus the equation of the graph is:

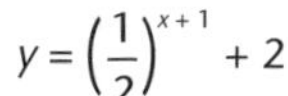

$$y = \left(\frac{1}{2}\right)^{x+1} + 2$$

Unit 12.4 Activity 11B: Graphs of the form $y = a^{x-b} + c$

1. Draw the graph of $y = 2^{x+1}$.
2. Draw the graph of $y = 2^x + 1$.
3. Draw the graph of $y = \left(\frac{1}{2}\right)^{x-1}$.
4. Draw the graph of $y = 3^x - 2$.
5. Find the equations of the following graphs:

a.

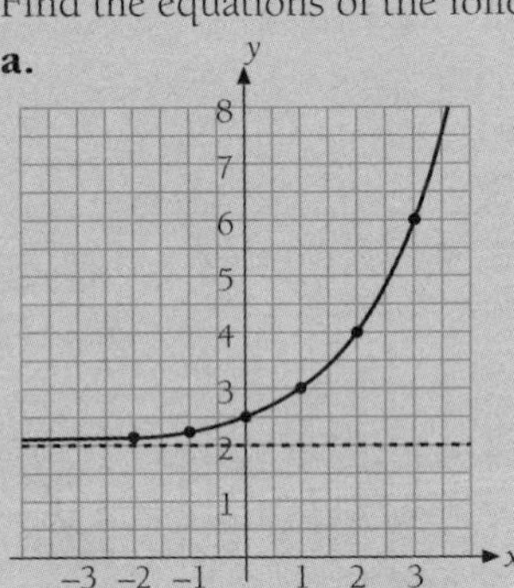

b.

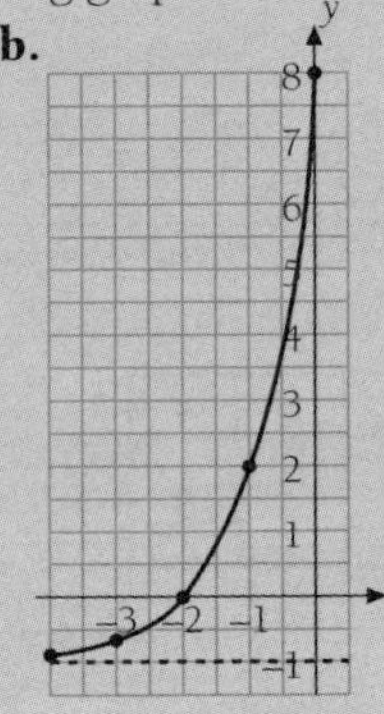

c.

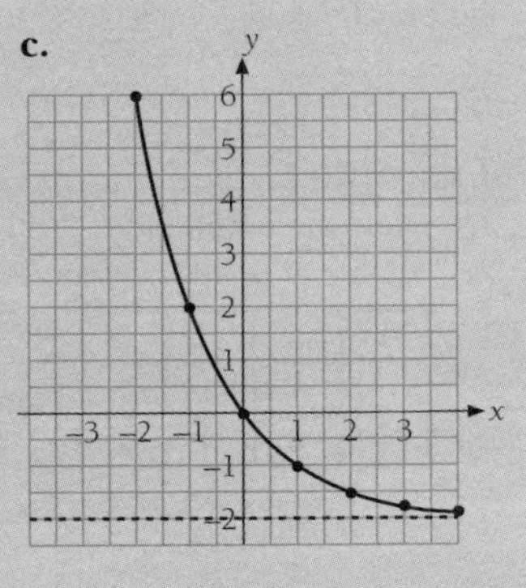

General logarithmic functions

If a is any positive number, $y = a^x$ and $y = \log_a x$ are **inverse functions**.

Note: $\log_a x$ is the logarithm of x to the base a.

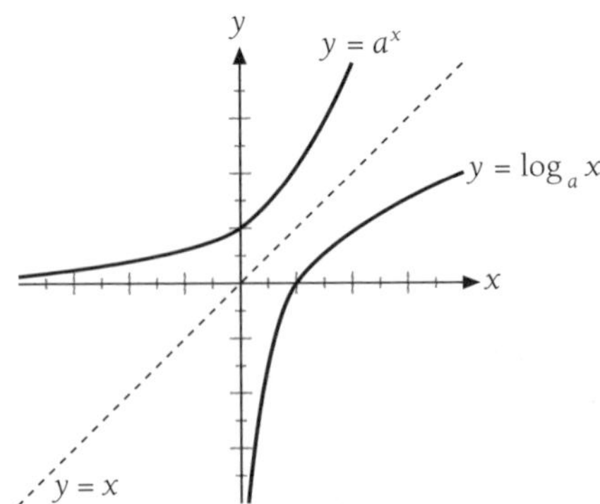

The graphs of $y = a^x$ and $y = \log_a x$ are **images** of each other under **reflection** in the line $y = x$, as shown in the graph.

The graph of $y = \log_{10} x$ is shown below ($\log_{10} x$ appears as  on most calculators).

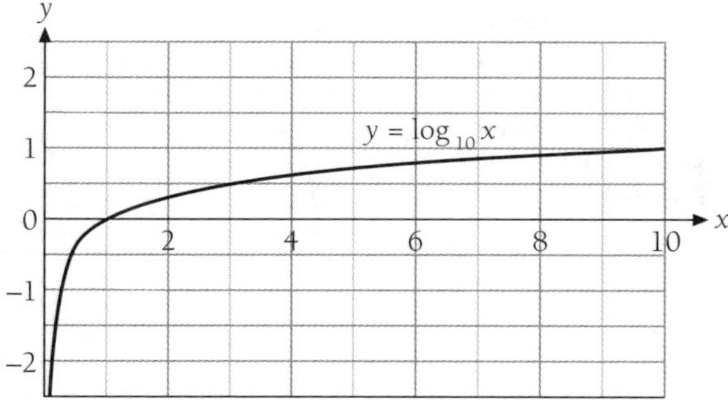

Note:

1. The y-axis is an asymptote for the curve $y = \log_{10} x$.
2. There is no horizontal asymptote for the curve $y = \log_{10} x$ (as $x \to \infty$, $\log_{10} x \to \infty$ also, although $\log_{10} x$ is not climbing steeply).

Logarithm rules can be used to evaluate the logarithms of large numbers.

Example F

Q. Find the value of: **1.** $\log_{10} 1\,000\,000$ **2.** $\log_{10} 10^{1\,000}$.

A. **a.** $\log_{10} 1\,000\,000 = 6$ [by calculator]

or $\log_{10} 1\,000\,000 = \log_{10} 10^6$ [since $10^6 = 1\,000\,000$]

$= 6 \log_{10} 10$ [since $\log x^n = n \log x$]

$= 6$ [since $\log_{10} 10 = 1$]

b. Although this cannot be calculated on most calculators currently (it causes a magnitude error), it can be evaluated using logarithm laws.

$\log_{10} 10^{1\,000} = 1\,000 \log_{10} 10$ [as $\log x^n = n \log x$]

$= 1\,000$ [since $\log_{10} 10 = 1$]

As can be seen from Example F, x has to be extremely large for $\log x$ to be large.

The change of base rule can be used to evaluate logarithms whose bases are not 10.

Example G

Q. Sketch the graph of $y = \log_4 x$.

A. Using the fact that $\log_4 x = \dfrac{\log_{10} x}{\log_{10} 4}$ the following table of values can be drawn up:

x	0.1	0.5	1	2	3	4	5
$\log_4 x$	−1.7	−0.5	0	0.5	0.8	1	1.2

The graph is as shown.

Note:

1. This graph is the inverse of $y = 4^x$. By reversing the ordered pairs of $y = 4^x$ the graph can be drawn.
2. $\log_4 16 = 2$ (since $4^2 = 16$) and $\log_4 64 = 3$ (since $4^3 = 64$) so (16, 2) and (64, 3) lie on the graph. It can be seen that the graph climbs slowly.

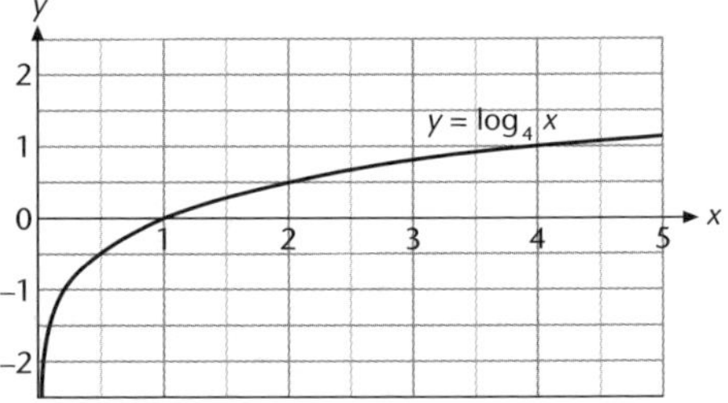

Example H

Q. Sketch the graph of $y = \log_2 (x - 1)$.

A. Using the fact that $\log_2 (x - 1) = \dfrac{\log_{10} (x - 1)}{\log_{10} 2}$ the following table of values for x and y can be drawn up.

x	1.25	1.5	2	3	5
y	–2	–1	0	1	2

This gives the graph shown alongside.

This is the graph $y = \log_2 x$ translated 1 unit to the right.

Note: The graph of $y = \log_2 x$ is the inverse of $y = 2^x$.

Example I

Q. Sketch $y = \log_3 x + 2$.

A. The table is:

x	$\frac{1}{9}$	$\frac{1}{3}$	1	3	9
y	0	1	2	3	4

The graph is as shown.

This is the graph of $y = \log_3 x$ translated 2 units up.

Note: The graph of $y = \log_3 x$ is the inverse of $y = 3^x$.

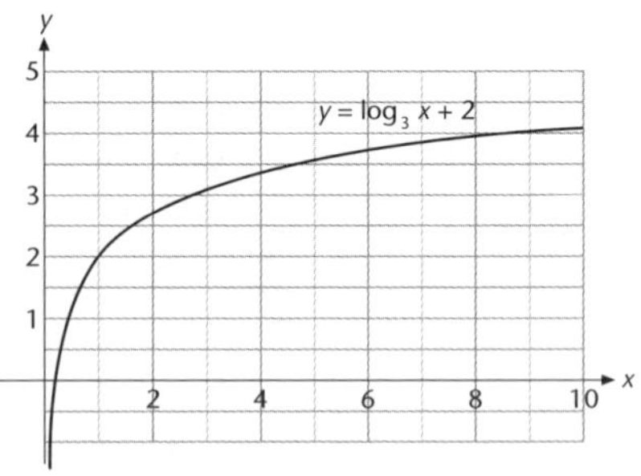

Unit 12.4 Activity 11C: Graphs of general logarithmic functions

1. Sketch the graphs of the following logarithmic functions:
   a. $y = \log_2 x$ **b.** $y = \log_3 x$ **c.** $y = \log_5 x$ **d.** $y = \log_8 x$
2. Investigate the claim that as a gets larger, the graph of $y = \log_a x$ gets flatter.
3. Investigate the claim that the graph of $y = \log_a x$ never crosses the line $y = x$.
4. Sketch the graphs of the following:
   a. $y = \log_2 (x - 2)$ **b.** $y = \log_2 x - 3$ **c.** $y = \log_3 (x - 1)$
   d. $y = \log_2 (x + 1)$ **e.** $y = \log_3 (x - 2) + 1$ **f.** $y = 1 - \log_5 x$

Exponential and logarithmic modelling (problems in context)

Exponential graphs model certain types of population growth (or decay).

Example J

Q. Toby is studying the rate of increase of a population of flies in a container. At the start of this project he has 4 flies in his container.

Each week, Toby finds there are 50% more flies in his container than there were in the previous week.

A good mathematical model for this is $F = 4(1.5)^n$, where F is the number of flies in the container after n weeks.

1. Sketch a graph of F against n, where n varies from 0 (at the beginning of his project) to 4 (when his project ends).
2. If Toby had had 8 flies at the start of his project, what effect would this have had on the graph?
3. If the number of flies had doubled each week (rather than increasing by 50%), what effect would this have had on the graph?
4. How does the number of flies vary as n increases?

A. 1. A graph is drawn from a table of values.

n	0	1	2	3	4
$4(1.5)^n$	4	6	9	13.5	20.25

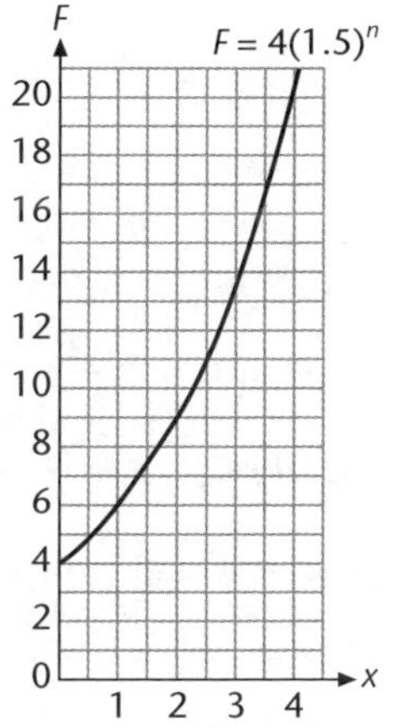

1. Every F-value on the graph would double for any particular week (eg week 2 would be $2 \times 9 = 18$).
2. The graph would have been steeper (eg week 2 would be $4(2)^2 = 16$).
3. As the number of weeks increases, the number of flies increases.

Unit 12.4 Activity 11D: Exponential and logarithmic modelling

1. A scientist studying an insect plague notices that the area (A) affected by the insects is given by $A = 1\,000(1.1)^n$, where n is the number of weeks after the first observation and A is measured in hectares.

a. Draw the graph of A against n.

b. Estimate the time taken for the infected area to reach 1 500 ha.

c. Explain why this model can only apply for a fixed length of time.

d. Explain in words what this model $A = 1\,000(1.1)^n$ means.

2. The number (N) of people in a district is modelled by the equation $N(t) = 100\,000 \times (1.01)^t$, where t is in years.

a. Sketch the graph of N against t.

b. How many people are in the district after 5 years?

c. Explain why this formula will no longer be valid after a certain time.

3. The population of a type of fish is given by $P(t) = 1\ 000\ (0.6)^{0.04t}$, where t is in months.

a. Sketch a graph of P against t.

b. How many fish of this type will there be after 10 months?

c. When will this population become extinct?

d. The graph of P against t appears to be linear. Why is a linear model inappropriate?

4. A model proposed for the population of insects in a field is $P = 50(1.2)^x$, where x is in weeks.

a. Draw the graph of P against x.

b. Are there any asymptotes for this model?

c. What happens to P as x increases?

5. a. Comment on the appropriateness of the model in question 4.

b. A refinement of this model proposes that this relationship holds true for the first 10 weeks and thereafter can be replaced by $P = 310 - \dfrac{50x - 500}{x - 1}$.
Sketch the graph of this new model.

c. What value does P get closer and closer to, as x gets larger and larger?

6. Scientists doing an experiment found that a quantity q is a function of time t given by $q = A\log_{10} t + B$. After 15.6 seconds, q is 47.897 and after 247 seconds, q is 65.890. Find A and B to the nearest integer.

7. A mathematical model for a function is $y = \log_a x - b$. If $y = -21.6429$ when $x = 105$ and $y = -21.1395$ when $x = 211$, find a and b to the nearest integer.

Unit 12.5 Applying Geometry in Papua New Guinean Arts

Topic 1: Polyhedra

Unit 12.5 focuses on mathematics that deals with shapes and patterns. The Syllabus (p. 26) states that mathematical skills should be developed and assessed mainly in an applied context and the local environment should be used as the context for all the application problems that students undertake. Students can examine, for example, the application of traditional patterns and measurement and expand their regular polygon properties.

The text and exercises in Topics 1 & 2 in this Unit are designed to give students mathematical ideas on which they can base their study of the traditional patterns and shapes from their local environment. Topic 1 deals with polyhedra, which are solid shapes with many sides ('poly' means many).

Prisms and pyramids

Prisms

We have already studied the geometry of prisms and some pyramids in Unit 11.1 Topics 8 and 9 (pp. 41 and 47).

Prisms are solids that have a constant cross-section that is a polygon, and hence rectangular sides.

Here are some prisms with cross-sections that are regular polygons:

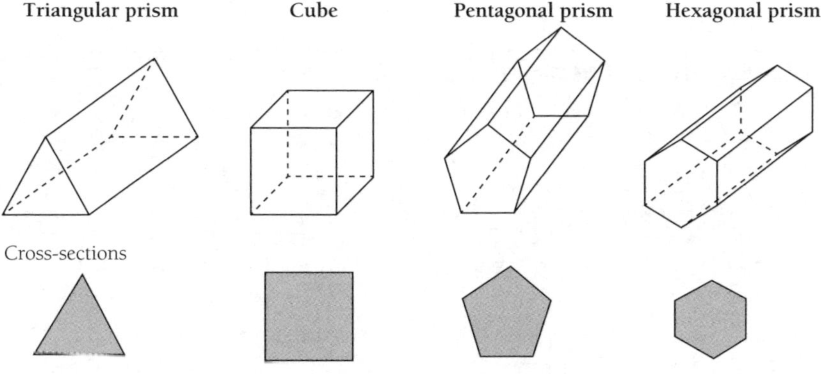

Some prisms that have cross-sections that are not regular polygons:

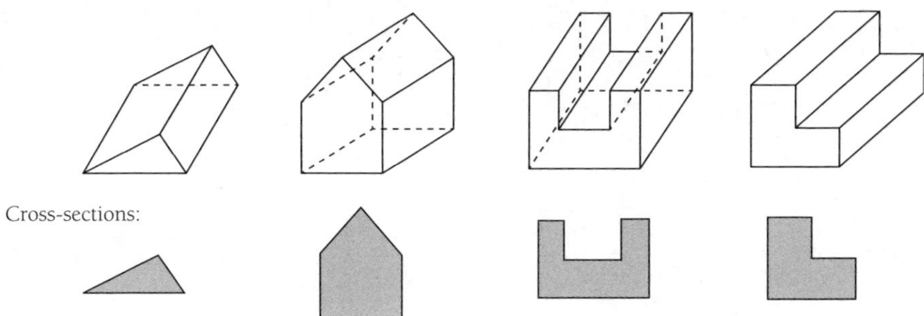

Pyramids

A pyramid is a solid that has a regular or non-regular polygon base and triangular sides. The sides meet at a point that is called the vertex.

A right pyramid has the vertex vertically above the 'centre' of the base. A non-right pyramid has the vertex not vertically above the centre of the base.The vertex can be vertically above a point that is not inside the base polygon.

Some examples of pyramids:

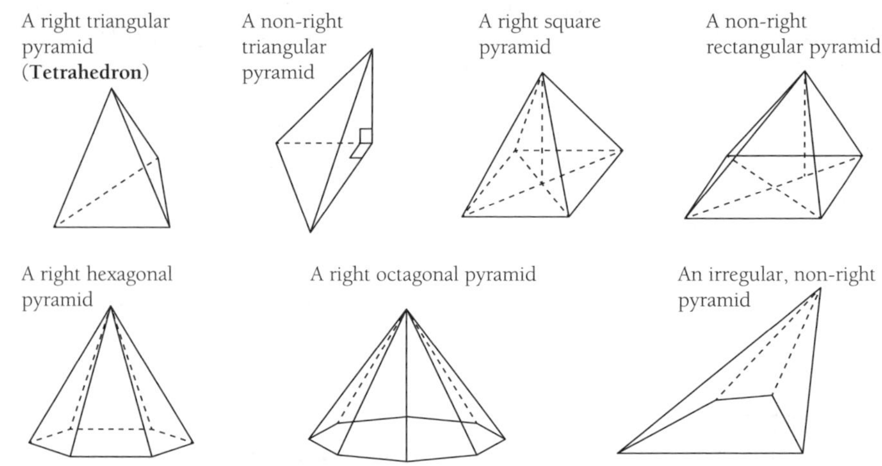

Unit 12.5 Activity 1A: Prisms and pyramids

1. Draw the cross-section of each of the following prisms:

a. 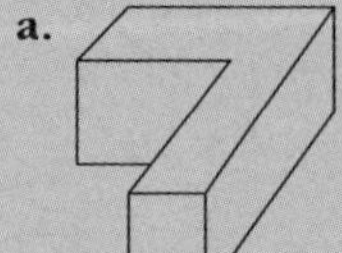

b. 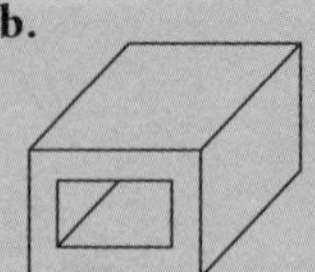

c. 

2. For each of the following construct a prism that has the given cross-section:

a. 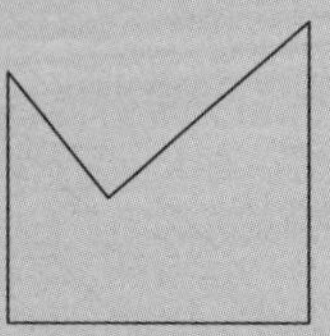

b. 

c. 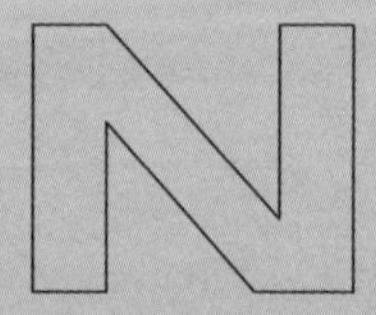

3. Construct each of the following pyramids:

a. A right rectangular pyramid.

b. A right pentagonal pyramid.

c. A non-right hexagonal pyramid.

4. Describe each of the following solids:

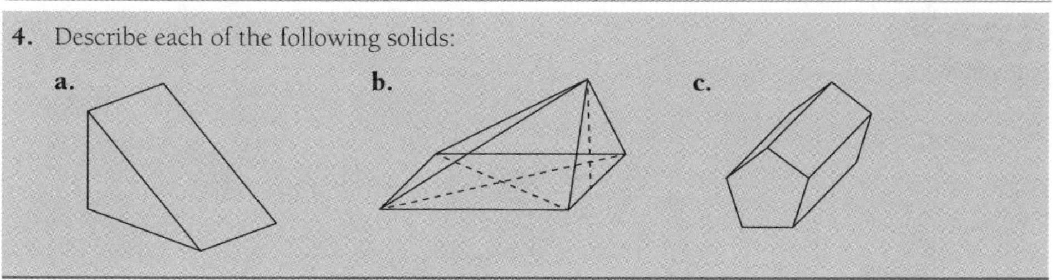

Constructing polyhedra

Polyhedra are solids that have polygons as their faces.

There are many polyhedra that can be constructed that combine regular and irregular polygons; these can be seen in books and on the Internet. Each of these solids can be deconstructed to show the underlying **plane** (flat) faces of the solid; this is called the **net** of the solid

Constructing polyhedra requires careful attention to measuring lengths and angles and it is advisable to draw a net that minimises the number of edges that need to be joined. Templates can be used to draw the net if there are many repeated shapes. The addition of tabs to glue the structure together also makes it easier to construct.

Platonic solids

Platonic solids are **polyhedra** that have only **congruent, regular polygons** as their faces and the same number of faces meet at each vertex.

There are only five platonic solids; the tetrahedron, the cube, the dodecahedron, the octahedron and the **icosahedron**.

Constructing a tetrahedron

A **tetrahedron** is a solid which has four faces that are equilateral triangles; four vertices and three faces meet at every vertex. To construct a tetrahedron:

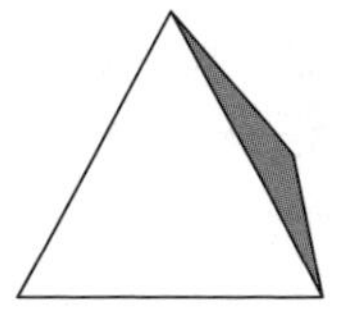

1. Using a compass, construct three intersecting circles of the same size radius so that the circumference of each of them goes through the centres of the other two circles.
2. Join the points of intersection on the circumference to form a large equilateral triangle. Join the points of intersection of the centres to form the smaller equilateral triangle.
3. Add some tabs for gluing your tetrahedron together, crease along the lines and glue together to form the tetrahedron. The best effect is achieved if the tabs are glued on the inside of the polyhedron.

Note: It is possible to make tabs that are not attached to the net. Make them in this shape and crease along the dotted line.

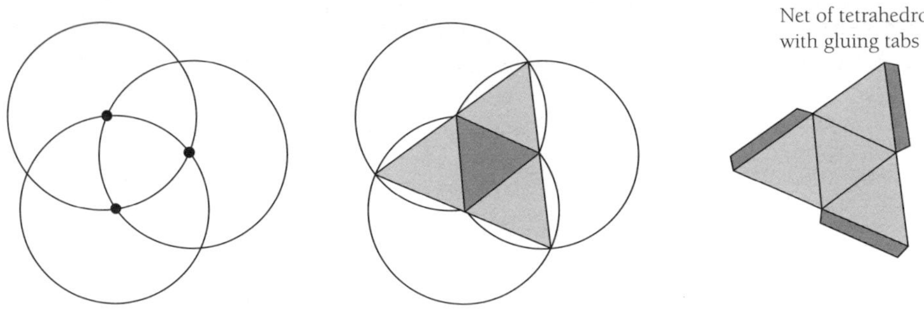

Constructing a cube

A **cube** has six faces that are squares. Eight vertices and three faces meet at each vertex.

Constructing a dodecahedron

A **dodecahedron** has twelve faces, all of which are regular pentagons. Twenty vertices and three faces meet at each vertex. To construct a dodecahedron:

1. Make an accurate template of a pentagon, carefully measuring the lengths of the sides and the angles.
2. Use the template to draw the net
3. Add tabs, cut out, crease along the lines and glue together.

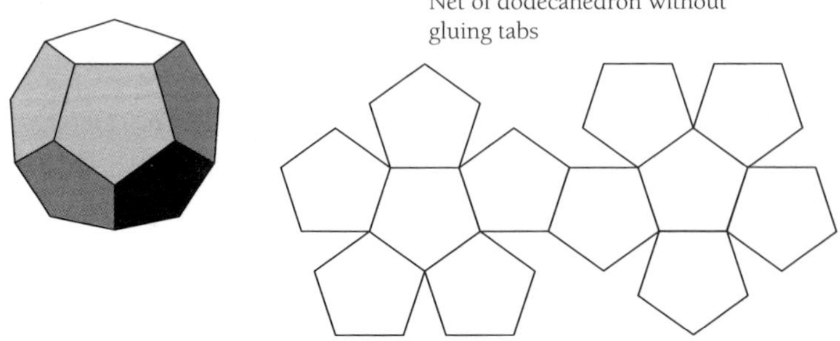

Constructing an octahedron

An **octahedron** has eight faces, of which all are equilateral triangles. Six vertices and four faces meet at each of the vertices.

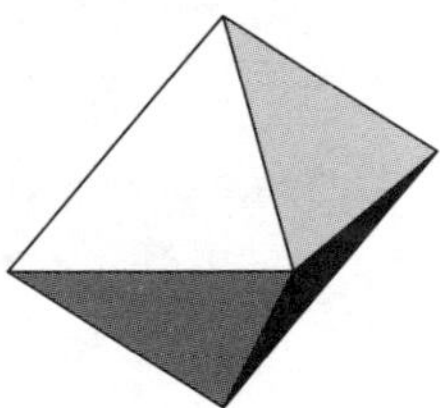

Net for octahedron without tabs

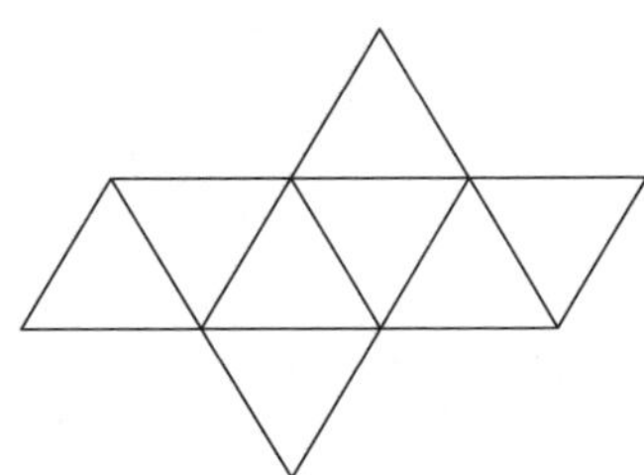

Constructing an icosahehron

An icosahedron has 20 faces, of which all are equilateral triangles. 14 vertices and five faces meet at each vertex.

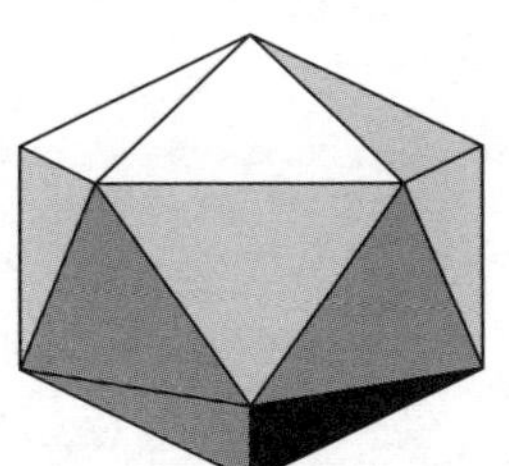

Net for an icosahedron, without tabs

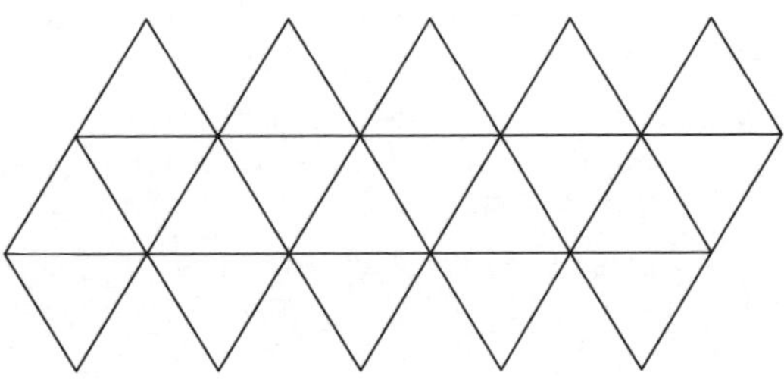

Polyhedra that are not platonic

A **polyhedron** combines several types of regular polygons or polygons that are not regular polygons.

1. The snub cube: a combination of squares and equilateral triangles.
Here is the net (without gluing tabs):

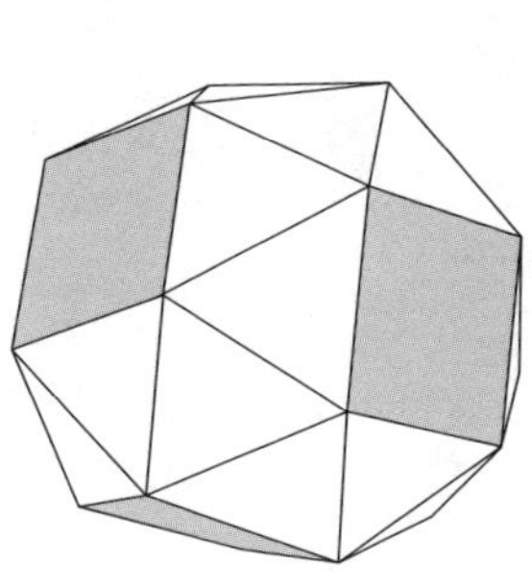

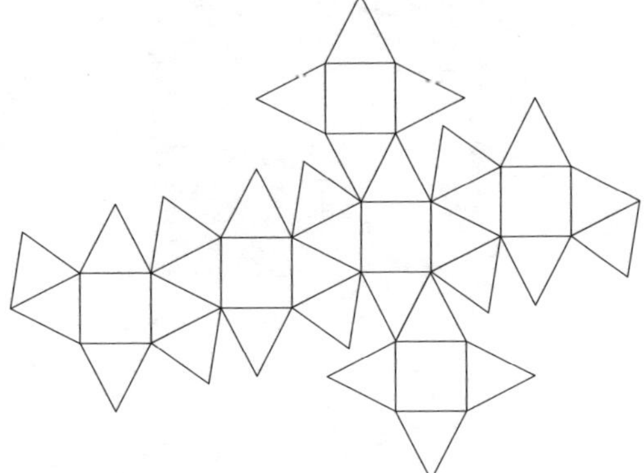

2. The great dodecahedron is made from identical isosceles triangles:

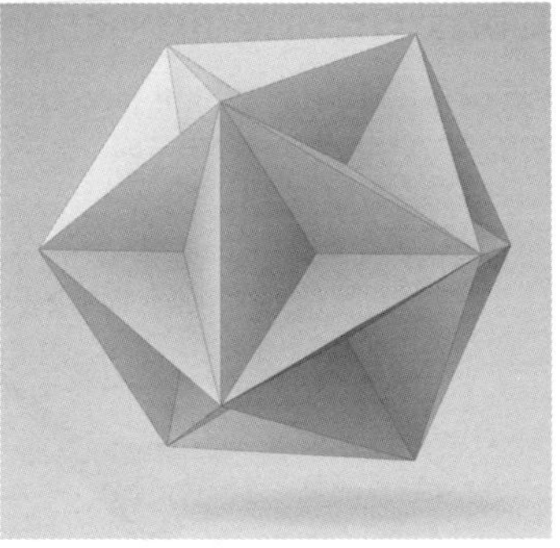

There are many other polyhedrons whose photos and nets can be found in books and on the internet.

Unit 12.5 Activity 1B: Constructing polyhedra

1. Construct a tetrahedron. For this task you will need ruler, compass, protractor, cardboard or thick paper, scissors or cutter, glue. Although the tetrahedron is the simplest of the polyhedra, it will require care and accuracy to make it look correct. Having completed this tetrahedron you will realise the skills that are needed to produce accurate and good-looking polyhedra.
2. Construct another polyhedron of your choice. You should draw the net of the polyhedron first. Write down the name of the polyhedron that you have constructed and the polygons that made up the solid.
3. Soccer balls have traditionally been made in the shape of a truncated icosahedron. This shape is made up of regular pentagons (the black shapes) and regular hexagons. When the shape is made from a flexible material and air is pumped into it and kept under pressure the polyhedron becomes rounded in shape. Make a template of a regular hexagon and a regular pentagon of the same side length and use these to construct a net for a truncated icosahedron. There are 12 black pentagons and 20 white hexagons in the net.

   Paper model of a truncated icosahedron:

4. Leonhard Euler (1707–1783) found that for any polyhedron, the number of edges plus 2 is always equal to the number of vertices plus the number of faces:

   $$e + 2 = v + f$$

   For each of the polyhedra that you have made, write down the values of e, v and f and show that Euler's rule applies.

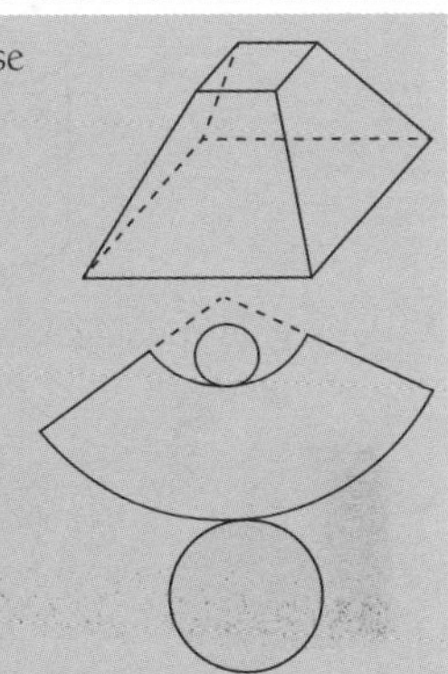

5. **a.** Draw the net for a truncated, right, square pyramid; you choose the size.
   b. Construct your truncated pyramid.
   c. Calculate the TSA and volume of your truncated pyramid.
6. A right truncated cone has a net as shown.
   a. Draw a net for the truncated cone; you choose the size. Show your calculations of the dimensions of the circles and the curved surface of the cone.
   b. Construct your cone.
   c. Calculate the TSA and volume of your truncated cone.

Symmetry

There are two types of symmetry relevant in geometry:
1. Bilateral symmetry. **2.** Rotational symmetry.

Bilateral symmetry

An object is said to have **bilateral symmetry** if, when the object is folded along a straight line, the two halves of the object would match. We can say: 'one half is the **mirror image** of the other half'. The straight line in this case is called the axis of symmetry.

Some objects have more than one axis of symmetry:

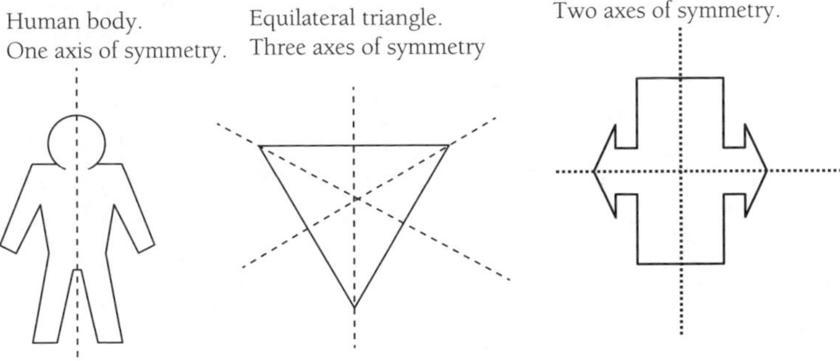

Rotational symmetry

An object has rotational symmetry if it looks the same after it has been rotated through an angle up to, and including, 360°.

The number of times the object can be rotated so that it matches the original object is called the **order of the rotational symmetry**. An object with rotational symmetry of one implies no true rotational symmetry since a full 360° **rotation** was needed to restore the object to its original position.

Some objects can have both bilateral symmetry and rotational symmetry:

1. The figure ABCD below has no bilateral symmetry but can be rotated 180° about a point so that it looks exactly like the original figure. This figure is said to have rotational symmetry of order two as there are only two rotations where the rotation fits over the original.

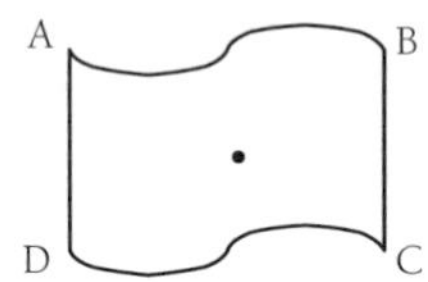

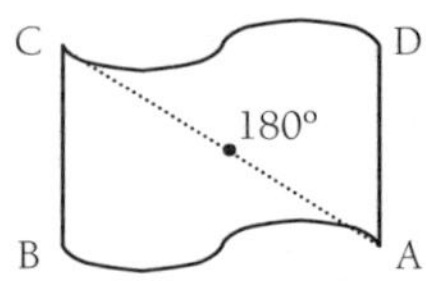

2. A regular pentagon has both bilateral symmetry (five axes of symmetry) and rotational symmetry of order five.

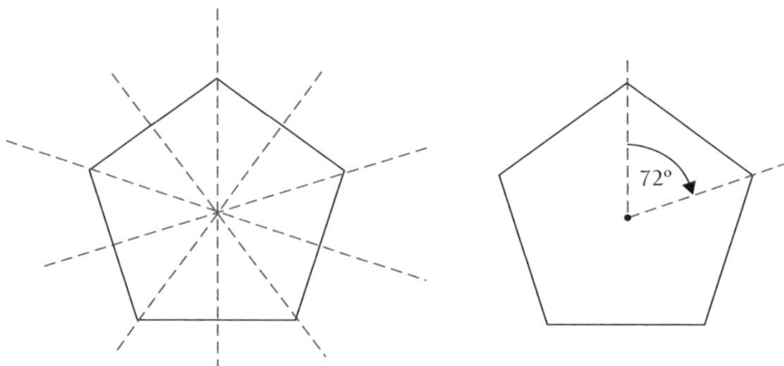

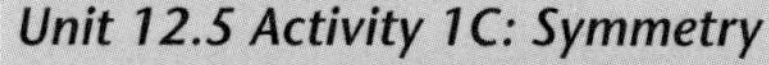

Unit 12.5 Activity 1C: Symmetry

1. State the number of axes of bilateral symmetry in each of the following figures:

a. 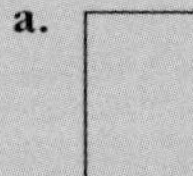 

b.  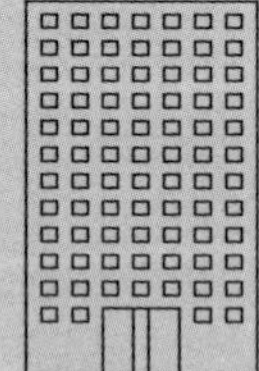

c. 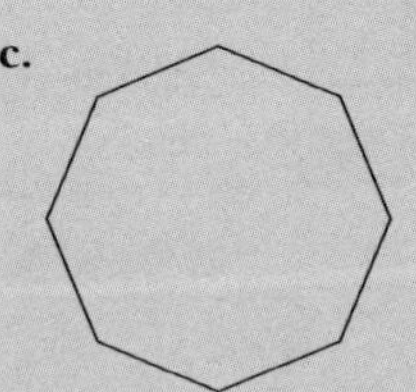

d. 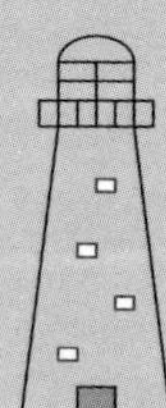

e. 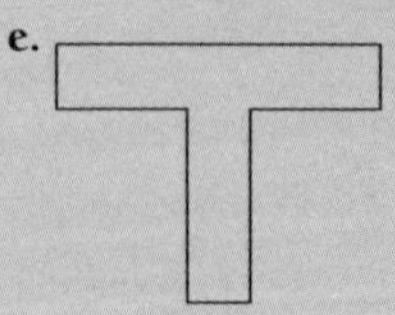

f. 

g. 

h. 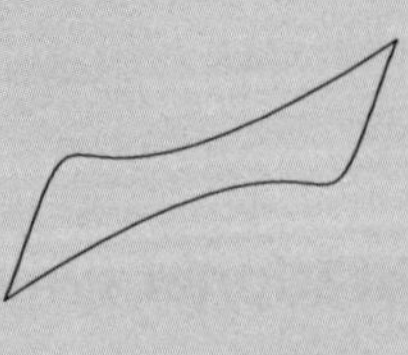

2. State the order of rotational symmetry in each of the following figures:

a. 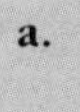 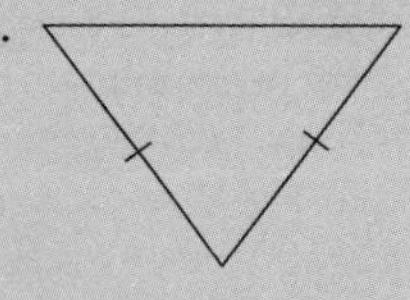

b. 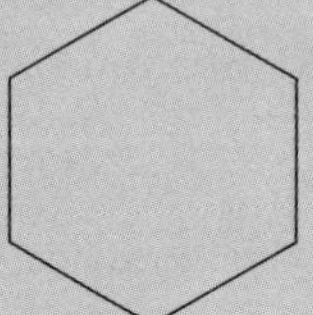

c. 

d. 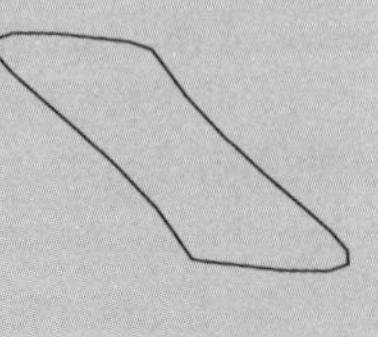

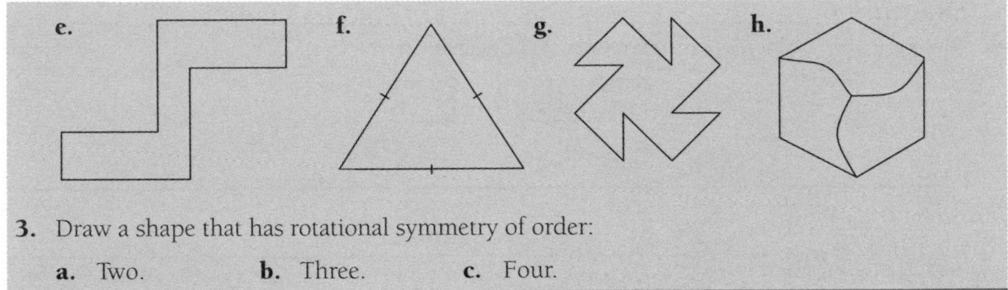

3. Draw a shape that has rotational symmetry of order:

a. Two. b. Three. c. Four.

Polyominoes

A polyonimo is a connected arrangement of squares. The squares are arranged in such a way that:

Polyomino | Not a polyomino

- Each connected square must be connected edge-to-edge.
- All squares must be in the same plane. Polyominoes are *not* three-dimensional arrangements of squares.
- Two polyominoes are the same if they can be translated (slid sideways, or up and down), reflected (as in a mirror) or rotated (turned around) so that they look the same.

For example, all the shaded polyominoes below are the same polyomino as the original:

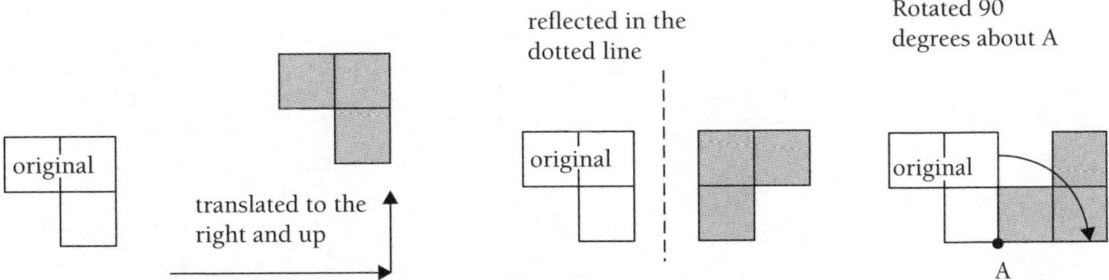

Polyominoes are named according to how many cells they have.

The table below shows the names and the number of different arrangements of the cells:

Number of cells (Order)	Name	Number of different arrangements
1	monomino	1
2	domino	1
3	triomino	2
4	tetromino	5
5	pentomino	12
6	hexomino	35
7	heptomino	108
8	octomino	369

Example A

There are only 2 different arrangements of three cells (triominoes):

Unit 12.5 Activity 1D: Polyominoes

1. There are five different tetrominoes (four cells). Draw each of these tetrominoes. If you have grid paper then you can shade in the squares to show each different tetromino.
2. There are twelve pentominos. Six of them have been drawn here. Find and draw the remaining six pentominoes.

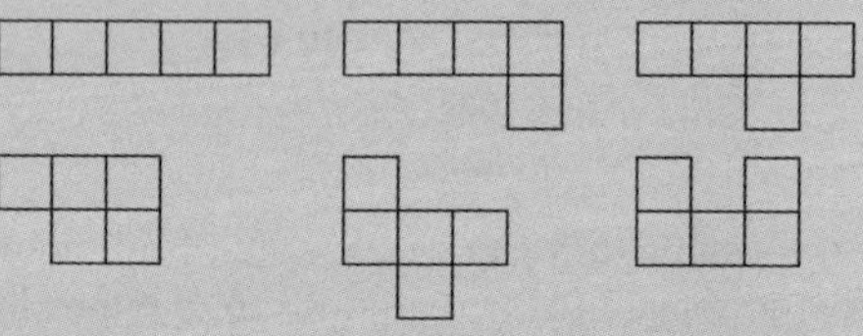

3. There are only nine of the distinct hexominoes drawn below. Match the pairs that are the same hexomino.

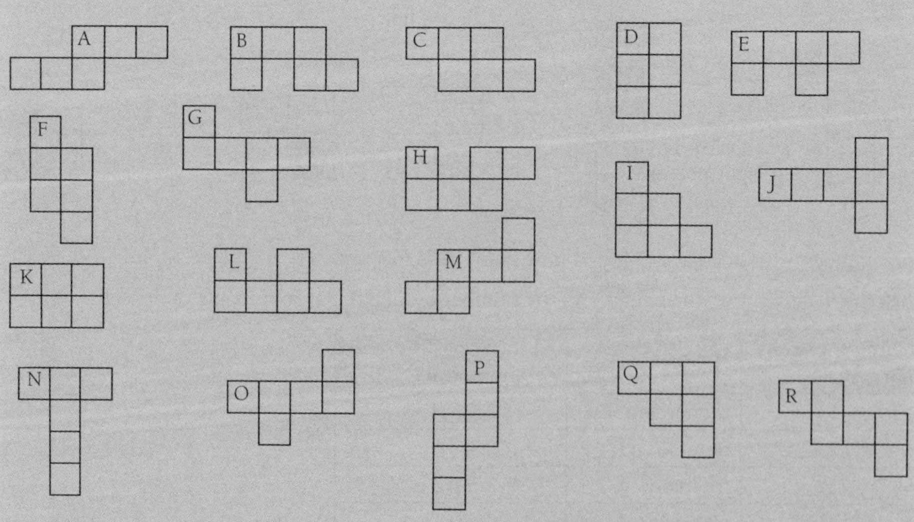

4. Nine of the 35 hexominoes are drawn in question **3** above. Find and draw six more distinct hexominoes.

5. Give a reason (translation, reflection or rotation) why both members of the following pairs of polyominoes are the same polyominoes.

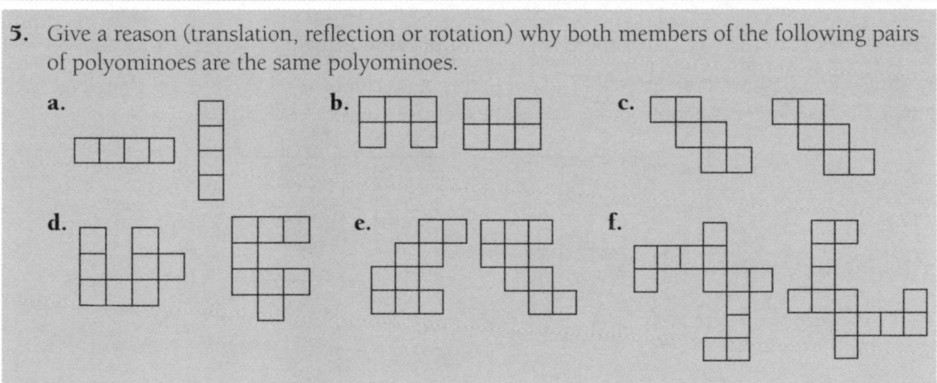

Tiling with polyominoes

The terms tiling and tesselating both mean covering an area, without leaving any gaps. There are two ways that polyominoes can be used in tiling:

1. Tiling a rectangle where the rectangle has dimensions whose units are the area covered by the size of a monomino. For example, a 3 × 4 rectangle would contain 12 monominos.
2. Tiling the plane where the tiles can cover any area, leaving no gaps, but the edges of the area are not all straight.

Tiling an area using a single polyominoe shape

A **monomino** can be used to cover any rectangular area and therefore any plane area:

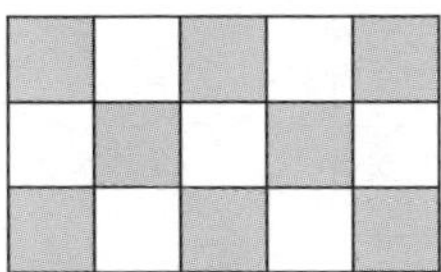

There is only one distinct shape for a monomino:

The rectangle can have any size dimensions where the unit is the side length of a monomino.

A **domino** can also be used to cover a rectangle or a plane area.

There is only one distinct shape for a domino:

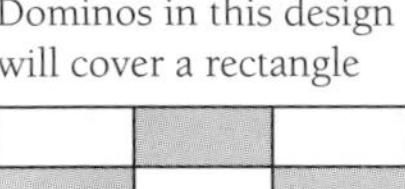
Dominos in this design will cover a rectangle

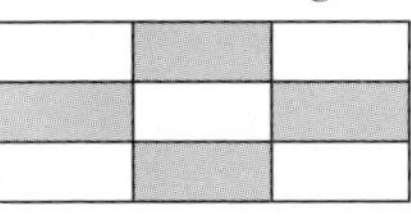

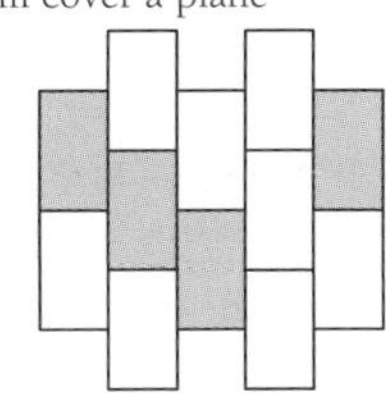
Dominoes in this design will cover a plane

Triominoes have three cells and there are two distinct arrangements:

In some tiling designs, both arrangements of tetrominoes can be used to cover a rectangle, although there are restrictions on the size of the rectangle that can be covered.

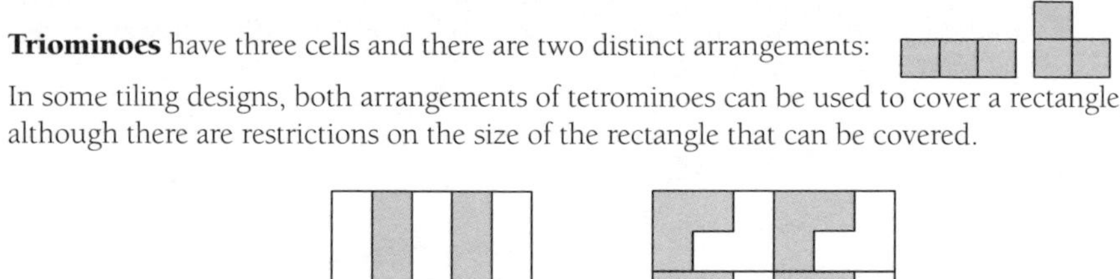

The tiling designs of triominoes (below) can be used to cover a plane area.

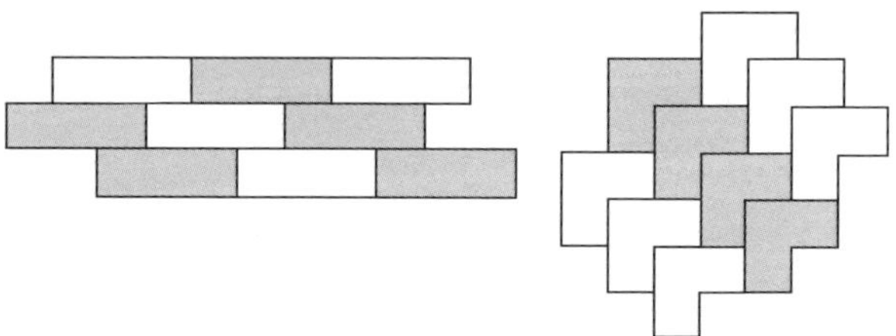

Tetrominoes have four cells. There are five distinct tetrominoes.

A tetromino of this shape can be used to cover a rectangular tile area:

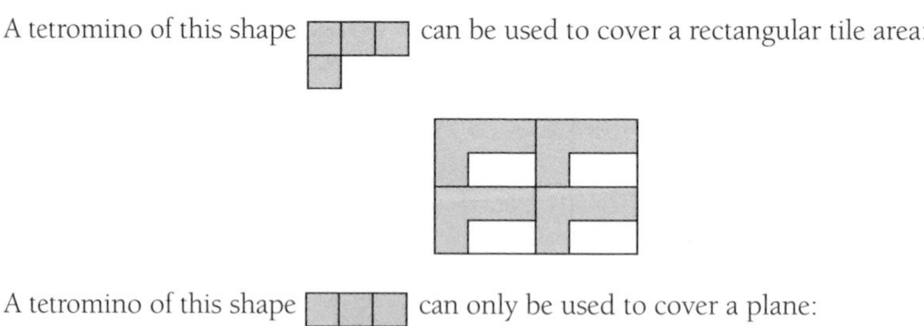

A tetromino of this shape can only be used to cover a plane:

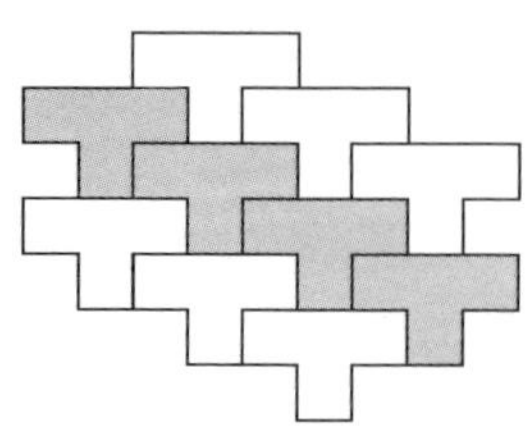

There are some polyominoes *that cannot tile a plane* using only the single polyomino shape, although it has been shown that all polyominoes of order 1 to 6 will tile a plane area.

Some tiling designs combine different polyominoes:

Example B

The tiling pattern below contains two types of tetrominoes, a pentomino and a heptomino.

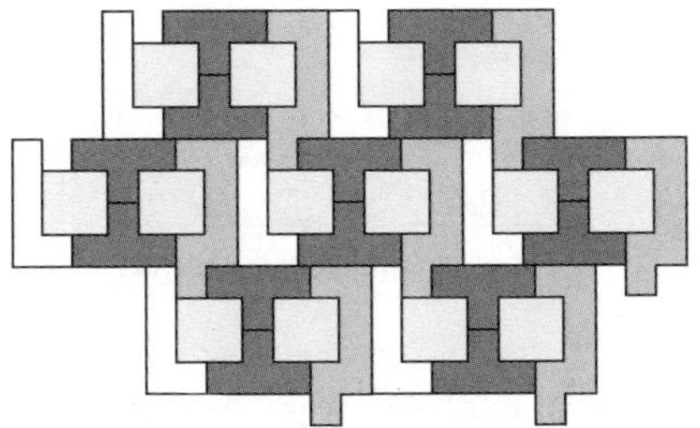

There is a whole body of knowledge, and applications, attached to the study of polyominoes. Further study can be done from books and from the internet.

Unit 12.5 Activity 1E: Tiling with polyominoes

The use of grid paper is recommended for this exercise.

1. Using only tiles in the shape of a domino, draw a pattern, different from the patterns already shown above, that:
   a. Can fit in a rectangle.
   b. Covers a plane area.
2. Use triominoes in the shape of [3-square straight triomino] to draw a tiling pattern, other than the patterns already shown above, that:
   a. Can fit in a rectangle.
   b. Covers a plane area.
3. Use tetrominoes in the shape of [L-tetromino] to draw a tiling pattern, that:
   a. Can fit in a rectangle.
   b. Covers a plane area.
4. Design a tiling pattern that combines both triominoes.
5. Find a tiling pattern that combines the tetromino [2×2 square tetromino] and the pentomino [pentomino]
6. It is possible to tile a plane area using all 5 tetrominos. Can you do this? It may help to cut out copies of the tetrominoes.

Polyiamonds

A polyiamond is a connected arrangement of equilateral triangles. The triangles are arranged in such a way that:

- Each connected triangle is connected side-to-side.
- All triangles must be in the same plane. Polyiamonds are *not* three-dimensional arrangements of triangles.

- Two polyiamonds are the same if they can be translated (slid sideways, or up and down), reflected (as in a mirror) or rotated (turned around) so that they look the same. For example, all the shaded polyiamonds below are the same polyiamond as the original:

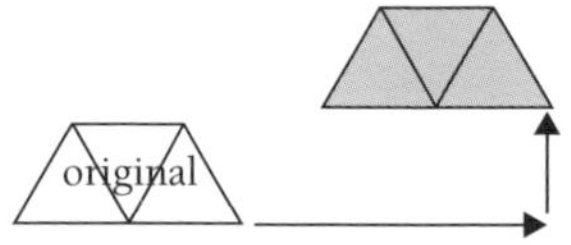

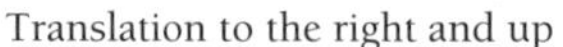
Translation to the right and up

Reflection as in a mirror

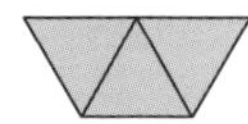
Rotation of 180 degrees

Polyiamonds are named according to how many triangles they have. The table below shows the names and the number of different arrangements of the triangles:

Number of triangles (order)	Polyiamond name	Number of different arrangements
1	moniamond	1
2	diamond	1
3	triamond	1
4	tetriamond	3
5	pentiamond	4
6	hexiamond	12
7	heptiamond	24
8	octiamond	66

There is only one each of moniamond, diamond and triamond, shown respectively below:

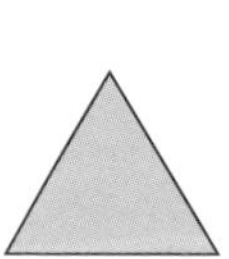
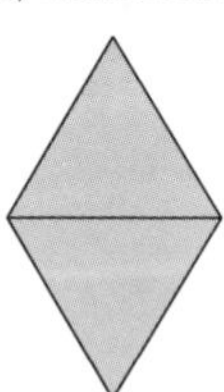
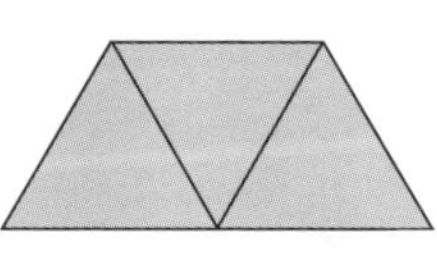

It is usual to shade polyiamond patterns on an **isometric** grid, or dot paper:

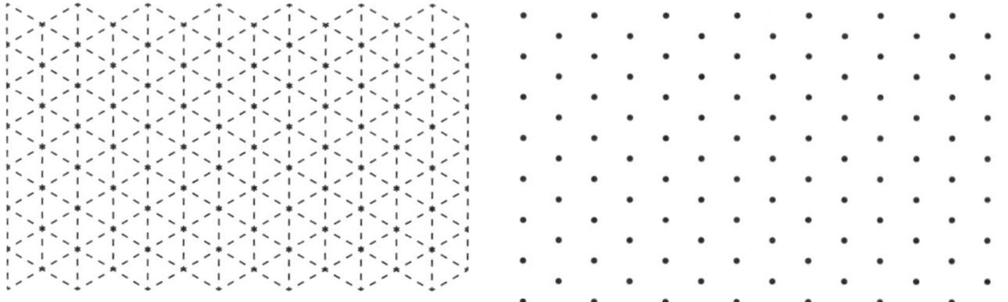

Unit 12.5 Activity 1F: Polyiamonds

1. There are three tetriamonds. Shade all of these tetriamonds on isometric paper.
2. There are four pentiamonds. Shade all of these pentiamonds on isometric paper.

3. There are 12 hexiamonds. Shade all 12 of these on isometric paper.
4. Match the identical pairs of polyiamonds shaded on the isometric paper below:

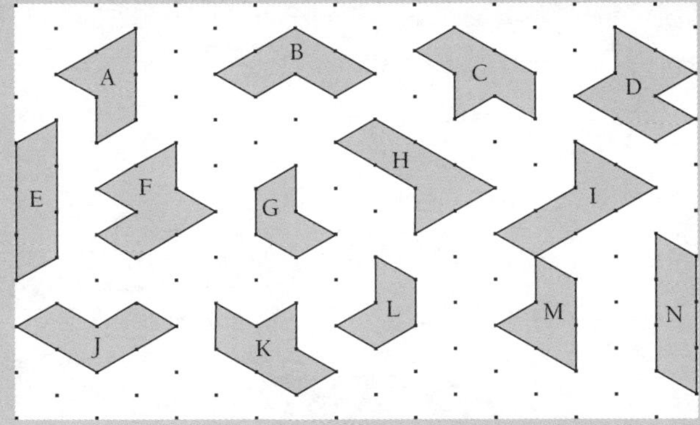

Tiling with polyiamonds

There are many tiling patterns, with no gaps, that can be produce using either a single pentiamond or a combination of pentiamonds. There are also puzzles such as fitting one of each of the hexiamonds into a parallelogram area. All polyiamonds of order 1 to 6 can tile the plane.

Example C

1. A pattern that tiles the plane using a single hexiamond:

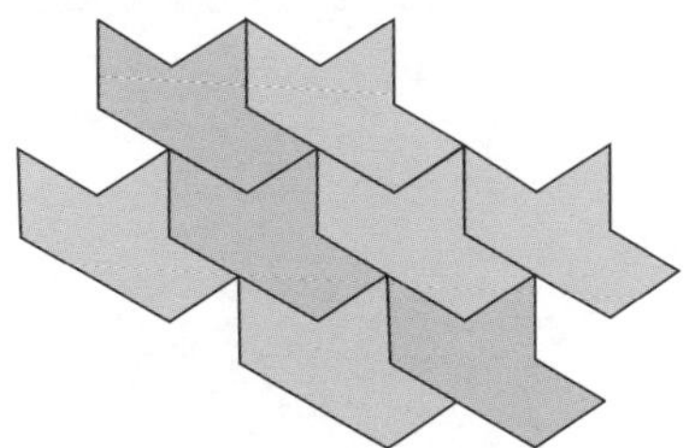

2. A pattern that tiles the plane with a combination of polyiamonds:

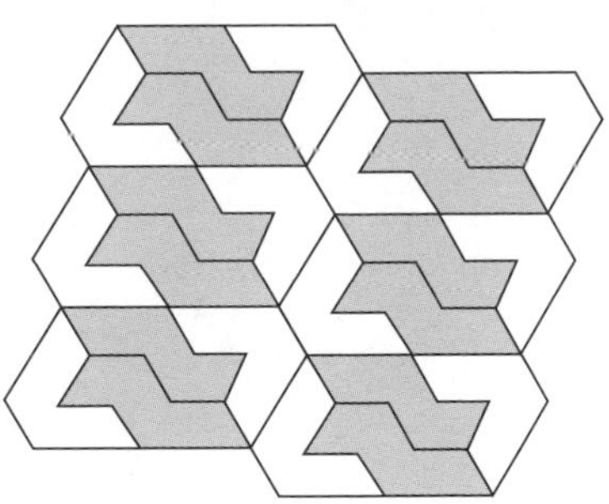

Unit 12.5 Activity 1G: Tiling with polyiamonds

It is recommended that you use isometric paper for this exercise.

1. Choose one of the pentiomonds and show how it can be used to tile the plane.

2. Show how this hexiomond 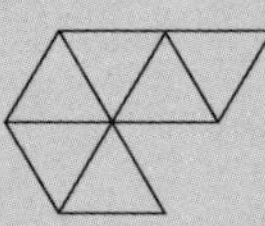 can tile the plane.

3. Show that the plane *cannot* be tiled with the heptiamond: 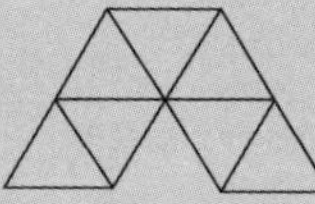

4. Below is one combination of all 12 hexiamonds. Count the number of triangles in this parallelogram and use this to explain why the twelve hexiamonds fit into this shape.

5. Can you fit all twelve hexiamonds into the parallelograms below?

a. 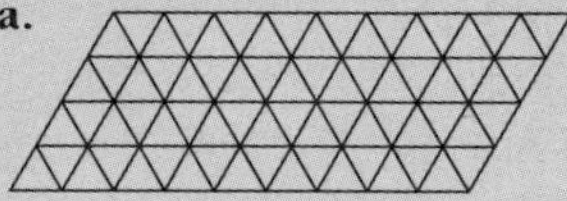

b.

Unit 12.5 Applying Geometry in Papua New Guinean Arts

Topic 2: Tessellations

Topic 2 continues the coverage of mathematics that deals with shapes and patterns. Having examined polyhedra in the previous Topic, we now look at tessellations. The word 'tessella' means 'small square' in Latin; and tessellation means 'tiling', which is a way of covering floors or walls with squares (or triangles or hexagons) of baked clay or carpet or other material. The text and exercises in this Topic should generate ideas for a study of patterns and shapes that are traditional in Papua New Guinea.

Tessellations (tiling)

A **tessellation** is produced when an identical shape is used to cover a plane surface without any gaps or holes. The pattern of tessellation should be able to continue indefinitely. Another word for a tessellation is a **tiling**.

There are only three tessellations using regular polygons: triangles, square and hexagons.

1. Using equilateral triangles:

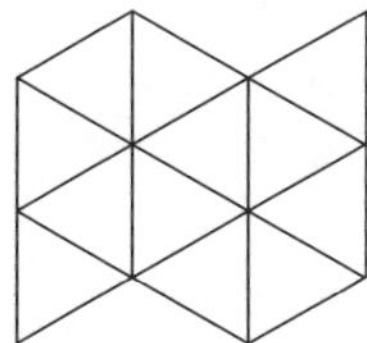

2. Using squares:

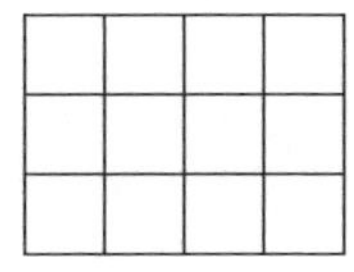

3. Using hexagons:

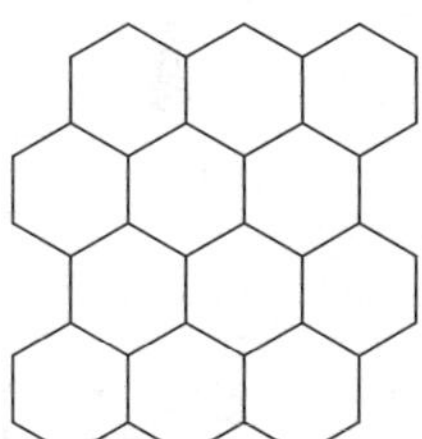

A regular pentagon cannot be used as a tessellation because of the gaps that are created in the pattern – see the diagram at right, with the gaps shown by the shaded areas.

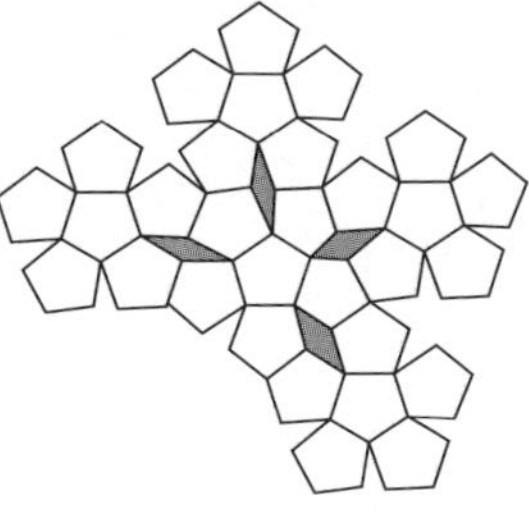

However, a pentagon in the shape  can be used as a tessellation:

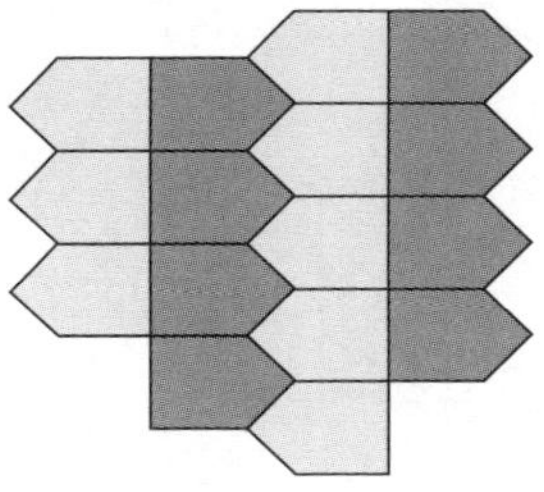

Creating a tessellation

Simple tessellating shapes can be created by starting with a basic shape, that tessellates, like a rectangle, altering this shape on one side and then translating this alteration to the other side.

Start with a rectangle.

Alter the top side; translate to the bottom side.

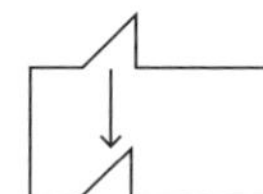

Alter the right side, translate to the left side.

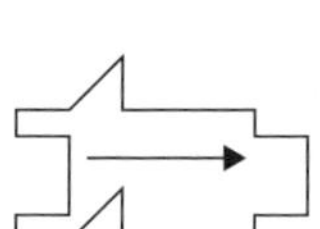

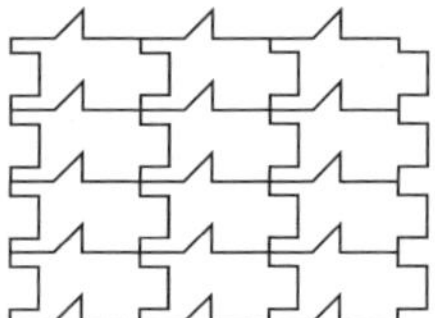

Creating more complex tessellations

There are many examples of complex tessellations to be found in books and on the internet.

The Dutch graphic designer M.C. Escher (1898–1972) was famous for his designs and art involving tessellations. He was inspired by the tiling patterns he saw on the walls and floors of a Muslim palace and fortress known as the Alhambra, in Spain, during a visit there in 1922.

The following tessellation relies on the rotational symmetry of the equilateral triangle and the hexagon:

Construct a hexagon and divide it into the six equilateral triangles. Find the centroid of one triangle; draw a curvy line from the centroid to the vertex and copy this line to the other vertices.

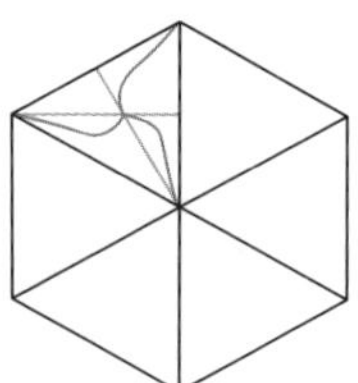

Rotate the set of curvy lines into the adjacent triangle.

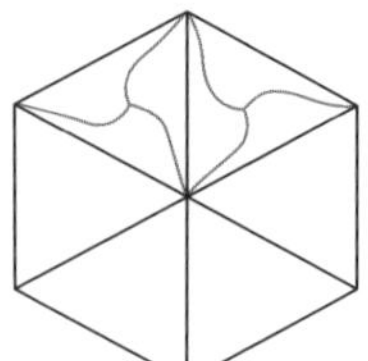

Continue rotating the set of curvy lines to fill the hexagon.

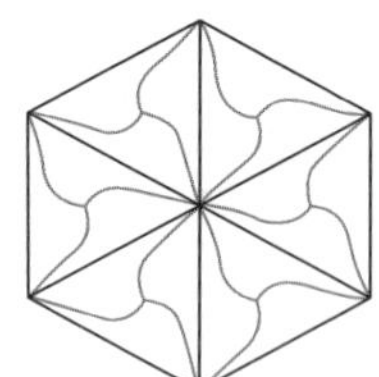

Copy the completed hexagon and join them to construct the tesselation.

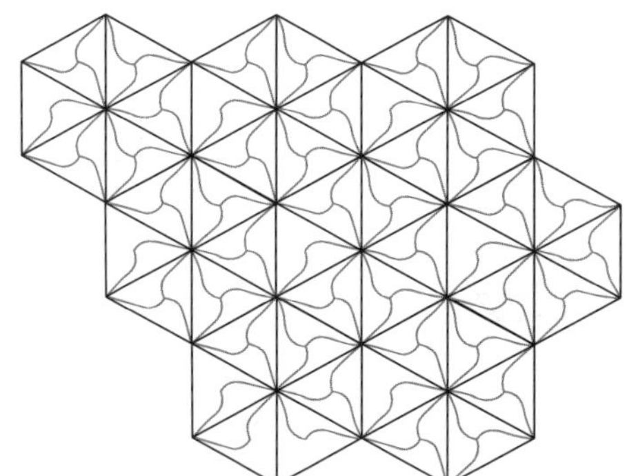

The same design without the construction lines:

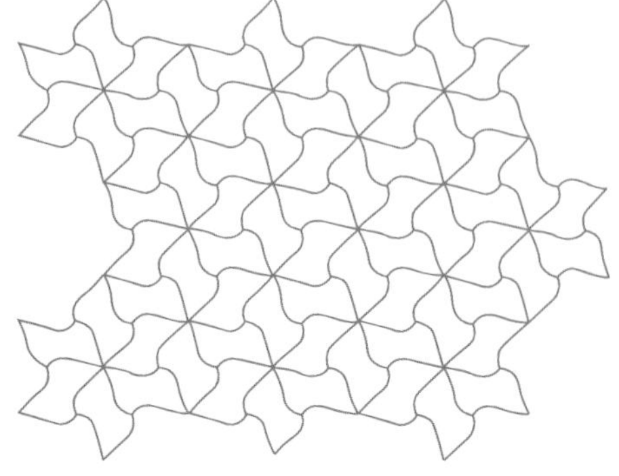

Unit 12.5 Activity 2: Tessellations

The use of grid paper or isometric paper is recommended for this exercise.

1. Which of the following figures tessellate?

a.

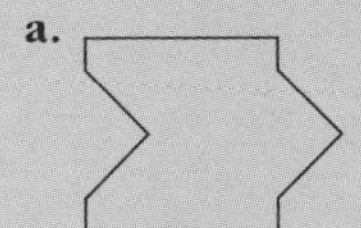

b.

c.

d.

2. Design a simple tessellation with straight edges, based on a rectangle. Show that it tessellates by constructing a plane area covered by your design.

3. Design a simple tessellation with curvy edges, based on a polygon that tessellates. Show that your design tessellates by constructing a plane area covered by it.

4. Design a more complex tessellation based on a regular hexagon as shown in the example. You could use tracing paper to rotate your lines in the hexagon. Show how your design tessellates over an area.

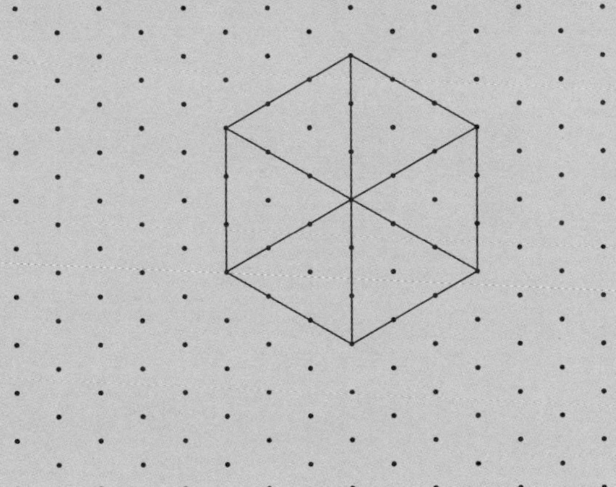

Unit 12.5 Applying Geometry in Papua New Guinean Arts

Topic 3: Application of patterns and shapes in traditional arts

As part of your assessment in Grade 12 General Mathematics you will be required to study an object of art from your local community. You will need to apply your knowledge of shapes and patterns to analyse this object and use mathematical terms to describe it. The examples and exercises in this Topic should help you and give you ideas.

Having covered polyhedra and tessellations in Topics 1 and 2, Topic 3 looks at how mathematical shapes are applied in the traditional patterns and measurement in PNG and uses mathematical language to describe them.

Patterns and shapes in traditional arts

Gather some examples of art from your home or your local community. Apply your knowledge of shapes and patterns to analyse these objects and use mathematical terms to describe them.

For example, you can look at the decorative pattern of a work of art and describe it in terms of polyominoes, poliamonds or other tessellations. You can identify lines of symmetry and reflections or rotations of the basic pattern. If the work of art is three-dimensional you can investigate volume or total surface area.

Graph paper or isometric paper can be used to help with illustrations of patterns.

Example A

Analysis of pattern

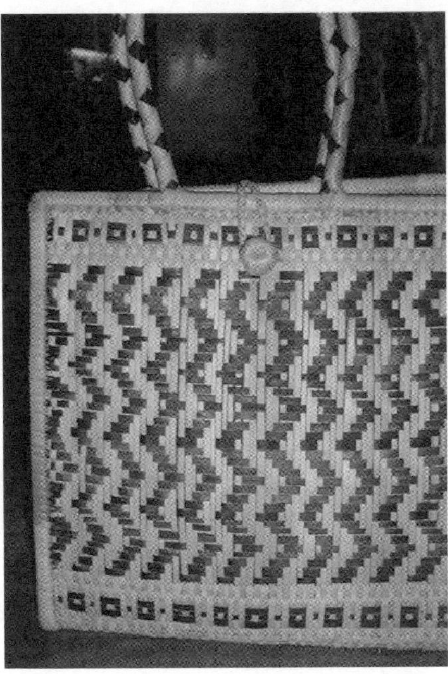

The main pattern on the bag is reproduced below:

The pattern is a combination of polyominoes:

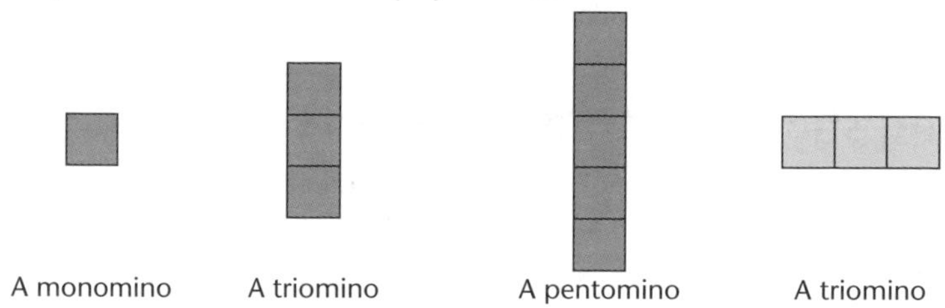

The lighter-shaded triomino, illustrated above, is a rotation of the darker-shaded triomino, so is the same trionimo.

The pattern combination of these polyominoes could not be used to tile a rectangle but can be used to tile the plane. This means that reproductions of the basic pattern, joined together, will always have ragged edges.

The basic unit of pattern is given below. There is only one axis of symmetry for the basic unit of pattern and, to tile the plane, reflections of the basic unit of pattern are required. This is illustrated on the diagram below, at right.

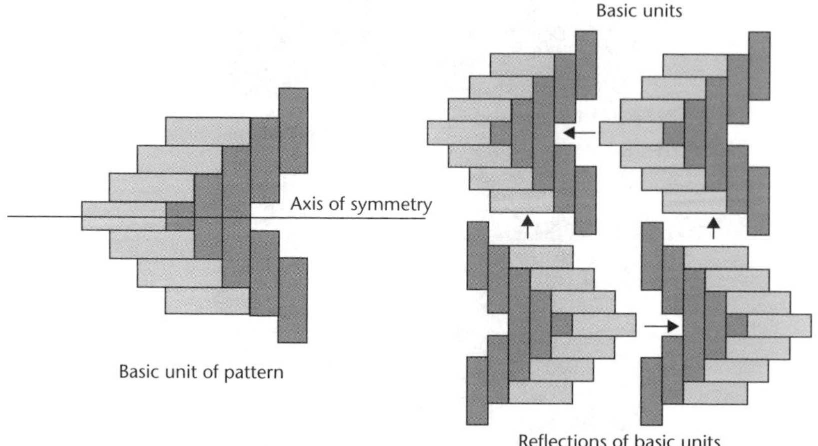

Further work could be done on the other patterns on the bag and handle.The capacity (volume) of the bag and the total surface area could also be investigated.

Example B

Analysis of shape

The basic shape of this basket is that of a truncated right cone. A truncated right cone has the centre of the upper circular edge vertically above the centre of the base circle.

Truncated means cut off, so this cone is inverted with the top cut off parallel to the base of the cone (this becomes the open top of the basket.)

top of cone that is removed

base of basket

open top of basket

Measuring the basket gives the following dimensions:

Diameter of top circular edge: 32 cm.

Diameter of base circle: 24 cm.

Height of the basket: 20 cm.

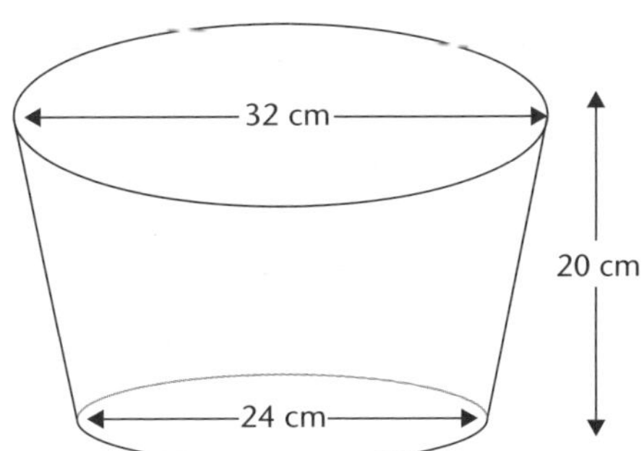

The capacity (volume) of the basket can be found using the formula:

$$\text{Volume} = \frac{1}{3}(B + B' + \sqrt{BB'}) \times h \text{ where:}$$

B is the area of the lower base.

B′ is the area of the upper base.

h is the vertical height of the truncated solid.

In this case:

1. The radius of the lower base is 12 cm, so the area of the lower base is $\pi \times 12^2 = 452.39 \text{ cm}^2$ so $B = 452.39$.
2. The radius of the upper base is 16 cm, so the area of the lower base is $\pi \times 16^2 = 804.25 \text{ cm}^2$ so $B' = 804.25$.
3. $h = 20$ cm.

$$\text{Volume of the basket} = \frac{1}{3}(452.39 + 804.25 + \sqrt{452.39 \times 804.25}) \times 20 = 12398.82\text{cm}^3$$

The capacity of the basket is 12.398 or approximately 12.4 litres.

(There are 1 000 cm^3 in one litre)

The **total surface area** (TSA) of a cone is given by the formula:
$\text{TSA} = \pi rs + \pi r^2$

where:

πrs is the TSA of the **curved surface** of the cone and πr^2 is the area of the circular base.

The basket that we are studying has a net given in the diagram at right:

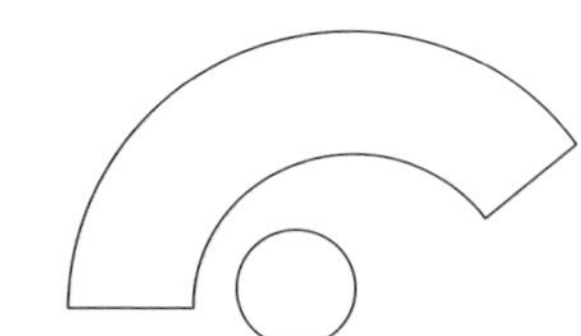

Using similar triangles and the 'whole cone':

$$(\tan\theta =)\frac{20}{4} = \frac{h}{16}$$

Giving: $h = \frac{20 \times 16}{4} = 80$ cm

For the whole cone $s = \sqrt{16^2 + 80^2} = 81.58$ cm.

Subtract from this 81.58 (the length of the sloping edge of the basket) gives 81.58 – 20.40 = 61.18 cm.

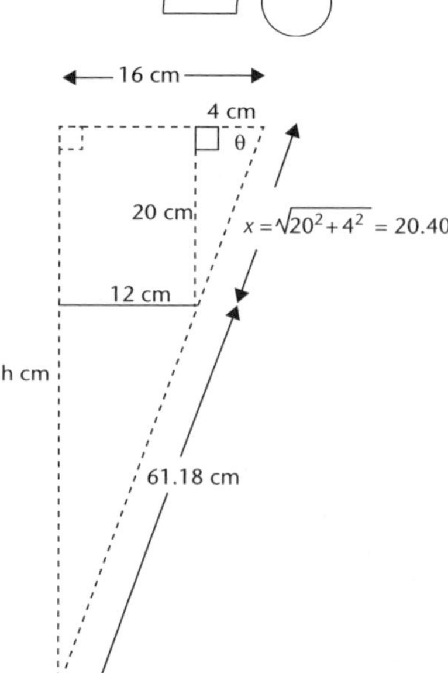

The TSA of the curved surface of the whole cone

$= \pi rs$

$= \pi \times 16 \times 81.58$

$= 4\,100.66 \text{ cm}^2$

The TSA of the discarded part of the cone:

$= \pi rs$

$= \pi \times 12 \times 61.18$

$= 2\,306.43 \text{ cm}^2$

Hence the TSA of the curved surface of the basket is:
$4\,100.66 - 2\,306.43 = 1\,794.23 \text{ cm}^2$

The area of the base of the basket $= \pi \times 12^2 = 452.39 \text{ cm}^2$

Adding these two areas together gives:

The TSA of the basket $= 2\,246.62 \text{ cm}^2$

Unit 12.5 Activity 3: Application of patterns and shapes in traditional arts

1. Apply your knowledge of shapes and patterns to analyse this bilum and use mathematical terms to describe it.

2. Apply your knowledge of shapes and patterns to analyse the shape of this tray. Use mathematical terms to describe it. The shape is part of a sphere where the diameter of the outside circle of the tray (excluding the handles) is 25 cm and the greatest depth is 3 cm.

3. Find objects from your own home or local community and analyse them in mathematical terms.

Appendix
Mathematical symbols and formulae

Area

$\frac{1}{2}bh$	triangle	a^2	square (side a)
bh	rectangle or parallelogram	$\frac{1}{2}(a+b)h$	trapezium
$\frac{1}{2}d_1d_2$	rhombus or kite	πr^2	circle
$\frac{\theta}{360} \times \pi r^2$	sector (θ in degrees)	$4\pi r^2$	sphere (surface area)
$\frac{1}{2}r^2\theta$	sector (θ in radians)	$2\pi r(r+h)$	closed cylinder (surface area)

Volume

l^3	cube (side l)	Ah	prism (A = base area)
lbh	cuboid	$\frac{1}{3}Ah$	pyramid (A = base area)
$\frac{1}{2}bhl$	triangular prism	$\frac{4}{3}\pi r^3$	sphere
πr^2h	cylinder	$\frac{1}{3}\pi r^2h$	cone

Geometry

$\angle$	angle
$\angle ABC$ or $A\hat{B}C$	the angle at B defined by the points A, B and C.
$\overline{AB}$	the line segment A to B.
//	parallel to
$\perp$	perpendicular to
Δ	triangle
$^\circ$	degree
π	Pi = 3.14159 (5 dp)

Probability

P(A)	the probability of event A occurring
$\overline{x}$	the mean of a sample
s.d. or sd	the standard deviation of a sample
μ	population mean
σ	population standard deviation

Co-ordinate geometry

Distance

$d = \sqrt{(x_1 - x_2)^2 + (y_1 - y_2)^2}$

Mid-point of a line

mid-point $\left(\frac{x_1 + x_2}{2}, \frac{y_1 + y_2}{2}\right)$

Gradient

$m = \frac{y_2 - y_1}{x_2 - x_1}$

Lines

$y = mx + c$

$y - y_1 = m(x - x_1)$

Perpendicular and parallel lines

$y = m_1x + c_1$ and $y = m_2x + c_2$ are parallel if $m_1 = m_2$ and perpendicular if $m_1 \times m_2 = -1$

Circles

$x^2 + y^2 = r^2$ circle with centre (0, 0) and radius r

$(x - a)^2 + (y - b)^2 = r^2$ circle with centre (a, b) and radius r

Perimeter

$C = \pi d$ or $2\pi r$ circumference of a circle

$\frac{\theta}{360} \times 2\pi r$ arc length of a sector (θ in degrees)

$r\theta$ arc length of a sector (θ in radians)

Trigonometry

$\sin\theta = \frac{o}{h}$ o = opposite

$\cos\theta = \frac{a}{h}$ h = hypotenuse

$\tan\theta = \frac{o}{a}$ a = adjacent side

$h^2 = a^2 + o^2$ Pythagoras' theorem

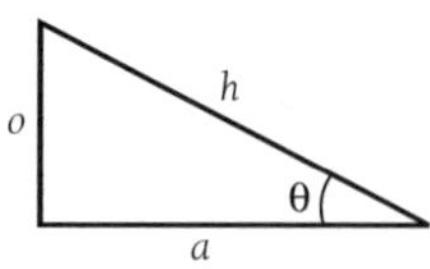

Special angles

θ	0°	30°	45°	60°	90°
$\sin\theta$	0	$\frac{1}{2}$	$\frac{1}{\sqrt{2}}$	$\frac{\sqrt{3}}{2}$	1
$\cos\theta$	1	$\frac{\sqrt{3}}{2}$	$\frac{1}{\sqrt{2}}$	$\frac{1}{2}$	0
$\tan\theta$	0	$\frac{1}{\sqrt{3}}$	1	$\sqrt{3}$	

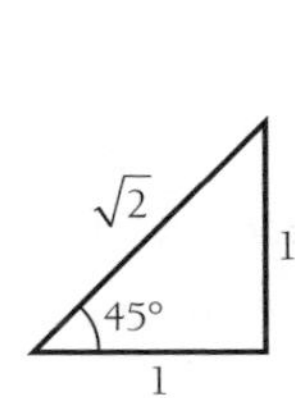

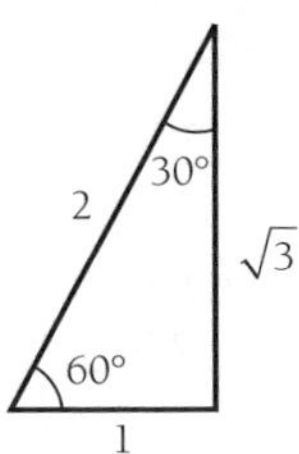

Trigonometric identities

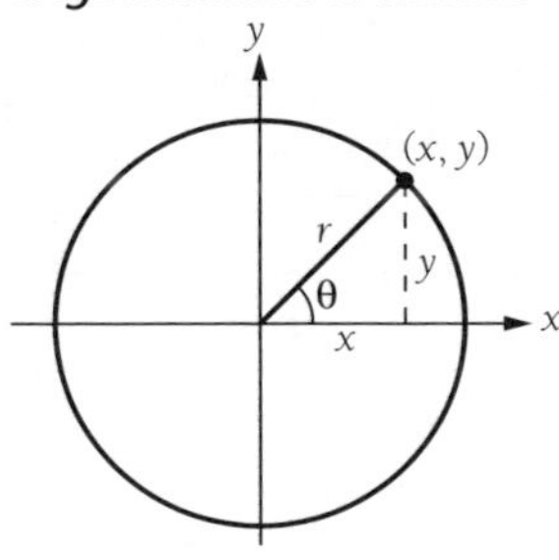

$\sin(90^\circ \pm \theta) = \cos\theta$

$\cos(90^\circ \pm \theta) = \mp\sin\theta$

$\sin(180^\circ \pm \theta) = \mp\sin\theta$

$\cos(180^\circ \pm \theta) = -\cos\theta$

$\sin(360^\circ - \theta) = -\sin\theta$

$\cos(360^\circ - \theta) = \cos\theta$

Triangles

Sine rule: $\frac{a}{\sin A} = \frac{b}{\sin B} = \frac{c}{\sin C}$ or $\frac{\sin A}{a} = \frac{\sin B}{b} = \frac{\sin C}{c}$

Cosine rule: $a^2 = b^2 + c^2 - 2bc\cos A$ or $\cos A = \frac{b^2 + c^2 - a^2}{2bc}$

Area of triangle: Area $= \frac{1}{2}ab\sin C$

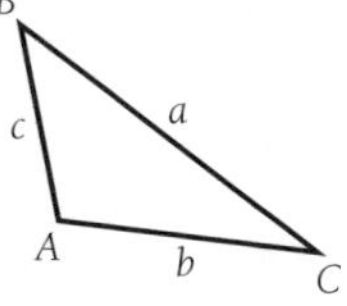

Sequences and series

Arithmetic series

$a + (a + d) + (a + 2d) + (a + 3d) + \ldots$

$t_n = a + (n - 1)d$

$S_n = \frac{n}{2}(2a + (n - 1)d)$

Geometric series

$a + ar + ar^2 + ar^3 + \ldots$

$t_n = ar^{n-1}$

$S_n = \frac{a(1 - r^n)}{(1 - r)}, r \neq 1$

$S_\infty = \frac{a}{1 - r}$ for $|r| < 1$

Probability

$P(A) = 1 - P(A')$

$P(A \text{ or } B) = P(A \cup B) = P(A) + P(B) - P(A \cap B)$

A and B are independent events $\Leftrightarrow P(A \cap B) = P(A) \times P(B)$

Mean, variance and standard deviation

Statistic	Sample of ungrouped data	Samples of grouped data
Mean	$\bar{x} = \dfrac{\sum x_i}{n}$	$\bar{x} = \dfrac{\sum x_i f_i}{\sum f_i}$
Variance	$s^2 = \dfrac{\sum (x_i - \bar{x})^2}{n}$ $= \dfrac{\sum x_i^{\ 2}}{n} - \bar{x}^2$	$s^2 = \dfrac{\sum (x - \bar{x})^2 f_i}{\sum f_i}$ $= \dfrac{\sum x_i^{\ 2} f_i}{\sum f_i} - \bar{x}^2$
Standard deviation	$s = \sqrt{s^2}$	$s = \sqrt{s^2}$

Answers

Answers for many questions include a 'Marking Guide':

- **A** ('Achievement' meaning 'satisfactory achievement')
- **M** ('Merit' meaning 'high achievement').
- **E** ('Excellence' meaning 'very high achievement').

The Marking Guide has been made by the authors and the publishers and is not an official guide but we hope it will be a help to students who are striving for the best possible results.

Unit 12.1 Measurement

Unit 12.1 Activity 1: Perimeter, circumference and area (page 5)

1. a. i. 26 cm **(A)** **ii.** 40 cm^2 **(A)** **b. i.** 24 m **(A)** **ii.** 36 m^2 **(A)**
c. i. 5.6 m **(A)** **ii.** 1.6 m^2 **(A)** **d. i.** 17.5 m **(A)** **ii.** 9.375 m^2 **(A)**
e. i. 34.2 m **(A)** **ii.** 42.84 m^2 **(A)** **f. i.** 326 cm **(A)** **ii.** 3 380 cm^2 **(A)**
2. a. 20 m^2 **(A)** **b.** 30 cm^2 **(A)** **c.** 144 cm^2 **(A)** **d.** 28.8 m^2 **(A)**
e. 100 m^2 **(A)** **f.** 45.98 m^2 **(A)**
3. a. i. 31.4 cm **(A)** **ii.** 78.5 cm^2 **(A)** **b. i.** 27.0 m **(A)** **ii.** 58.1 m^2 **(A)**
c. i. 37.7 cm **(A)** **ii.** 113.1 cm^2 **(A)** **d. i.** 77.3 cm **(A)** **ii.** 475.3 cm^2 **(A)**
e. i. 3.1 m **(M)** **ii.** 4.7 m^2 **(M)** **f. i.** 3.1 cm **(M)** **ii.** 6.3 cm^2 **(M)**
4. a. 3.27 cm^2 or 327 mm^2 **(M)** **b.** 9.96 cm^2 or 996 mm^2 **(M)**
5. a. 1.235 ha **(A)** **b.** 19 sections **(A)** **6.** 7.96 m **(A)** **7.** 10 m^2 **(M)**
8. 2 500 cm^2 or 0.25 m^2 **(M)** **9.** 315 m^2 **(M)** **10.** 79.4 g **(M)**
11. 509 kg (3 sf) or 510 kg (2 sf) or 500 kg (1 sf) **(E)** **12.** 77 m^2 **(E)**

Unit 12.1 Activity 2: Solving triangles (page 12)

1. a. 49° **b.** 104° **c.** 44°
2. a. 20 cm **b.** 13.86 cm **c.** 5.81 cm
3. a. 11.49, 9.64, 36.13 cm **b.** 14.58, 23.16, 55.74 cm **c.** 15.78,18, 51.78 cm
4. a. 55.51 m **b.** 20.00 cm **c.** 27.80 cm **d.** 33.57 cm
5. a. 46.84° **b.** 32.21° **c.** 57.40°

Unit 12.1 Activity 3: Area of a triangle (page 19)

1. a. 44.64 cm^2 **b.** 250 cm^2 **c.** 128.06 mm^2 **d.** 23.72 cm^2
e. 12.84 m^2 **f.** 61.64 m^2 **g.** 594.26 mm^2 **h.** 30 cm^2
2. a. 1 141.77 mm^2 **b.** 38.42 cm^2 **c.** 123.51 m^2
d. 586.15 cm^2 **e.** 227.48 cm^2 **f.** 578.76 m^2
3. d.

Unit 12.1 Activity 4: Area of a polygon (page 22)

1. a. 352.38 cm^2 **b.** 88.24 m^2 **c.** 39.77 m^2
d. 41.94 m^2 **e.** 88.94 m^2 **f.** 737.24 cm^2
2. a. 45.94 cm^2 **b.** 49 m^2 **c.** 75.99 m^2 **d.** 494.22 m^2
3. 1478.28 m^2 **4.** 267.5 m^2

Unit 12.1 Activity 5: Regular polygons (page 27)

1. **a.** 60° **b.** 120° **c.** 150°
2. **a.** 90°, 90°, 90°, 45° **b.** 72°, 72°, 108°, 54° **c.** 45°, 45°, 135°, 62.5°
3. **a.** 61.94 m^2 **b.** 120.71 cm^2 **c.** 2 895.75 m^2 **d.** 363.39 cm^2
4. **a.** **i.** 72° **ii.** 108° **iii.** 36° **b.** 50.11 m^2
5. 18 sides

Unit 12.1 Activity 6: Volume and surface area (page 32)

1. **a.** 280 cm^3 **b.** 1 131 cm^3 **c.** 100 m^3 **d.** 15 268 cm^3
e. 4 051 cm^3 **f.** 600 cm^3 **g.** 1 414 m^3 **h.** 3
2. **a.** 900 cm^2, 1 800 cm^3 **b.** 1 100 cm^2, 2 771 cm^3 **c.** 942 cm^2, 2 094 cm^3
d. 636 cm^2, 864 cm^3 **e.** 3 376 cm^2, 13 764 cm^3 **f.** 283 cm^2, 314 cm^3
g. 80 cm^2, 40 m^3
3. 9 cm^3, 26 cm^2 **4.** 159 cm **5.** 129 L
6. **a.** 2.66 m **b.** 26.15 m^2 **7.** **a.** 672 **b.** 304
8. **a.** 204 **b.** 249 **9.** **a.** 851 **b.** 529
10. 30 min ***(M)***
11. **a.** 104 min ***(E)*** **b.** K244.40 ***(M)***
c. Part **a** depends on a very steady water flow without interruption. Also, a rate of 2 600 L may only be correct to 2 sf, ie the water may be flowing at any rate between 2 550 L/min or 2 650 L/min which would give times between 78.5 min and 81.6 min. The answer to part **b** depends on the accuracy of the water meter. If it registers water in any way that differs from the 208 m^3 or 208 000 L of the calculation in **a**, then the changes would be different. ***(E)***
12. 22 cm

Unit 12.1 Activity 7: Similar figures and areas (page 40)

1. 9:25 **2.** 4:25 **3.** 450 **4.** 1 963.6 cm^2 **5.** 28.88 hectares **6.** 1:8
7. **a.** 34.75 cm^2 **b.** 75 m **c.** 1 954.54 m^2

Unit 12.1 Activity 8: Similar solids, length, area and volume (page 44)

1. **a.** 1:2 **b.** 1:4 **c.** 1:8
2. **a.** 3:4 **b.** 9:16 **c.** 27:64
3. **a.** 3/5 **b.** 9/25 **c.** 27/125
4. **a.** 5:4 **b.** 125:64
5. 4:7 **6.** 50p **7.** **a.** 4:3 **b.** 1 181.25 cm^2 **c.** 64:27
8. 5 mL = 5 cm^3 **9.** 120 m^3 **10.** **a.** **i.** 8:27 **ii.** 2:3 **b.** 36 cm
11. 4:5 **12.** 27:125 **13.** **a.** 4:25 **b.** 64 cm^2
14. **a.** 168.75 cm^2 **b.** 1:26 **15.** 810 mL
16. **a.** 9:64 **b.** 9 cm **c.** 27:485

Unit 12.1 Activity 9: Truncated solids (page 51)

1. **a.** 105 m^3 **b.** 145.86 m^2 **2.** **a.** 8 377.58 cm^3 **b.** 2 513.27 cm^2
3. 1 406 mL **4.** 2 780.57 cm^3
5. **a.** 15.59 cm **b.** 763.41 cm^2 **c.** 3.82 L **d.** 1 017.88 cm^2

6. **a.** 15.20 cm **b.** 725.71 cm^2 **c.** 12.47 L **d.** 2 111.15 cm^2
7. **a.** 2 m **b.** 52 m^3 **c.** 57.69 m^2

Unit 12.1 Activity 10: Estimation (page 57)

1. **a.** Measure line with a piece of string and measure length of string. Multiply length of string, in cm, by 40 to get distance in metres. ***(M)***
b. 320 m ***(M)***
c. The distance measured is 'as the crow flies', it does not account for the uphills and downhills. ***(M)***
2. **a.** 40–50 km ***(M)*** **b.** 85–95 km^2 ***(M)*** **3.** 31 000 to 34 000 m^2 ***(M)***
4. 0.18 to 0.22 km^3 or 180 000 000 to 220 000 000 m^3 ***(M)*** **5.** 5.0 to 5.4 m^3 ***(E)***
6. Answers will vary. For example:
a. Weigh 100 grams of rice. Divide answer by 100 to get mass of individual grain. ***(E)***
b. Put 100 grains in water. Measure volume displaced (mL). Convert to cm^3 and divide by 100 to get volume of individual grain. ***(E)***
7. **a.** 26 500 cm^3 ***(E)*** **b.** 37 cm ***(E)***
c. Many objects cannot be immersed in water without change/damage to the object. A large container of water may be needed if the object is large and this is difficult to arrange. The method is, however, accurate if the water displaced can be read reasonably accurately (using a suitable scale). ***(E)***
8. **a.** 0.58 m^3 ***(E)***
b. 3.1 loads (3 or 4 loads, depending on how particular Bunu is about topsoil depth). ***(E)***
9. **a.** 39 cm (assuming sand forms a square-based pyramid) ***(E)***
b. Answers will vary but will need to be a container which will hold approximately 0.4 m^3. For example, a cylinder of diameter 1.85 m and depth 18 cm would hold all the sand to within 3 cm of the top of the sandpit. ***(E)***
10. **a.** Approximately 1.3 m^3.
b. Assumption – sand forms a pyramid from the top of the sides of the trailer.
c. **i.** A pyramid shape may not be a close approximation to the true shape of the soil above the sides of the trailer.
ii. Soil can be packed firmly to take up less space (volume) and so a trailer can carry more 'volume' of soil from a pile if firmly packed. ***(E)***

Unit 12.1 Activity 11: Using pace length for distance and area (page 62)

1. **a.** 37.5 × 17.86 m **b.** 670 m^2
2. **a.**

Paces	Distance(m)
12	10.4
13	11.27
14	12.13
15	13
16	13.87
17	14.73
18	15.6

b. 227 m^2

3. a.

Paces	Distance(m)
7	5.38
8	6.15
9	6.92
11	8.46
13	10
16	12.31
18	13.85
25	19.23

b. 71 m^2 **c.** 29.26 m^2 **d.** 100.26 m^2

4. a. 0.64 m **b.** 9 paces = 5.77 m, 11 paces = 7.05 m **c.** 12.96 m^2

5. a. 0.714 m **b.** 141 paces = 100.71 m, 238 paces = 170 m **c.** 14 944 m^2

6. Present your investigation to the class and discuss your answers with your teacher.

Unit 12.1 Activity 12: Offset surveys (page 68)

1. a.

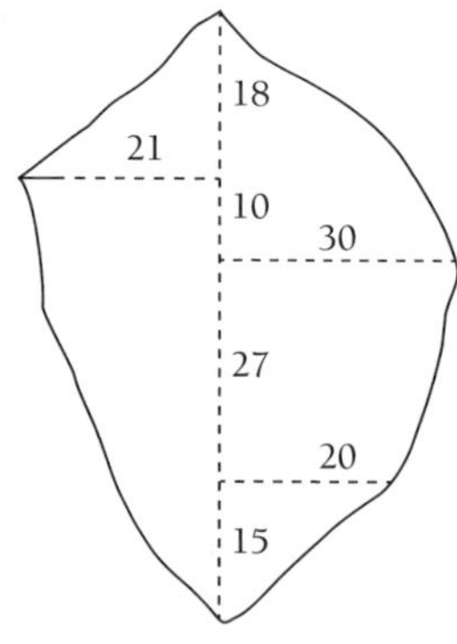

b.

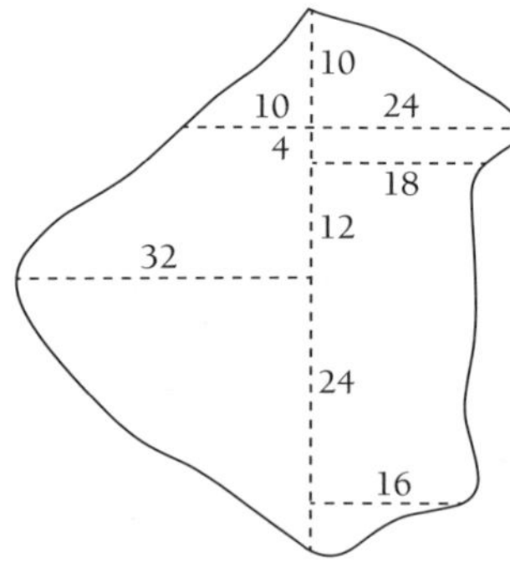

2. a. 3 266 m^2, 233.77 m **b.** 4 110 m^2, 260.04 m

c. 1 980 m2, 178.57 m **d.** 1 826 m2, 176.17 m

3. a. 1 106.5 m^2 **b. i.** 21.95 m **ii.** 39.40 m

4. a. Measurements in metres **b.** Area ≈ 2 930 m^2, Perimeter ≈ 105 m

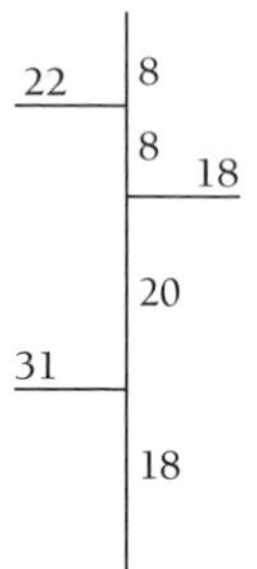

Unit 12.1 Activity 13: Estimating areas with irregular boundaries (page 74)

1. 17 750 m² **2.** 2.58 m² **3.** 91 m²
4. a. 693 m² **b.** 512 m² **c.** 582 m²
5. a. 17 505m² **b.** 5 940 m²

Unit 12.1 Activity 14: Surveying around obstacles (page 79)

1. 105.83 m **2.** 46.10 m **3.** 60 m **4.** 18.39 m **5.** 163.33 m

Unit 12.2 Managing Money 2

Unit 12.2 Activity 1A: Percentages – revision (page 83)

1. a. i. 1.2 **ii.** 1.1 **iii.** 1.03
b. i. 0.92 **ii.** 0.88 **iii.** 0.75
2. a. 20% **b.** 12.5% **c.** 7% **d.** 14%
3. a. K19.96 **b.** K84
4. a. K11.13 **b.** K14.50
5. 8% discount, K1.80
6. a. K332 **b. i.** K174.30 **ii.** 47.5%
7. a. K76 **b.** 17%
8. a. K51.37 **b.** K77.23
9. 3.36% **10. a.** K252 000 **b.** K250 000

Unit 12.2 Activity 1B: Simple interest (page 88)

1. a. K240 **b.** K656.60 **c.** K2 428.80
2. a. K170 **b.** K522.33 **c.** K14 105
3. K3 913.22 **4.** K5 081.25 **5. a.** K436.64 **b.** 9 562.50
6. a. K34 440 **b.** K1 722 **7. a.** K12 600 **b.** K350
8. K1 006.58 **9. a.** 224 **b.** K397.68 **10.** K5 731.23 **11.** K1 710.85

Unit 12.2 Activity 1C: Transposing the simple-interest formula (page 92)

1. a. K9 850 **b.** K36 500 **c.** K3 750
2. a. 6.25% **b.** 5.8% **c.** 8.4%
3. a. 2.5 years **b.** 17 months **c.** 6 months
4. K691.20 **5.** 6.8% **6.** 15 days **7.** 26 months
8. K8 450 **9.** 7.2% **10.** K21 900 **11.** 8.33%
12. 8.57% **13.** 3.33% **14.** K40 000 **15.** K2 080

Unit 12.2 Activity 2: Interest and balances (page 97)

1. $3 940.54 **2.** $3 418
3. a. $2.50 **b.** $1.82 **c.** $19.27
4. a. $6.31 **b.** $10.93 **c.** $13.19
5. a. $162 **b.** $163.70 **6. a.** $3.87 **b.** $4.34
7. a. K5 785 **b. i.** $139.05 **ii.** $143.68

Unit 12.2 Activity 3: Compound interest (page 104)

1. a. K2 805.10 **b.** K152 124.31 **c.** K11 181.62 **d.** K11 810.07 **e.** K29 542.98
2. a. K5 851.76 **b.** K852.45 **c.** K7 447.20
3. K9 149.12 **4.** K648, K672.59, simple interest **5.** K10.12 **6.** 6.08%
7. a. K2 599.35 **b.** K1 662.91
8. a. K4 016.35 **b.** K3 197.39 **c.** K2 341.51
9. a. Compound **b.** 1 555.99 **c.** 1 469.67 **d.** 9 185.45

Unit 12.2 Activity 4A: Inflation (page 108)

1. K122 **2.** K31 763 **3. a.** K55 645 **b.** 8
4. Profit of K18 110 **5.** K479 **6.** K74 889 **7.** 2.2%

Unit 12.2 Activity 4B: Appreciation (page 110)

1. a. K36 501 **b.** 5 years **2. a.** K44 958 **b.** 7 years
3. K42 020 **4.** K2 974 **5.** K210 219

Unit 12.2 Activity 4C: Depreciation (page 112)

1. a. K240 **b.** K1 280 **c.** K560
2. a. K900 **b.** 6 years **3. a.** K975 **b.** K2 250
4. a. 17.5% **b.** K2 280
5. a. Flat rate **b.** K3 400 **c.** K1 000 **d.** K400
e. Book value = K3 400 – 400 × Age

Unit 12.2 Activity 4D: Reducing-balance depreciation (page 115)

1. a. K850 **b.** K5 019.17 **c.** K4 065.52
2. a. K30 750 **b.** K40 359 **c.** K17 908
3.

4. 13%, greater than 13%
5. a. 16% **b.** 26% **c.** K301
6. a. K8 730 **b.** K7 159 **c.** K21 928 **d.** 8 years
7. K1 312.20 **8.** 7 years **9. a.** 5 years **b.** 6 years
10. a. 14 years **b.** 5 years

Unit 12.2 Activity 4E: Unit-cost depreciation (page 117)

1. a. i. 8 toea **ii.** K80 **iii.** K20 000 **b.** K13 600
2. a. 83 toea **b.** 32.2%
3. a. i. 12.4 toea **ii.** K124 **b. i.** K31 260 **ii.** K20 960
c. i. 5 years **ii.** 10.75%
4. a. K39 000 **b.** K25 500
5. a. K9 000 **b.** 325 000 copies **c.** 375 000 copies **d.** Profit of K300
6. a. II or III **b.** I **c.** II **d.** II **e.** III **f.** I **g.** I or IV **h.** I

Unit 12.2 Activity 5A: Credit cards (page 120)

1. 1.5% **2.** K3.85 **3.** K0, K3.65
4. a. K3.57 **b.** K302.75 **5.** K270.75
6. a. K2 868.32 **b.** K2 860.21 **c.** K2 892.36
d. K2 884.61 **e.** K2 876.74

Unit 12.2 Activity 5B: Hire-purchase (page 123)

1. K480 **2.** K400; 20% **3.** K127.50 **4. a.** K454.17 **b.** K315.28
5. 25% **6.** 26% **7. a.** 13 **b.** K420 **c.** 19%
8. Option A 31.7%; Option B 23.3%; Option B is preferable.

Unit 12.2 Activity 5C: Reducing-balance loans (page 126)

1. a.

Year	Interest (K)	Balance(K)	Repayment (K)	Amount owing (K)
0				5 000.00
1	450.00	5 450.00	1 200.00	4 250.00
2	382.50	4632.50	1 200.00	3432.50
3	308.93	3741.43	1 200.00	2541.43
4	228.73	2770.16	1 200.00	**1570.16**

b. K1 370.16

2. a. 2.1%

b.

Quarter	Interest(K)	Balance (K)	Repayment (K)	Amount owing (K)
0				15 000.00
1	315.00	15 315.00	1 000.00	14 315.00
2	300.62	14 615.62	1 000.00	13 615.62
3	285.93	13 901.54	1 000.00	12 901.54
4	270.93	13 172.47	1 000.00	**12 172.47**

c. K1 172.47

3. a.

Month	Interest (K)	Balance (K)	Repayment (K)	Amount owing (K)
0				780.00
1	11.70	791.70	50.00	741.70
2	11.13	752.83	50.00	702.83
3	10.54	713.37	50.00	**663.37**

b. K33.37
c. K152.26

4. a.

Quarter	Interest (K)	Balance (K)	Repayment (K)	Amount owing (K)
0				7 500.00
1	180.00	7 680.00	200.00	7 480.00
2	179.52	7 659.52	200.00	7 459.52
3	179.02	7 638.54	200.00	7 438.55
4	178.53	7 617.08	200.00	7 417.08

b.

Month	Interest (K)	Balance (K)	Repayment (K)	Amount owing (K)
0				200 000.00
1	1 150.00	211 150.00	2 400.00	198 750.00
2	1 142.81	199 892.81	2 400.00	197 492.81
3	1 135.58	198 628.39	2 400.00	196 228.39
4	1 128.31	197 356.70	2 400.00	194 956.70
5	1 121.00	196 077.70	2 400.00	**193 677.70**

5.

Month	Interest (K)	Balance (K)	Payment (K)	Investment amount (K)
0				300 000.00
1	2125.00	302 125.00	2 400.00	299 725.00
2	2123.05	301 848.05	2 400.00	299 448.05
3	2121.09	301 569.14	2 400.00	299 169.14
4	2119.12	301 288.26	2 400.00	298 888.26

Unit 12.2 Activity 5D: Using the annuities formula (page 132)

1.

	P	R	Q	n
a.	6000	1.081	500	4
b.	15 000	1.006	250	36
c.	850	1.027	120	6
d.	200 000	1.0031	800	260
e.	82 000	1.0016	180	520

2. a. K5 936.82 **b.** K8 591.96 **c.** K226.95
d. K1 28 223.86 **e.** K42 460.33
3. K964.92 **4. a.** K10 553.14 **b.** K8 142.57
5. a. K5 767.87 **b.** K15 297.93 **c.** K83 487.49
6. a. K846.86 **b.** K281.14
7. a. K174.03 **b.** K646.08 **c.** K1 905.43 **8.** K334.85

Unit 12.2 Activity 5E: Using the annuities formula (page 133)

1. i. a. $A = 84\,000 \times 1.007^{60} - \frac{650(1.007^{60} - 1)}{1.007 - 1}$ **b.** K79 396.62

ii. a. $0 = P \times 1.019^{40} - \frac{5\,000(1.019^{40} - 1)}{1.019 - 1}$ **b.** K13 9207.18

iii. a. $5\,000 = 100\,000 \times 1.006^{120} - \frac{Q(1.006^{120} - 1)}{1.006 - 1}$ **b.** K885.71

2. a. C **b.** D **c.** B **d.** A
3. a. K101 000 **b.** K114 000 **4. a.** K409.90 **b.** K53 287.50
5. a. K264 540 **b.** K305 994
6. a. I **b.** II **c.** III **d.** IV
7. a. K275.97 **b.** K521.25 **c.** K203 000 **d.** K40 **e.** K99 462.03

Unit 12.2 Activity 5F: Service fees and charges (page 137)

Discuss your reports with your teacher and share them with your class.

Unit 12.2 Activity 6A: Types of investments (page 141)

1. a. K6 491.37 **b.** K7 096.97 **c.** K7 702.58
2. a. K6 680.27 **b.** K9 385.16 **c.** K12 090.06 **3.** K4 683.56
4. a. K554.10 **b.** K3 865.67 **5. a.** K124 **b.** K129.02
6. a. K353.96 **b.** K163.10 **7. a.** K7 789.64 **b.** 63 months

Unit 12.2 Activity 6B: Investing in real estate (page 145)

1. a. K597.40 **b.** K2 040 **c.** K4 800 **d.** K18 000 **e** K72 500
2. a. K14 500 **b.** K20 425 **c.** K31 400 **d.** K45 750
3. a. K56 000 **b.** K1 764
4. a. K19 300 **b.** K96 500 **c.** K6 140.51 **d.** K5 16 555 **e.** K22 500
5. a. K55 000 **b.** K57 520 **c.** K194 988 **d.** K12 988 loss
6. a. i. A = 0, R = 1.003, n = 520 **ii.** K723 592 **b.** K916 667
7. Discuss your investigation with your teacher and share your results with the class.

Unit 12.2 Activity 6C: Investing in the stock market (page 149)

1. a. i. $1 500 **ii.** $37.50 **b. i.** $8 554 **ii.** $171.08
c. i. $9 248 **ii.** $184.96
2. a. i. $3 895 **ii** $77.90 **iii.** K8 156.38
b. i. $899.25 **ii.** K65 **iii.** K1 856.52
c. i. $6 675 **ii.** $133.50 **iii.** K13 977.88
3. a. $8.49 **b.** $4 796.85 **c.** $95.94 **d.** $47.97
e. $4 652.94 **f.** $7 617.90 **g.** 38.9%
4. a. $6.17 **b.** $12 031.50 **c.** Profit of $1 696.50
5. 14 423 shares.

Unit 12.2 Activity 6D: Assessing the suitability of companies that sell shares (page 151)

1. a. $5 200 **b.** $725 **c.** $156
2. a. 9 cents **b.** 4.68% **c.** 9 cents per share **d.** 20.6
3. a. Yes **b.** 6.8% **c.** 14.9 years
d. It will take 14.9 years for the purchase price of the shares to be returned by the dividends.
4. a. 2.64% **b.** 23.4 years
5. a. 4.4% **b.** 9.2 years
c. The dividend yield of the similar company is 7.3% compared to 4.4% with the original company. This means that more income is generated by the similar company.
6. a. 3.36% **b.** 14.8 years **c.** About the same **d.** 11.6%

Unit 12.2 Activity 7A: Life insurance (page 155)

1. a. K4.62 **b.** K0
c. K22 000 of insurance represents the best value at K8.73 per K1 000 of cover.
d. i. K22 000 **ii.** K0 **iii.** K0
e. i. K144 **ii.** K8.31 **iii.** K24 000

2. **a.** K24.62 **b.** K273 **c.** K31 436.66 **d.** K12.09
3. **a.** K656.1 **b.** K28.18 **c.** **i.** K50.58 **ii.** K553.80 **iii.** K67 017.10 **iv.** K21.34
4. **a.** 582.40 **b.** K10 000 **c.** K7 571.20

Unit 12.2 Activity 7B: Health insurance (page 157)

1. **a.** K12.92 **b.** K77.19 **c.** K190.40 **d.** **i.** K23.42 **ii.** K500
2. **a.** Conrad, his wife, his children aged 12 and 15 and his father.
   b. K51.15 **c.** K0 **d.** K0 **e.** K1 235.20
3. Discuss your investigation with your teacher and share your results with the class.

Unit 12.2 Activity 7C: Household insurance (page 159)

1. **a.** K557.10 **b.** K80.22 **c.** K24.51 **d.** K141 600
2. K14 000 **3.** **a.** K78.08 **b.** K1 300
4. **a.** K76 577 **b.** K603.20
5. **a.** The excess is higher with the Smarta cover. **b.** **i.** K4.84 **ii.** K850
   c. **i.** The Superior policy has a relatively low excess and higher valuables cover.
   ii. K470.87 **iii.** K381.40 **iv.** K7 300
   d. **i.** K120.86 **ii.** K865 **iii.** K1 000 **iv.** K23 000
6. Discuss your investigation and your findings with your teacher, and compare your findings with other students.

Unit 12.2 Activity 7D: Motor vehicle insurance (page 162)

1. **a.** K316 **b.** K5 180.50 **c.** 16.61% **d.** K135.25
   e. There are three parties in a situation where this insurance applies. The first two parties are the insurer and the insured and the third party is the 'party' who has been injured.
2. K0 to Gavin and K2 700 to the other driver. Gavin will be expected to pay K500 (excess) to the other driver.
3. **a.** **i.** K1 679 **ii.** K504 **iii.** K1 194
   b. **i.** K560 **ii.** K7 280 **iii.** K403
   c. **i.** K1 250 **ii.** K2 340 **iii.** K125
4. Discuss your report with your teacher and share it with your class

Unit 12.3 Probability and Statistics

Unit 12.3 Activity 1A: Probability (page 167)

Note: *Fractions have generally been left in unsimplified form to make checking easier.*

1. **a.** $\frac{16}{120}$ **(A)** **b.** $\frac{46}{120}$ **(A)**
   c. There should be approximately equal frequencies for each number but there is a higher number of 3's than expected, and a lower number of 4's. **(E)**
   d. Repeat the experiment (with more throws if possible). **(A)**
2. $\frac{59}{75} \approx 0.79$. Petra's claim is exaggerated. **(A)**
3. **a.** $\frac{1}{8}$ **(A)** **b.** $\frac{3}{8}$ **(A)** **c.** $\frac{5}{8}$ **(A)**
4. **a.** $\frac{2}{10}$ **(A)** **b.** $\frac{3}{10}$ **(A)** **c.** $\frac{5}{10}$ **(A)**

5. **a.** $\frac{1}{2}$ **(A)** **b.** $\frac{1}{4}$ **(A)** **c.** $\frac{1}{8}$ **(A)** **d.** $\frac{7}{8}$ **(A)**

6. **a.** $\frac{2}{5}$ **(A)** **b.** $\frac{3}{5}$ **(A)** **c.** $\frac{1}{5}$ **(A)**

7. **a.** $\frac{1}{2}$ **(A)** **b.** 1 **(A)** **c.** 0 **(A)**

8. **a.** 10 **(A)** **b.** $\frac{4}{10}$ **(A)**

Unit 12.3 Activity 1B: Probabilities for statistical data (page 170)

Note: *Fractions have generally been left in unsimplified form to make checking easier.*

1. **a.** $\frac{23}{40}$ **(A)** **b.** $\frac{27}{40}$ **(A)** **c.** $\frac{5}{40}$ **(A)**

2. **a.** **i.** $\frac{4}{30}$ **ii.** 0 **iii.** $\frac{7}{30}$ **iv.** $\frac{19}{30}$ **(A)** **b.** **i.** $\frac{2}{12}$ **ii.** $\frac{6}{12}$ **(A)**

3. **a.** $\frac{10}{50}$ **(A)** **b.** $\frac{22}{50}$ **(A)** **c.** $\frac{11}{50}$ **(A)**

4. **a.**

	Factory	Office	Total
Male	88	27	115
Female	52	38	90
Total	140	65	205

b. **i.** $\frac{115}{205}$ **ii.** $\frac{140}{205}$ **iii.** $\frac{38}{205}$ **(A)**
c. $\frac{52}{90}$ **(A)** **d.** $\frac{88}{140}$ **(A)**

5. **a.** **i.** $\frac{235}{600}$ **ii.** $\frac{280}{600}$ **iii.** $\frac{95}{600}$ **(A)**
b. $\frac{90}{280}$ **(A)** **c.** $\frac{125}{235}$ **(A)** **d.** $\frac{195}{320}$ **(A)**

6. **a.** $\frac{12}{48}$ **(A)** **b.** $\frac{22}{48}$ **(A)** **c.** $\frac{10}{48}$ **(A)** **d.** $\frac{14}{22}$ **(A)**

e. $P(\text{win at home}) = \frac{17}{26} \approx 0.654$ $P(\text{win away}) = \frac{14}{22} \approx 0.636$

A win during a home game is more likely (but only just). **(M)**

f. $P(\text{loss on a weekday}) = \frac{6}{24}$ $P(\text{loss on a weekend}) = \frac{4}{24}$

A loss on a weekday is more likely. **(M)**

7. **a.** $\frac{2}{30}$ **(A)** **b.** $\frac{15}{150}$ **(A)** **c.** $\frac{13}{31}$ **(A)** **d.** $\frac{3}{15}$ **(A)**

Unit 12.3 Activity 2A: Probabilities for combinations of events (page 174)

1. **a.** $\frac{8}{12}$ **(A)** **b.** $\frac{9}{12}$ **(A)** **c.** $\frac{7}{12}$ **(A)**

2. **a.** $\frac{2}{9}$ **(A)** **b.** $\frac{6}{9}$ **(A)** **c.** $\frac{7}{9}$ **(A)**

3. **a.** (H, 3) (H, 4) (H, 5) (H, 6) (T, 1) (T, 2) (T, 3) (T, 4) (T, 5) **b.** $\frac{3}{12}$ **(A)**

4. **a.** See table in Example B, page 164. **b.** 36 **c.** $\frac{6}{36}$ **(A)** **d.** $\frac{6}{36}$ **(A)**
e. 0 **(A)**

5. **a.** **i.** The two numbers are the same and even.
ii. The two numbers are the same or even or both.
iii. The two numbers are different.

b. i. $\frac{3}{36}$ **(*M*)** **ii.** $\frac{12}{36}$ **(*M*)** **iii.** $\frac{30}{36}$ **(*M*)**

6. $\frac{2}{3}$ **(*M*)**

Unit 12.3 Activity 2B: Probability trees (page 179)

1. a. $\frac{5}{9}$ **(*M*)** **b.** $\frac{25}{81}$ **(*M*)** **c.** $\frac{40}{81}$ **(*M*)**

2. a. $\frac{6}{15}$ **(*M*)** **b.** $\frac{2}{15}$ **(*M*)** **c.** $\frac{7}{15}$ **(*M*)**

3. a.

Start	1st child	2nd child	3rd child	Outcome
Start	B	B	B	BBB
			G	BBG
		G	B	BGB
			G	BGG
	G	B	B	GBB
			G	GBG
		G	B	GGB
			G	GGG

b. i. $\frac{3}{8}$ **ii.** $\frac{4}{8}$ **iii.** $\frac{4}{8}$ **(*M*)**

4. a. 0.357 **(*M*)** **b.** 0.048 **(*M*)**

5. a. $\frac{2}{8}$ **(*M*)** **b.** $\frac{2}{8}$ **(*M*)**
c. $\frac{4}{64}$ **(*M*)**

6. a. 0.14 **(*M*)** **b.** 0.53 **(*M*)** **c.** 0.135 **(*M*)**
7. a. 0.18 **(*M*)** **b.** 0.46 **(*M*)** **c.** 0.272 **(*M*)**
8. 0.902 **(*E*)** **9. a.** 0.22 **(*M*)** **b.** 0.4 **(*M*)** **10.** 0.4 or 40% **(*E*)**

Unit 12.3 Activity 3: Probabilities of one event and/or another event (page 185)

1. 0.92 **2.** 0.77 **3.** 0.5 **(*M*)**
4. a. 0.55 **(*M*)** **b.** 0.15 **(*M*)**
5. 0.37 **6.** 0.55 **7.** 0.6
8. a. 0.7 **b.** 0.3 **c.** 0.21 **d.** 0.06
e. 0.27 **f.** 0.76 **g.** 0.79
9. a. A = {4, 5, 6}; B = {1, 2, 3}; C = {5}; D = {2, 4, 6}; E = {1, 3, 5}; F = {3}
b. i. Y **ii.** N **iii.** N **iv.** N
v. Y **vi.** N
c. i. $\frac{1}{2}$ **ii.** $\frac{1}{2}$ **iii.** $\frac{1}{6}$ **iv.** $\frac{1}{6}$
v. 1 **vi.** $\frac{1}{2}$ **vii.** $\frac{2}{3}$ **viii.** $\frac{1}{2}$
10. a. i. Y **ii.** N **iii.** N **iv.** Y
b. i. $\frac{1}{4}$ **ii.** $\frac{1}{4}$ **iii.** $\frac{1}{13}$ **iv.** $\frac{1}{2}$
c. i. $\frac{1}{13}$ **ii.** $\frac{1}{2}$ **iii.** $\frac{1}{13}$ **iv.** 0
v. $\frac{2}{13}$ **vii.** $\frac{1}{13}$ **viii.** $\frac{1}{2}$

Unit 12.3 Activity 4A: Independent events (page 187)

1. **a, b** **2.** none are independent **3.** There are many possible answers for each.

4. **a.** **i**, **ii** are independent **b.** **i.** $\frac{4}{100}$ **ii.** $\frac{16}{100}$

5. The probability of A and B occurring is not equal to the product of the probability of A and the probability of B.

Unit 12.3 Activity 4B: Conditional probability (page 189)

1. **a.** 0.3 **(M)** **b.** 0.95 **(M)** **c.** 0.6 **(M)** **d.** 0.035 **(M)**

2.

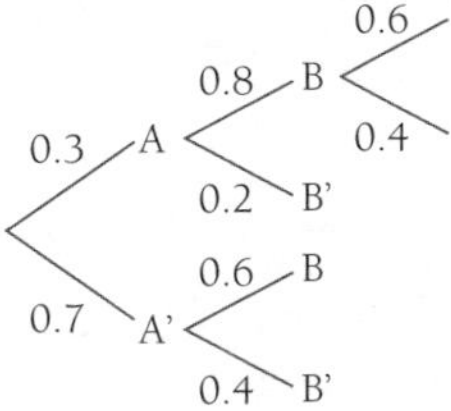

(M)

a. 0.4 **(M)**
b. 0.4 **(M)**
c. 0.42 **(M)**
d. 0.144 **(M)**

3. **a.**

0.3 M: 0.8 P, 0.2 P'
0.7 M': 0.1 P, 0.9 P'

(M)

b. 0.9 **(M)**
c. 0.59 **(M)**

4. **a.** 0.1 **(M)** **b.** 0.4 **(M)** **c.** 0.9 **(M)** **d.** 0.81 **(M)** **e.** 0.06 **(M)**

5. **a.** 0.51 **(M)** **b.** 0.4 **(M)**

6. **a.** 0.45 **(M)** **b.** $\frac{32}{45}$ **(M)** **c.** 0.18 **(M)** **d.** $\frac{18}{55}$ **(M)**

Unit 12.3 Activity 4C: Probability and expected values (page 192)

1. **a.** 0.02 **(M)** **b.** 0.28 **(M)** **c.** 14 **(M)**

2. **a.** $\frac{1}{4}$ **(M)** **b.** $\frac{5}{18}$ **(M)**

3. **a.**

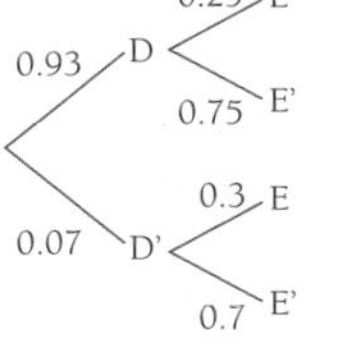

(M)

b. 0.2325 **(M)** **c.** 0.2535 **(M)**

4. **a.**

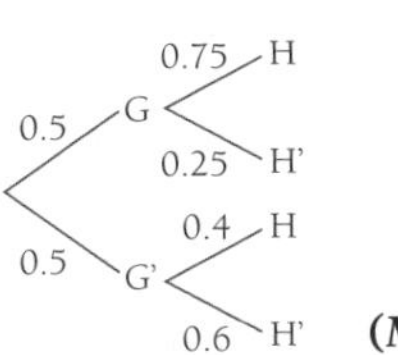

(M)

b. 0.375 **(M)** **c.** 0.3 **(M)**

Unit 12.3 Activity 5: Scatter diagrams (page 196)

1. a. i.

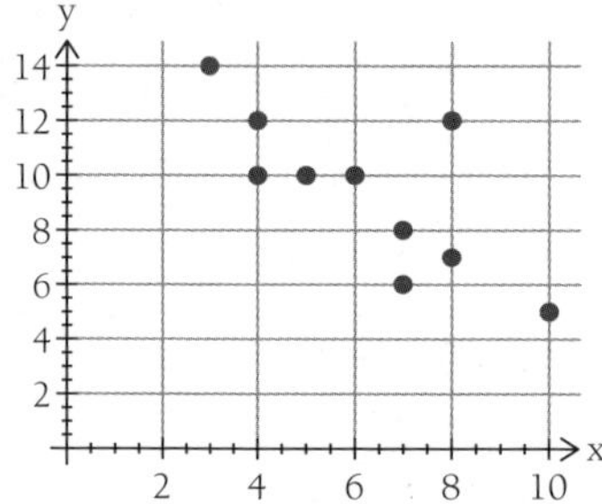

ii. Moderate, negative, linear with one possible outlier, the point (8, 12).

iii. As x increases, y decreases.

b. i.

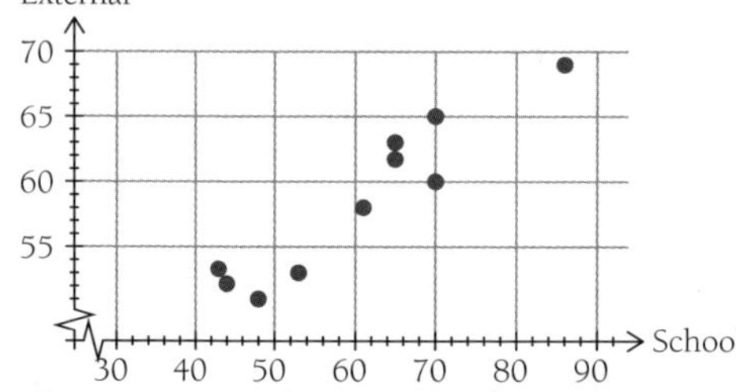

ii. Strong, positive, linear; no outliers.

iii. As the school results increase, the external result also increases.

c. i.

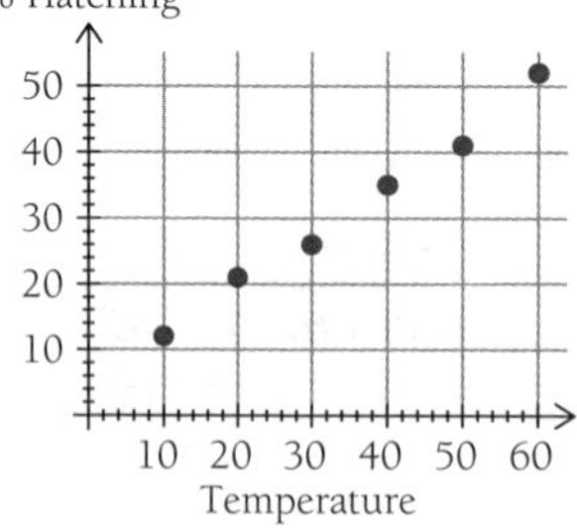

ii. Strong, positive, linear; no outliers.

iii. As the temperature increases, the percentage of eggs hatching also increases.

d. i.

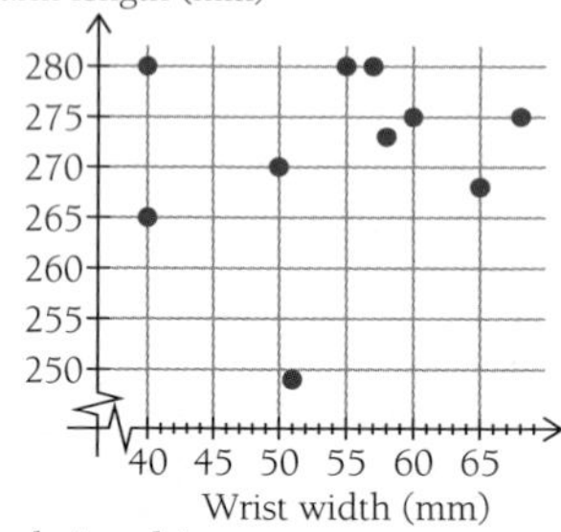

ii. No relationship

iii. The forearm length has no correlation with the width of the wrist.

Unit 12.3 Activity 6: Correlation analysis (page 201)

1. **a.** **i.** –0.33 **ii.** Weak, negative
b. **i.** 0.6 **ii.** Moderate, positive
c. **i.** 0.45 **ii.** Weak, positive

2. **a.** **i.** –0.71 **ii.** Moderate, negative **b.** **i.** 1 **ii.** Strong, positive
c. **i.** 1 **ii.** Perfect positive **d.** **i.** 0.2 **ii.** No relationship

3. **b.** and **e.**

Unit 12.3 Activity 7: Regression (page 206)

1. **a.** See graph (at right). **b.** 1
c. *Physics mark* = 0.875 × *Advanced Maths mark* + 1

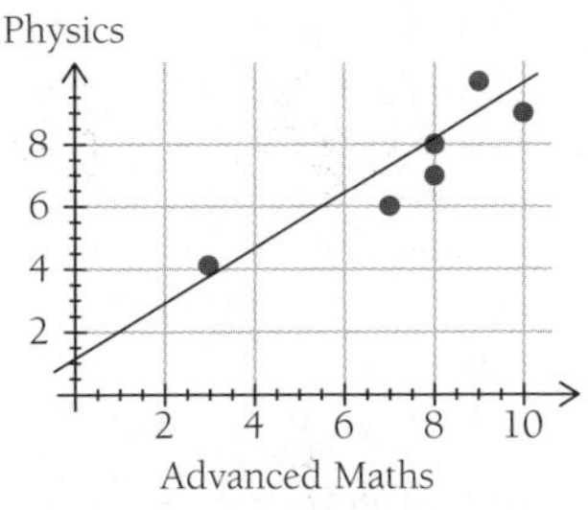

2. **a.** See graph (at right).
b. 0.5; moderate, positive
c. *Accounting* = 0.475 × *Economics* + 40.48
d. See graph (at right).

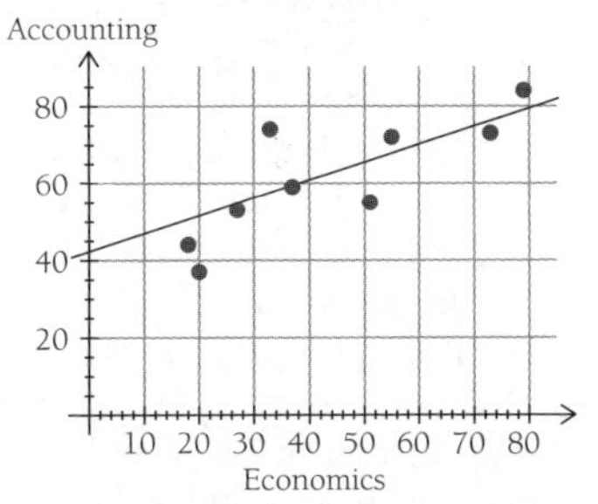

3. **a.** See graph (at right).
b. weak, positive
c. *Height* = 0.614 × *Weight* + 135
d. See graph (at right).

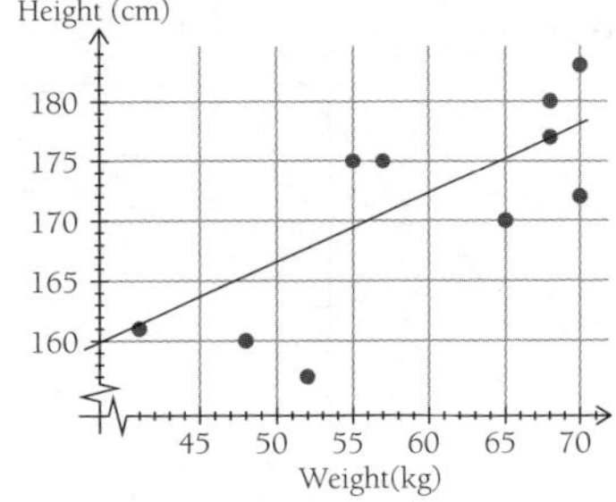

4. **a.** See graph (at right).
b. strong, positive
c. $y = 0.58 \times x + 0.03$
d. See graph (at right).

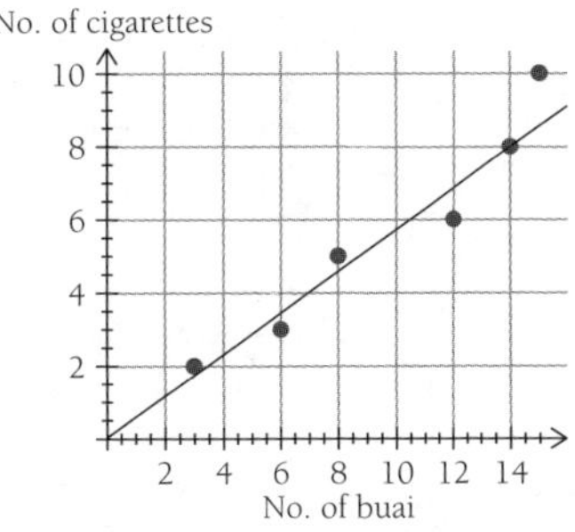

Unit 12.3 Activity 8: The three-median regression line (page 212)

1. a. The median (middle value) of the x-values is 2, the median of the y-values is 6, so the point is (2, 6).
   b. The median (middle value) of the x-values is the average of 2 and 3 (= 2.5), the median of the y-values is the average of 6 and 8 (= 7). The point is (2.5, 7).
   c. The median (middle value) of the x-values is 3, the median of the y-values is 6, so the point is (3, 6).
2. a. The data is split 3–3–3 and the median points found on the graph as shown, marked x.
   To draw the equation place a ruler on the two outside points and move 1/3 of the way towards the middle point.
   Using the points (2, 17) and (8, 13) to find the equation.
   The equation is: $y = -\frac{2}{3}x + \frac{55}{3}$

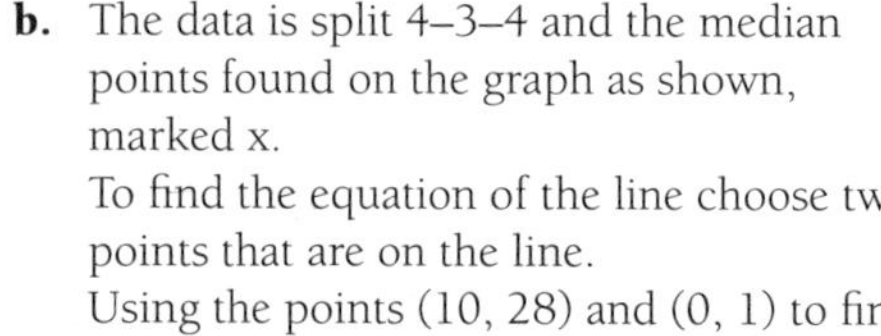

   b. The data is split 4–3–4 and the median points found on the graph as shown, marked x.
   To find the equation of the line choose two points that are on the line.
   Using the points (10, 28) and (0, 1) to find the equation.

   c. The equation is $y = 2.7x + 1$

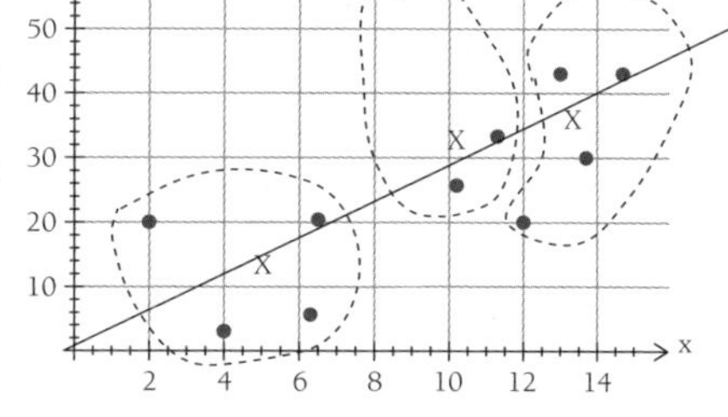

3. There appears to be an outlier at the point (8.3, 57). A three-median line would not be distorted by this outlier.

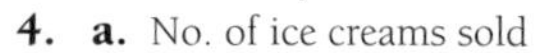

4. a. No. of ice creams sold

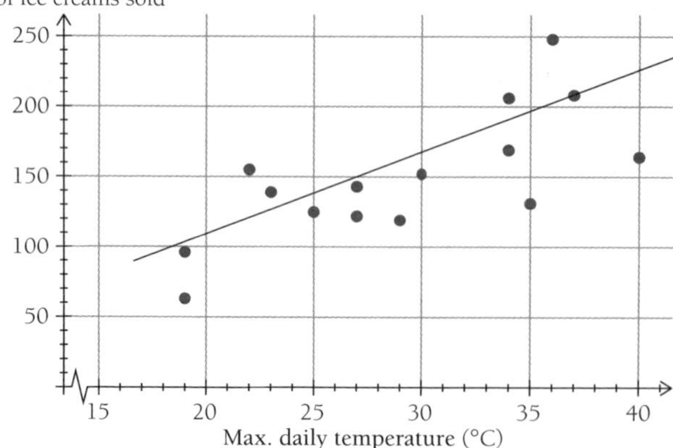

   b. Using the points (40, 225) and (27.5, 150) the equation is:
   Number of ice creams sold = –15 + 6 × *Maximum daily temperature*.
   c. *Number of ice creams sold* = –9.8 + 5.78 × *Maximum daily temperature*.
5. a. *Score on test* = 28 + 0.17 × *Minutes spent preparing*

b. 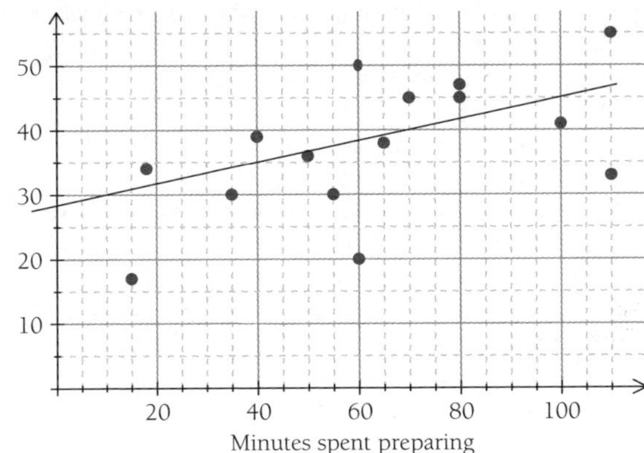

6. a. 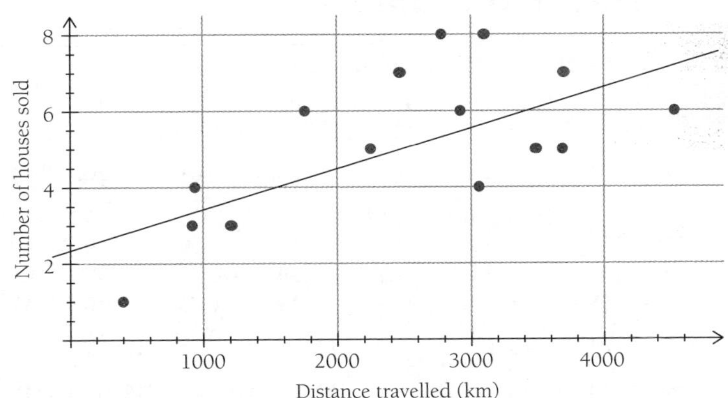

b. Using the points (2000, 4.5) and (3000, 5.5) the equation is:
Number of houses sold = 2.5 + 0.001 × *Distance travelled*

7. Answer **b.**

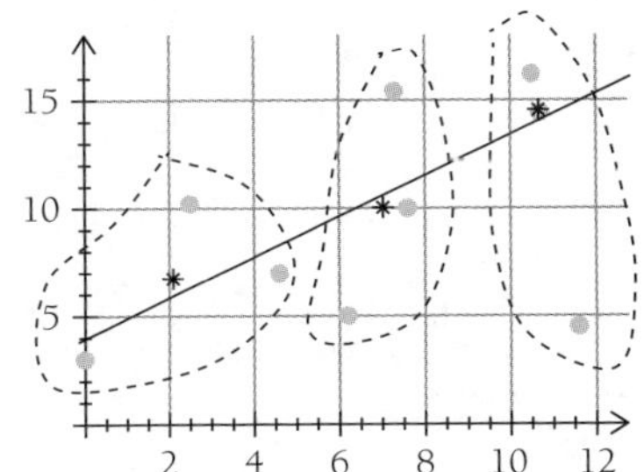

Unit 12.3 Activity 9: Predicting using the regression line (page 218)

1. a. i. 11.13 **ii.** 28.38 **b.** interpolation

c. For an increase of one in the value of x the value of y increases by 0.21.

d. When the value of x is zero the value of y is 5.04.

2. a. i. 64.65 ii. 7.8 b. extrapolation
   c. For each increase of one year in age the pulse rate increases by 1.668 beats/minute.
3. a. i. 27.1 ii. 44.02 b. interpolation
   c. For each increase of one year in age the BMI increases by 0.1045.
   d. When the age of a person is zero the BMI is 22.4.
4. a. i. 4 ii. 9 b. i. 1 ii. 6 c. **b** part **i**.
5. a. i. 59 ii. 76 b. i. 50 ii. –28
   c. Predicting an economics score for an accounting score of 20 is a case of extrapolation where the prediction is too far from the data group.
6. a. The slope of the regression line suggests that for each increase of a minute spent preparing for the exam the score on the test increased by 0.203 of a mark. The intercept suggests that if a student spends zero minutes preparing for the test then their mark on the test will be 24.5.
   b. i. 33 ii. 45 c. 15.4% d. (–)9.75%

Unit 12.4 Algebra and Graphs

Unit 12.4 Activity 1A: Linear equations (page 223)

1. 5 **2.** 15 **3.** 5 **4.** 9 **5.** –7 **6.** 1.68 **7.** 21 **8.** –40
9. 1.6 **10.** 12.1 **11.** 6 **12.** 3.6 **13.** 8 **(A)** **14.** 7.5 **(A)** **15.** 20 **(A)**
16. 18 **(A)** **17.** 8 **(A)** **18.** –3 **(A)** **19.** $10\frac{1}{3}$ **(A)** **20.** –2 **(A)**
21. 14.5 **(A)** **22.** –3.25 **(A)** **23.** 2.125 **(A)** **24.** 65 **(A)** **25.** 36 **(A)**
26. 24 **(A)** **27.** –9 **(A)** **28.** 16 **(A)** **29.** 14.4 **(A)** **30.** 22.5 **(A)**
31. 5 **(A)** **32.** 5 **(A)** **33.** 3.3 **(A)** **34.** 3 **(A)** **35.** 7 **(A)**
36. –31 **(A)** **37.** –2 **(M)** **38.** 3.6 **(M)** **39.** 4.5 **(M)** **40.** 2 **(A)**
41. 8.5 **(A)** **42.** 22 **(A)** **43.** 22 **(A)** **44.** $9\frac{2}{3}$ **(A)** **45.** $-5\frac{1}{6}$ **(A)**
46. $2\frac{1}{3}$ **(A)** **47.** 10 **(A)** **48.** 1 **(A)** **49.** 1.7 **(M)** **50.** 6.25 **(M)**

Unit 12.4 Activity 1B: More advanced linear equations (page 225)

1. a. $\frac{1}{8}$ b. $2\frac{5}{8}$ c. $1\frac{1}{19}$ d. $\frac{1}{4}$ e. $9\frac{1}{4}$ f. $-1\frac{3}{8}$
   g. $\frac{8}{11}$ h. $2\frac{1}{3}$ i. 1.6 j. $\frac{-2}{3}$ k. $-12\frac{3}{4}$ l. $\frac{1}{7}$ (A)
2. a. $3\frac{3}{7}$ b. $\frac{2}{5}$ c. 3 d. 1 e. $2\frac{3}{4}$ f. $-2\frac{1}{2}$
   g. –42 h. $-1\frac{6}{7}$ i. $1\frac{12}{19}$ j. 1 k. $1\frac{3}{4}$ l. 2 (A)

Unit 12.4 Activity 1C: Solving word problems (page 226)

1. $5x + 6(x + 1) = 32.4$, K2.40 (**M**)
2. $x + 2(x + 1) + 6x = 15.5$, T K1.50; B, D K2.50, rest K3 each (**M**)
3. $2x + x - 20 = 86.5$, K35.50, K15.50 (**M**)
4. $2x + 2(x - 3.3) = 50$, 14.15m (**M**)

5. $4x + 6 + 110$, 26 (**M**)

6. $17.35x = 1\,000 - 687.70$, K18 (**M**)

7. $\frac{x+2}{2} = \frac{117}{2}, \frac{x+2}{2} = \frac{117}{26}$, K7 (**M**)

8. $x - 20 = \frac{7x}{11}$, 55 (**M**)

9. $12(x - 5) = 5x + 45$, 15 (**M**)

10. $1.15(x + 5) = 13.8$, K7 (**M**)

11. $36(1 + \frac{x}{100}) = 40.86$, 13.5% (**M**)

12. $\frac{x}{x+44} = \frac{11}{17}$, 143 (**M**)

13. $x - 30 = 0.25x$, 40 (**M**)

14. $4x - 12 = x + 12$, K8, K32 (**M**)

15. $\frac{x + 3\,000}{x + 5\,000} = \frac{13}{17}$, K3 500 (**M**)

Unit 12.4 Activity 1D: Linear inequations (page 228)

1. **a.** ..., –1, 0, 1, 2, 3, 4 (**A**) **b.** 14, 15, 16, ... (**A**) **c.** ..., 3, 4, 5, 6, 7 (**A**)
   d. ..., –4, –3, –2, –1 (**A**) **e.** 5, 6, 7, ... (**A**) **f.** ..., –1, 0, 1, 2 (**A**)
2. **a.** $x \leq -6$ (**A**) **b.** $x > 2$ (**A**) **c.** $x \geq -2$ (**A**)
   d. $x > -7$ (**M**) **e.** $x \geq -4$ (**M**) **f.** $x < \frac{-7}{6}$ (**M**)
   g. $x > -2$ **h.** $x \geq -5$ **i.** $x < -17$
   j $x \leq -43$ **k.** $a \geq 3\frac{2}{11}$ **l.** $x > 1\frac{4}{5}$
   m. $x \geq -\frac{11}{25}$ (**A**)
3. –4 (**M**) 4. –2 (**M**) 5. 3, 4 (**M**) 6. 3 (**M**)

Unit 12.4 Activity 2A: Writing formulae (page 230)

1. $100D$ toea
2. **a.** $\frac{x}{20}$ hours **b.** $180x$ seconds
3. **a.** $(1\,000x + y)$ grams **b.** $(x + \frac{y}{1\,000})$ kg
4. **a.** $100LW$ cm^2 **b.** $\frac{LW}{100}$ m^2 **c.** $(2\,000L + 20W)$ mm
5. **a.** $800L$ cm **b.** $8\,000L$ mm
6. **a.** $\frac{1}{2}ab$ **b.** $\sqrt{a^2 + b^2}$
7. **a.** $\frac{\pi L^3}{4}$ **b.** $\frac{3\pi L^2}{2}$
8. **a.** $2y + 2z$ **b.** $yz - xy + x^2$
9. **a.** $2(L + M) + 8x$ **b.** LMx
10. **a.** $16L^3$ **b.** $40L^2$

Unit 12.4 Activity 2B: Changing the subject of formulae (page 232)

1. $\frac{T}{3}$ (**A**) 2. $\frac{T}{A}$ (**A**) 3. PT (**A**) 4. P^2 (**A**)
5. $AP + BP$ (**A**) 6. $\frac{5P}{2}$ (**A**) 7. $\frac{T^2}{5}$ (**A**) 8. $A - B$ (**A**)
9. $A + B$ (**A**) 10. $2B$ (**A**) 11. $2B^2 + 3B$ (**A**) 12. B^2 (**A**)
13. AB^2C (**A**) 14. $B(C + 3)$ (**A**) 15. $\frac{AC}{B}$ (**A**) 16. $\pm\sqrt{A}$ (**A**)
17. $\pm\sqrt{\frac{A+B}{C}}$ (**A**) 18. $\frac{B^2 + D}{A}$ (**A**) 19. $\frac{B^2}{A - D}$ (**A**) 20. $\frac{ABC}{A + B}$ (**A**)

21. $\dfrac{ED + A}{1 - E}$ (**A**) **22.** C^2 (**A**) **23.** C^2D^2 (**A**) **24.** $\dfrac{9L^2G}{4A^2}$ (**A**)

25. $\dfrac{A^2G}{C^2X^2}$ (**A**) **26.** $\sqrt{\dfrac{D}{3 - A}}$ (**A**) **27.** $\dfrac{D^2}{(A - B)^2}$ (**A**) **28.** $\dfrac{(A + C)^2}{E^2}$ (**A**)

29. $\dfrac{4y - 5}{By - A}$ (**A**) **30.** $\sqrt[3]{\dfrac{Fy - E}{4y - 3}}$ (**A**)

Unit 12.4 Activity 2C: Using formulae (page 233)

1. a. 45.08 m^2 **b.** 71.13 m^2 **c.** 3.44 m **d.** 1.94 m **e.** 10.82 m

2. a. 817.24 cm^3 **b.** 106.55 cm^3 **c.** 3.44 cm **d.** 5.70 cm **e.** 5.33 cm

3. a. K2 750 **b.** 8.5% **c.** 4 years

Unit 12.4 Activity 3A: Drawing graphs by plotting points (page 237)

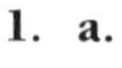

1. a.

x	−2	−1	0	1	2
y	−3	−1	1	3	5

b.

x	−2	−1	0	1	2
y	−8	−5	−2	1	4

c.

x	−2	−1	0	1	2
y	4	3	2	1	0

d.

x	−2	−1	0	1	2
y	7	6	5	4	3

e.

x	−2	−1	0	1	2
y	6	5	4	3	2

f.

x	−2	−1	0	1	2
y	9	8	7	6	5

2. a.

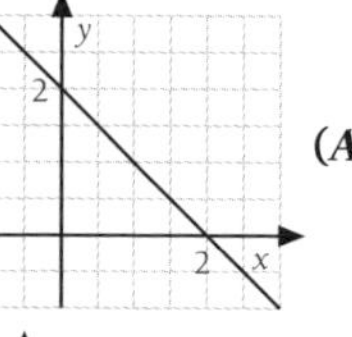

(**A**)

b.

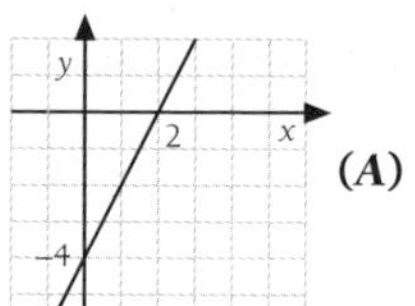

(**A**)

c.

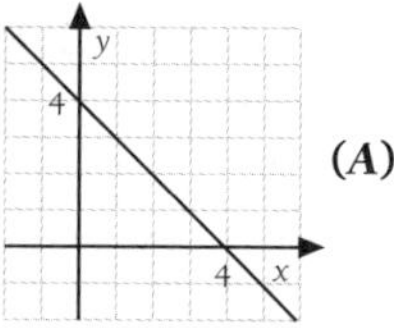

(**A**)

d.

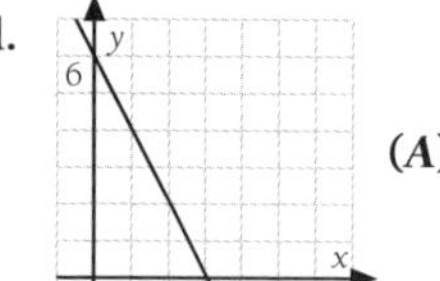

(**A**)

e.

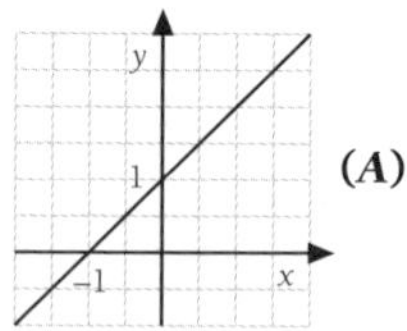

(**A**)

f.

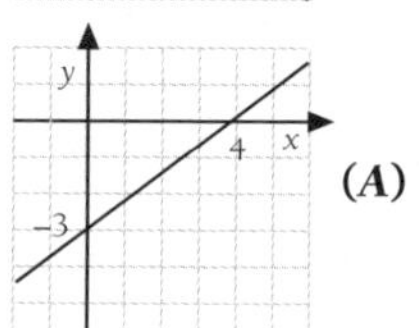

(**A**)

Unit 12.4 Activity 3B: Linear graphs (page 242)

1. a. a, b, d **b.** e **c.** c **d.** c, e **e.** f **f.** f

2. a. i. 1 **b. i.** 2 **c. i.** $\dfrac{2}{3}$

ii. 2 **ii.** −3 **ii.** 1

iii.

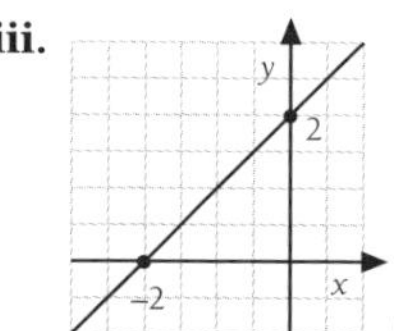

(**A**)

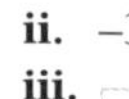

iii.

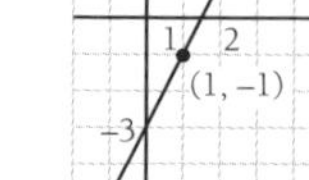

(**A**)

iii.

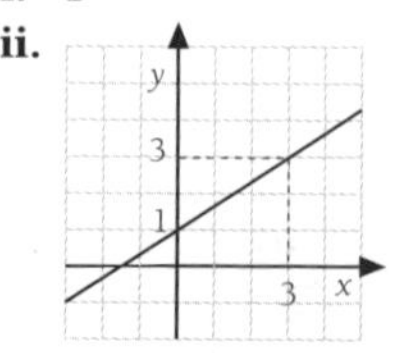

(**A**)

d. i. −1 **e. i.** 3 **f. i.** −2

ii. 6 **ii.** 0 **ii.** 0

iii.

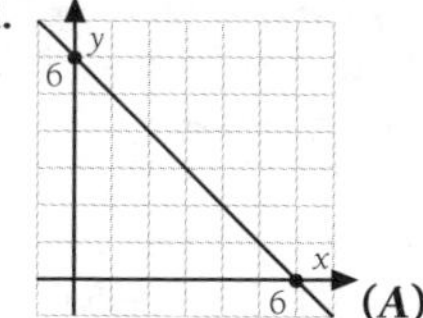

iii.

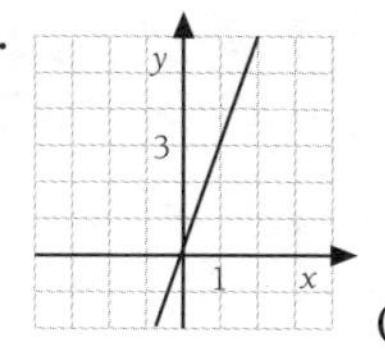

iii.

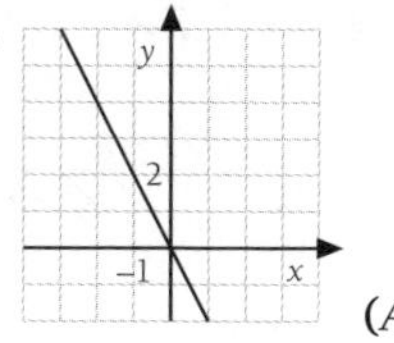

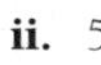

g. **i.** –2

ii. 5

iii.

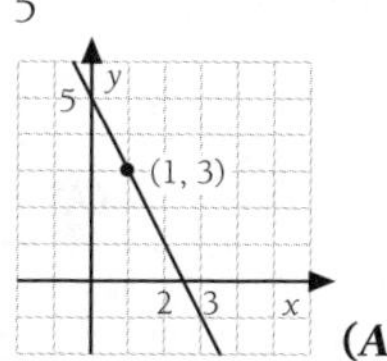

h. **i.** $\frac{3}{2}$

ii. 3

iii.

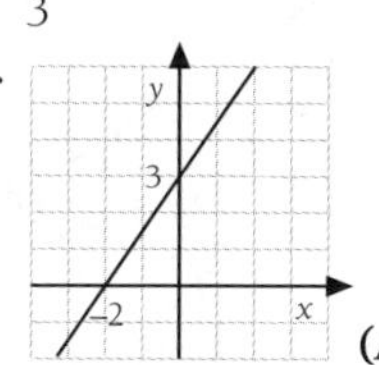

i. **i.** $-\frac{2}{3}$

ii. 2

iii.

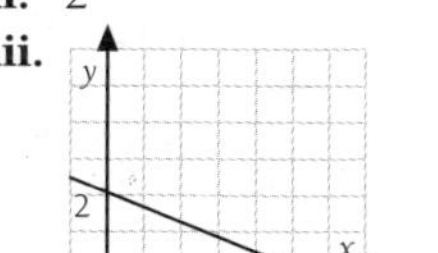

j. **i.** $\frac{3}{4}$

ii. –4

iii.

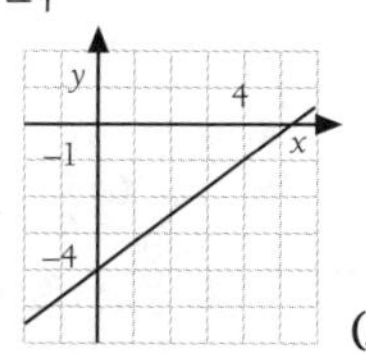

k. **i.** 0

ii. 3

iii.

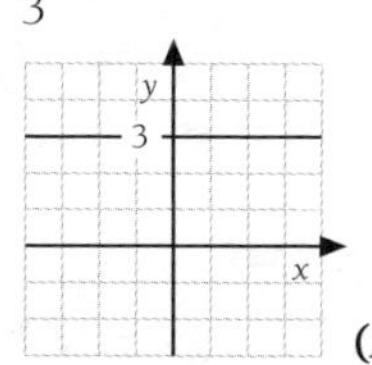

l. **i.** 0

ii. –2

iii.

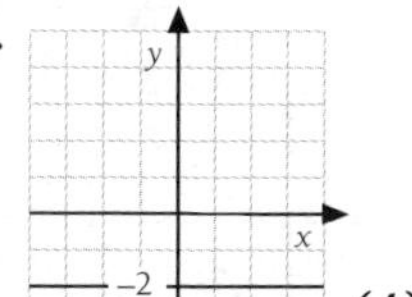

m. **i.** Undefined.

ii. None.

iii.

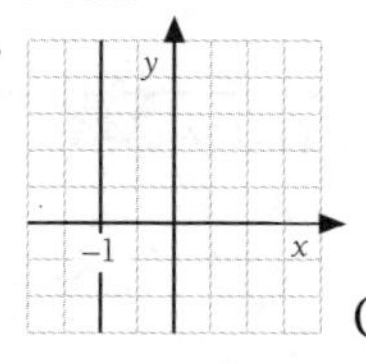

n. **i.** Undefined.

ii. None.

iii.

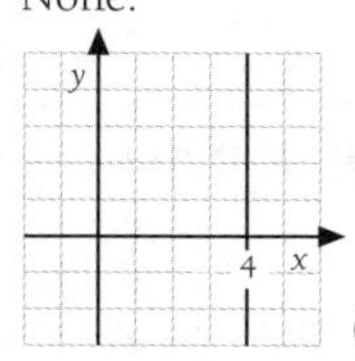

o. **i.** 4

ii. 2

iii.

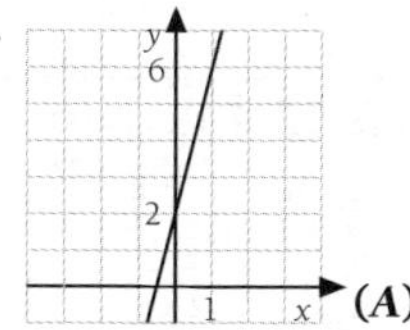

p. **i.** –1 **ii.** 1 **iii.**

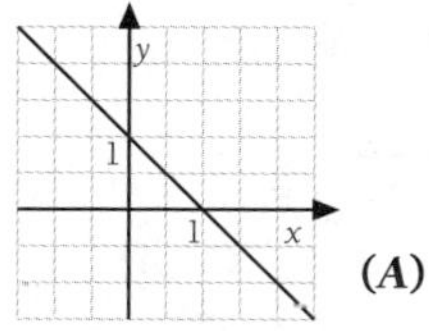

3.

	i.	ii.	iii.
a.	1 **(A)**	2 **(A)**	$y = x + 2$ **(M)**
b.	2 **(A)**	1 **(A)**	$y = 2x + 1$ **(M)**
c.	–1 **(A)**	4 **(A)**	$y = -x + 4$ **(M)**
d.	$\frac{2}{3}$ **(A)**	–1 **(A)**	$y = \frac{2}{3}x - 1$ **(M)**
e.	$-\frac{3}{4}$ **(A)**	5 **(A)**	$y = -\frac{3}{4}x + 5$ **(M)**
f.	Undefined. **(A)**	None. **(A)**	$x = 4$ **(M)**
g.	–1 **(A)**	0 **(A)**	$y = -x$ **(M)**
h.	0 **(A)**	–1 **(A)**	$y = -1$ **(M)**
i.	–1 **(A)**	–1 **(A)**	$y = -x - 1$ **(M)**

4\. a.

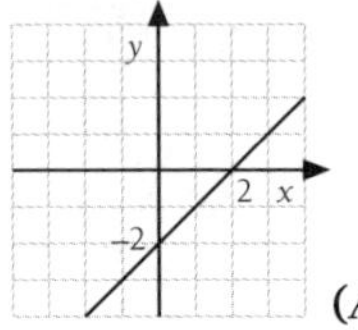

(A)

b.

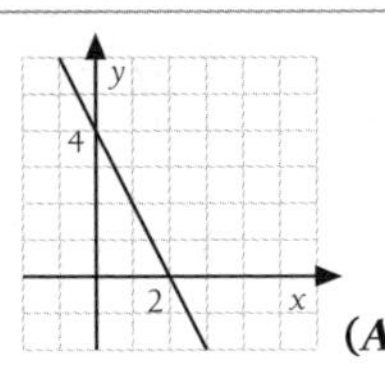

(A)

c.

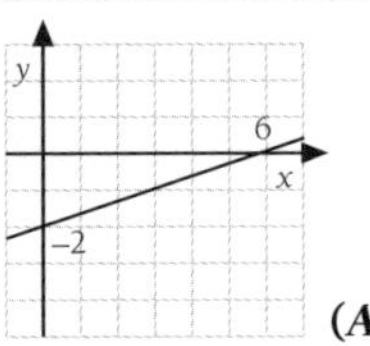

(A)

d.

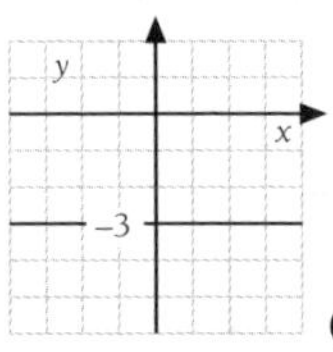

(A)

e.

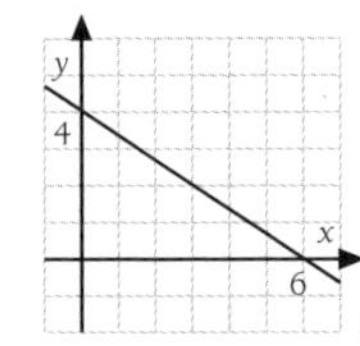

(A)

f.

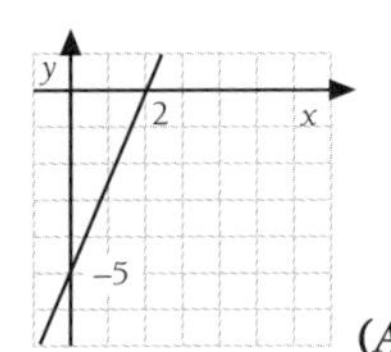

(A)

g.

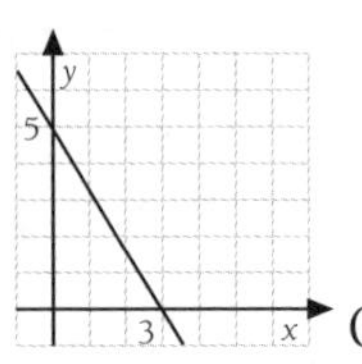

(A)

h.

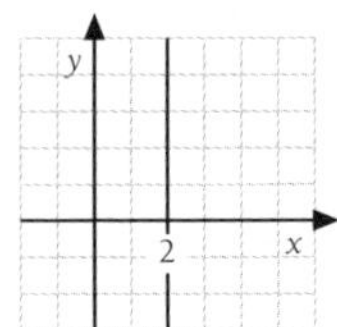

(A)

5\.

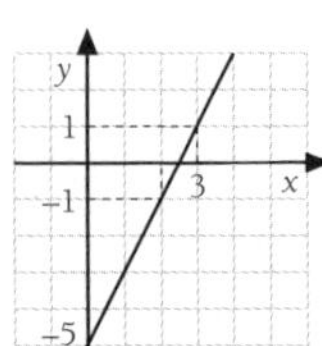

a. 2.5 (A)
b. −5 (A)
c. $y = 2x - 5$ (M)

6\.

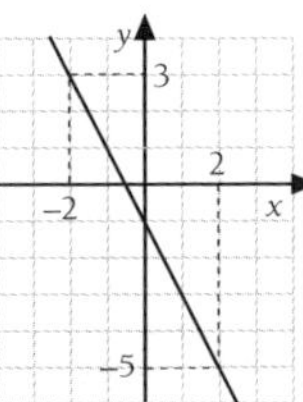

a. −1 (A)
b. −2 (A)
c. (−0.5, 0) (A)
d. $y = -2x - 1$ (M)

Unit 12.4 Activity 3C: Linear graphs in practical contexts (page 246)

1\. a., b.

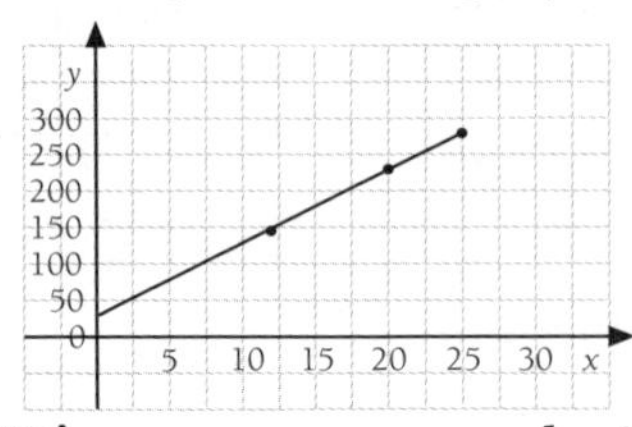

(A)

c. i. K205 (A) ii. 16 (A)
d. i. K25 (A) ii. K10 (A)
e. $d = 10h + 25$ (M)

2\. a.

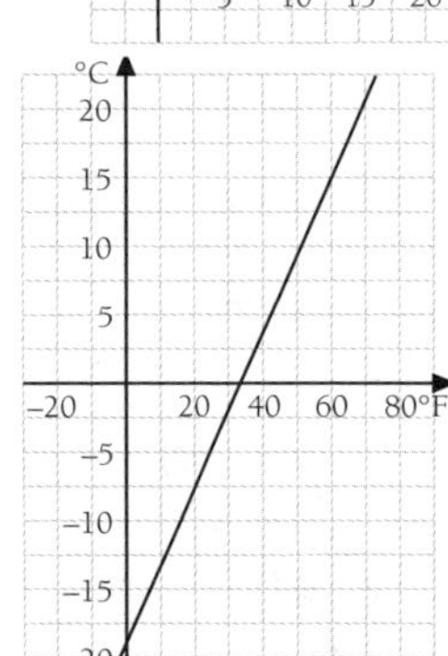

b. i. 5 °C (A) ii. 14 °F (A)

c. $C = \frac{5}{9}F - 18$ (y-intercept to nearest whole number) (M)

3\. a.

x	2	4	6	8	10
y	6	9	12	15	18

b. i., ii., iii.

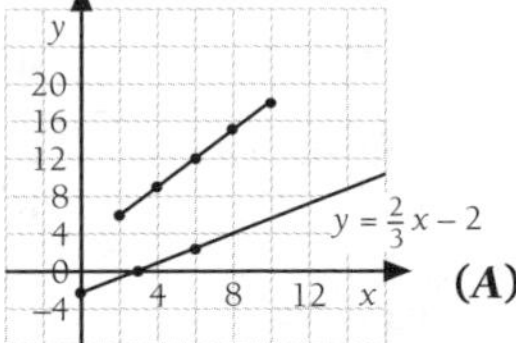

(A)

iv. $y = \frac{3}{2}x + 3$ **(*M*)**

c. See graph.

4. a. $y = 18x + 15$ **(*M*)**

b.

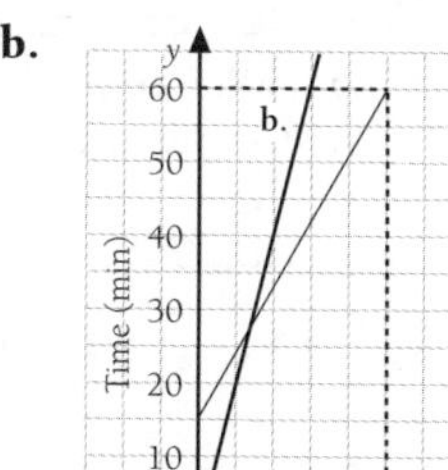

c. 680 g (27 min). It is the point of intersection of the two graphs. **(*A*)**

5. a. $y = 25x + 100$ **(*M*)**

b. Growing at *about* 25 cm a year. **(*A*)**

c. Grows at about 25 cm in 1 year. $\frac{100}{25} = 4$. **(*A*)**

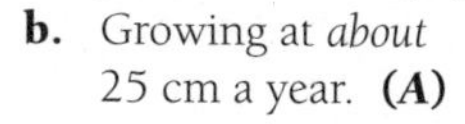

6. a.

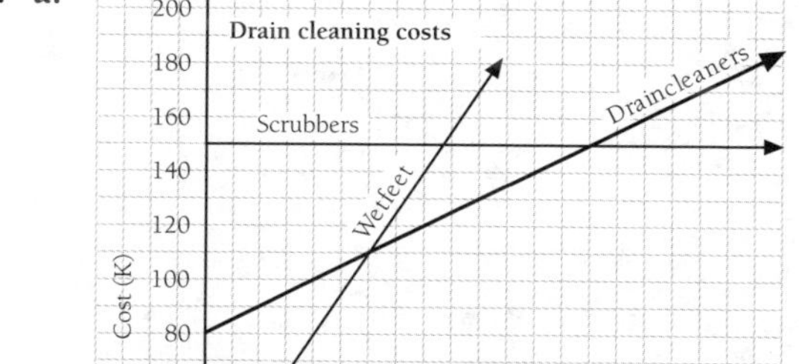

(*A*)

b. i. $C = \frac{3}{2}l + 20$ **(*M*)**

ii. $C = 150$ **(*M*)**

c. i. Scrubbers.

ii. Largest intercept on Cost axis. **(*A*)**

d. i. Wetfeet.

ii. Steepest (or largest) gradient. **(*A*)**

e. i. 140 m

ii. Intersection of two graphs. **(*A*)**

7. a.

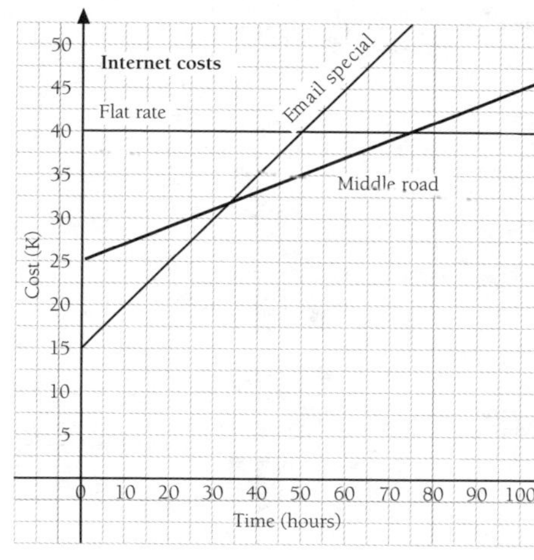

b. i. $C = \frac{1}{2}t + 15$ **(*M*)**

ii. $C = 40$ **(*M*)**

c. i. Flat rate.

ii. Largest intercept on Cost axis. **(*A*)**

d. i. Email special.

ii. Steepest (or largest) gradient. **(*A*)**

e. i. 50 hours

ii. Intersection of two graphs. **(*A*)**

8. a.

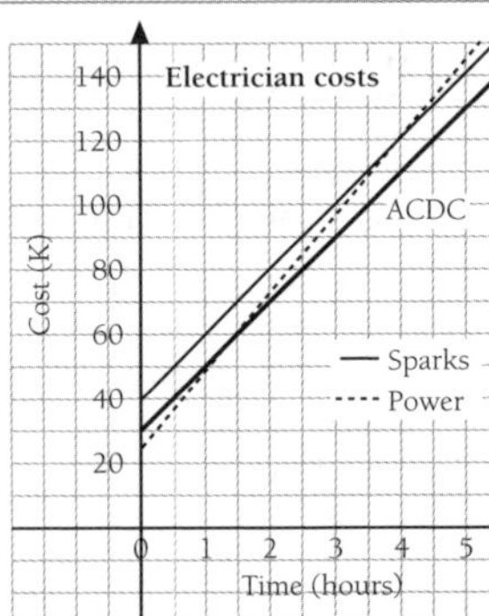

b. i. 3.75 hours.
ii. Where the two graphs intersect. **(A)**

c. i. Sparks and ACDC. **(A)**
ii. Same gradient or lines parallel. **(A)**

d. i. K20 per hour.
ii. Gradient of graph. **(A)**

e. i. $C = 20h + 40$ **(M)**
ii. $C = 24h + 25$ **(M)**

Unit 12.4 Activity 4: Straight lines and linear inequations (page 254)

1. a, b, f, g, i.

2. a.

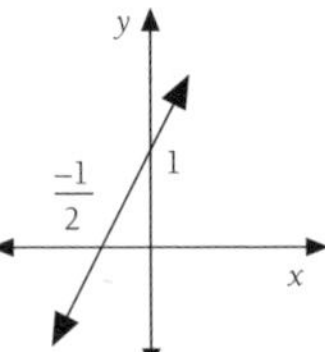

b.

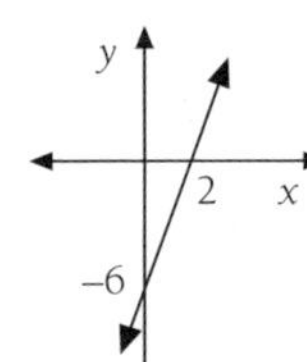

c.

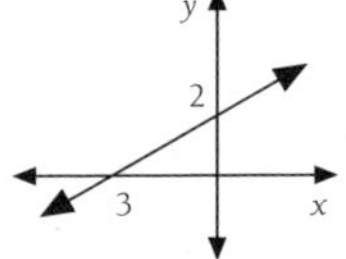

d.

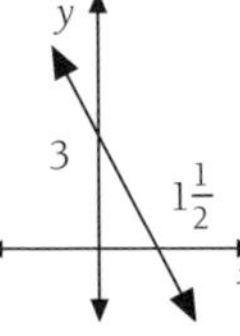

e.

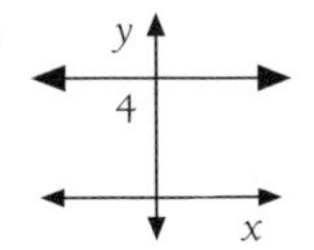

f.

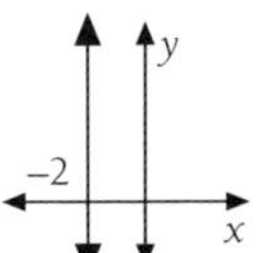

g.

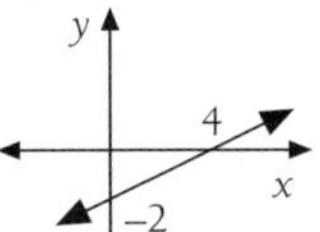

h.

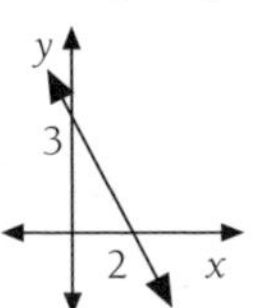

i.

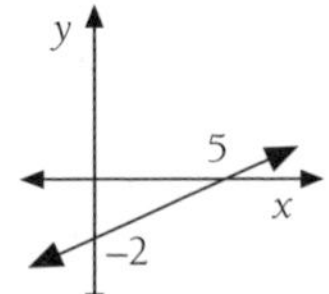

j.

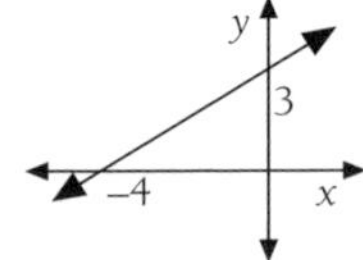

3. a.

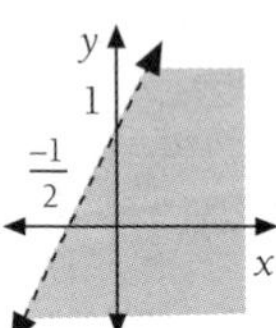

b.

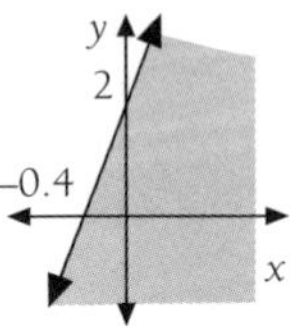

c.

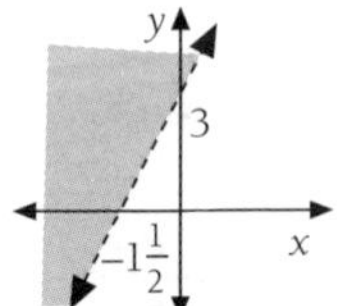

d.

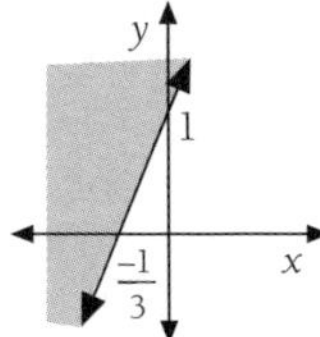

e.

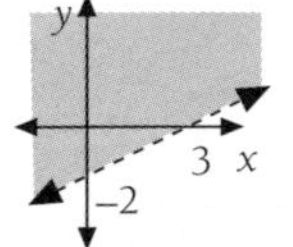

f.

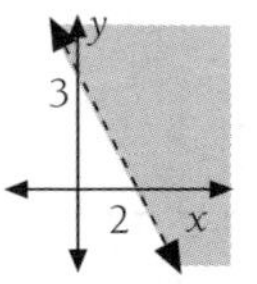

g.

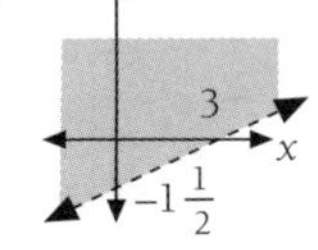

h.

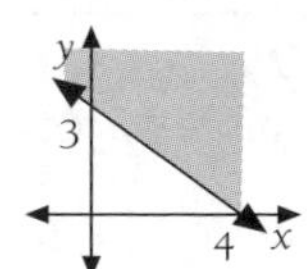

i.

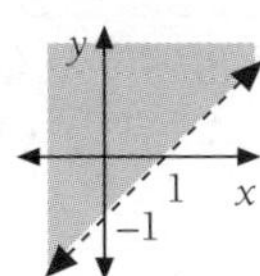

j.

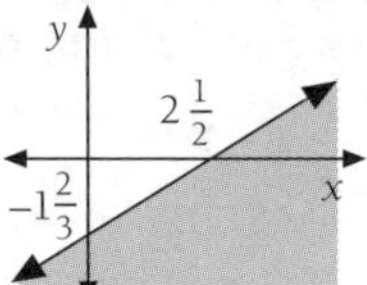

4. a.

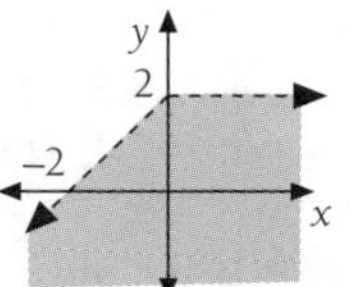

b.

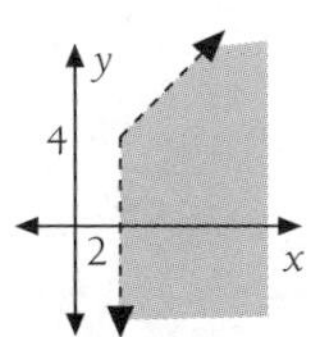

c.

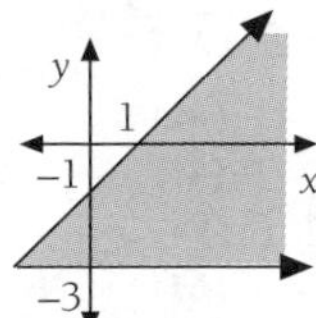

d.

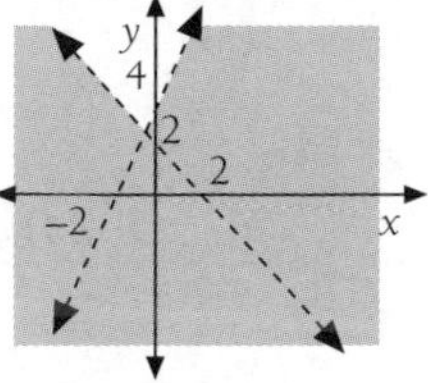

e.

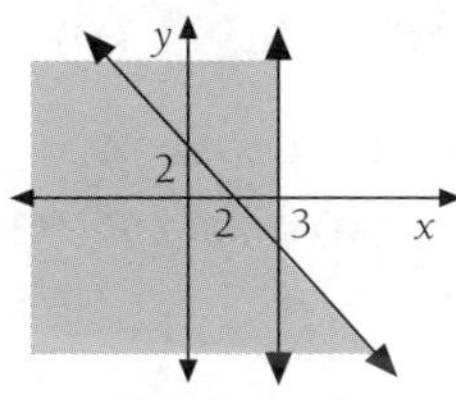

f.

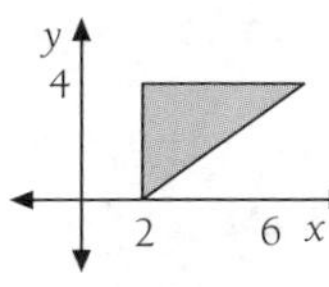

g.

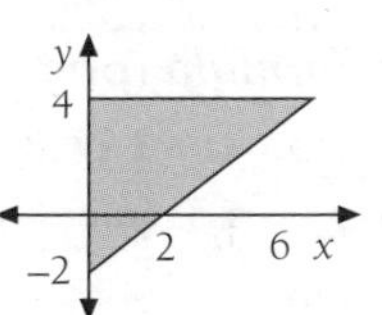

h.

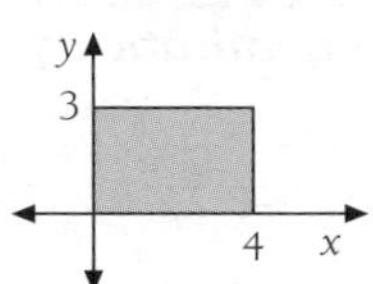

i.

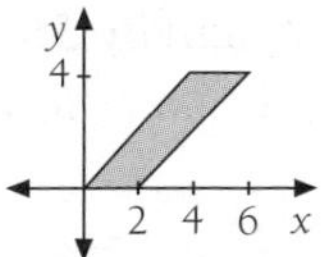

j.

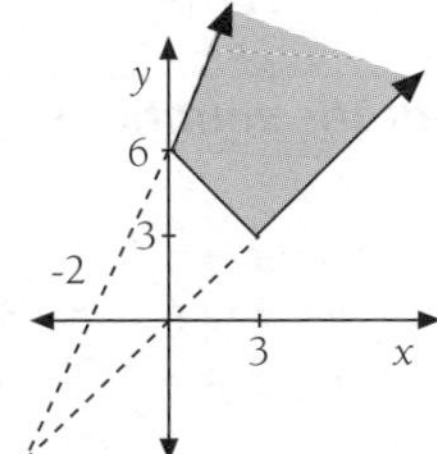

5. a.

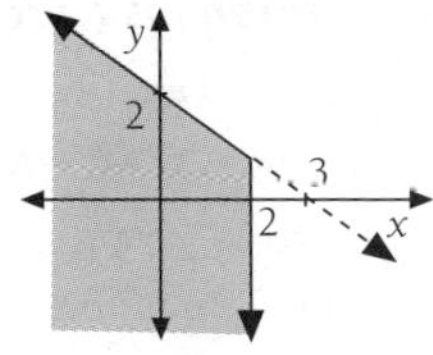

b.

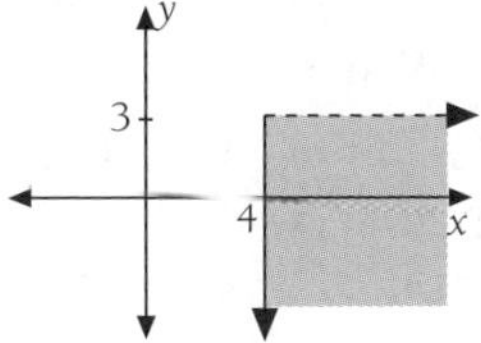

c.

d.

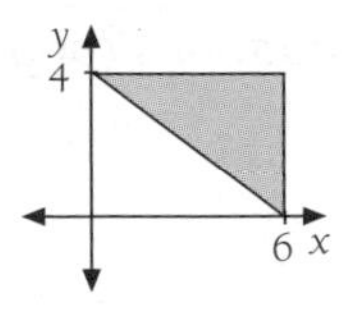

e.

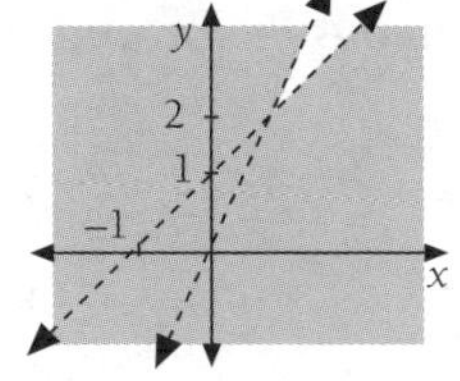

Unit 12.4 Activity 5A: Solving quadratic equations (page 257)

1. a. –2, –3 (**A**) b. 2, 4 (**A**) c. –1, 3 (**A**) d. 5, –2 (**A**)
   e. 0, 4 (**A**) f. 0, –5 (**A**) g. $\frac{3}{2}$, –6 (**A**) h. $-\frac{1}{2}$, 1 (**A**)
   i. $-\frac{3}{5}, -\frac{5}{2}$ (**A**) j. $\frac{3}{4}, \frac{9}{2}$ (**A**) k. $\frac{4}{3}$, 5 l. $\frac{2}{3}, -\frac{2}{3}$ (**A**)
2. a. 0, –3 (**M**) b. 0, 5 (**M**) c. –3, 2 (**M**) d. –1, –4 (**M**)
   e. 8, –5 (**M**) f. 3 (**M**) g. 2, 6 (**M**) h. –5, 2 (**M**)
   i. ±2 (**M**)
3. a. 5, –3 (**M**) b. –6, 3 (**M**) c. 1, 3 (**M**) d. 0, 6 (**M**) e. 0, $\frac{5}{2}$ (**M**)
   f. ±7 (**M**) g. –2, 4 (**M**) h. ±2 (**M**) i. $\frac{3}{2}$, –2 (**M**)
   j. –1, $\frac{1}{3}$ (**A**) k. $-1\frac{1}{2}, -\frac{1}{2}$ (**A**) l. ±2 (**A**) m. ±2 (**A**)
4. a. ±8 (**M**) b. –3, 9 (**M**) c. ±12 (**M**) d. ±5 (**M**)
   e. $\frac{-9}{2}$, 4 (**M**) f. $\frac{-19}{6}$, 2 (**M**)

Unit 12.4 Activity 5B: Solution of quadratic equations by formula (page 259)

1. a. –3.73, –0.27 b. –7.61, –0.39 c. 3.35, 0.15 d. –2.83, 0.83 e. –2.27, 1.77
   f. 0.23, –1.33 g. 1.81, –0.29 h. –8.24, 0.24 i. 1.55, –0.22 j. –4.44, –0.06 (**M**)
2. a. –5.83, –0.17 b. 0.18, 2.82 c. 2.29, –0.29 d. –2.30, 1.30 e. –2.26, 0.59
   f. 0.70, 4.30 g. –1.87, 5.87 h. –0.58, 0.58 i. –0.83, 4.83 j. 1 (**M**)

Unit 12.4 Activity 5C: Solving word problems with quadratic equations (page 260)

1. a. Sam's age. b. Sam = 14, Sam's sister = 11, Sam's brother = 16. (**M**)
2. Day 3 and Day 9. (**M**)
3. 12 m by 32 m (**M**)
4. 2 m square (**M**)
5. 5 (**M**)
6. 8 cm (**M**)
7. 4, 14 or –4, –14 (**M**)
8. 16 cm square (**E**)

Unit 12.4 Activity 5D: More word problems (page 261)

1. Width is 9 m 5 cm, length is 11 cm 5 cm (**M**)
2. Width is 10 m, length is 20 m (**M**)
3. 3.58 m (**M**)
4. 7.83 m (**M**)
5. 24, 53 (**M**)
6. 61, 62 (**M**)
7. 2.25 m (**M**)
8. 9.44 (**M**)
9. 0.73 m, 1.73 m, 2.73 m (**M**)
10. 4.79 or 0.21 (**M**)
11. a. 31 b. 45 (**M**)
12. Height = 3.28 m, base = 2.28 m (**M**)
13. K8.50 (**M**)
14. 4.3 m (**M**)

Unit 12.4 Activity 5E: Simultaneous linear and non-linear equations (page 263)

Students should show all working.

1. (0, –1), (1, 0) (**M**)
2. (0, 1), $(-\frac{4}{5}, -\frac{3}{5})$ (**M**)
3. (0, 2), $(-\frac{4}{3}, \frac{2}{3})$ (**M**)
4. (0, –3), $(\frac{12}{5}, \frac{9}{5})$ (**M**)
5. (0, 3), $(-\frac{36}{13}, -\frac{15}{13})$ (**M**)
6. (–2, –1), (1, 2) (**M**)
7. (5, 2), (–2, –5) (**M**)
8. (2, –2), (72, 12) (**M**)
9. (4, 3), $(\frac{9}{4}, \frac{16}{3})$ (**M**)

10. $(\frac{3}{2}, 4), (2, 3)$ (**M**) **11.** $(-3, -2), (2, 3)$ (**M**) **12.** $(2, 1), (1, 2)$ (**M**)

13. $(4, 2), (-2, -4)$ (**M**) **14.** $(\frac{16}{3}, 4), (\frac{4}{3}, -2)$ (**M**) **15.** $(1, \frac{1}{2}), (-\frac{2}{3}, 3)$ (**M**)

16. $(4, 3), (-4, 3)$ **17.** $(4, 3), (1.5, -0.75)$ **18.** $(-3, 1), (4.4, -0.48)$

Unit 12.4 Activity 6A: Quadratic graphs (page 269)

1. a.

x	–2	–1	0	1	2
x^2	4	1	0	1	4
$y = x^2 - 2$	2	–1	–2	–1	2

b.

x	0	1	2	3	4
$x - 2$	–2	–1	0	1	2
$y = (x - 2)^2$	4	1	0	1	4

c.

x	–5	–4	–3	–2	–1
$x + 3$	–2	–1	0	1	2
$y = (x + 3)^2$	4	1	0	1	4

d.

x	–2	–1	0	1	2
x^2	4	1	0	1	4
$y = 2 - x^2$	–2	1	2	1	–2

e.

x	–2	–1	0	1	2
x^2	4	1	0	1	4
$y = x^2 + 3$	7	4	3	4	7

f.

x	–2	–1	0	1	2
x^2	4	1	0	1	4
$y = 2x^2$	8	2	0	2	8

2. a. $y = x^2 + 3$ **b.** $y = (x + 3)^2$ **c.** $y = 2x^2$ **d.** $y = 2 - x^2$
e. $y = x^2 - 2$ **f.** $y = (x - 2)^2$

3. a.

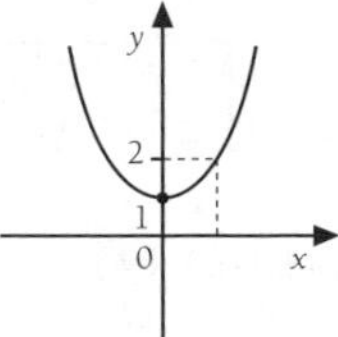
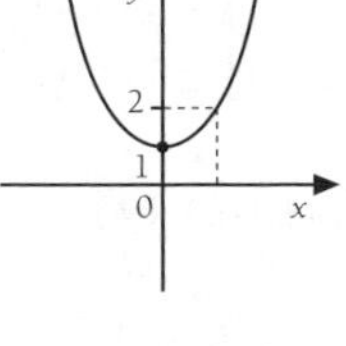

(*A*)

b.

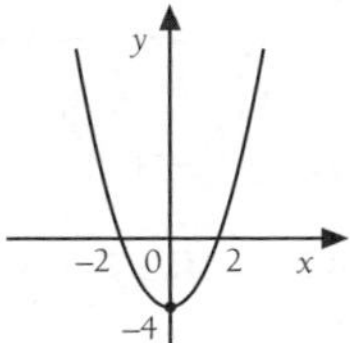

(*A*)

c.

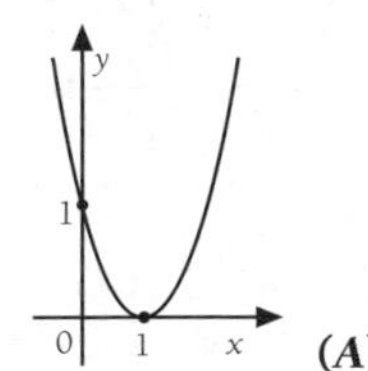

(*A*)

d.

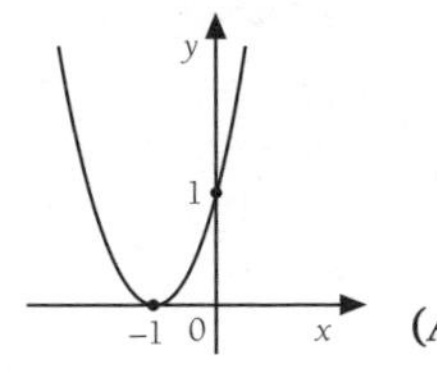

(*A*)

e.

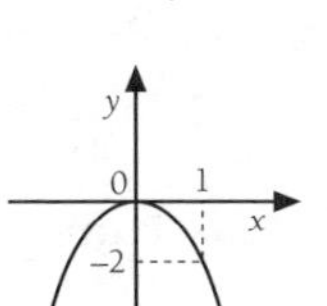

(**M**)

f. y, 4, –2, 0, 2, x (*A*)

4. a.

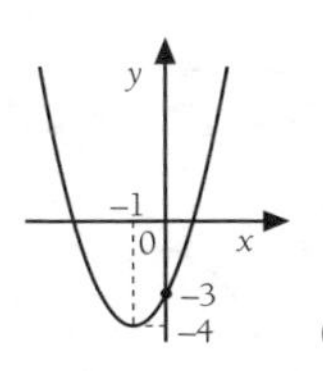

(**M**)

b.

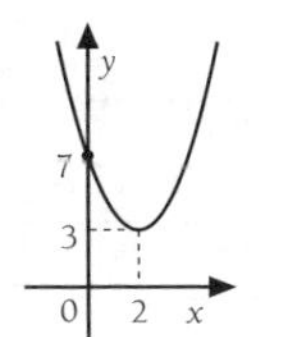

(**M**)

c. y, 5, 1, –2, 0, x (**M**)

d.

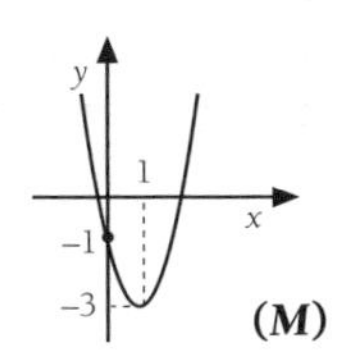

(**M**)

e.

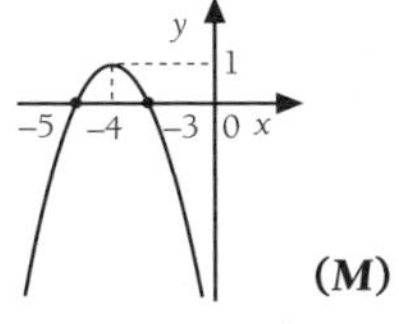

(**M**)

f. y, 2, x, –2, –4 (**M**)

Unit 12.4 Activity 6B: Sketching quadratic graphs (page 273)

1. a.

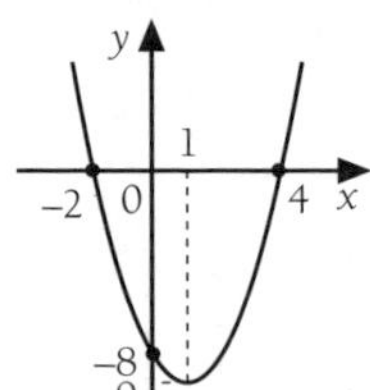

(A)

b.

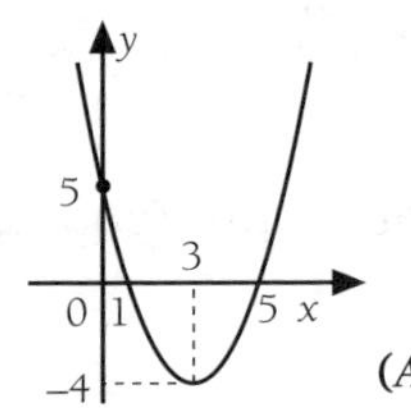

(A)

c.

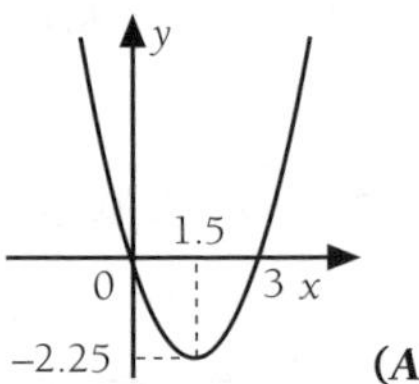

(A)

d.

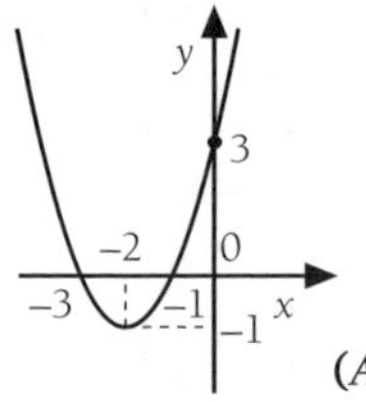

(A)

e.

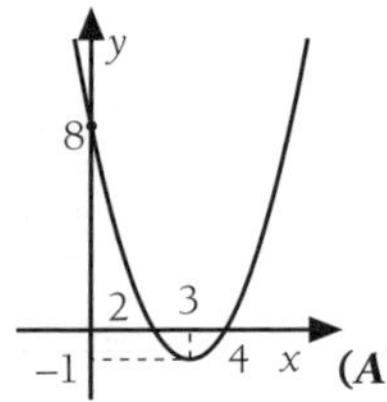

(A)

f.

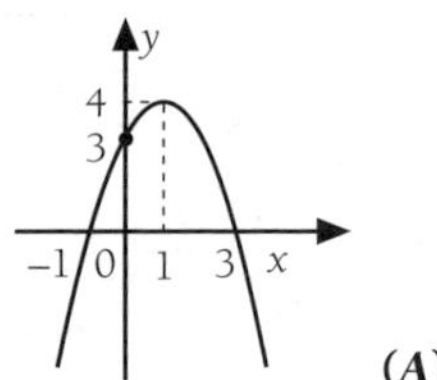

(A)

g.

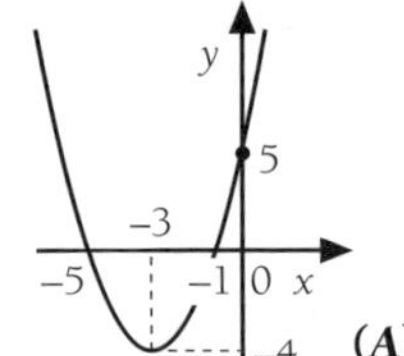

(A)

h.

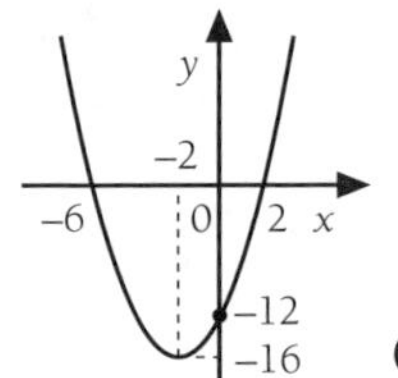

(A)

i.

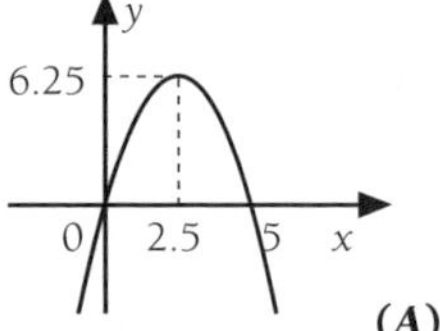

(A)

2. a. i. (–4, 0) **(A)** **ii.** (2, 0) **(A)** **iii.** (–1, 0) **(A)** **iv.** (–1, –9) **(A)**

b. –9 **c.** $y = (x + 1)^2 - 9$ **(M)**

3. a. $y = (x - 15)(x - 55)$ *or* $y = (x - 35)^2 - 400$ **(E)**

b. $y = 3(x - 5)(x - 9)$ *or* $y = 3(x - 7)^2 - 12$ **(E)**

c. $y = 2(3 - x)(x - 15)$ *or* $y = -2(x - 3)(x - 15)$ *or* $y = -2(x - 9)^2 + 72$ **(E)**

d. $y = \frac{-1}{2}x^2 + 800$ *or* $y = \frac{-1}{2}(x + 40)(x - 40)$ **(E)**

e. $y = -(x - 2)^2 + 9$ *or* $y = -(x + 1)(x - 5)$ **(E)**

f. $y = \frac{1}{3}(x - 9)^2 + 3$ **(E)**

Unit 12.4 Activity 6C: Using quadratic graphs to solve problems (page 276)

1. a.

x	0	1	2	3	4	5	6	7	8	9	10	11	12
A	0	22	40	54	64	70	72	70	64	54	40	22	0

b.

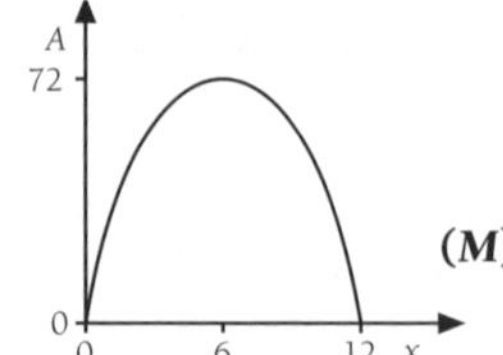

(M)

c. 72 m^2 **(M)**

d. Cannot have a negative length. **(A)**

e. 3.6 m or 8.4 m **(M)**

2. a.

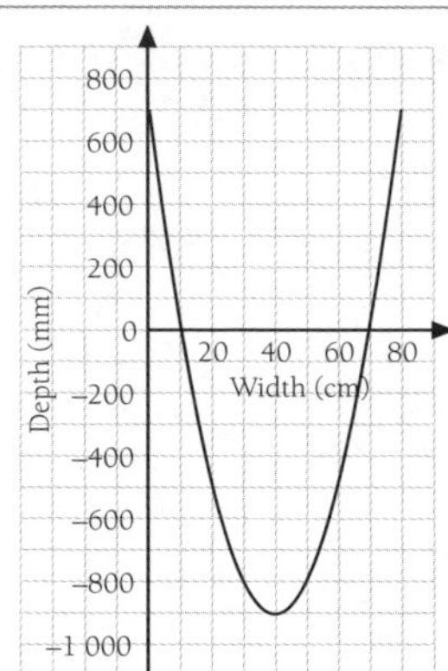

(*A*)

b. i. 40 cm **(*A*)**
ii. 900 mm **(*A*)**
c. i. 60 cm **(*A*)**
ii. Distance between intercepts on *x*-axis. **(*A*)**
d. 20 cm and 60 cm **(*A*)**
e. For $10 \le W < 70$ (a depth of +700 cm at the pole is inapplicable). **(*E*)**
f. Cross-section of the drain may not exactly 'fit' the parabola and the sides and bottom of the drain may be uneven. **(*E*)**

3. a.

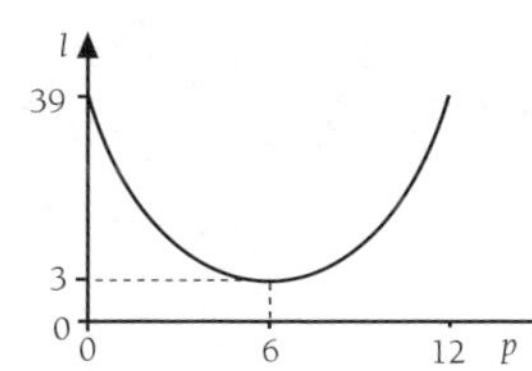

(*A*)

b. 3 m **(*A*)**
c. i. 12 m **(*A*)**
ii. Symmetry and distance from first pole to the lowest point is 6 m. **(*A*)**

4. a. $y = \frac{1}{8}(x - 2)^2 + 1.5$ *or* $y = \frac{1}{8}x^2 + 1.5$

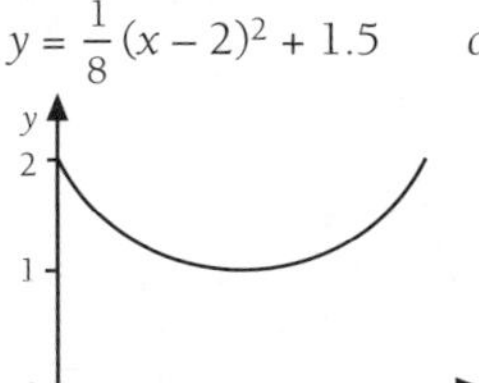

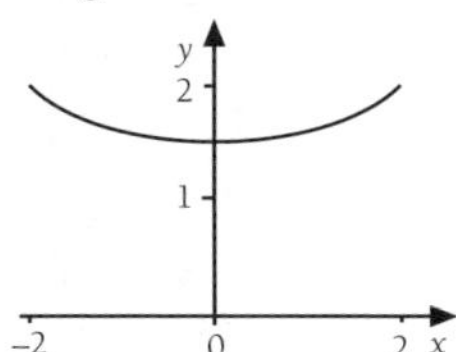

(*E*)

b. No, height of string is 1.68 m. **(*E*)**

5. a. $h = 50d + 10$ **(*M*)**
b. $h = -40(d - 4)(d - 10)$ or $h = -40(d - 7)^2 + 360$ or $h = -40d^2 + 560d - 1\,600$ **(*E*)**

6. a.

b. 1.25 m **(*A*)**
c. $y = 5(x - 1)(2 - x)$ or $y = -5x^2 + 15x - 10$ **(*E*)**
d. 0.9375 m **(*A*)**

7. 7.5 cm **(*E*)**

Unit 12.4 Activity 7: Cubic graphs (page 281)

1. a. (*A*)

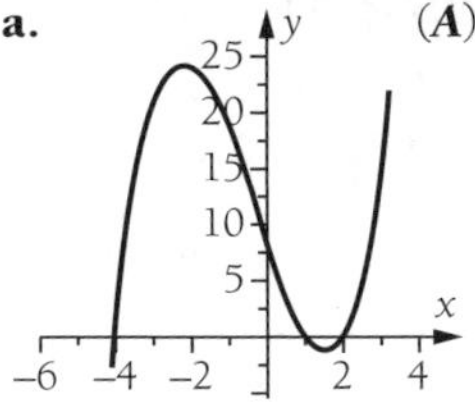

b. (*A*)

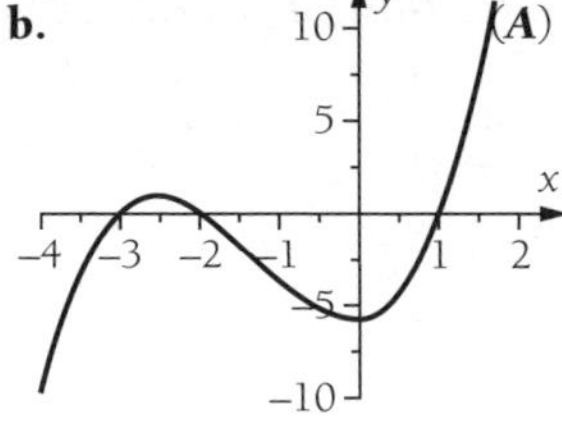

c. (*A*)

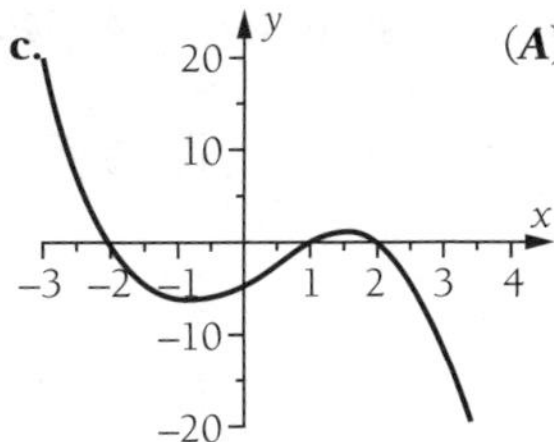

d. (***A***)

e. (***A***)

f. (***A***)

cuts y-axis at -12

g. (***A***)

cuts y-axis at 2

h. (***A***)

2. a. $y = (x + 2)(x - 1)(x - 3)$ (***M***)

b. $y = (x + 2)(x + 1)(x - 2)$ (***M***)

c. $y = -(x + 1)(x - 1)(x - 3)$ (***M***)

d. $y = (x + 3)(x - 2)^2$ (***M***)

3. (***E***)

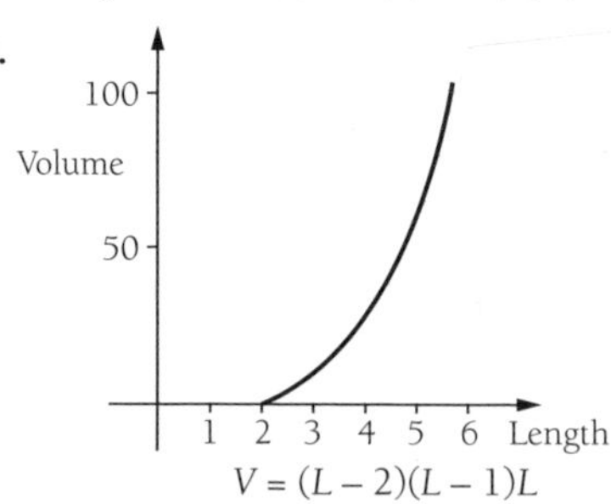

$V = (L - 2)(L - 1)L$

4. (***E***)

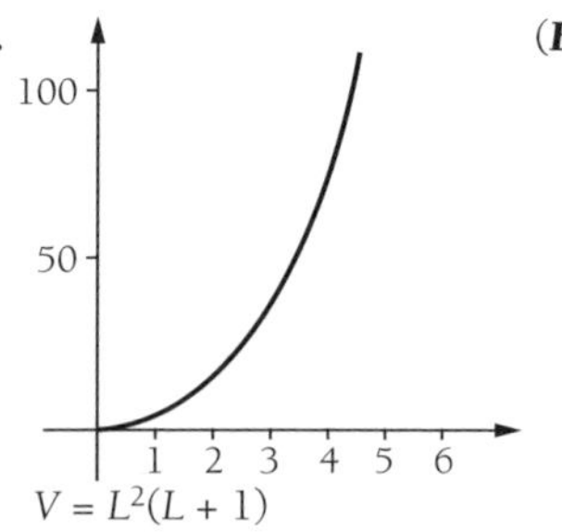

$V = L^2(L + 1)$

5. a. April, November 1 (***M***)

b. K10.50 (***M***)

c. Before February, then early July until nearly the end of September. (***M***)

d. $y = -0.1(x - 3)^2(x - 10)$

Where x is the number of months since the beginning of the year, y is the Kvalue above K1.50 of 1 kg of the commodity. (***E***)

Unit 12.4 Activity 8A: Multiplication of rational expressions (page 284)

1. a. $\frac{3a}{4b}$ **b.** $\frac{5ac}{bd}$ **c.** $\frac{3ac}{4db}$ **d.** $\frac{a^2}{bc}$ **e.** $\frac{b}{2}$ **f.** $3p$

g. $15a$ **h.** $\frac{a}{6}$ **i.** $\frac{2}{3}k^2$ **j.** $\frac{13b}{8}$ **k.** $\frac{y^3}{x^3}$ **l.** $\frac{y}{3x}$

2. a. $\frac{x}{x + 1}$ **b.** $\frac{1}{x - 1}$ **c.** $\frac{x - 2}{x + 2}$ **d.** $\frac{4}{x + 4}$ (***A***)

Unit 12.4 Activity 8B: Division of rational expressions (page 285)

1. **a.** $\frac{2c}{3a}$ **b.** $\frac{abe}{cd}$ **c.** $\frac{3x^2}{y^2}$ **d.** $\frac{a}{5}$ **e.** $\frac{b}{c}$ **f.** $\frac{3b}{a}$

g. $2a$

2. **a.** $\frac{2a+2}{15}$ **b.** $\frac{5(x^2-1)}{7x}$ **c.** $\frac{4b-4}{9}$ **d.** $\frac{2x+2y}{3y}$ **e.** $\frac{9x-3}{4}$ **f.** $\frac{6x}{5x+5}$ (**A**)

3. **a.** 5 **b.** $\frac{x+2}{x+1}$ **c.** $(x+1)(x-2)$ **d.** $\frac{x+3}{x+1}$ **e.** $\frac{x-2}{x+5}$ (**A**)

Unit 12.4 Activity 8C: Addition and subtraction of rational expressions (page 286)

1. **a.** $\frac{3a+b^2}{3b}$ **b.** $\frac{3a}{8}$ **c.** $\frac{a+1}{b}$ **d.** $\frac{4a+b}{4b}$ **e.** $\frac{4a-3b}{12}$ **f.** $\frac{a^2-bc}{ac}$

g. $\frac{a+2}{a^2}$ **h.** $\frac{2a+3}{3b}$ **i.** $\frac{10+b}{4}$ **j.** $\frac{21p+2q}{7}$

2. **a.** $\frac{2a+3}{2}$ **b.** $\frac{3b+4a}{3}$ **c.** $\frac{7a+17b}{6}$ **d.** $\frac{4a+4}{3}$ **e.** $\frac{15b+31}{6}$ **f.** $\frac{9a+2}{2a}$

3. **a.** $\frac{a+7}{12}$ **b.** $\frac{13b+6}{15}$ **c.** $\frac{-p-7q}{12}$ **d.** $\frac{11p+8q}{20}$ **e.** $\frac{14a-12}{15}$ **f.** $\frac{21-5p}{6}$

g. $\frac{26+5p}{10}$ **h.** $\frac{1+7r}{12}$ **i.** $\frac{6p-4q}{3}$

4. **a.** $\frac{2x+3}{(x+1)(x+2)}$ **b.** $\frac{3p+5}{(p+3)(p+1)}$ **c.** $\frac{3p-5}{(p+1)(p-3)}$ **d.** $\frac{5R+2}{(R+4)(R-2)}$

e. $\frac{2R+9}{R(R+3)}$ **f.** $\frac{R-A-1}{(A+3)(R+2)}$ **g.** $\frac{3A-1}{(A-3)(A-1)}$ **h.** $\frac{2A^2+2A-2}{(A+1)(A-1)}$

i. $\frac{3x-2}{6(x+1)}$ **j.** $\frac{xy-3x^2-4y^2}{12(x+y)(x-y)}$ **k.** $\frac{5x+9}{(x+1)(x+2)(x+3)}$ **l.** $\frac{x+8}{x(x+1)(x+2)}$

m. $\frac{3x-y+1}{(x-y+2)(x+y-3)}$ **n.** $\frac{x-5y+10}{x(x+y-2)(x-y+2)}$ **o.** $\frac{4y-3x}{xy(x-y)}$ (**A**)

5. **a.** $\frac{a}{3b^3}$ **b.** $\frac{4y^2z^4}{9x^2}$ **c.** $\frac{3a(a+c)}{c^2}$ **d.** $\frac{x+2}{x(x-3)}$ **e.** $\frac{(u-6)(x-1)}{(x+4)(u-1)}$

f. $\frac{4(a+2)}{3(a-2)}$ **g.** $\frac{5}{2x-2y}$ **h.** $\frac{5x+1}{x^2-1}$ **i.** $\frac{x+6y}{x^2-4y^2}$ **j.** $\frac{x}{x^2-y^2}$

k. $\frac{2x}{(x+1)(x+2)}$ **l.** $\frac{2}{(a-4)(a-5)(a-6)}$ **m.** $\frac{4x+2}{(x-1)(x+1)(x+3)}$

n. $\frac{(a-1)(a-2)(7a-12)}{12}$ (**A**)

Unit 12.4 Activity 9A: Using formulae (page 290)

1. a. 3 **(A)** **b.** 8 **(A)** **c.** 21 **(A)** **d.** 25 **(A)** **e.** $\frac{1}{2}$ **(A)**
f. 0 **(A)** **g.** 3 **(A)** **h.** –5 **(A)** **i.** 12 **(A)**

2. a. 0 °C **(A)** **b.** 100 °C **(A)** **c.** 148.9 °C **(A)**

3. a. 68 °F **(A)** **b.** 212 °F **(A)** **c.** 5 °F **(A)**

4. 904.8 cm^3 **(A)** **5.** 2.84 sec **(A)** **6. a.** K115 **(A)** **b.** 3 hr **(A)**

7. a. 102.8 cm **(A)** **b.** 12.8 m **(A)**

8. a. K47.65 **(A)** **b.** 4.2 kg **(A)** **c.** 515 km **(A)**

9. a. 584.8 cm^2 **(A)** **b.** 11.1 cm **(A)** **10.** K2 000 **(A)**

11. –40 °F **(M)** **12.** 160 °C **(M)**

Unit 12.4 Activity 9B: Rearranging formulae (page 292)

1. a. $u = v - at$ **(M)** **b.** $t = \frac{v - u}{a}$ **(M)** **c.** $r = \frac{C}{2\pi}$ **(M)**

d. $R = \frac{100I}{PT}$ **(M)** **e.** $w = \frac{P}{2} - l$ **(M)** **f.** $b = \sqrt{a^2 - c^2}$ **(M)**

g. $u = \frac{s}{t} - \frac{1}{2}at$ **(M)** **h.** $I = \frac{V}{R}$ **(M)** **i.** $l = \frac{2s}{n} - a$ **(M)**

j. $a = \frac{b^2}{c}$ **(M)** **k.** $a = \frac{2(S - ut)}{t^2}$ **(M)** **l.** $a = \frac{v^2 - u^2}{2d}$

m. $C = \frac{5}{9}(F - 32)$ **(M)** **n.** $u = \sqrt{v^2 - 2ad}$ **(M)** **o.** $a = \frac{2A}{h} - b$ **(M)**

2. a. $v = 3wu^2 - 1$ **(M)** **b.** $w = \frac{v + 1}{3u^2}$ **(M)** **3.** $k = \frac{4(T - ax^2 - c)}{ax}$ **(M)**

4. $r = \sqrt{\frac{A}{\pi}}$ **(M)** **5.** $F = \frac{9C}{5} + 32$ or $F = \frac{9C + 160}{5}$ **(M)**

Unit 12.4 Activity 10A: Graphs of $y = \frac{b}{cx}$ (page 297)

1. a.

Point	x	y	xy
A	6	1	6
B	4	1.5	6
C	3	2	6
D	2	3	6
E	–1	–6	6

Value of the constant is 6.

b. See graph

c. See graph

d. Hyperbola

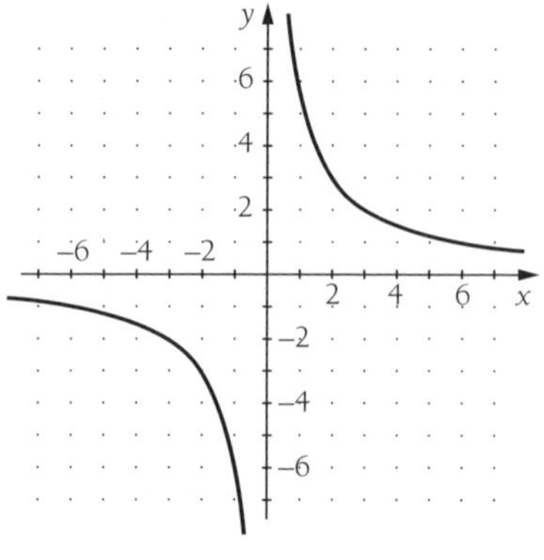

(A)

2. a. (***A***)

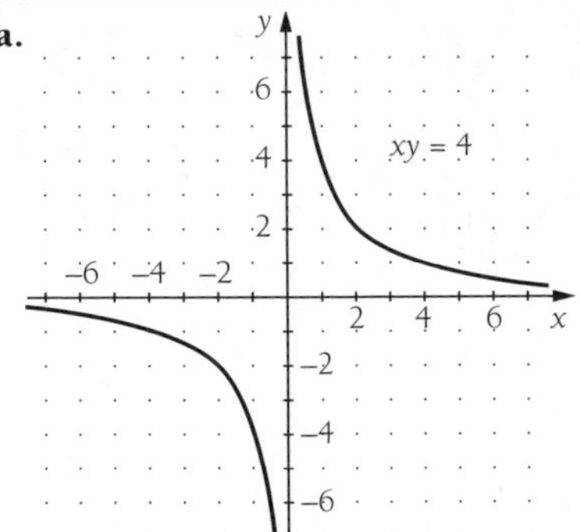

b. (***A***)

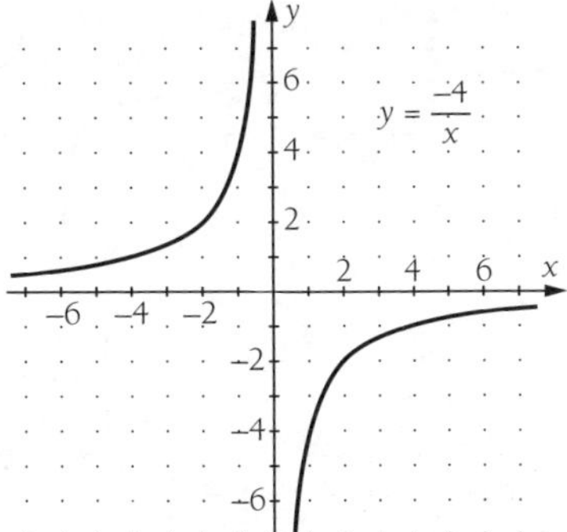

3. a. (***A***)

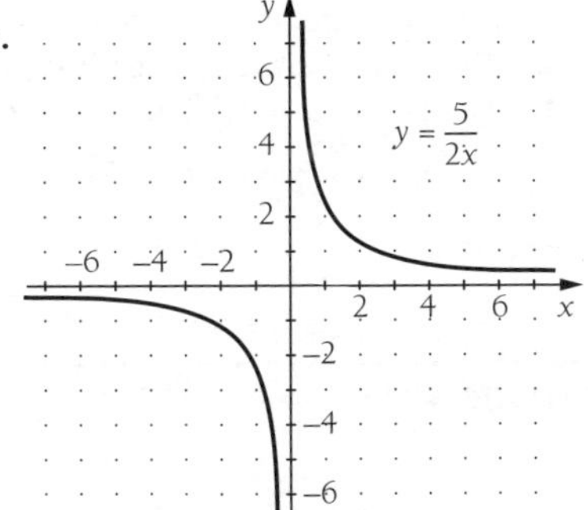

b. (***A***)

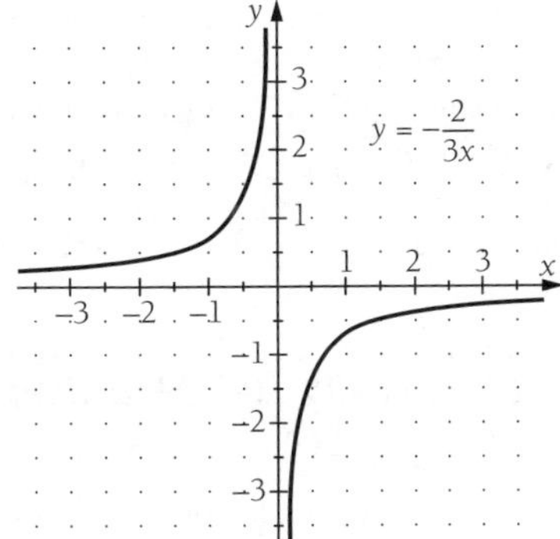

4. a. $xy = 4$ or $y = \frac{4}{x}$ **b.** $xy = -2$ or $y = \frac{-2}{x}$ (***M***)

5. a. $xy = 12$ or $y = \frac{12}{x}$ **b.** $xy = -12$ or $y = \frac{-12}{x}$ **c.** $xy = -10$ or $y = \frac{-10}{x}$ (***M***)

Unit 12.4 Activity 10B: Graphs of $y = \frac{a}{x-c} + b$ (page 300)

1. a.

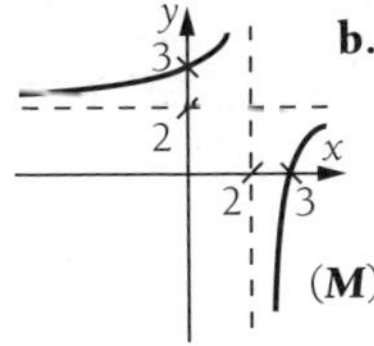

b.

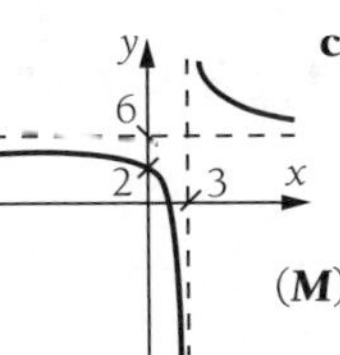

c.

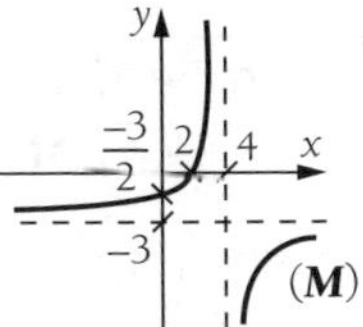

d.

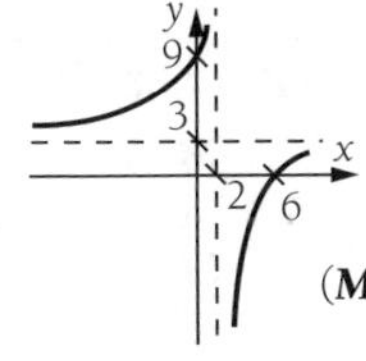

e.

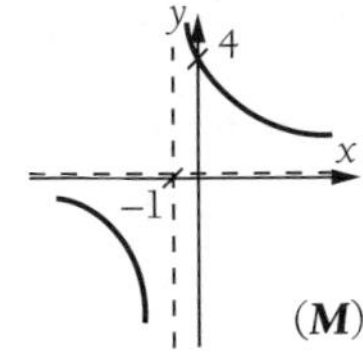

2. a. b. 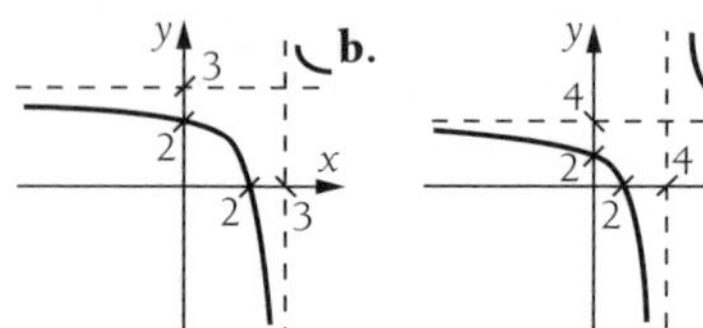

c. 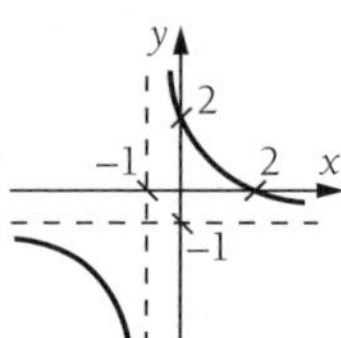

d. 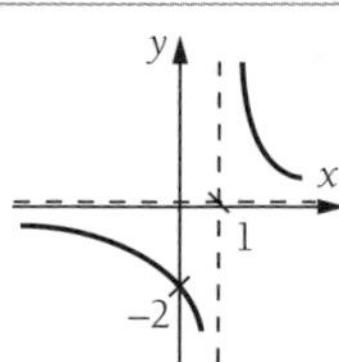

e. 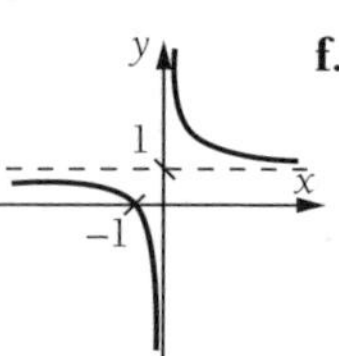

f. 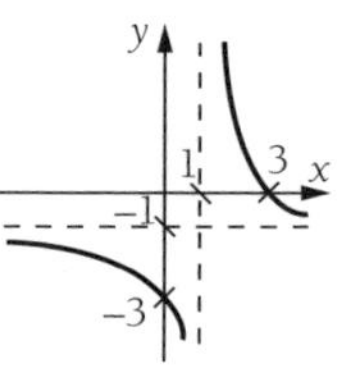

all (***M***)

3. a. Domain $\{x: x \neq 3\}$ Range $\{y: y \neq 3\}$
   b. Domain $\{x: x \neq 4\}$ Range $\{y: y \neq 4\}$
   c. Domain $\{x: x \neq -1\}$ Range $\{y: y \neq -1\}$
   d. Domain $\{x: x \neq 1\}$ Range $\{y: y \neq 0\}$
   e. Domain $\{x: x \neq 0\}$ Range $\{y: y \neq 1\}$
   f. Domain $\{x: x \neq 1\}$ Range $\{y: y \neq -1\}$

4. a. $y = \frac{12}{x+3}$ (***M***)
   b. $y = 3 - \frac{6}{x}$ (***M***)
   c. $y = -1 + \frac{3}{x}$ (***M***)
   d. $y = 2 + \frac{2}{x+2}$
   e. $y = -2 - \frac{14}{x-5}$

Unit 12.4 Activity 10C: Modelling problems (page 301)

1. a. 140, 105, $\frac{4\,200}{n}$
   b. The graph should continue through (30, 140), (40, 105), (50, 84), (60, 70), (70, 60), (100, 42), (120, 35).
   c. hyperbola

2. a. $x + 10$
   b. $(x + 10)(y + 5)$
   c. $y = \frac{500}{x+10} - 5$ (***A***)
   d. 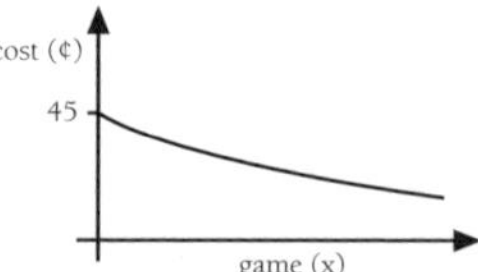

3. a. $d = \frac{A}{10A + 2\,000}$
   b. Only one branch of a hyperbola since only positive values of A are allowed. (***E***)

Unit 12.4 Activity 10D: Intersection of a hyperbola and a line (page 303)

1. $(\frac{1}{3}, 3)$ (***M***)
2. $(\frac{1}{5}, 5)$ (***M***)
3. $(2, \frac{1}{2})$ (***M***)
4. $(4, \frac{1}{4})$ (***M***)
5. (1, 1), (–1, –1) (***M***)
6. (0.71, 1.41), (–0.71, –1.41) (***M***)
7. (1, 1) (***M***)
8. $(\frac{1}{2}, 2)$, (–1, –1) (***M***)
9. (1.62, 0.62), (–0.62, –1.62) (***M***)
10. no intersection (***M***)
11. (0, 0), (2, 2) (***M***)
12. no intersection (***M***)
13. $(\sqrt{2}, \sqrt{2}), (-\sqrt{2}, -\sqrt{2})$ (***M***)
14. (0, 0) (***E***)
15. (–5, –5), (1, 1) (***M***)
16. (1, 5), (–5, –1) (***M***)

17. a.

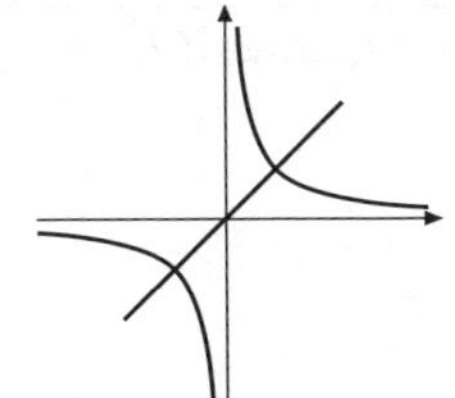

b.

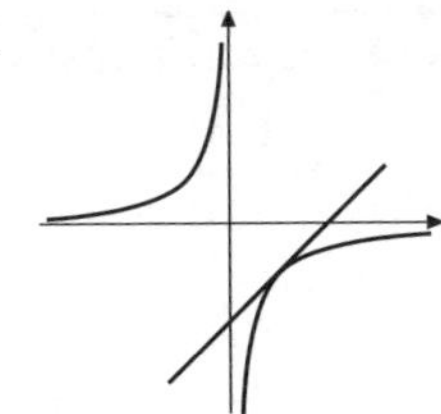

c.

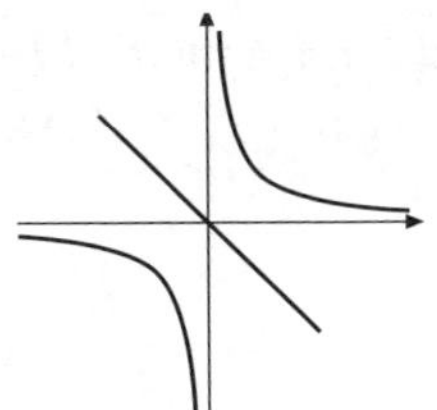

Unit 12.4 Activity 11A: Basic exponential graphs (page 306)

1. a. See graph.

b. See graph.

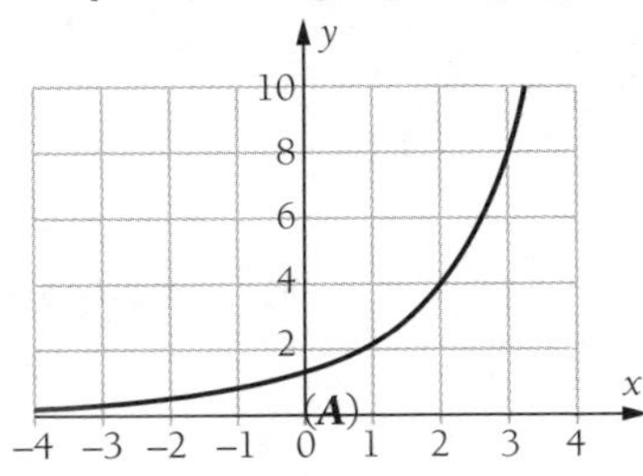

2. a.

x	–2	–1	0	1	2	3
y	0.44	0.67	1	1.5	2.25	3.38

b. (***A***)

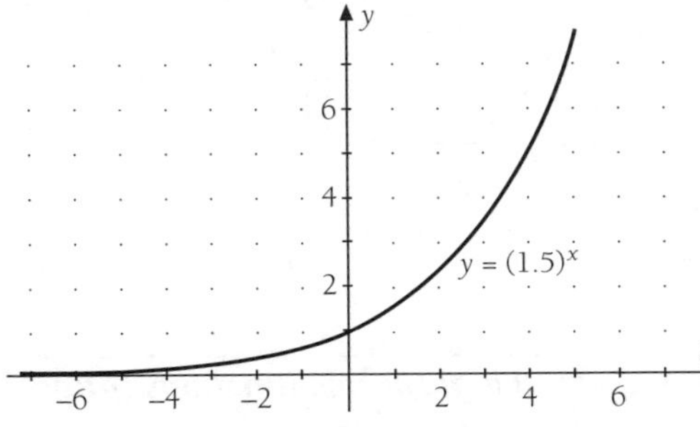

3. a. (***A***)

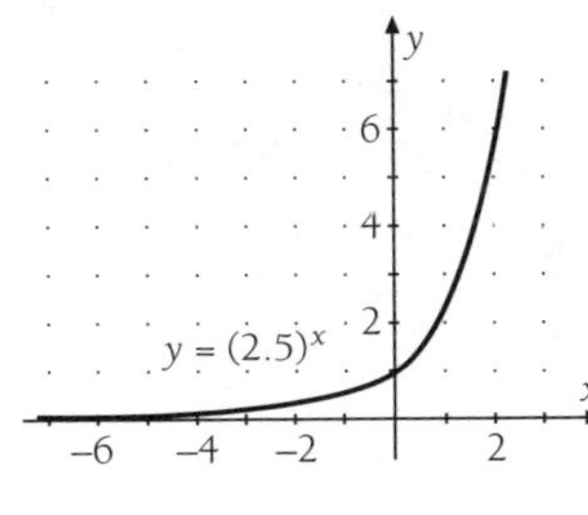

b. (***A***)

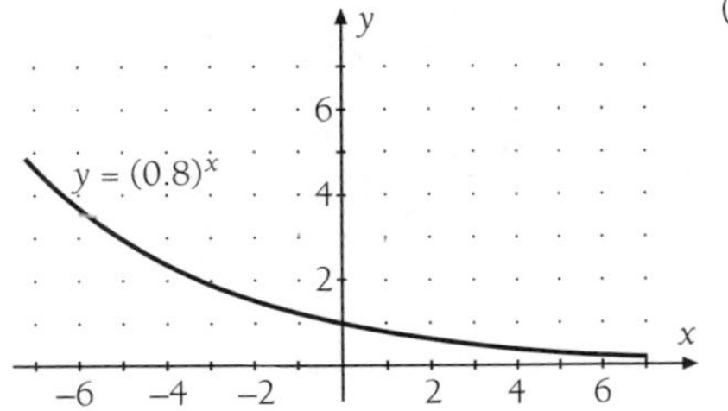

4. a. $y = 2^x$ (**M**) **b.** $y = 3^x$ (**M**) **c.** $y = \left(\frac{1}{3}\right)^x$

5. $a = 3$ (**M**) **6.** $a = 5$ (**M**)

Unit 12.4 Activity 11B: Graphs in the form of $y = a^{x-b} + c$ (page 308)

1. Graph of $y = 2^{x+1}$ (**M**)

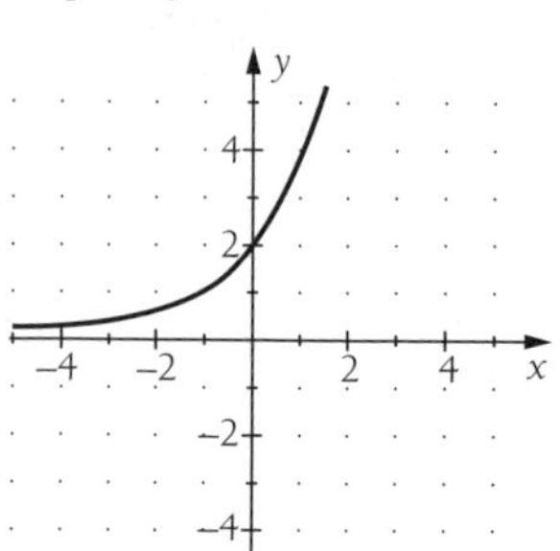

2. Graph of $y = 2^x + 1$ (**M**)

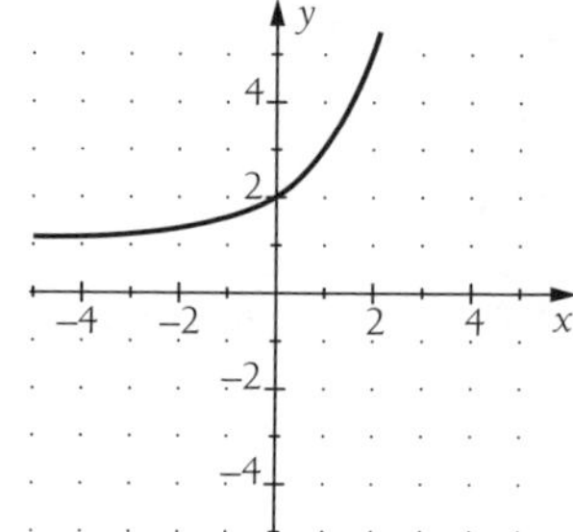

3. Graph of $y = \left(\frac{1}{2}\right)^{x-1}$ (**M**)

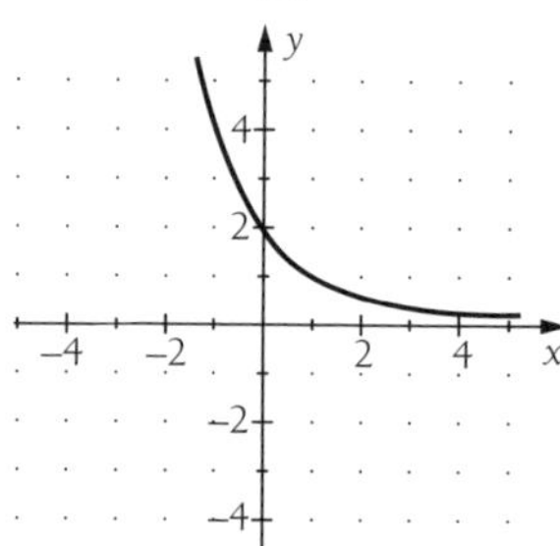

4. Graph of $y = 3^x - 2$ (**M**)

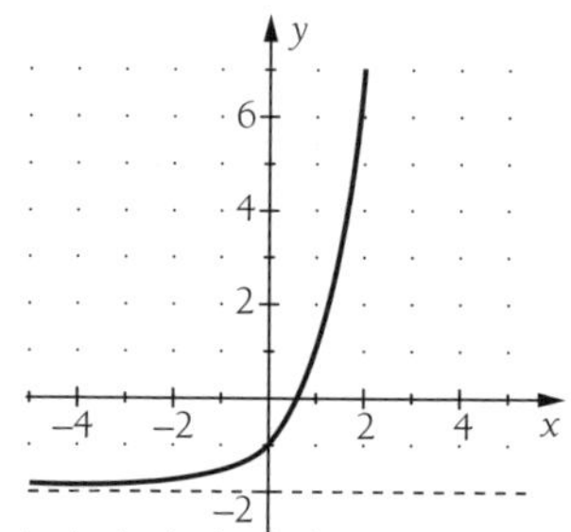

5. a. $y = 2^{x-1} + 2$ **b.** $y = 3^{x+2} - 1$ **c.** $y = \left(\frac{1}{2}\right)^{x-1} - 2$

Unit 12.4 Activity 11C: Graphs of general logarithmic functions (page 311)

1. a. (***A***)

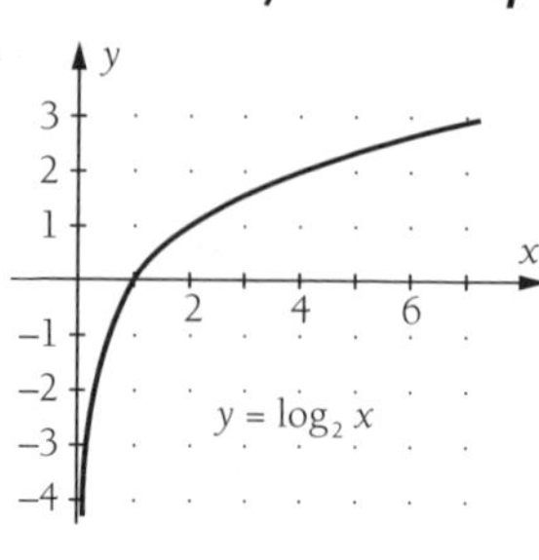

b. (***A***)

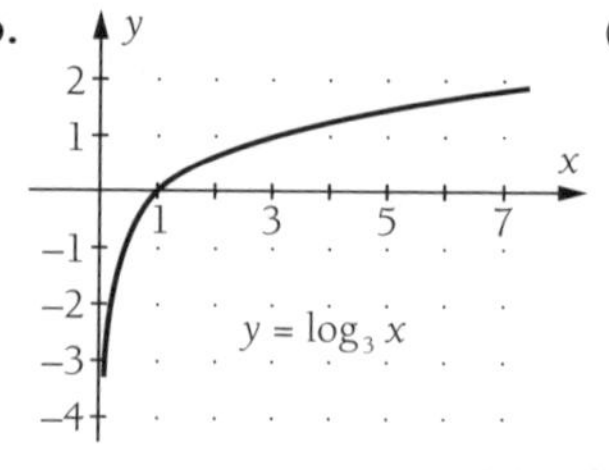

c. (*A*)

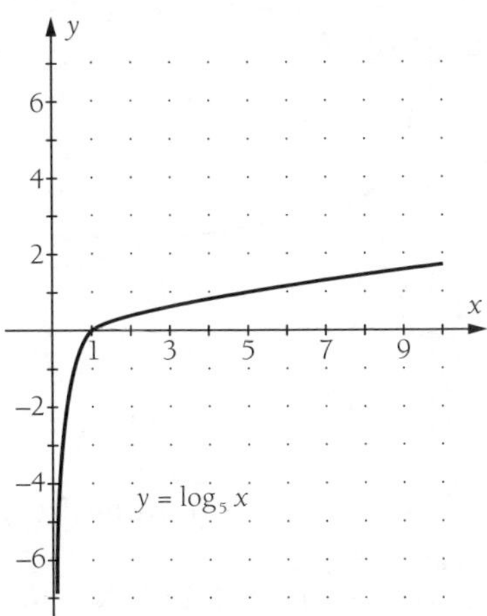

d. (*A*)

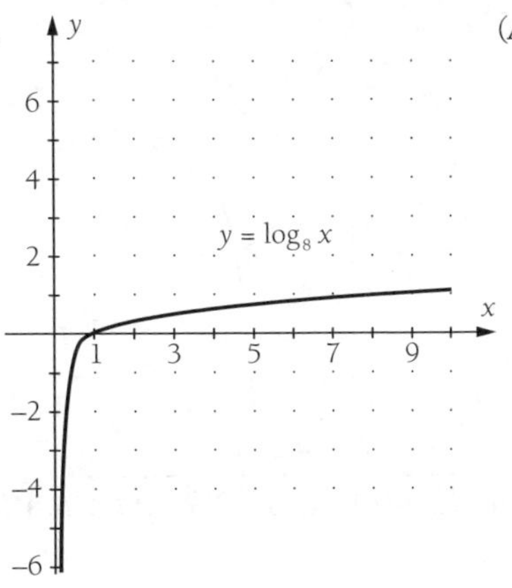

2. The claim is true.

3. The claim is false, eg on graph of $y = \log_{1.1} x$, the line $y = x$ would cut the graph.

4. a. (**M**)

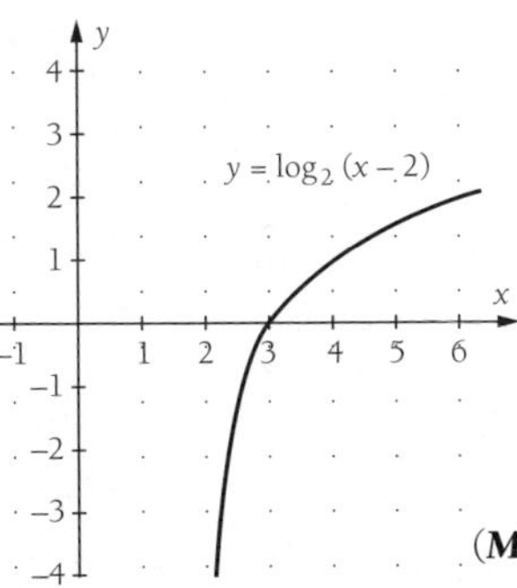

b. (**M**)

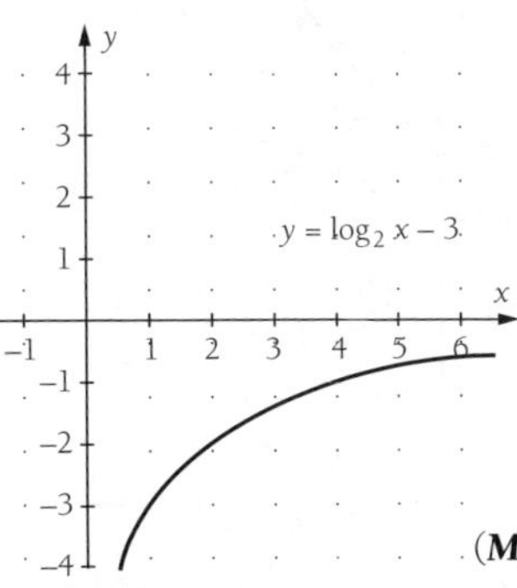

c. (**M**)

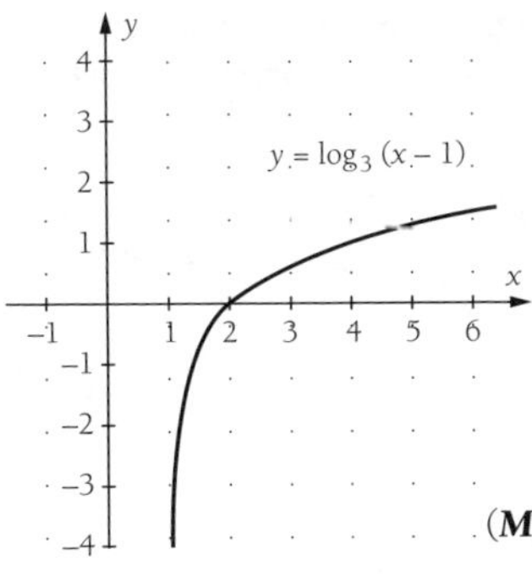

d. (**M**)

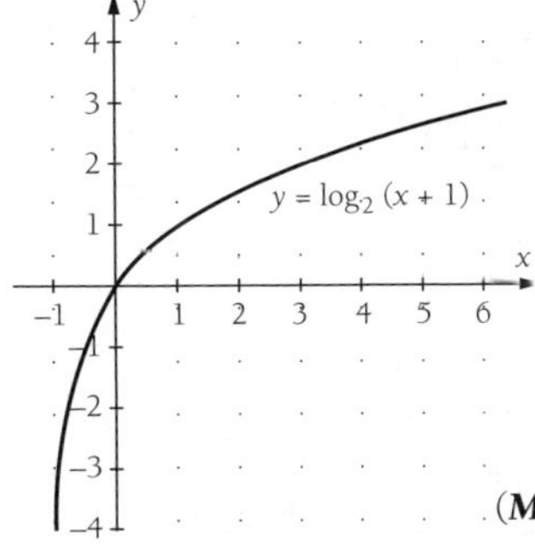

e.

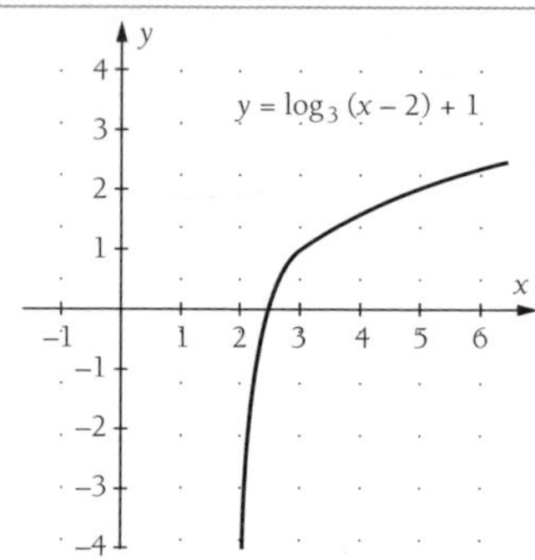

f.

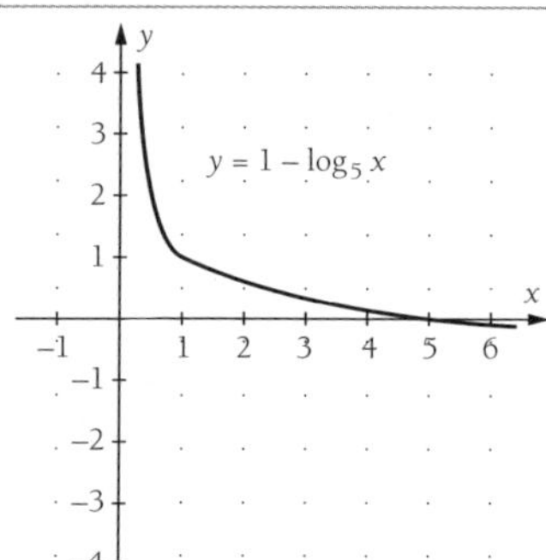

Unit 12.4 Activity 11D: Exponential and logarithmic modelling (page 312)

1. a.

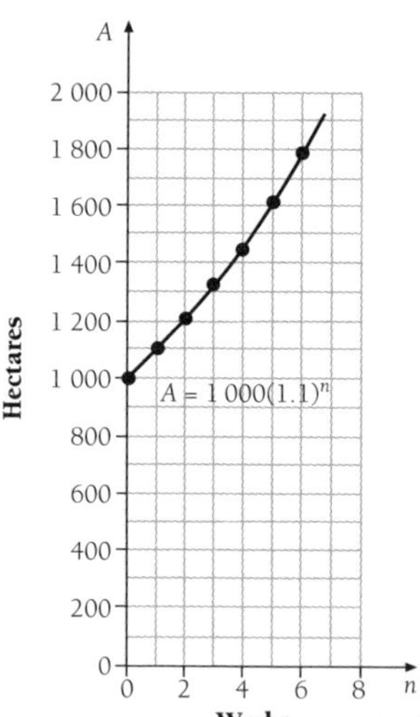

(*M*)

b. After about 4.3 weeks.

c. If this law held indefinitely the area infested would exceed the land available, which is impossible.

d. The area infested increases by 10% each week. (***M***)

2. a.

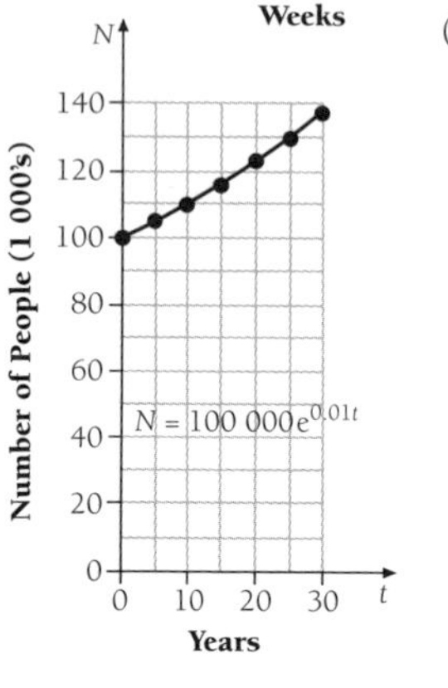

b. about 105 000. (***M***)

c. The number of people will exceed the ability of the district to support them. (***E***)

3. a. (**M**)

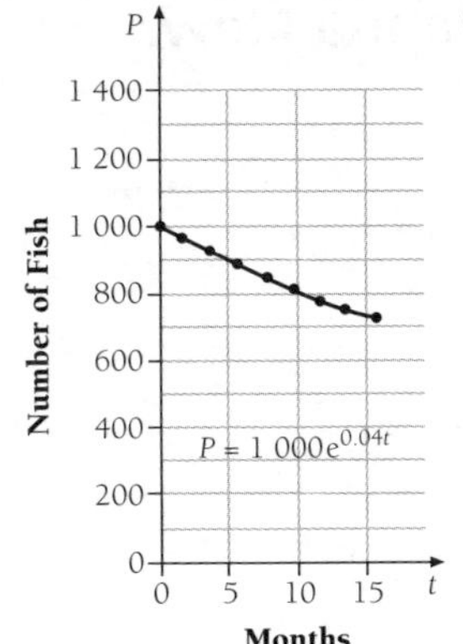

b. A few more than 800. (**M**)

c. Population extinct if $P < 1$.
This occurs when $t = 339$ weeks. (**M**)

d. As t increases, the graph of $P = 1\,000(0.6)^{0.04t}$ will be clearly non-linear. (**E**)

4. a.

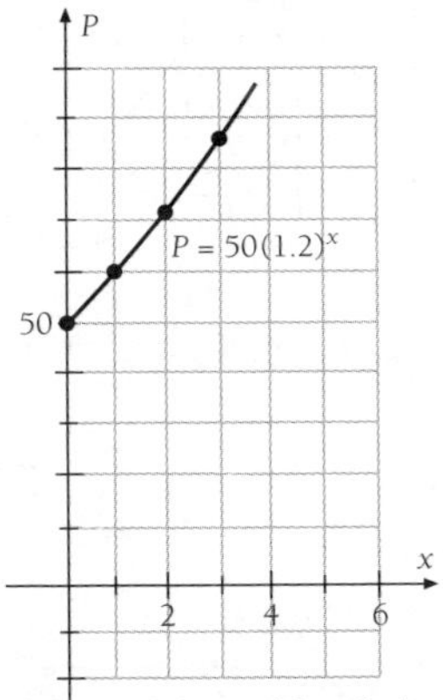

b. no asymptotes

c. P increases by 20% each week and will become very large as x increases. (**M**)

5. a. This model would only be appropriate for a limited time as the population would eventually outstrip the resources available to it.

b.

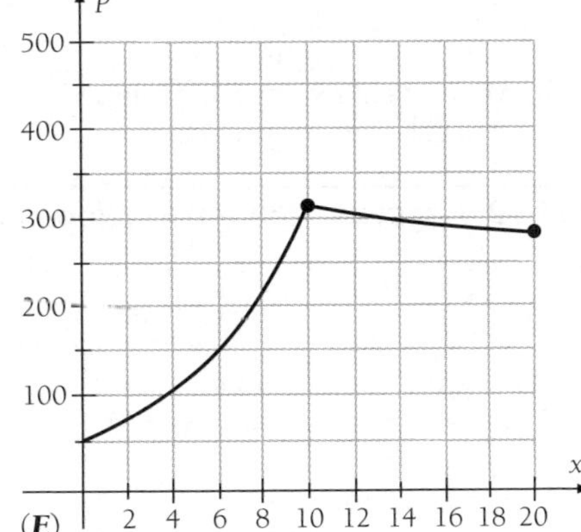

(**E**)

c. as $x \to \infty$, $P \to 260$

6. $A = 15$, $B = 30$ (**E**)

7. $a = 4$, $b = 25$ (**E**)

Unit 12.5 Applying Geometry in Papua New Guinean Arts

Unit 12.5 Activity 1A: Prisms and pyramids (page 314)

1. a. 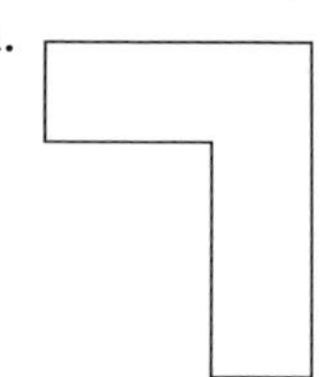
   b. 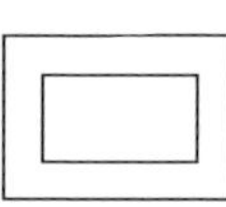
   c. 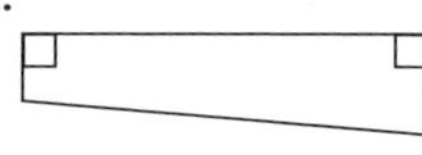

2. a. 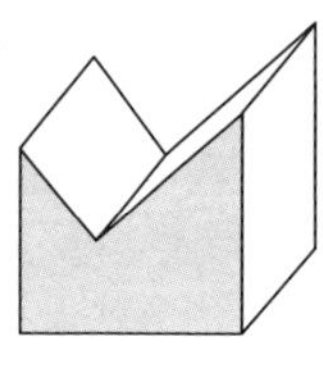
   b. 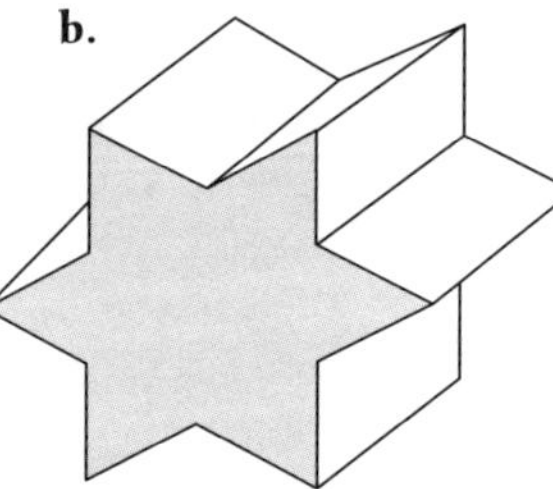
   c. 

3. a. 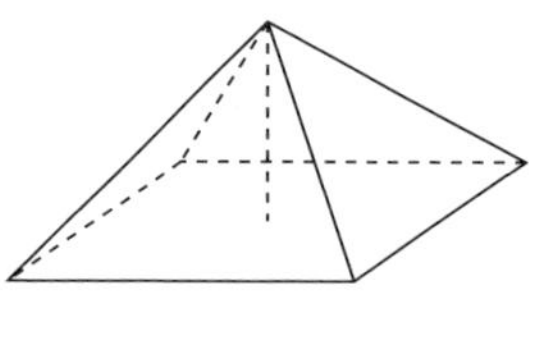
   b. 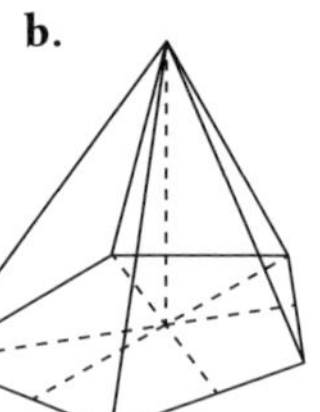
   c. 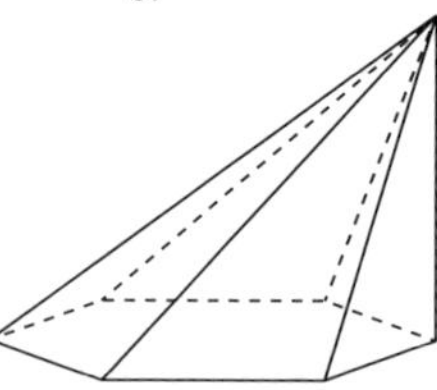

4. a. A triangular prism
   b. A non-right rectangular pyramid
   c. A regular pentagonal prism

Unit 12.5 Activity 1B: Constructing polyhedra (page 318)

1. Discuss your tetrahedron with your teacher and share your findings with your class.
2. Discuss your polyhedron (and its net) with your teacher. Share your findings with your class. How many different polyhedral were constructed?
3. Discuss your templates and nets with your teacher. Share your findings with your class.
4. A tetrahedron has six edges ($e = 6$), four faces ($f = 4$) and four vertices ($v = 4$).
   Verifying Euler's rule $+2 = v + f$:
   $e + 2 = 6 + 2 = 8$; $v + f = 4 + 4 = 8$ so $e + 2 = v + f$
5. Discuss your nets, truncated pyramid and calculations of TSA and volume with your teacher.
6. Discuss your nets, truncated cone and calculations of TSA and volume with your teacher.

Unit 12.5 Activity 1C: Symmetry (page 320)

1. **a.** 4 **b.** 1 **c.** 4 **d.** 0
   e. 1 **f.** infinite number **g.** 3 **h.** 0
2. **a.** 1 **b.** 6 **c.** 2 **d.** 2
   e. 2 **f.** 3 **g.** 4 **h.** 3
3. **a.** a shape based on a rectangle **b.** a shape based on an equilateral triangle
   c. a shape based on a square.

Unit 12.5 Activity 1D: Polyominoes (page 322)

1.
2.

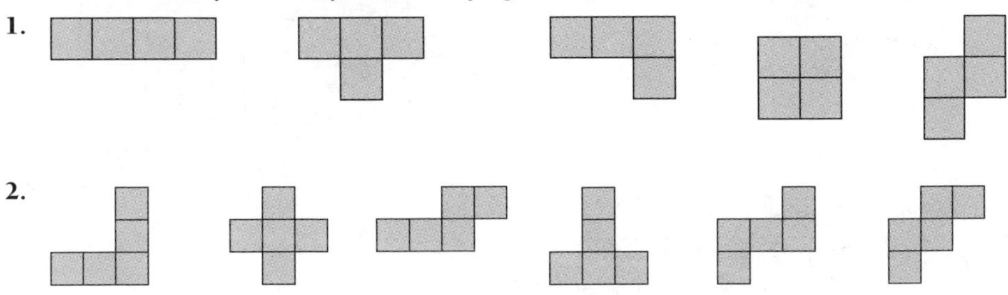

3. A & P, B & H, C & F, D & K, E & L, G & O, I & Q, J & N, M & R.
4. Here are six other hexominoes, but there are others:

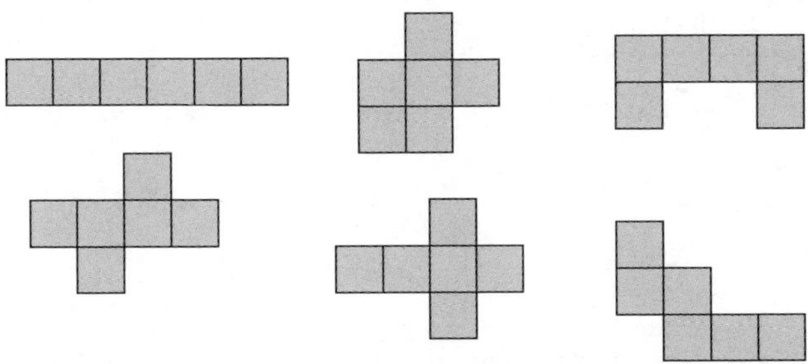

5. **a.** Rotation **b.** Rotation or reflection **c.** Translation
   d. Rotation **e.** Reflection **f.** Rotation

Unit 12.5 Activity 1E: Tiling with polyominoes (page 325)

1. Here are two designs, but there are others:

a.

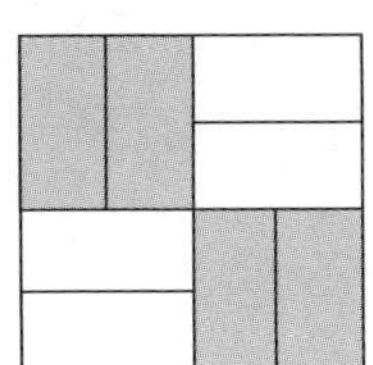

b.

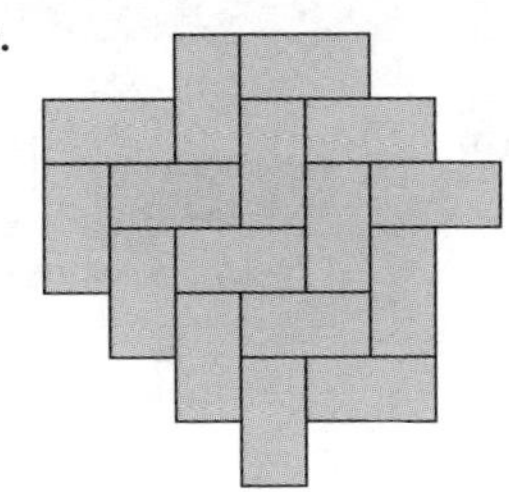

2. **a.**

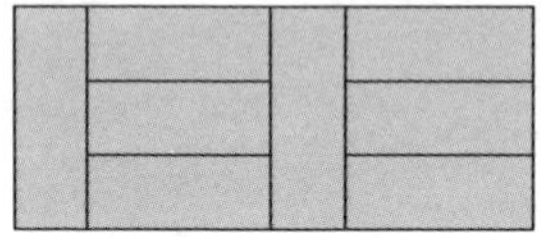

b.

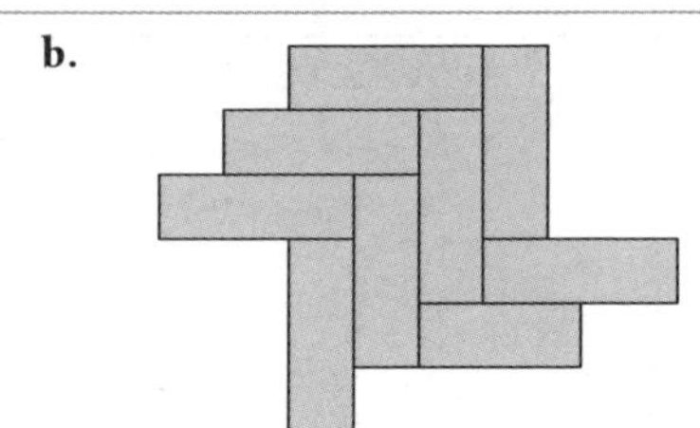

3. Other designs possible:

a.

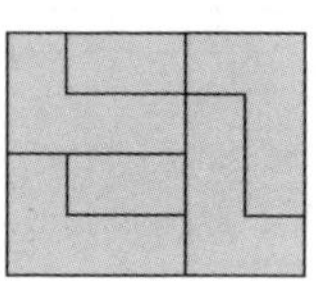

b.

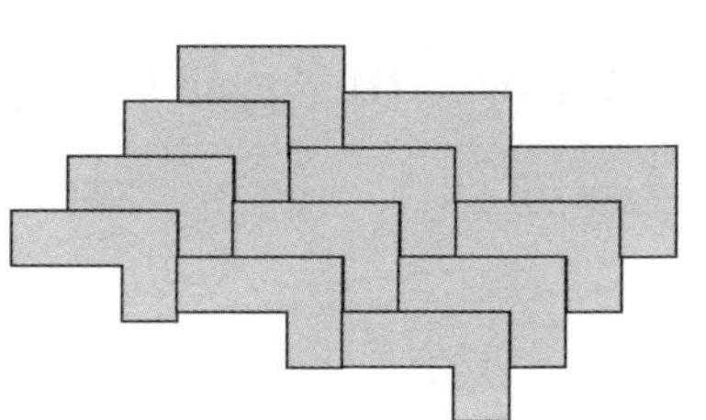

4. There are other designs also:

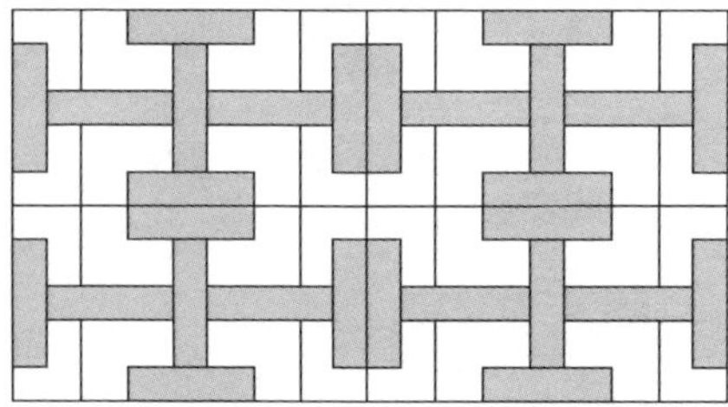

5. There are other designs:

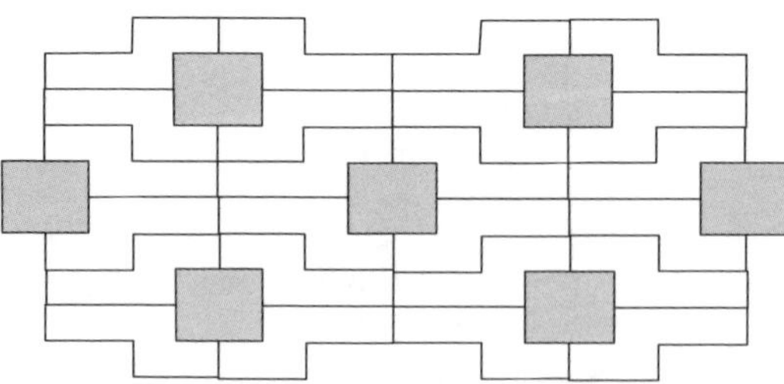

6. Yes.

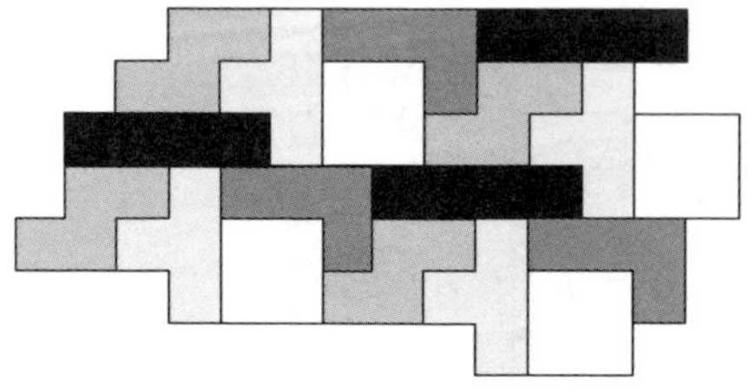

Unit 12.5 Activity 1F: Polyiamonds (page 326)

1.

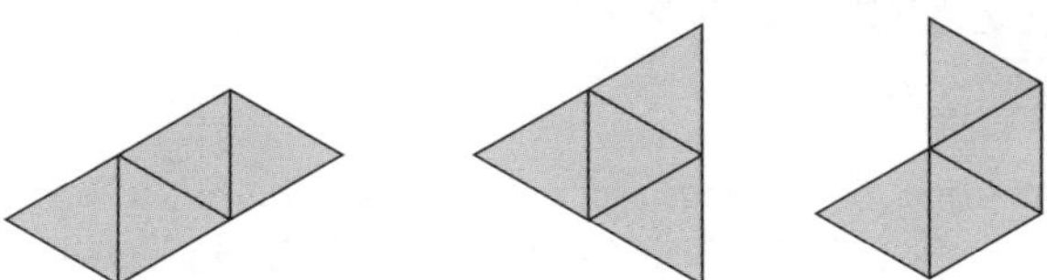

2.

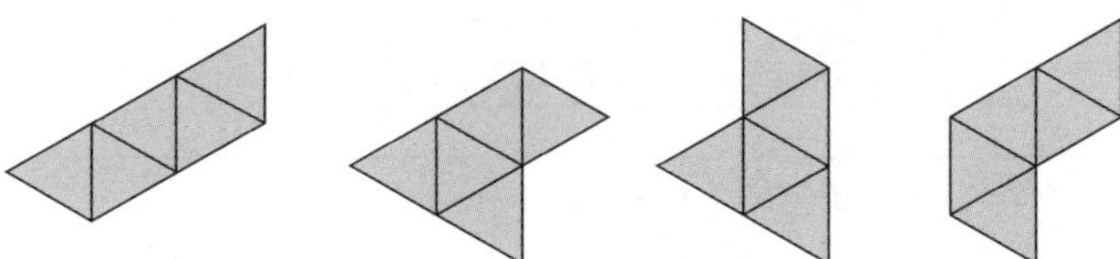

3.

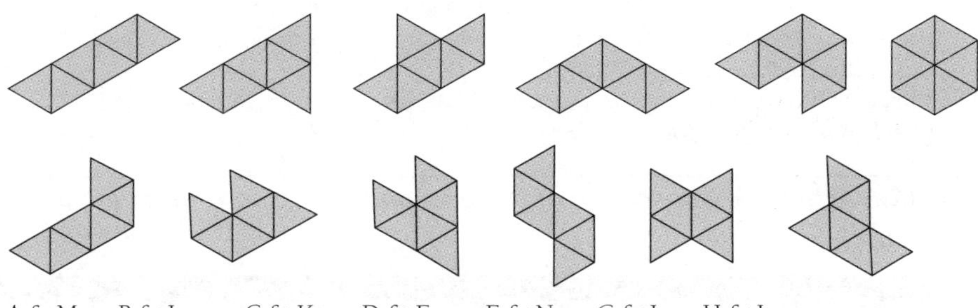

4. A & M, B & J, C & K, D & F, E & N, G & L, H & I.

Unit 12.5 Activity 1G: Tiling with polyiamonds (page 328)

1. Other tilings are also possible.

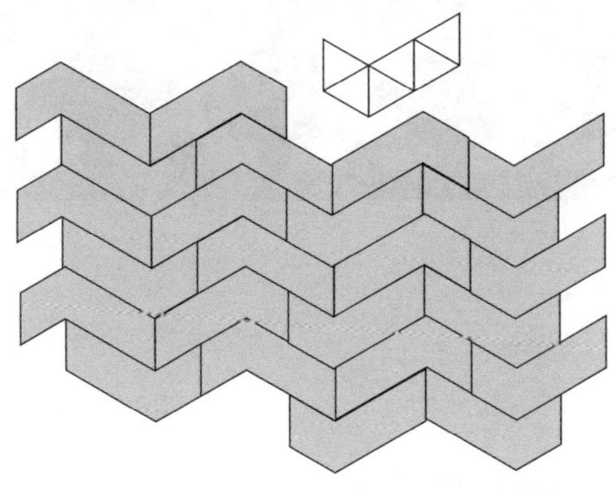

2.

3. There are always 'gaps' in the tiling.

4. There are 72 triangles and each of the 12 hexiamonds has 6 triangles. $12 \times 6 = 72$

5. a.

b. One of the hexiamonds has the shape shown shaded on the parallelogram below. This shape isolates one triangular space so it cannot be filled, therefore this parallelogram cannot be used.

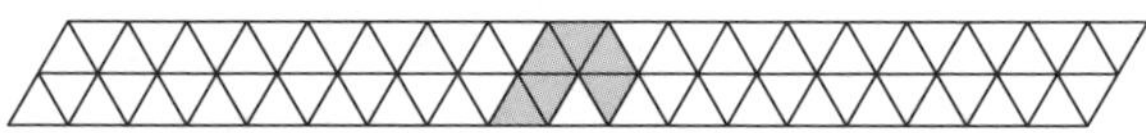

Unit 12.5 Activity 2: Tessellations (page 331)

1. **a.** and **d.**
2. Discuss your tessellation with your teacher and share with your class.
3. Discuss your tessellation with your teacher and show that your design tessellates by constructing a plane area covered by it.
4. Discuss your tessellation with your teacher. Make sure it is more complex than your other tessellations and show how your design tessellates over an area.

Unit 12.5 Activity 3: Application of patterns and shapes in traditional arts (page 337)

1.

The pattern on this bilum is a combination of polyominos.

The dominant pattern is a combination of

monominos, triominos, and pentominos.

This dominant pattern has four axes of bilateral symmetry and has rotational symmetry of order four.

The central border

is a combination of pentominoes and triominoes .

This border pattern has one axis of bilateral symmetry and rotational symmetry of order one.

The 'fill-in' areas 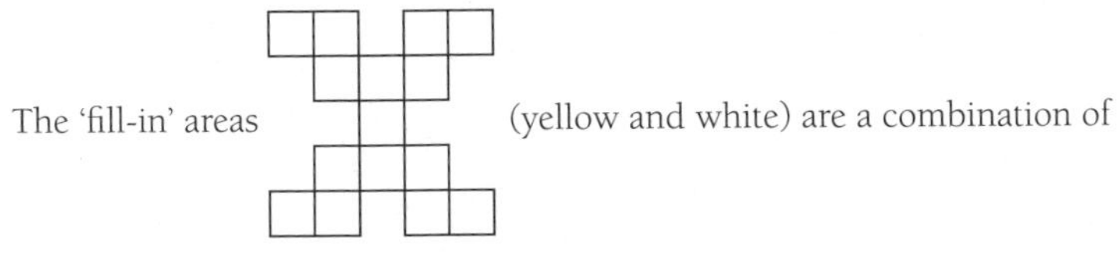 (yellow and white) are a combination of

dominoes , triominoes and monominoes

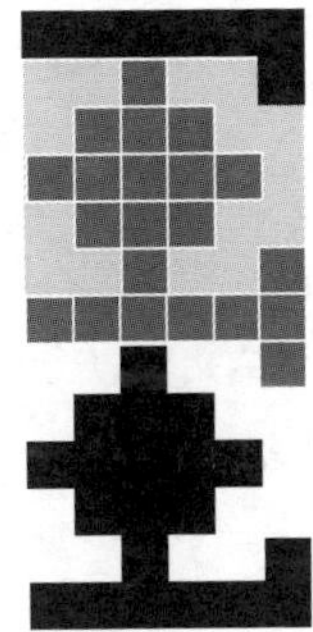

The combinations will be different if the basic pattern (below) is being analysed. This fill-in area has two bilateral axes of symmetry and rotational symmetry of order two.
The basic pattern combination (see basic pattern at right) of these polyominoes could be used to tile a rectangle and therefore can be used to tile the plane. This means that reproductions of the basic pattern, joined together, will always have straight edges.
The basic pattern combination shown does not have any axes of bilateral symmetry if colour is considered, but there is one if colour is ignored.
The basic pattern shown has rotational symmetry of order one.

2. The shape of this object is a sector of a sphere.

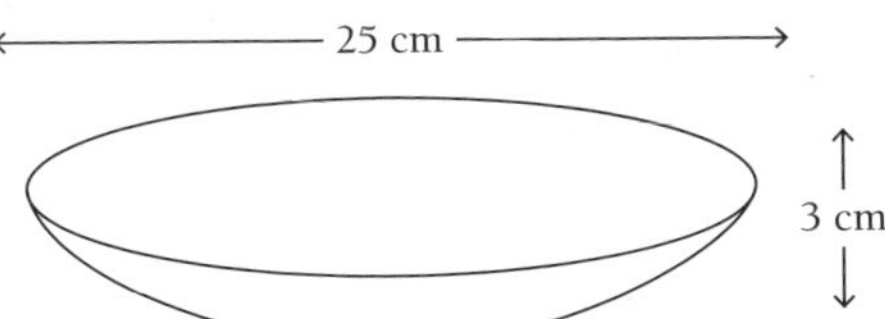

R–3

R

12.5 cm

The formula for the volume of a spherical segment is:

$$V = \frac{\pi h^2}{3}(3R - h)$$

$h = 3$

We need to find the value of R, the radius of the whole sphere. Using Pythagoras:

$R^2 = (R-3)^2 + 12.5^2$

$R^2 = R^2 - 6R + 9 + 12.5^2$

$6R = 165.25$

$R = 27.54$ cm

Volume = volume $= \frac{\pi \times 3^2}{3}(3 \times 27.54 - 3) = 750.45\ \text{cm}^3$

(The shape holds approximately 750 mL)

The formula for the surface area of a spherical segment is: Surface area = $2\pi Rh$

The surface area of the shape is $2 \times \pi \times 27.54 \times 3 = 519.12\ \text{cm}^2$

Glossary/Index

addition (+): An arithmetic operation. Adding rational expressions **(285)**.

adjacent side (10, 340): The side of a triangle, not the hypotenuse, which is next to a particular angle.

algebra (221–312): The use of numbers and *variables* to solve problems.

algebraic equation (221): An *equation* with at least one *variable*.

algebraic expression (221–312): A collection of algebraic (and maybe *numerical*) *terms*, linked by + and –.

algebraic fractions (283): Fractions containing *variables*. Also called *rational expressions*.

algebraic method (221, 255): Solving a problem using algebraic modelling and techniques.

altitude (15, 30): A line from the *vertex* of a triangle that meets the opposite side at *right angles*.

analysis (197): The process of deriving information from *data* in a *statistical investigation*.

and (173, 183, 184): The event *A and B* means the outcomes that are in *both* event *A* and event *B*.

angle (∠ or ^) (4, 9): A turning, usually measured in degrees.

approximate (2, 53, 73): A less accurate value which is close to the actual value.

arc (4): Part of the circumference of a circle; arc length **(4, 340)**.

area (2–3): The amount of 2-dimensional space an object covers. Units are m^2, cm^2 or *hectare*; Area of a *sector* **(4)**. Estimation of area **(71–76)**.

area rule (15): A relationship used to find the area of any triangle given the length of two sides and the angle between them.

asymptote (296, 298, 305): Line to which a graph gets closer and closer, but never touches.

average (56, 61, 71): A measure of the central or middle value of a *data* set, eg the *median*, *mean* and *mode*.

axis (plural axes) (193, 235): One of a pair of *number lines* that intersect at right angles on a two-dimensional graph.

axis of symmetry (265, 319): A line about which an object may be folded so that one half sits exactly on the other half. Also called a mirror line.

base: The number being raised to a power in an index expression, eg in 5^2, the number 5 is the base. The unequal side of an *isosceles triangle* **(9)**.

base angles (9): The equal angles on the *base* of an *isosceles triangle*.

base unit: The fundamental units (eg metre) from which other units (eg centimetre and millimetre) are derived.

bivariate data (193, 203, 209): Data involving pairs of variables.

brackets (( ) or [ ]) (139, 222, 285): Punctuation used to establish the order of basic operations.

calculators, use of with: logarithms **(309)**; circles **(2)**; least-squares regression line **(206)**; correlation coefficient **(197)**.

cancelling (283): Simplification of a fraction by dividing the numerator and denominator by the same factor.

capacity (29, 334, 336): Amount of liquid a container holds, measured in *litres*, *millilitres*, etc.

Cartesian plane (235, 251): The x, y plane in which graphs are drawn and points represented by ordered pairs (x, y).

causal relationship (200): There is a causal relationship between two variables if a change in one variable causes a change in the other variable.

Celsius (245): See *degrees Celsius.*

cent (c) (150): A unit of Australian *money*. 100 c = $1.

centigrade: See *degrees Celsius.*

centimetre (cm): A smaller unit of length than the metre. 100 cm = 1 m.

centre (25, 314, 335, 340): The middle of a *circle* or polygon.

certain (142, 165): Having a 100% chance of occurring. A certain *event* has a *probability* of 1.

changing the subject (231, 291): Rearranging a *formula* to make a different *variable* the *subject.*

chord: A straight line joining any two points on the circumference of a circle.

circle (2): A set of points equidistant from a fixed point (its *centre*). equation of **(340)**; intersecting with a line **(263)**.

circumference (2, 315, 340): The *perimeter* of a *circle.*

coefficient (195, 228, 269, 271, 280): A number that multiplies a variable in a term.

combinations (173): Subsets of a set, in which order is not important. Probabilities of combinations (**173**).

common denominator (285): A denominator common to two or more fractions.

common factors (37, 41, 86, 100, 283): Factors that divide into each term of an expression without remainder.

complement (A') (173): A' is the set of elements *not* in set A.

complementary (174): The complement of an event, A, is written A' and is the set of all elements in the sample space which are outside A.

composite (1, 4, 31): A composite figure is made up of two or more basic shapes. Also called *compound.*

compound (1, 5): See *composite.*

compound interest (99–105, 123, 151): Interest that is added to the capital amount and therefore also attracts interest.

conditional probability (178, 188): The probability of an event occurring given that another event has already occurred. The conditional probability of A occurring given that B has occurred is written $P(A \mid B)$.

cone (30, 39, 41, 47, 229, 335, 339): A circle-based *pyramid.*

congruent (315): Identical in shape and size.

constant (2, 31, 110, 240, 241, 257, 265, 295, 313): A fixed amount that does not vary.

constraints: Restrictions or conditions.

conversion (3): Changing from one unit to another in the metric system.

conversion graph (245): A graph use to change from one unit to another, eg dollars to kina.

coordinates (203, 235, 279): *Ordered pair* of numbers (x, y) describing the position of a point on a *cartesian plane*.

correlated (200): A linear (or other) relationship between two variables. Positive linear correlation has a trend line with positive gradient ($R > 0$); negative linear correlation has a trend line with negative gradient ($R < 0$).

correlation (193–201): An association between variables (a change in one variable is accompanied by a change in the other variable).

correlation coefficient, r (195, 197–200, 205): A measure (between –1 and 1) of the linearity of a trend line fitted to a set of data points.

cosine (cos) (10): A *trigonometric ratio*. Relative to a particular angle it is the ratio $\frac{\text{adjacent side}}{\text{hypotenuse}}$.

cosine rule (10, 19, 341): A relationship between three sides and one angle of a triangle.

cross-section (5, 31–32, 48, 276, 313): The view when a plane 'cuts' an object at right angles to the base.

cube (313, 315, 316, 339): Dice-shaped solid with 6 congruent (equal) square faces. Also called a regular *hexahedron*.

cubic centimetre (cm^3) (29): Unit of measurement for *volume*. $1\ cm^3 = 1\ 000\ mm^3$.

cubic functions (279): An equation of the form y = cubic expression.

cubic metre (m^3) (29): Base unit of measurement for *volume*. $1\ m^3 = 1\ 000\ 000\ cm^3$.

cubic millimetre (mm^3) (29): Unit of measurement for *volume*. $1\ 000\ mm^3 = 1\ cm^3$.

cubics: Drawing graphs of **(279)**; writing equation of **(279)**.

cuboid (29): A solid with 6 rectangular faces meeting at right angles (opposite faces congruent). Also known as a rectangular *prism*.

cylinder (30): Circle-based solid with uniform cross-section. Surface area **(30, 339)**. Volume **(30, 41, 339)**.

data (169, 193, 199, 203): Pieces of information; collected by observation, interviews or questionnaires.

decagon (21): A 10 sided *polygon*.

decimal (2): A number that has a *decimal point*. Percentages and decimals **(165)**.

decimal places (18): The number of *digits* after the decimal point.

decimal point: A full stop used to separate the whole part from the fractional part in a *decimal*.

degrees (295): Unit used to measure angles. 1 degree (1°) is $\frac{1}{360}$ revolution

degrees Celsius (°C) (245): A unit for *temperature*. Water freezes at 0°C and boils at 100°C. Also called degrees centigrade.

degrees Fahrenheit (°F) (245): A unit for *temperature*, no longer widely used except in North America. Water freezes at 32°F and boils at 212°F.

denominator (222, 283, 286, 298): The integer or variable below the horizontal line in a fraction.

diagonal (3): A line segment joining two vertices of a polygon and which is not a side.

dial: A circular *scale* used in measurement.

diameter (2, 54, 335): A straight line passing through the *centre* of a circle and joining two points on the *circumference*.

die (plural dice) (165, 167, 174, 187): A 6-sided cube usually labelled 1, 2, 3, 4, 5, 6.

difference (130): The result of subtracting two terms or numbers.

digits: The ten symbols, 0, 1, 2, 3, 4, 5, 6, 7, 8, 9, used in numbers.

discontinuity (296): A hole or break in a graph.

division (÷) (284, 298): An arithmetic operation.

divisor (284): The term 'doing' the division.

dodecagon: A 12-sided *polygon*.

dodecahedron (316, 318): *Solid* with 12 faces, each a regular *pentagon*.

dollar ($): A unit for money in Australia, New Zealand, the USA and some other countries. $1 = 100 cents.

domain (296, 298): The set of first elements in the ordered pairs of a bivariate data set or a relation.

equally likely outcomes (167, 173, 176): *Outcomes* which all have an equal chance of occurring.

equation: A mathematical sentence with an equals sign in it.

equation of a straight line (236, 240): Gradient intercept form is $y = mx + c$. General form $ax + by + c = 0$.

equidistant (72, 73): The same distance away from another point or line.

equilateral triangle (9, 315, 319, 325, 329): A *triangle* in which all three sides are of equal length.

equivalent fraction (285): A different way of representing the same *fraction*.

estimate (53–57, 61, 65, 166): An approximate value. A number *rounded* to fewer *significant figures* than it originally had.

even numbers (165, 173): *Multiples* of 2.

event (165–167, 173–191): A set of *outcomes* of interest (a subset of the *sample space*).

expanding (222, 259, 273): Removing *brackets* from an expression using the rule $a(b + c) = ab + ac$.

expansion (271): Removing brackets from an expression by multiplying.

expected number or value (191): The average number of occurrences of an event after a set number of trials.

experimental probability (166): The proportion of times an event occurs in a planned investigation where observations are recorded.

exponent: A superscript (raised) number or variable used to indicate repeated multiplication of a factor. Also called a *power* or *index*

exponential function (305): A function of the form $y = a^x$, whose inverse is the logarithmic function $y = \log_a x$.

exponential graph (100, 305–311): Graph of an exponential function.

exterior angles (25–26): The outside angles of a polygon formed by extending the sides of a *polygon*.

extrapolation (218): Using data from a sample to make future predictions (outside the domain of the sampled data).

factor (17, 37, 144, 167, 257, 272, 280): Any *natural number* (or *algebraic expression*) which divides into another number or expression without a remainder.

factorise (231, 256–272, 283, 286): Rewriting a mathematical expression as a product of simpler factors.

formula (formulae, plural) (1, 9–14, 229–234, 289–294, 339): An algebraic equation in which the mathematical relationship between different variables is expressed. Changing the subject (231).

fraction (4, 165, 222, 230, 283, 285): A number expressed as a quotient. The top number is called the numerator, the bottom the denominator. An algebraic fraction or rational expression is a fraction with variables.

frequency (124, 165): How often something occurs.

function (265–278): A relation in which each element of the domain appears in only one of the ordered pairs of the relation.

geometric sequence (128): A sequence where each term gives a constant value when divided by the previous term.

gradient (193, 203, 206, 218, 238–247, 340): The ratio of the $\frac{\text{vertical change}}{\text{horizontal change}}$ of a straight line that describes the steepness of the slope of a line.

gradient-intercept form (236, 240): The $y = mx + c$ form of the equation of a straight line.

gram (g) (217): The base unit of *mass* in the *metric system*. 1 000 g = 1 kg.

graph (85, 100): Visual display of information, often involving axes and a grid of quadratic functions (parabola) **(265–278)**; of hyperbolae **(295–304)**; linear graphs **(235–250)**; of linear equations **(251–254)**; exponential and logarithmic graphs **(305–312)**; cubic graphs **(279–282)**.

graphical method (209): Solving a problem using a graph.

grouped data (342): *Data* values that are put into groups or classes.

GST (Goods and Services Tax) (159, 161): A tax of 10% added to all goods and services.

hectare (ha) (2, 40, 65): A unit for *area* in the *metric system*. 1 ha = 10 000 m^2.

hexagon (21, 25, 313, 329, 330): A 6-sided *polygon*.

hexahedron: Solid with six faces. A regular hexahedron is a *cube*.

horizontal asymptote (298–300, 309): A horizontal line that a graph approaches but does not cross. The value that is excluded from the range of a hyperbola.

horizontal line (197, 239, 241): Line parallel to the horizon.

hyperbola (hyperbolae, plural) (295–303): The graph of a function of the type $y = \frac{Ax + B}{Cx + D}$ **or** $y = \frac{A}{x - B} + C$.

hypotenuse (10, 22, 340): The side of a right-angled triangle opposite the right angle.

icosahedron (315, 317): *Solid* with 20 faces, each an *equilateral triangle*.

impossible (165): Having no chance of occurring. An impossible *event* has *probability* zero.

independent (187–192): Independent events have no effect on the occurrence (or not) of each other. Events A and B are independent iff (if and only if) $P(A \cap B) = P(A) \times P(B)$. Also A and B are independent iff $P(A \mid B) = P(A)$ (and $P(B \mid A) = P(B)$). Independent random variables (247).

independent variable (193, 199, 204, 218): In an experiment or modelling, the variable that represents an input; can be tested to see if it is the cause.

index (indices, plural): Exponent or power (of a base number or variable).

inequality symbols ($<, \leq, >, \geq$) (227): Symbols used in inequations.

inequation (227, 251–253): A mathematical expression which has an *inequality* instead of an equals sign.

inflation rate (107): Rate of change of the CPI (Consumer Price Index).

integers (I) (238): The set of *whole numbers* and their opposites: $\ldots, -3, -2, -1, 0, 1, 2, 3, \ldots$

intercept method (236, 237): A method of drawing a *linear graph*.

intercepts (237, 269–276, 279–280, 298–300, 307): Points where a graph cuts the *axes*.

interior angles (25–26): The angles inside a *polygon* and between adjacent sides.

interpolation (218): The process of using a model to predict values within the domain of a data set.

interpret (150, 194, 199, 218, 235, 243, 259): Make sense of. Understand the implications of.

intersection (183, 244, 252, 302, 315): The set containing the elements common to two or more other sets; a point where the graphs of two (or more) functions meet. The intersection of two sets A and B is written $A \cap B$.

inverse function: The function that reverses the effect of another function.

inverse operations (2, 231, 291, 292, 310): Pairs of operations that reverse each other, such as addition and subtraction.

inverted (266, 271, 335): Turned upside down.

irrational numbers (2): Numbers that are not *rational*, ie that cannot be written as a *fraction*.

isometric (326, 333): An isometric drawing or diagram has equal parallel sides of a figure drawn equal and parallel.

isosceles triangle (9, 11, 25, 26, 318): A triangle with two sides of equal length.

kilogram (kg): A larger unit for *mass* than the *gram*. 1 kg = 1 000 g.

kilometre (km) (2, 29, 65): A longer unit of length than the metre. 1 km = 1 000 m.

kite (339): A *quadrilateral* with two pairs of equal adjacent sides.

length (1, 3, 9): The distance from one end of an object to the other. The base unit is the *metre*.

limit (298): The value a function approaches.

line of best fit (203): Fairest line for a *scatter diagram* showing a narrow linear band pattern of points.

line segment (339): Part of a line lying between two points.

linear (3, 108, 140, 193, 194, 198, 209): Pertaining to the straight line.

linear equation (221–223): An algebraic equation in which the power of the variable is 1.

linear graphs (235–246): Straight line graphs.

linear inequation (251–253): An algebraic inequation in which the power of the variable is 1.

litre (L) (29, 57, 336): A unit for the *capacity* (or *volume*) of an object. Related to volume by the rule 1 L = 1 000 cm^3.

logarithm (log) (308–311): The inverse of $y = a^x$ is $y = \log_a x$, a function with domain $(x > 0)$.

lowest common multiple (LCM) (222, 285, 286): The smallest possible *multiple* shared by two or more numbers.

m: Commonly used symbol for the gradient of a straight line.

magnitude (15): Size.

map (53, 57, 61, 65): A representation of a geographical area, drawn to scale often with grid lines for reference.

mass (58, 247): The amount of material in an object. Often referred to as weight.

mathematical modelling (225, 229–300): The process where problems described in words are changed into equations.

mathematical sentence (221): A description of a problem using numbers, *variables* and *operations*.

maximum (266, 273): The *y* value for a function that has a peak at its turning point.

mean (204, 339): Central value in data, often called the *average*. The mean is found by adding all the scores and then dividing by the number of scores.

measurement (1, 5, 1–80): The process of assigning a numerical value to properties of an object, such as mass.

median (197, 200, 209): The central (or middle) value found by *ranking* data and selecting the middle number.

metre (m) (2, 29): The base unit for *length* in the metric system.

midpoint: The point halfway between the end-points of a line segment.

milligram (mg): A unit of *mass* smaller than the *gram*. 1 000 mg = 1 g.

millilitre (mL) (29): A unit of *capacity* (or *volume*) smaller than the *litre*. 1 000 mL = 1 L.

millimetre (mm) (29): A unit of *length* smaller than the centimetre and metre.
1 000 mm = 1 m, 10 mm = 1 cm.

minimum (266, 275): The *y* value for a function that has a trough at its turning point.

minimum turning point (268): A turning point where a function changes from decreasing to increasing.

model (203, 209): A mathematical equation representing a relationship between variables.

money: Currency used to exchange goods and services. The basic unit in PNG is the kina.

multiplication principle (175): A multiplying technique to work out the number of ways a sequence of events can occur.

multiplying (×): An arithmetic operation.

multiplying factor (81–82, 144): The amount by which you multiply a value.

multivariate statistical data (169): Data with two or more characteristics being measured.

mutually exclusive (184): When two events or sets have no common elements.

natural numbers (N) (226, 262): The set of counting numbers: 1, 2, 3, . . .

net (29): Flat shape that folds up to make a three-dimensional figure.

non-linear (279): Not linear (not a straight line). Non-linear equations **(262)**.

numerator (283): The integer or variable above the horizontal line in a fraction.

objective function (251): A function to be optimised (maximised or minimised).

octagon (21): An 8-sided *polygon*.

octahedron (317): *Solid* with 8 faces, each an *equilateral triangle*.

odd numbers (167): The numbers 1, 3, 5, 7...

operation (221): A process which changes one or more elements to other elements, eg addition.

or (173): The event *A or B* means the outcomes in *either* event *A or* event *B or* in both events.

order of rotational symmetry (319): The number of times an object will map onto itself in one turn.

ordered pairs (235): Two numbers (*a*, *b*) which give the position of a point relative to axes in the *cartesian plane* (the order of the numbers is important).

origin (235): The point (0, 0), where the *x*- and *y*-axes intersect on the *cartesian plane*.

outcome (165): The result of a trial in a statistical experiment.

outliers (195, 209): Data values that do not fit a general pattern, eg values that are extra large or small.

parabola (235, 265): The curve which is the graph of a *quadratic equation*.

parallel lines (239): Two lines are parallel if they are always the same distance apart.

parallelogram (3): A *quadrilateral* in which the opposite sides are parallel to each other.

pentagon (21): A polygon with 5 sides.

percentage (%) (81): A percentage means 'out of every 100'.

perfect square (280): A *square number* such as 25 or an algebraic expression such as $(a + b)^2$, formed by multiplying a factor by itself.

perimeter (1): The distance around the outside of an object.

perpendicular lines (74): Lines at right angles to each other.

plane (29, 47, 315): A flat (2-D) surface.

plotting (193): Transferring the ordered pairs of a relation accurately onto a graph.

point symmetry (279): A graph which can be rotated through 180° about a point onto itself has point symmetry.

polygon (21): A closed figure with many straight sides.

polyhedron (plural polyhedra) (317): *Solid* figure with flat surfaces.

polynomial (279): A function which is the sum of several terms. Each term is of the form ax^n, where n is a whole number.

population (188): Total group of interest in a statistical investigation.

population mean (μ) (339): A measure of central tendency in a population.

population standard deviation (σ) (339): A measure of spread of a population (the square root of the variance).

predict (217): Use patterns to work out future values.

prism (31, 313): A solid object with the same *cross-section* throughout.

probability (165): A measure of how likely it is that an event will occur.

probability tree (176): A branching diagram showing combinations of events and probabilities.

product (72, 187): The result of multiplying two or more terms together.

proof (15): Statements showing that a relationship is correct.

proportion (165): A comparison of different amounts. The proportion in a group with a certain property is calculated by dividing *the number in the group with the property* by *the total number in the group*.

proportional (78): Of equal proportions; in the same ratio.

pyramids (30): Solid objects with all but the base face meeting at the same point (the *apex*).

Pythagoras' theorem (10): A relationship ($a^2 + b^2 = c^2$) involving the three sides of a *right-angled triangle*.

quadrant (197, 296): One of four regions defined by the x- and y-axes.

quadratic equation (255): An equation which can be expressed in the form $ax^2 + bx + c = 0$.

quadratic formula (258): A formula for solving a quadratic equation.

quadratic function (265): A function with equation of the form $y = ax^2 + bx + c$.

quadratic graphs (235, 274): Parabolas.

quadrilateral (21): A 4-sided *polygon*.

radius (plural radii) (2): A straight line from the *centre* to a point on the *circumference* of a *circle* or *sphere*.

random selection (169): A method for collecting a sample in which every member of the population has the same chance of being selected.

range (218, 295): The difference between the highest and lowest scores in a set of data.

ratio (10, 38): A comparison of two or more quantities.

rational expression (283): Any quotient consisting of algebraic expressions.

real numbers (R) (228): The set of all possible numbers on the number line.

rearranging (221, 231, 291): Changing the order of terms in an expression, often so that another variable becomes the subject.

reciprocal (284): The multiplicative inverse. The reciprocal of x is $\frac{1}{x}$. The reciprocal of $\frac{a}{b}$ is $\frac{b}{a}$.

rectangle (3): A *quadrilateral* in which all the *interior angles* are right angles.

reflection (308): A *transformation* which maps objects across a *mirror line*.

region (251): Another expression for 'a set of points' on a graph.

regression (203): A technique for fitting a trend line to data.

regular (25): A regular *polygon* has the lengths of all sides equal and the sizes of all *interior angles* equal. A regular polyhedron has all faces made up of the same regular polygons.

relative frequency (165): The relative frequency of an event is the *proportion* of times it occurs.

rhombus (3): A *parallelogram* in which all the sides are equal.

right angle (⊾) (10): An angle of 90°.

right-angled triangle (10): A *triangle* which has a *right angle*.

root (221, 258): A *solution* to an equation.

root (231): The inverse of raising a base to the power of n is taking the nth root of the base.

rotation (319): A *transfomation* in which an object is turned about a point.

rotational symmetry (319): A shape has rotational symmetry if there is a point about which the shape rotates onto itself.

rounding (2, 233): The writing of a number with fewer significant figures.

rule: See *formula*.

sample space (188): The set of all possible outcomes of a statistical experiment.

satisfy (221): Values which, when substituted, make an equation true, are said to satisfy the equation.

scale (53): A line of values at set intervals marked on a measuring device.

scale factor (SF) (37): The *ratio* of the image size to the object size.

scalene triangle (9): A *triangle* which has no equal sides and no equal angles.

scatter diagram (193): Graph used with paired data, to investigate relationships between the variables.

sector (4): A fraction of a whole *circle* formed by two radii and the arc in between. Area **(4)**.

segment (50): The region of a *circle* enclosed by a chord and an *arc*.

semicircle (315): A shape enclosed by a diameter of a circle and half the *circumference* of the circle.

sequence (341): A function which associates a real number with each natural number. Arithmetic sequences, pp. 341; Geometric sequences, pp. 128, 341.

series (139): The sum of the successive terms of another sequence.

sign (+ or –) (231, 239): The positive or negative nature of a number.

similar figures (37): Similar figures have the same shape (angles) and corresponding sides are in proportion.

simplest form (228): *Simplified* as far as possible.

simplify (1, 283): To write an expression in a briefer form (by combining like terms, etc). Simplifying algebraic fractions **(283)**.

simultaneous equations (262): Two or more equations, each of which contains the same variables, which hold true at the same time.

sine (sin) (10): A *trigonometric ratio*. Relative to a particular angle it is the ratio $\frac{\text{opposite side}}{\text{hypotenuse}}$.

sine rule (10): The relationship between one side of a triangle and its opposite angle, and a second side and its opposite angle.

sketch (58, 61): A freehand drawing showing the main features of an object.

slope (206, 238): Another expression for gradient.

SOH CAH TOA (10): Mnemonic relating the *trigonometric ratios* to the sides of a *right-angled triangle*.

solution (221): A value which substitutes into an *equation* to make it true; an answer.

solving (221): Finding the values of the variables in an equation or inequation which make the equation or inequation true.

speed (v): The positive rate of change of distance with respect to time. (Also called velocity.)

sphere (30): Completely rounded three-dimensional figure; ball-shaped.

spread (203): A description of the *distribution* of data values between their highest and lowest values.

square: A *quadrilateral* in which all four sides are equal and meet at *right angles*.

square centimetre (cm^2) (2): Unit of measurement of *area* in the *metric system*. 10 000 cm^2 = 1 m^2.

square kilometre (km^2) (2): Unit of measurement of *area* in the *metric system*. 1 km^2 = 1 000 000 m^2.

square metre (m^2) (2): Base unit of measurement of *area* in the *metric system.*

square millimetre (mm^2): Unit of measurement of *area* in the *metric system*. 100 mm^2 = 1 cm^2.

square root (231): A number which when squared gives the number under the $\sqrt{\ }$ sign.

standard deviation (s or σ) (339, 342): A measure of spread. The square root of the variance.

statistics: The collection, display and analysis of data.

straight angle (26): An angle on a straight line, ie 180° or a half turn.

subject (89, 231, 289): The variable appearing on its own on one side of the equals sign in an equation or formula.

substitute (262): To replace variables with numbers or other variables.

substitution (289): The process of replacing variables with numbers or other variables.

subtracting (–): An arithmetic operation.

sum: The result of adding two or more numbers or like terms.

surface area (1): The sum of the *areas* of the faces of a solid.

survey (61): A statistical investigation in which data are collected.

survey line (71, 77): Straight line dividing a survey region into two parts: a regular region and an irregular region.

table (3): Information arranged in columns and rows.

tangent (tan) (10): A trigonometric ratio. Relative to a particular angle it is the ratio $\frac{\text{opposite side}}{\text{hypotenuse}}$.

temperature (245, 289): How hot or how cold an object is. The unit used is *degrees Celsius* or *centigrade*.

term (81, 84, 89, 99, 119, 128, 136): Length of time of an investment or loan.

tetrahedron (314, 315): *Solid* with 4 faces, each an *equilateral triangle*.

tonne: A unit of *mass*. 1 000 kg = 1 tonne.

transformation (266, 268, 272, 279): A change made to an object. *Reflection* **(308, 326, 333)**. *Rotation* **(319)**.

transformed (267): Changed in position or size.

transforming (291): See *rearranging*.

translation (267): A *transformation* which changes the position of an object by sliding.

trapezium (3): A *quadrilateral* that has one pair of parallel sides.

trend (247): A pattern of changes over time.

trend line (247): See *line of best fit*.

trial (166): One repetition of an experiment.

triangle (3): A 3-sided *polygon*.

trigonometric ratios (10): Relationships between side lengths and angle sizes in a *right-angled triangle*. The three basic trigonometric ratios are *sine, cosine* and *tangent*.

turning point (265): A point on a graph where it reaches a *maximum* or *minimum*. See also *vertex*.

unbiased (165): An unbiased sample is representative of the population from which it is selected.

ungrouped data (342): Data for which each possible value is listed individually.

union, symbol ∪ (183): The union of two (or more) sets is the collection of all outcomes in the sets (or in their intersection). The union of two sets A and B is written $A \cup B$.

variable (221): A letter or symbol representing an unknown number.

variance (σ^2 or s^2) (342): A measure of spread.

vector (272): A quantity that has both direction and magnitude.

velocity (symbol v): The instantaneous rate of change of distance (displacement) with respect to time.

Venn diagram (183): A rectangular diagram used to show relationships between sets (represented as ovals or circles inside the rectangle).

vertex (plural vertices) (265): The *turning point* of a parabola. The junction where two sides of a polygon *intersect* **(9)**.

vertical asymptote (298): A vertical line which a graph approaches but does not cross. The value which is excluded from the domain of a hyperbola.

vertical line (241): Line at *right angles* to a *horizontal line*.

volume (29): The amount of 3-dimensional space an object occupies. Units are m^3, cm^3 etc. Volume and *capacity* **(29)**.

whole numbers (W) (32): The set of numbers 0, 1, 2, 3, 4, 5, 6, …

word problem (225): A description of a mathematical problem in words.

x-axis (235): Horizontal axis in *cartesian plane*.

x-coordinate (235): First number of a point's *coordinates* on a graph.

x-intercept (175): The point on the x-axis where a graph cuts.

x-intercept (237): The point where a graph cuts the x-axis.

y-axis (235): Vertical axis in *cartesian plane*.

y-coordinate (235): Second number of a point's *coordinates* on a graph.

y-intercept (237): The point on the y-axis where a graph cuts.

ZFS (257): A mnemonic for the strategy required to solve quadratic equations (zero, factorise, solve).